1 MONTH OF FREE READING

at

www.ForgottenBooks.com

By purchasing this book you are eligible for one month membership to ForgottenBooks.com, giving you unlimited access to our entire collection of over 1,000,000 titles via our web site and mobile apps.

To claim your free month visit:

www.forgottenbooks.com/free1193686

ISBN 978-0-331-48948-4
PIBN 11193686

S–R Q 608 In85 v.7

The Inventive age and
Industrial review

The Inventive Age

AND INDUSTRIAL REVIEW

A JOURNAL OF MANUFACTURING INDUSTRY AND SCIENTIFIC PROGRESS

Seventh Year.
No. 1.

WASHINGTON, D. C., JANUARY, 1896.

Single Copies 10 Cents.
$1 Per Year.

THE NEW CONGRESSIONAL LIBRARY.

A Magnificent Addition to the Nation's Structures.

At the Capitals of all governments are erected buildings in which to transact the nation's business. These vary in solidity, utility, and magnificence of structure, in proportion to the importance, financial standing and artistic taste of each nation, and while the Republic of the United States is comparatively young in years, yet in the magnificence and stability of its public buildings is among the first. The Capitol building is said to be equal to, if not exceeding in its architectural magnificence, that of any other nation. Washington contains many public build-

24th, 1800, entitled "an act to make further provisions for the removal of the seat of government." etc. This act was passed in Philadelphia, before the seat of government was removed to Washington, which occurred in October of that year.

The first Librarian was John Buckley of Virginia, appointed January 29, 1802. In 1809 the total appropriation on account of the library, including the purchase of books and expenses connected with its management, was $900. The battle of Bladenburg was fought on August 24th, 1814, and in the evening of that day the British took possession of the Capitol building. The library, which was stored therein, was totally destroyed by fire. On the 21st day of September, following this destruction, Thomas

Congress, might be removed to Washington so as to be accessible at once. They were then at Monticello his home in Virginia, which is some three hours run by rail from the city. He said, speaking of the delivery of the books, "Eighteen or twenty wagons would place it in Washington at a single trip in a fortnight." When delivered the collection consisted of 6,700 volumes for which Congress paid $23,950, which by Mr. Jefferson's biography is claimed to have been a "measly sum," though generally considered to be their full value. From the date of the purchase of the Jefferson collection to 1864 the Congressional Library may be said to have existed rather as a necessity than from any national pride. Appropriations were insignificant and its

THE NEW CONGRESSIONAL LIBRARY BUILDING, WASHINGTON, D. C.

ings the equal if not superior of the best European governments. Among those of greatest note outside of the Capitol building, is that in which the State, War and Navy Departments have headquarters. This building will hereafter take second place, as the new Library Building, fast nearing completion, is easily entitled to first—if not first of all the public buildings in the world. It is called the "Library Building," and intended for the accommodation of what is known as the "Library of Congress." It is separate from the House and Senate libraries, they being used exclusively by members of the two branches of Congress, and not intended to contain books on general topics.

The Library of Congress is distinctively a storehouse for the preservation of the nation's literature. The first mention of the Library of Congress by legislative enactment was the act approved April

Jefferson made a proposition to Congress to sell his library consisting, as he stated in the proposition, of about 9,000 volumes. Knowing the financial condition of the government at that time he proposed to take such a price as appraisers should fix and pay at such time and in such manner as might be satisfactory to the purchasers. His proposition provoked much discussion and Daniel Webster voted against its acceptance. Cyrus King, a member of Congress from Massachusetts, proposed that before accepting the books they be delivered and inspected, and that all of any atheistic tendency be returned to Mr. Jefferson without expense to him. When the collection was finally received it was found to contain many rare and valuable works on theology. One of the clauses in Mr. Jefferson's offer is particularly worthy of note, and sounds really odd at this date. He was anxious that the books, if desired by

general management inefficient. The present librarian, Mr. A. R. Spofford, was appointed by President Lincoln on December 31st, 1864, and has since been in charge (and probably will be until he dies). In September after his appointment the first complete alphabetical catalogue of the library was gotten out by him and consisted of eight volumes, in all 1,236 pages. The library at that time did not exceed 70,000 volumes. The copyright law, which originally provided for registration in the office of the clerk of the district court of the several states, was in 1870 changed and the entire copyright business transferred to Washington and placed under the charge of the Librarian of Congress. It provided that all persons who procured copyright should, with their application for such, place in the library two copies of the work for which copyright was to

(Continued on page 4.)

The Inventive Age

Established 1889.

INVENTIVE AGE PUBLISHING CO.,

8th and H Sts., Washington, D. C.

ALEX. S. CAPEHART. MARSHALL H. JEWELL.

The INVENTIVE AGE is sent, postage prepaid, to any address in the United States, Canada or Mexico for $1 a year; to any other country, postage prepaid, $1.50.

Correspondence with inventors, mechanics, manufacturers, scientists and others is invited. The columns of this journal are open for the discussion of such subjects as are of general interest to its readers.

Technical matter is particularly desired. We want practical information from practical men.

Nothing will be published in the editorial columns for pay.

The INVENTIVE AGE is thoroughly independent.

Advertising rates made known on application. Special facilities for furnishing cuts of any patented article together with descriptive article. Business specials 25 cents a line each insertion, 7 words to the line. No advertisement less than 50 cents.

Address all communications to THE INVENTIVE AGE, Washington, D. C.

Entered at the Postoffice in Washington as second-class matter.

WASHINGTON, D. C., JANUARY, 1896.

THE horseless carriage is so much of a success that an experiment in the shape of a regular time schedule transportation line is to be inaugurated in Cleveland, the fare to all parts of the city being 2½ cents.

ELECTRICITY of December 4th last, contains an interesting article on the forthcoming electric light and railway exhibit in New York. Elsewhere in this issue will be found an illustration of the building in which the exposition will be held.

A recent report shows that about 80 per cent of the work of digging the great Chicago drainage canal has been completed. Nearly $18,000,000 has already been expended on the enterprise and the present estimate of the entire cost is $28,000,000.

INVENTORS and others desiring comprehensive cuts made of their inventions from photographs—either line cuts or half tones—or reproductions from Patent Office drawings, can obtain them through the INVENTIVE AGE at a low rate, and any kind of commercial printing will also be furnished promptly and reasonably.

So many patents are obtained for inventions wherein the specifications are deficient and the claims so narrow that the inventor has really a patent in name only—and of no commercial value—that we take this and every occasion to advise that inventors be extremely cautious in selecting attorneys to prosecute their cases. In this connection the INVENTIVE AGE desires to add that all business entrusted to its Patent Bureau receives the most careful attention, and inventors may depend upon honest opinions as to patentability and upon the broadest most comprehensive claims consistent with the state of the art in the class of the invention.

ABOUT the most ludicrous thing that has happened lately was the recent banquet given by a few individuals connected with the Bureau of Engraving and Printing in Washington in honor of the "near completion" of the diplomas and medals awarded at the World's Fair in Chicago two years ago. Here is an exemplification of Rip Van Winkle's experience sure. And all this jollification because it is "expected" the diplomas will be ready for distribution about April—probably the 1st, to be in keeping with the succession of events in the past. The contract of the Atlanta exposition management is quite refreshing. It is announced that they will deliver the diplomas and medals at once, in fact some of them were nearly ready before the close of the exposition. The total number of medals awarded is 1543 and the number of diplomas 1546. These include 164 grand prize gold medals and diplomas, 414 diplomas of honor with gold medals, 414 silver medals and diplomas of excellence, and 495 diplomas of honorable mention and bronze medals, in addition to 56 diplomas of grateful recognition to exhibiting States and foreign countries accompanied by gold medals, and seven similar diplomas without medals.

THE activity of invention is without relapse, and in this age is more marked than ever, says Age of Steel. In fact it never tires. It has its spurts of course, as new directions open for its ingenuity. These may be in logical sequence to such epoch making discoveries as steam power and electric energy; some new application of mechanical laws or forces, or in the persistent and never halting demand for labor-saving appliances, but be the impulse what it may, scientific or economic, the current no matter its course is incessantly in motion. No observer of the constant changes in a machine shop can question the perpetual movement of invention. It has no vacation in its vigilance, and is seldom or ever blind to its opportunities. In every department of industry from cutting a log to planing a plank; from a gimlet to a drill, from a wagon brake to that of an express train, from an iron casting to a steel shaft; in spinning cotton and crushing sugar cane, and in making a nail or a sub-marine cable, in short everywhere and in everything, modern inventive genius is the vital breath of industrial success. It may be overdone and in some cases it may precipitate serious disturbances in the equilibrium of labor, but in its total results it is helpful as it is irrepressible and necessary.

MR. ALEX KLINZER, now an old man, writes from Silverton, Col., congratulating the INVENTIVE AGE on its vigorous policy in the protection of inventors and reciting an instance in his own experience of the theft of a valuable invention by a "heartless corporation." Mr. Klinzer claims to have been the inventor of the simple hook now in use on nearly all shoes manufactured away back in 1872 and that the patent was infringed on with impunity by eastern manufacturers of shoes; that he was poor and could not find any one to take interest enough in the matter to assist him in fighting for his rights; that after a few years a New York patent lawyer wrote him offering the small sum of $150, which he accepted and as the Patent Office records will show, finally after three transfers reached the real parties Foster and Paul, who, according to the long history of the transaction appearing in the New York Sun in 1890, made the factories using the patent, come down, and pay them royalties amounting to over $300,000. All of this and the real inventor, Mr. Klinzer, still a poor man and well along in years. Mr. Klinzer argues that there should be some system whereby the government would, under such circumstances step in and protect the inventor. While it frequently happens that the inventor of a valuable device does not reap the reward he is entitled to still it is difficult to see wherein the government, having given an inventor the patent his invention and specifications call for, could after that protect his rights. After the patent is issued the inventor has redress for infringement but he must needs go to the courts, and prove his claims. Lack of business sagacity as often as lack of funds to prosecute infringers, prevents the inventor from the enjoyment of the fruits of his inventive genius. The government cannot furnish these requisites.

WORK in the Patent Office is reported to be about up to date. That is well. "Another thing is wanted," says *Iron Industry Gazette*. "In addition to the prompt granting of patents there should be assurance that a patent granted is granted in earnest, is irrefragable, is absolutely protected by the government from all suits, all infringements and all other expensive and senseless forms of defense. Prompt patenting is a desirable thing, but *safe and unbreakable patenting* is the one great thing needed."

The keynote is somewhat in the last sentence. Unbreakable patents—and they can only be obtained through reliable and competent patent attorneys. The anxiety of "patent or no pay" attorneys to get some kind of a patent—so that they may gain their fee has resulted in thousands of worthless patents—patents in name only, that will not stand the test of infringement and mechanical and scientific investigations. Through the incompetency of patent attorneys hundreds and thousands of inventors have become possessed of patents of no commercial value whatever. It was with a view of protecting inventors against fraud and incompetency that THE INVENTIVE AGE Patent Bureau was started. Through it patents on inventions can be obtained at the lowest rates consistent with conscientious and efficient work, and where the inventors cannot obtain rights that justify expense of applying for or obtaining a patent the INVENTIVE AGE Patent Bureau does not hesitate to so inform the inventor. An investigation extending over a period of six years has convinced the INVENTIVE AGE that the inventor needs protection against unscrupulous patent attorneys as well as against fraudulent patent sellers. The columns of the INVENTIVE AGE, advertising or reading, are not open to announcements from any but thoroughly reliable attorneys or patent agents.

NOTES.

Gas Pipes from Paper—European papers state that parties in Germany have succeeded in making gas pipes from paper which are claimed to be cheap, gas tight and serviceable. The method of manufacture is to pass strips of manilla paper through molten asphalt, the material being then molded under heavy pressure. After cooling, the pipes receive another water proof coating.

* * *

A Boom for Angle-Worms.—The importance of angle-worms in agriculture has been demonstrated by Professor Wollny, of Munich. Peas, beans, potatoes, and other vegetables were grown in wooden boxes, with and without worms, and in every case the presence of the worms gave an increase of crop, varying from 25 per cent. in the peas to 94 per cent. in the rye.

* * *

Substitute for Rubber.—An excellent opportunity in the field of invention is offered in finding a substitute for rubber. An European chemist offers the following compound which may furnish a clue as to the proper way to begin investigations: Tar, 1 part; paraffine, 10 parts; dissolve together at 120 degrees and then add caoutchouc, 2 parts. Keep at this temperature until a homogeneous mass results.

* * *

Amazing Speed of Electricity.—Under most favorable conditions, the speed of electricity is, so it is estimated, 180,000 miles per second. An eminent scientist gives the following illustration: Suppose that a row of telegraph posts 25,000 miles long were erected around the earth at the equator, and that a wire be stretched upon these posts for the circuit of 25,000 miles, and then continued around a second time, a third time, etc. In fact, let the wire be wound round this great globe seven times. We should then find that an electric signal sent into the wire at one end would accomplish the circuit of seven convolutions in one second of time.

BOOKS AND MAGAZINES.

The December number of Power completed its fifteenth volume. It was a particularly interesting edition containing many illustrations and much valuable matter relating to the exhibits in machinery hall at the Atlanta exposition.

* * *

Cassier's Magazine has gathered into its January number some of the best available electrical talent on both sides of the Atlantic, and the result is what has very appropriately been termed an electrical number, of exceptional interest and importance to the engineering profession. The latest developments in applied electricity, the latest realizations of electric power transmission and utilization, and the possible achievements of the near future have all received attention.

* * *

The Indian Textile Journal published in Bombay, is indeed, as it claims a representative publication for the textile and engineering industries, and the first and only journal of the kind in the East. It is about to print a "vernacular" edition, for the large number of native mill-owners, factory managers, engineers and others who do not know English.

* * *

No one ever thought of introducing so expensive a feature as lithographic color work in the days when the leading magizines sold for $4.00 a year and 35 cents a copy. But times change, and the magazines change with them. It has remained for The Cosmopolitan, sold at one dollar a year, to put in an extensive lithographic plant capable of printing 320,000 pages per day (one color). The January issue presents as a frontispiece a water-coloring drawn by Eric Pape, illustrating the last story by Robert Louis Stevenson, which has probably never been excelled even in the pages of the finest dollar French periodicals. The cover of The Cosmopolitan is also changed, a drawing of page length by the famous Paris artist Rossi, in lithographic colors on white paper takes the place of the manilla back with its red stripe. Hereafter the cover is to be a fresh surprise each month.

That Disastrous Detroit Boiler Explosion.

A great deal has been published about the great boiler explosion that occurred at Detroit on the morning of November 6th last, but our readers will be interested in the illustration herewith presented through the courtesy of *Power*, of New York. By this explosion thirty-seven lives were lost and some eighteen or twenty other persons were severely injured. The explosion occurred about 9 o'clock in the forenoon, the boiler being located in the building occupied by the Detroit Journal. The illustration shows the condition of the collapsed building immediately after the explosion. Steam for power and heating purposes was furnished by two horizontal tubular boilers five feet in diameter, fourteen feet long with three-inch tubes. These were connected by a five-inch pipe from which a 3½-inch supply pipe went to the elevator pump, and other feeds to the heating mains, etc. We have no knowledge of the feed connections or what possible communication existed between the boilers below

was waiting for the pressure to get nearly equal before opening the valve in the pipe which connected them. The boilers had been inspected by the city inspector and allowed 90 pounds. The safety valves were set at 80.

When the debris had been cleared away it was found that the east boiler had exploded. It had parted just behind the dome and the tubes were broken off clean and square at about the center of their length. The other boiler did not go out, but the force of the explosion carried it out off its setting and through the solid stone foundation into the adjoining cellar. The force of the blow injured the boiler near the dome, parting the seams and bending the tubes. This was the boiler under which the fire had just been started, and which had 15 pounds of steam when the engineer last saw it before the explosion. The injector was found closed as was also the valve in the five-inch steam pipe connecting the boilers.

"We do not want to pose as alarmists," says *Power*, "but is it not a self-evident fact that with this increasing tendency in the direction of high steam pressure, the number of accidents due to boiler explosions and the death-rate from this cause

The Erhardt Process.

The Erhardt process for the production of hollow articles of iron, steel, or other metals from billets is a German invention promising important results the manufacture of hollow articles. The process comprises making tubes of all kinds, gun barrels, hollow projectiles, acid flasks, parts of machinery which require the utmost lightness with adequate strength in fact, an infinite variety of hollow metallic articles by thrusting a steel core or mandrel through a billet of hot steel, wrought iron, or other metal, which is held firmly in a matrix of such shape and calibre as to give the required outward form to the completed object. The mandrels operated by hydraulic pressure; they are guided by proper machinery to pierce the exact centre of the billet, to compress the yielding metal outwardly into the space between the mandrel and matrix, producing a hollow body, the shell of which is everywhere of exact and uniform thickness. A touch upon the lever controlling the hydraulic press starts the mandrel, which with one noiseless thrust pierces the billet, drives it through the matrix, draws it out like

BOILER EXPLOSION AT THE DETROIT JOURNAL BUILDING IN WHICH THIRTY-SEVEN LIVES WERE LOST.

the water-line. Oil was used as fuel. About two weeks before the accident the gasket of the manhole of the west boiler blew out, and that boiler had consequently been out of service. On Tuesday, the day before the explosion, fired this boiler up to 90 pounds and the safety valve blew off. Wednesday he started to run with the east boiler as before, but the demand for steam was such that he started oil burners under the other also. At the time of the explosion the engineer was in one of the upper stories talking with a stereotyper. In his testimony before the coroner's jury he says:

"I was just on the point of getting down, to go into the boiler room when something gave me a crack on the side of my neck that knocked me galley west. Then there was a sound all around my head that reminded me of a lot of firecrackers exploding under an empty can. When I got up I saw George trying to get from under a lot of brick. I took hold of his arm, but the skin peeled of his arm and off my hands when I tried to pull. A fireman then came up and pushed me away."

The gauge in the east boiler at the time showed 65 pounds and the one on the west 15 pounds and he

is bound to increase, unless steps are taken to insure their intelligent supervision and use. A boiler never exploded yet unless the pressure got above that which the boiler was competent to support, or the integrity of the boiler became impaired so as to render it incapable of maintaining its accustomed pressure. Intelligent and careful supervision is all that is required to forestall either contingency. The public has a right to demand that these menaces to the public safety be placed in charge of men who are competent to determine positively their ability to stand the pressure imposed upon them, and who are not only competent to maintain them in that condition and detect any deterioration from it, but who may be depended upon from their known habits of sobriety, faithfulness and painstaking interest, to operate them within the lines of safety, to stay by them when in operation, and see to it that safe operative conditions are maintained."

It is said that in Japan "earthquakes are so frequent that the only attention paid to them is to stop shaving until the shock is over."

To be perfectly proportioned a man should weigh 28 pounds to every foot of his height.

wax, spreading the pliant material over the end of mandrel like the finger of a glove, shoves the end of it into a die which tapers it to a point, and within two seconds of time the process is complete.

The Russian government is about to undertake some very elaborate and thorough census work, and will do its counting by means of the ingenious electrical apparatus of Dr. Herman Hollerith, of Washington, which was used in the last American and Canadian censuses with great success. It is said, however, that perhaps the Russian government may get its Hollerith machines from Austria, where the system was also adopted in 1891, but where the inventor's latest ideas have not yet been introduced. For this reason, the order may still come direct to this country.

A new fuel made in France is of coal dust compressed into bricks and soaked with chemicals, which make it last a long time in a glow when once alight.

India rubber used for erasing pencil-marks was known in England as early as 1770. A cube of it ½ in. square costs 3s.

THE NEW CONGRESSIONAL LIBRARY.

(Continued from first page.)

issue, and since that time every literary production, protected by copyright, has been deposited, and is now on file. This includes more than 150,000 musical productions. Of all the men in public life since the formation of the government none have more completely filled the niche into which the inmuthble law of circumstances have placed them than has Librarian Spofford. The phrase of "walking encyclopedia" fits him with better grace than any to whom we have ever heard it applied. In 1873 the books in the library consisted of 246,345 volumes, of which 29,373 belonged to the law library. In 1882 Dr. J. H. Toner, a Washington physician, donated to Congress his private library consisting of 30,000 books among which were a very complete collection of literal transcripts of all the letters, journals and writings of President Washington. According to the report of the Librarian of December 2nd, 1895, the Library of Congress now contains 725,000 volumes. (The library of France, the largest in the world, contains 3,120,000.) Felly one-half of these are either stored in vaults almost inaccessible, or are piled on the floor about the library rooms now in use. The proposition to build more commodious quarters for the library has been long considered and many different plans proposed. Among them one to extend the west front of the Capitol building 100 feet to hold the books, another was to project the eastern front 250 feet, another scheme was to turn the rotunda into one vast book repository with rows of shelves around its sides, one upon the other, from floor to apex. Fortunately none of these ill-advised plans met with sufficient favor to secure their adoption. A separate building for the library, to be erected on the ground now occupied by the Botanical Gardens was at one time strongly urged. At another the present site of the Department of Justice was talked of, but was abandoned because of its distance from the Capitol. Librarian Spofford, as early as 1872, foresaw not only the necessity of a commodious building, but also at that time outlined almost exactly what has since been done. Congress provided by law for a new library building by act of 1886, and which was amended in 1889. The cost of the building is limited to $6,000,000. The site selected consists of ten acres, located just east of the Capitol grounds, the front of the Library Building facing the east front of the House of Representatives. The architectural plans for the new building were prepared by Smithmeyer and Pelz, but have been considerably modified in constructing the building which is being done under the supervision of the Chief of Engineers of the Army, General Thomas L. Casey. The outside walls are of white New Hamshire granite and the inner courts of Maryland granite of a slightly darker hue. Its dimensions are 470 by 340 feet, covering about three and one-half acres of ground. It has four large inner courts 150 feet in length by 75 to 100 feet in width. The outer walls have a frontage on four streets. The building has nearly 2,000 windows and is said to be the best lighted library building in the world. There are three floors, comprising a basement, level with the ground; a first story, 19 feet high, and a second story 29 feet high. The walls are 69 feet to the roof, and the apex of the dome 195 feet from the ground. The solid massiveness of the granite walls are relieved by numerous windows with their casing treated in high relief and 16 ornate pillows and capitals in the central front, besides 15 additional columns in each of the corner pavilions. Upon the keystones of 33 of the window arches on the four sides of the edifice are carved human heads, types of so many races of men. The central pavilion on the west front has four colossal figures, each representing Atlas, and is surmounted by a pediment with two sculptured American eagles as the center of an emblematic group in granite. Over the arches of the three entrance doors are carved three spandrels in relief, each representing two female figures emblematic of Art, Science and Literature. The roof is throughout of sheet copper, the dome is gilded by a thick coating of gold leaf 23 carats fine, costing about $3,800. The crest of the dome terminates in a gilded final representing the torch of science ever burning. The reading room is an octagonal or nearly circular hall 100 feet in diameter and 125 feet high, lighted by eight large semi-circular windows 32 feet wide. The room is designed to seat 250 readers, furnishing each a desk with four feet or more to work in, and isolated from his neighbors by a light screen or curtain. In the center of this room will be the desk of the Superintendent and his assistants, and from his desk leads 28 speaking tubes giving him communication with every part of the building in which books are stored and also with the Capitol building. The interior walls of the reading room are finished with variagated marble with eight massive pillars rising 40 feet to the concave ceiling and are of dark chocolate colored Tennessee marble at the base, surmounted by two shades of lighter Numidian marble, on which rests shades of heroic size. There are 34 arches on the library floor intercalated with

pilasters, architraves, carved in classic sculpture, and above these on the gallery floor are 53 more arches in continuous succession, surrounded by running balustrades, extending entirely around the reading hall. Next to the reading room in the center is the grand entrance hall or vestibule lined throughout with fine Italian marble, highly polished. On the sides rise lofty rounded columns with elaborately carved capitals of Corinthian design, while the graceful arches are adorned with marble rosettes, palm leaves and foliated designs of exquisite finish and delicacy. The height of the entrance hall is 72 feet to the skylight, has vaulted ceilings, and the grand double staircase with its white marble balustrades lead up on either side; exhibiting an architectural effect rarely if ever surpassed. The upper staircases are ornamented with 26 miniature marble figures carved in relief. This beautiful and spacious hall has been described as "a vision in polished stone." The statuary for the reading room comprises eight colossal emblematic figures representing art, history, philosophy, poetry, science, law, commerce and religion. Two representative men for each subject are cast in bronze statues of heroic size to be arranged in groups around the galleries of the rotunda. Philosophy is represented by Plato and Bacon. history by Herodotus and Gibbon, poetry by Homer and Shakespeare, art by Michael Angelo and Beethoven, science by Newton and Henry, law by Solon and Kent, commerce by Columbus and Fulton and religion by Moses and St.

NEW YORK INDUSTRIAL BUILDING.

Paul. The crown of the dome will be decorated with allegorical groups in fresco. The hollow concave of the dome is enriched with ornamental stucco work in which figures in high relief of cherubs, vases, birds. rosettes, griffins and flowers in geometrical designs are harmoniously distributed. Next to the reading room there opens out on either side extensive book magazines or repositories called stacks. Each tier of shelves is seven feet in height. All the floors of the stack rooms are of white marble; the shelves of rolled steel, easily adjusted to any height. The grand courts into which these stack rooms look on both sides are lined from ground to roof with enamelled brick of the color of ivory or porcelain with 200 windows on each side. Each stack has a shelving capacity of 800,000 books, and the building when complete will accomodate 4,000,-000 books, which, judging from accumulation in other libraries, will be sufficient for the necessities of the government for only a little over a century; this, when we come to think of the centuries the British Museum has existed seems but a brief space of time comparatively. In the basement of the building will be located a book bindery for the proper care and preservation of books, pamphlets and periodicals. On the ground floor there is on each of the four sides a wide corridor finished in marble, each of a different color. The second floor contains an art gallery measuring 274 by 35 feet with a glass roof for the exhibition of works of graphic art, and an-

other room of the same size for the exhibition of maps. Numerous lifts for the carriage of books and seven elevators for passengers are distributed throughout the building, and time saving machinery, both vertical and horizontal, for the quick transmission of books from point to point in the Library Building, and from the Library to the Capitol will be supplied. The appliances for conveyance of books from the Library building to the Capitol will be constructed in a tunnel 4x6 feet in diameter, and through this channel persons in the Capitol building will be enabled to be supplied with books from the Library building in from four to six minutes after request is made. There are separate rooms for the accommodation of members of both branches of Congress, and also for the accommodation of persons engaged in special studies, or the investigation of subjects of particular importance. The Librarian of Congress is appointed by the President, and he has charge under a joint committee of House and Senate of the Library, the selection of books, pamphlets and periodicals and the general management of all the details connected with the Library.

This committee also has charge of the Botanical Gardens, and the selection of all works of art for decoraiton in the Capitol and surroundings. The chairman of the joint committee has been for a number of years one of the prominent Senators of the United States. Among those who may be mentioned in connection with the position are Senators Sherman, Hoar, Sewell, Holly, Howe, Everett, Quay and Mills. The present chairman of the committee is the Hon. H. C. Hansbrough. senior Senator from North Dakota. He is a practical printer and journalist and therefore well versed in the making and proper care of books. He is besides well versed in literature, and will undoubtedly take special interest in the organization of the new Library so as to materially promote its efficiency. It is expected that the building will be fully completed and occupied by 1897.

Most Powerful Battery in the World.

The recent test of the heavy battery of dynamite guns commanding the entrance to the Golden Gate proved in every way successful. This battery is the largest and most powerful of its kind in the world. It consists of three 15-inch and five 5-inch dynamite guns. The range of the 8-inch guns, which carry shells 340 pounds in weight. was found to be 4500 yds., the guns being fired with an air pressure of 1000 pounds. Under similar pressure the 15-inch guns threw 500-pound shells a distance of 2000 yards. This battery is considered by experts to afford an absolutely complete defense against hostile fleets essaying to enter the harbor of San Francisco.

The Southern California Railway Company has now eighteen locomotives in regular service burning oil, and the results are quite satisfactory.

A Commercial Museum for New York.

Preparations have been for some time quietly making for the establishment in New York City of a permanent International Commercial Museum, and a number of well-known business men and manufacturing firms are interesting themselves in the project. The museum will be located in the Industrial Building, at Lexington avenue and Forty-third street, which has been leased by the managers for a period of ten years. The plan is to establish there an exhibition where the foreign buyer can see samples of American manufactured goods and where the American buyer can see specimens of foreign manufactured goods. A large number of firms, it is stated, have already applied, for space for exhibits of their goods, including some of the best known manufacturing concerns in the United States.

This exhibition will contain a library of information as to industries, habits, products, climate, soil, etc., of the different countries, with full particulars as to their currency, weights and measures, duties, etc. Complete business directories of all foreign countries will be kept on file. There will also be found in the library technical works on the principal industries of the world, dictionaries of technology, catalogues of museums of arts and manufactures, also copies of all statistical reports issued by the United States and other countries. There will also be kept on file the prominent trade journals of the world, and especially such as contain notices of proposed contracts for all kinds of material and labor products. All documents furnishing information in regard to the articles on exhibition will be regularly filed and numbered, so that by consulting the catalogue and the number of the document any article may at once be found by the persons desiring to examine it and know its commercial value.

We give in this issue an exterior view of this building and also an interior view taken at the time of one of the great flower shows therein. Interest just now is centering in the great electrical exhibit to be held in this building next spring. It is proposed to make this exhibit one of international importance. The building is peculiarly adapted for this class of expositions being very large, fire proof and substantially built.

The ground space covered by the building is 200 x 275 feet, and including its roof garden and auditorium it affords over 425,000 square feet of floor surface. Every foot of this is perfectly lighted—by natural light by day and by innumerable electric lamps and gas jets by night.

The central section of this grand auditorium, which will easily accommodate 8,000 people, is surmounted by a glass dome 65 feet high and containing 12,000 square feet of glass, and its construction is such that it is one of the strongest and most perfect light-giving domes ever built.

Very Light Petroleum Motor.

According to the *American Machinist*, a Kane-Pennington petroleum motor, weighing only 17.5 pounds gave 4.75 indicated horsepower when run at 700 revolutions and 10 horse-power at 2000 ! A speed of over a mile a minute was said to have been obtained with this motor on a bicycle. The inventor is about to experiment with the application of this motor to flying machines, some preliminary experiments having given very promising results.

An important invention, now making a place for itself in textile manufacturing, is a loom which feeds the bobbin into the shuttle automatically. As this hitherto has been one of the chief duties of the weaver, the new device promises to supplant a considerable amount of labor. Five persons can take care of 80 of these looms, where now six looms are most commonly consigned to a single operator. They can be run an hour or two after everybody has left the factory at night, and throughout the noon hour, when the operatives are at dinner. How important the invention is may be judged from the fact that where tried it is saving one-half the labor cost in weaving and about one-fourth the whole labor cost in manufacturing.

Science in Warfare.

Thomas A. Edison has a good many interesting ideas. One of his latest is for the [illegible] York by lining the bottom of [illegible] out with a network of electric cables [illegible] could be strung at intervals deadly torpedoes which could be discharged into the bottom of hostile ships. Ships coming to the attack of New York [illegible] certain to pass over some of the cables and means of a range-finder located in an observatory on the shore the proper moment could be determined for discharging any of the torpedoes by means of an electrical connection. Thus the whole [illegible] navy could be blown out of the water and destroyed without the loss of a man on our side. All other coast cities could be defended in the same way at an expense insignificant. Mr. Edison claims, when compared with the cost of building battle ships.

Another of the great inventor's ideas is the possibility of defending a fort by means of streams of electrically charged water. These streams could be played upon an advancing enemy through hose, and he claims that 25 men, armed with such a hose, could successfully defend a fort against 10,000 or any number of men. The whole advancing line could be drenched with water and prostrated. The water is a good conductor of electricity. The moment the stream would strike any part of the body, even the smallest exposed portion of the face, the body would complete the circuit and the man would fall as one who accidentally touches a "live" wire.

The Western Strawboard Company is the name of a new corporation organized in Chicago to manufacture a new kind of straw board. The plant is now being installed. The board will be manufactured from the contents of the paunches of cattle killed at the stock yards. It is believed by manufacturers under the present system that the board made from this material will lack strength and the odor will be objectionable.

NEW YORK INDUSTRIAL BUILDING—VIEW OF AUDITORIUM.

Taylor Patent Upright Water-tube Boiler.

The manufacturers of this boiler, which is a marked departure from the general form of water-tube boiler, claim for it that the pipes being upright, conduct the steam directly to the dome, naturally, as fast as generated. It forces no considerable quantity of water ahead of it as in horizontal or inclined tubes, to enter the dome in the form of spray, causing the steam to be wet and weak, but being comparatively free from water is of the dryest and most elastic quality.

Neither do the pipes, being upright, accumulate ashes to retard the absorption of heat, but receive full benefit of the fire. The manifolds over the fire, while close enough together to form a crown sheet, forcing the volume of heat to follow the course indicated by arrows in the illustration, have apertures between them sufficient for the force of the draft to keep this crown sheet free from any accumulation of ashes, even on its top, and every pipe, and every part of the heating surface is subjected to the nearest possible approximate to the same degree of heat.

The cut illustrates a 300 H. P. boiler, in which the longest distance through pipe from bottom manifolds to dome is six feet. There are four large manifolds at the base, from which arise 128 1¼-inch pipes, or 32 from each, making three separate fire holes. There are 32 manifolds over the fire, each entered at the bottom by 4 of the ¼-inch up-flow pipes, and

removed with a brush, the pipes being in straight rows and spaces ample for the purpose.

The pipes being straight and upright, and all being subjected to the same degree of heat, and having the same water level, the variation in expansion and contraction is reduced to the vanishing point. It is the boast of the manufacturers that not a cent has ever been expended for repairs to any pipe or any fitting in any boiler ever made under these patents.

These boilers are made in two general styles. One for boats desiring the maximum of speed without regard to fuel consumption. The other for boats desiring good speed with proper regard to coal bills, also economy of floor space. The two forms are covered by separate patents. The latter form while running 15 miles per hour, show a stack temperature of less than 600 degrees (same style as illustrated here.)

The Detroit Boat Works were first to put the Taylor boiler into yachts. It was first put into the launch "Dream," 63 ft. long, 9 ft. beam. The "Dream" as originally built had a first class water tube boiler of the usual form but when called on for much greater speed, could only maintain it for short distances, but when the Taylor boiler was put into the "Dream," the engine made more than 200 additional revolutions and the boat made more than two miles per hour additional speed, and can maintain it indefinitely, the same hull, engine, engineer, fuel and same pressure being used, (and with all a saving of fuel.) The showing was so good that the Detroit Boat Works ordered one for the yacht

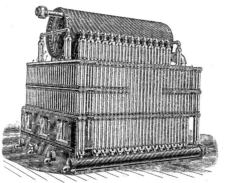

THE TAYLOR PATENT WATER-TUBE SECTIONAL BOILER.

at the top of each are 29 openings for ¼-inch pipes. These 32 manifolds each having 29 upright pipes, makes 928 1¼-inch upright pipes between all of which the products of combustion must pass. The heat must follow the course indicated by the arrows in cut, up the vertical flue to fire brick baffle, across among all of the pipes to center where a number of the second series of manifolds are left out to form an upright flue, thence upward and again each way across among all of the pipes, under the upper manifold to the vertical flue, thence upward under the dome and feed-water pipes to the stack, which may be placed central or otherwise at will.

Experience indicates that the steam and water are separated mainly in the upper manifolds, but that sufficient water is carried into the dome to form a constant column in the two 5-inch down-flow pipes at the back of boiler, keeping the hot water about the fire in constant circulation toward the dome, where it delivers up its steam.

At each end of all upper horizontal manifolds there are brass plugs which may be removed for examination or cleaning the manifolds. At the caps of each of the 5-inch bottom manifolds are 1½-inch blow-off pipes, controlled by valves. There are also brass plugs in the caps for the purpose of cleaning, washing or inspecting. There is a 7-inch hand-hole in the front end of dome. The casing is made in sections any of which can be readily removed. A section removed from the front exposes all of the upright pipes, and if there should from any cause be any accumulation of ashes or other matter, they could be

"Azalea" 115 ft. long and later for the "Cynthia."

The Detroit Boat Works made the following, report to the manufacturers of the Taylor Boiler of its performance in the two splendid yachts built by them last year.:

"Azalea" 104 ft. on water line, 115 ft. over all, 16 ft. beam, cylinders 9¼-14-24 x 14-inch stroke, 300 revolutions per minute, 325. H. P. (max.) made 17 miles per hour, (250 lbs steam pressure.) "Cynthia" 111 ft. on water line, 132 ft. over all, 17½ ft. beam, cylinders, 10-16-26-inch x 16-inch stroke, 280 revolutions per minute, 425 H. P. (max.) made 17 miles per hour, (250 lbs steam pressure.)

The stacks of both of these beautiful yachts are painted white and after a season's use show no discoloration from heat, showing that the fuel spent its force where it would do most good.

One by One They Fall.

Last year a lady inventor of Washington called on the INVENTIVE AGE to ascertain if an alleged patent selling firm in San Francisco were reliable. Enquiry was made through an attorney in San Francisco and information received that where the firm of R. McGregor & Co., advertise themselves as a saloon stands and—no patent seller. Doubtless all "advance fees" sent them reached headquarters all right.

One-quarter of all the people born die before six years, and one half before they are sixteen.

The Patent Committees and Legislation.

In the Senate the new committee on patents consists of seven Senators, as follows: O. H. Platt of Connecticut, J. C. Pritchard of North Carolina, C. D. Clark of Wyoming, Geo. P. Wetmore of Rhode Island, W. Call of Florida, R. Q. Mills of Texas, and J. H. Berry of Arkansas.

Senator Platt, who is chairman, has served in the same capacity for many years, is thoroughly posted on every thing relating to the subject, and has shown himself to be one of the staunchest friends the patent system ever had. His speech in the Senate March 31, 1884, is probably the most complete statement showing the benefits to the country of the patent system, ever made.

Of the other six members, three, Messrs Call, Mills and Berry were on the committee before, but as the committee was not called upon to act on any bill, it is not known what their feeling on the subject is, except that Mr. Call has always avowed himself as the friend of the inventor. All are lawyers.

In the House the committee consists of thirteen members, as follows:

W. F. Draper, Mass., chairman. J. D. Hicks, Pa., E. Sauerhering, Wis., B. L. Fairchild, N. Y., W. M. Treloar, Mo., C. A. Sulloway, N. H., E. D. Cooke, Ill., W. S. Kerr, Ohio, R. J. Tracewell, Ind., J. C. Hutcheson, Texas, T. J. Strait, S. C., G. A. Robbins, Ala., and J. A. Walsh of N. Y.

Of these, five were on the committee in the last Congress, viz: Messrs. Draper, Hicks, Hutcheson, Strait and Robbins.

The chairman, Hon. W. F. Draper is a manufacturer, and like Senator Platt, is thoroughly posted on the subject and is a man of marked ability. Of the others, ten are lawyers, one a pharmacist, and one a graduate of a medical college.

From their location, it will be seen that five are from the eastern states, four from the west, and four from the south, a very fair distribution territorially. In the Senate three are from the northeast one from the extreme west, and four from the south; the southern quota having been increased of late years, since the "New South," has engaged so largely in manufactures.

These committees will be called upon to act on a bill prepared by a special committee of the National Bar Association, to amend the patent law, the most important feature of which is to relieve American inventors from the injurious effect of section 4887, as finally construed by the Supreme Court.

As the law now stands it punishes the American inventor, if he tries to protect his invention abroad, because it provides that his U. S. patent shall expire at the same time as his shortest prior foreign patent.

If he will patent his invention in the United States only, he can have a full term seventeen year patent, all foreign nations of course being free to use his invention from the beginning. But if he tries to secure a patent abroad—and, unless he patents it abroad first he cannot secure a valid patent in most foreign countries—then his patent here can run only as long as his shortest foreign patent.

The injustice of this to American inventors has long been seen, and many attempts made to correct the evil, but so far in vain. Now that the Supreme Court, twenty-five years after the enactment of that section, has finally decided what it does mean, it is agreed on all hands that it should be amended so that the American inventor who files his application in the United States before he does abroad, shall have a full term patent here, regardless of the date or issue of his foreign patents.

That will enable our inventors to secure valid patents at home and abroad, and will prevent our foreign competitors from having the free use of inventions for which our own manufacturers have to pay royalty, thus benefiting our manufacturers as well as our inventors.

It is a remarkable fact, that this provision, in a somewhat different form, was first introduced in 1839 to enable aliens to secure patents in the U. S. notwithstanding the fact that they had previously

patented the invention abroad, the object of course being to induce foreigners to introduce their inventions here, so that our people could have the benefit of them. In order however, that our people should not be compelled to pay a royalty on these foreign patents after they had expired and become public property abroad, the law provided that the patents should expire here at the same time they did abroad.

No one then dreamed of its applying to American inventors, as at that time Americans did not take patents abroad. But in the meantime, things have so changed that of late years Americans desire to patent all valuable inventions in Europe as well as at home; and as the act of 1870 was worded the Supreme Court held that it applies to American inventors.

The bill as presented contains a few other amendments, but none of anything like the importance of this.

What ought to be done, is to reorganize the Patent Office, separate it from the Interior Department, give it the use of the Patent Office building, built largely with money paid by the inventors, and give it enough of the $4,500,000 surplus patent fund now in the U. S. Treasury, to furnish it with the force and appliances necessary to enable it to promptly and properly perform its duties. Every dollar of that money has been paid by the inventors, in addition to the cost of running the bureau, and these inventors have in addition paid their proportion for the support of the government, the same as all other citizens. The Patent Office is the only self-supporting bureau that the government has for ever had, and it seems an outrage to deprive it of the use of its own funds in such a manner as to cripple it, as it now is, and for years has been.

Then, there should be a special court established to hear appeals from the Patent Office, and try all patent causes; so that there might be speedy and uniform decisions, by judges who know something of mechanism as well as of patent law, which Justice Story said was "the metaphysics of law."

There is no question but that our patent system has done as much for the growth and prosperity of our agricultural as well as our manfacturing industries, as the tariff has, and yet Congress spends nearly every session fighting over the tariff, while it seldom devotes an hour to the patent system.

It is not expected that these improvements can be brought about at the present time, but it is to be hoped they may before many years pass.

It is a very important subject, requiring far more time and attention than any one seems disposed to give it at present.

Death of Alfred E. Beach.

The INVENTIVE AGE pays its tribute of praise due to the memory of this truly useful man. His loss we greatly feel, and so will the whole country. He possessed a mind of great versatility of thought and his perceptive qualifications were of a very quick and far reaching order; his industry knew no bounds. At one time about 1853-6 he ventured outside of the Scientific American, upon a great, independent enterprise in New York City—the publication of "The Illustrated People's Art, Literary, and Scientific Journal," but finally merged the same with the Scientific American, and threw all of his energies in the latter named journal, associating himself as a partner with the other proprietors, Messrs. O. D. Munn and Salem H. Wales.

For 40 years, he ceaselessly devoted his brain and energy in the cause that he had espoused, promoting science and invention. Such a man as Alfred Ely Beach is seldom found in the walks of life. Large in his sympathies, wide in his views of everything, simple as a child in his daily living; habits strictly temperate, and diet or food of the very simplest kind, and kind to his family; and being faithful to his God and fellow men, and full of charity, he has gone to his external reward.

On one occasion 1857-8, his heart being deeply touched by the fact that an old, worthy, colored man named "Tasko," and about 60 years of age, who was employed to clean up the branch office of the Scientific American in Washington, D. C. was still a slave, he conferred with Mr. Fenwick, who at that time had the conduct of the branch office in his charge, and the finale of the matter was the raising of the sum of $300 by Mr. Beach in New York and Washington, paying the same to Tasko's owner, and Tasko became a free man during the remainder of his life. This incident soon brought Mr. Beach to the attention of others, and a little, fair faced, colored girl, about 12 years of age, named "Pinkie" was carried, through Mr. Beach's influence, to New York, and members of the Plymouth Church also purchased her freedom.

Mr. Beach was the inventor of the Beach Hydraulic Shield, the first example of this machine which is now in common use by engineers in all parts of the world. The Beach Hydraulic Machine was used in the construction of t e great railway tunnel under the St. Clair river at Port Huron and also for the tunnels in London, Glasgow and under the Hudson river. Mr. Beach founded, soon after the close of the war, the Beach Institute, at Savannah, for the education of freedmen, and has always been noted for his private charities. He leaves a widow, one son and a daughter.

Wonderful Discovery in Photography.

A dispatch from London says: The noise of wars alarms should not distract attention from the marvelous triumph of science which is reported from Vienna. It is announced that Prof. Routgen, of the Wurzburg university, has discovered a light which, for the purposes of photography, will penetrate wood, flesh and most organic substances. The professor succeeded in photographing metal weights which were in a closed wooden case; also a man's hand which shows only the bones, the flesh being invisible. The Chronicle correspondent says the discovery is simple. The professor takes a so-called crook and pipe, viz., a vacuum glass pipe, with an induction current going through it, and by means of rays which the pipe emits photographs on ordinary photographic plates. In contrast with the ordinary rays of light these rays penetrate organic matter and other opaque substances just as ordinary rays penetrate glass. He has also succeeded in photographing hidden metals with a cloth thrown over the camera. The rays penetrated not only the wooden case containing the metals, but the fabric in front of the negative. The professor is already using his discovery to photograph broken limbs and bullets in human bodies.

The $1 Edison Motor.

The cut shown herewith represents one of the most satisfactory and perfect of all small dynamos, capable of running toys, fans, light models and for innumerable uses. To the youth scientifically inclined, it offers great satisfaction and benefit.

It is a marvel in price, and if not made in large quantities would cost from $5 to $10. We can fur-

nish the motor for $1 and will send one free to any person forwarding us three new subscribers.

Horseless carriages have now become a feature of the boulevards of Paris. Their introduction in American cities is no longer a novelty.

Meeting of A. A. I. & M.

THE annual meeting of the American Association of Inventors and Manufacturers which is to be held at the Board of Trade Rooms, in this City, on the 21st instant, promises to be an unusualy interesting and important one. In addition to a few informal papers on subjects of importance, it is expected that there will be short pointed talks by members present and letters from absentees on a variety of live topics. The Executive Council has sent out a schedule of subjects on which discussion is invited and members are free to select other subjects if they desire. The Council's list of subjects is given below:

1. The propriety of a Federal Statute of Limitations in Patent Causes.

2. The importance of having American Patents granted for a definite fixed term without regard to the date of expiration of corresponding foreign patents.

3. Shoud not a patent become void by failure to manufacture and use the invention in such manner that it may be brought to the attention of the public within a fixed period from the date of the grant?

4. Would it not be beneficial to our Patent System to so amend the Patent laws that innocent users of devices and processes covered by old and unworked patents, of doubtful validity, may not be compelled to pay tribute to purchasers of such patents?

5. Should not the Commissioner of Patents be clothed with authority to appoint, on application, Special Examiners, for taking testimony in Interference Proceedings the same as Special Examiners are appointed by the Courts in Equity Cases?

6. Are the New Patent Office Rules abridging the time for response to official action desirable and sustainable under the statute?

7. Should or should not the United States Patent Office be made a separate and distinct department and be provided with a building of its own?

8. In view of the necessity, already arisen, of storing a part of the models elsewhere than in the Patent Office should not models be placed, all together, in a suitable building specially designed for that purpose?

9. How can the rules of procedure in the Patent Office be amended or changed to facilitate and improve the character of the work which it is intended to accomplish?

They embrace topics which are of vital interest to the great army of inventors throughout the country, and some of them have already been discussed editorially and by letter in this paper, which is always found in the front rank whenever reforms are to be instituted. During the last year, a large number of very prominent inventors, manufacturers and others have joined the society and a large attendance at the meeting is anticipated.

While the foregoing subjects are deemed to be proper ones for discussion by the Association it is not intended to discourage the presentation of papers on other subjects relating to patents and papers are invited from members of the Association upon any subjects within the Association's recognized field of work.

A full report of the proceedings will be printed in the INVENTIVE AGE.

The letters in the various alphabets of the world vary from twelve to 202 in number. The Sandwich Islanders' alphabet has twelve, the Tartarian, 202.

Electricity Applied to Bennett Amalgamator.

A somewhat unique application of electricity and one of interest to the mining industry has recently been made in the adaptation of electric motors to the operation of a Bennett Amalgamator. The Amalgamator consists essentially of the following parts : A truck and frame for supporting the larger part of the machine—arranged to be run forward on a track as the work progresses ; a turn table supporting a boom and dipper : the boom and dipper for excavating the dirt and a revolving cylinder or screen with a hopper. Into this earth is thrown from the revolving screen, the finer material passing through into the amalgamator, the coarse material being discharged at the end of the drum into the tailings carrier

The amalgamator is a large trough, in which the fine material after passing through the screen is amalgamated. In the bottom of the amalgamator is an agitator, moving backwards and forwards to keep the material well stirred up, and a wheel for raising and discharging the fine tailings after the removal of the gold.

The four motors used in operating this machine have been supplied by the General Electric Company and are so constructed as to be thoroughly protected from water or dust. One of these motors is mounted on the frame of the machine between the turn table carrying the cab and the revolving screen. It is geared to a shaft connected by a sprocket chain

THE BENNETT AMALGAMATOR OPERATED BY G. E. MOTOR.

to the revolving cylinder, the tailings carrier, the agitator in the amalgamator and the wheel for discharging the tailings at the outer end of the amalgamator. A clutch is provided to throw it into gear with the trucks for moving the entire machine backwards and forwards.

The second motor is placed in the cab and operates the dipper by means of a fine wire rope, passing around the drum to which the motor is geared.

The third motor is mounted on the dipper boom and is geared to a drum placed just above it. Fine wire ropes pass around this drum and are attached to both ends of the dipper handle. The dipper may thus be thrust out or in according to the requirements of the work.

The fourth motor occupies a place in the cab and drives by means of a beveled gear and large sprocket chain the turn table. This motor is also used to swing the dipper boom around from the front of the amalgamator to a position which allows of the dirt being thrown into the hopper of the revolving screen. The motor operating the cylinder, the tailings carrier and the agitator in the amalgamator run continuously when the machine is in operation and is provided with a simple starting rheostat only. The other three motors are controlled by reverse rheostats in the cab and are handled by one man.

The " Ferris " wheel erected at Earls' Court in London, does not seem to be a very successful piece of engineering work. Its proportions have been criticised, and the method of driving is a failure. It seems that common chains resting in smooth grooves it was thought would give traction enough, but the wheel ran back with a test load, the chain slipping

in the grooves. It does not matter much. There is neither beauty, science, nor safety in such a hideous structure, and it should not be set up permanently anywhere. It is better to go up on an elevator and look out from the roof of a high building.—*Industry.*

Value of Acetylene.

When it was first proved that by heating a mixture of coal or lime or charcoal and chalk in an electric furnace, a compound resulted which upon being thrown into water evolved acetylene, the gas companies were naturally elated. They looked upon the new discovery as a cheap and easily procurable substance for mixing with their own product and thus raising its illuminating power. Acetylene, however, is turning out to be an even more extraordinary substance than has been recognized. It has so many strings to its bow that its exploiters are presumably abandoning the problem of enriching gas and are actively engaged in demonstrating that by means of it the cheaper manufacture of insumerable substances which are used in the arts, but which up to the present have been the products of pure chemistry, can be achieved. The acetylene on being passed through an iron tube heated to dull redness turns rapidly and completely into benzine. This is a product of prime importance and is the base of thousands of organic substances. In illustration of the transmutations which can be effected, it may be

pointed out that if the resultant benzine vapor be passed into strong nitric acid it is transformed into nitro-benzine, and this, on treatment with hydrochloric acid and iron fillings, go into aniline, whereby the road is opened for the prediction of the immense series of dye substances, of which aniline is the starting point. Instead of transforming acetylene into aniline, however, it may be changed into carbolic acid—thence it is but a step to picric acid, the formation of the modern high explosives. Or it may be made into aniline and then boiled with acetic acid, when it is transformed into anti-fibrin, the well-known fever specific. Again, by passing it through a tube heated to bright redness, napthalene is produced, which is also the starting point of a legion of valuable chemicals. It would seem as though almost all the needs of man were able to be satisfied by this protean substance. The further investigation is pushed into its possibilities, the more astounding and bewildering they become. By the action of nascent hydrogen acetylene becomes ethyphuric acid and water becomes alcohol, which, apart from its other uses, is absolutely necessary to the production of an enormous number of economic substances. In a similar ways we can get such deadly poisons as oxalic acid and prussic acid, while acetylene is a cheap source of the aldehyde so much used in the production of artificial essences and the manufacture of mirrors. When, therefore, it is considered that from acetylene can be derived whole systems of dyes, medicines, essences, perfumes, poisons, explosives,not to mention cheap whiskey, it will be seen that the latest product of the electric furnace has a utility out of all proportion greater than that which can be derived from its peculiar light-giving

powers. In speaking of the wonders of calcium carbide and acetylene gas *Invention*, of London, says :

Although known for many years as a laboratory product, its rediscovery as a commercial element comes from T. L. Willson. A brief history of the discovery may be of interest to our readers, as the new product is more than likely to become a principal illuminating factor in the future. In addition to its illuminating powers it is more than probable that it will become the principal fuel of the future, either by itself or in some combination with hydrogen and oxygen as yet undiscovered. Mr. Willson was trying to manufacture cheap aluminum at his works in the hamlet of Spray, North Carolina, United States. He conceived the idea of decomposing lime and separating the metal calcium, and further bringing calcium into contact with ordinary alumina or clay by fusing them together by the heat of the electric arc. Mr. Willson in the course of his experiment brought together lime (calcium oxide) and finely divided or powdered carbon in the form of coke dust. He placed the mixture between the poles of an electric arc in a furnace, and the intense heat caused it to melt. Instead of however, finding what he hoped for, the metal calcium, he was in possession of 100 pounds of a brownish crystalline substance. Treating this as refuse it was thrown aside, and but for an accident might have slowly disintegrated and practical acetylene gas have been still in the future. By a chance almost as strange as that which befell the French scientist Gramme when, through the stupidity of a workman, he discovered the reversion of the Gramme ring which gave the electro-motor to the world, by means of a little water accidentally thrown on the disagreeable looking slag, acetylene gas blazed out as a new luminant to the world. Providence is kind to the earnest seeker in the field of science; and sometimes guides his erring hand as a father guides the hand of a child in its first tentative effort. In this case Mr. Willson did not succeed in his efforts to obtain calcium, but his experiment has given to the world what is infinitely more valuable. By this accidental discovery, as it may be termed, an amazing future is opened up to the world. A new element is added to our civilization more potent in its possibilities than any other discovery made since science passed from its purely empirical stage.

It is not often that the greatest discoveries come from the most profound schoolmen ; in fact, this is but rarely the case. A comparatively unknown experimentalist is the discoverer of the commercial value of the new agent. It remains for the profoundly learned minds of trained scientists to work out the why and wherefore of the discoveries; but it is for the comparatively untrained minds to push into unexplored fields of knowledge, just as the discovery of gold is due not to the geologist but to the ignorant prospector. After the discovery is made, and the riches of the hidden mines are opened up to the world, the trained geologist comes along, the trained chemist, the scientific engineer, and they make practicable the discovery of the more ignorant mind. The lesson to be drawn from this is, that it is the bounden duty of everyone, to turn his attention, his mind, his efforts, his experience, to searching for new inventions, for new methods, or new designs. The prize that awaits the successful inventor or discoverer is twofold—fame and pecuniary betterment. One should not despise therefore, the day of small things, nor turn his attention from the little things of life, but constantly strive to work out some idea, however vague, into its concrete form, and give to the world some new impulse of civilization.

The original models of the first sailing vessel and the first steamboat ever built in the Pacific Northwest are now on exhibition at the Washington State horticultural rooms in Tacoma. They were placed there by Captain Grounds, designer of the schooner and owner of both models. The steamer was called the Columbia. She was built in this city in 1849 by Frost & Adair. She is of the sidewheel type, and not unlike the sidewheelers to be seen on the Columbia today. She was used for carrying freight and passengers on the river. The schooner was called Calaipooia. She was built in Oregon City in 1845 by John Cook. She was 65 feet long, 19 feet beam, and 5 feet hold, being flat-bottomed. The two models are attracting considerable attention. They are not of the finest type, but superior in build to many steamers of the present generation. Capt. Grounds now resides in Tacoma.—*Portland Oregonian.*

An evidence of the striking uniformity of size among the Japanese is found in the fact that recent measurements taken of an infantry regiment showed no variations exceeding two inches in height or twenty pounds in weight.

Cats die at an elevation of 13,000 feet, even though they are reputed to have " nine lives " when on a level with the ocean. Dogs and men can climb the greatest known natural elevations.

A Water Cycle.

The accompanying illustration represents a very clever invention in the shape of a water cycle, now being manufactured by the International Water Cycle Company of Milwaukee, Wis. The boats consist of two half shells or floats made preferably of copper, supplied with four water-tight compartments making them non-sinkable. On top of each float are five adjustable slides fastened firmly which makes the frame work adjustable on these slides fore and after. On top of these aforesaid five slides there rest the three top cross adjustments which hold the two boats together and on which the frame work rests. The frame rests on the first and second cross pieces and the third supports the propeller shaft at the stern, or near the stern. The gear hanger is made of miter iron and is made adjustable for adjustment of the propeller up or down. This can readily be done. The propeller shaft is made up out of bicycle tubing with a pin or piece of iron brazed and pinned into each end of this shaft, one of which is to hold the bevel gear and ball bearing box at one end and the propeller at its other end. The shaft is supported by two hangers which are ball bearing throughout where there is any friction. The frame is made similar to the regulation bicycle except that where the wheels are supposed to be are placed supports or braces. The

front fork is stationary and acts as a brace. The steering is done by the handle bar, which runs down through the head and terminates into a T piece, on which T piece the rudder ropes are fastened, which by turning the handles turns both rudders at once. The whole machine weighs less than 90 pounds.

Money in Simple Inventions.

The idea of copper toed shoes was patented Jan. 5, 1858, by a Maine genius, who made $100,000 out of it. Another similar invention which made a great deal of money was the metal button fastener for shoes invented and introduced by Heaton, of Providence. At the time it was considered a fine invention, for the old sewed button was continually coming off. It has gradually grown in popularity since its introduction in 1869 until now very few shoes with buttons are manufactured without the Heaten appliance.

Placing the actual daily consumption of coal for the entire world as low as 1,600,000 tons, a solid cube of coal more than 100 yards on a side is burned every day.

The climatic limit to the cultivation of wheat is not so much the cold of winter as the heat of summer.

The brain is not affected by the movements of the body, even though these are sometimes very violent,

because it rests on a basis of soft cushions between bones of the spine.

A Hydraulic Air-Compressor.

The *Buffalo Express* states that C. H. Taylor has invented a hydraulic air-compressing system whereby power can be developed at much less cost than by turbines. The air is compressed by the direct action of falling water, without the aid of any moving machinery. The water is conveyed to a compressor by an open flume or through a pipe supplying a tank or stand-pipe round the headpiece of the compressor, where it can attain the same level as the water in the dam or source of supply. There are a number of small, horizontal airpipes around the headpiece. These draw their supply of air through larger vertical pipes, which extend above the surface of the water and open to the atmosphere. As the water enters this down flow pipe and passes the ends of small air-pipes, it draws in the air constantly, in the form of small uniform globules, which become entangled in the descending water, are carried down to an air chamber at the bottom of the pipe, compressing the air by weight of the water surrounding the globules, according to their depth below the tail-race water level, until they reach the point of separation. The air chamber at the bottom is sufficiently large to allow the air to rise to the surface of the water therein. Thence the air is taken through a pipe for transmissions to be utilized as power or for other purposes.

The idea of having the gases leave a boiler at a high temperature, in order that it may be more effectively used by heating the feed water in an economizer, is reported by Schmidt, a German engineer, to have been successfully applied by him in producing a very economical engine by extraordinary heating. The gases are represented as leaving the boiler at a temperature sufficiently high to permit of superheating the steam to over 650°. It is thought by experts, however, that though by this action the economy of the boiler must be reduced, the question presents itself whether it is not preferable to permit of less economy in the boiler, in order that the engine may be more economical—a point, of course, of special practical moment.

Dr. Berliner's gramophone, an instrument similar to the phonograph, but simpler, will soon be put upon the market. Its records of human speech and music are indestructible and can be cheaply multiplied to an indefinite extent by simple mechanical means. The entire machine is of such simple construction that it will cost less than $20 to manufacture it.—*Invention.*

A prize of 12,000 marks has been offered by the German Hygienic Association for a paper on the Efficiency of Electric Heaters.

Engineering Tools at Pompeii.

Under the title of "Things of Engineering Interest Found at Pompeii," Professor Goodman lately gave his inaugural lecture to the engineering department of the Yorkshire College, Leeds. The lecturer remarked that he had recently visited Pompeii, and was not only charmed by the great beauty of the works of the ancient Romans, but also by their extreme ingenuity as mechanics—in fact, it was a marvel how some of the instruments and tools they were in the habit of using could possibly have been made without such machinery as we now possess. After explaining the situation and destruction of Pompeii by showers of ashes and mud, not lava, as is usually supposed, in the year 79 A. D., Professor Goodman showed a series of about fifty lantern slides, prepared from photographs taken by himself in Pompeii last Easter. The streets he explained were used as waterways to carry off the surface water, and probably sewerage from the houses. The pavements were raised about a foot above the streets, and stepping stones were provided at intervals for foot passengers. The horses and chariot wheels had to pass between, and in many places deep ruts have been worn by the chariot wheels in the stone paved streets. The water supply of Pompeii was distributed by means of lead pipes laid under the streets. There were many public drinking fountains, and most of the large houses were provided with fountains, many of most beautiful design. The amphitheater, although a fine structure, capable of seating 15,500 people, was small compared with many in Italy. The bronzes found at Pompeii reveal great skill and artistic taste. The bronze brazier and kitchener were provided with boilers at the side and taps for running off the hot water. Ewers and urns have been discovered with internal tubes and furnaces precisely similar to the arrangement now used in modern steam boilers. Several very strong metal safes, provided with substantial locks, have been found. The locks and keys were most ingenious, and some very complex. On looking at the iron tools found in Pompeii, one could almost imagine he was gazing into a modern tool shop, except for the fact that the ancient representatives have suffered severely from rust. Sickles, billhooks, rakes, forks, axes, spades, blacksmiths' tongs, hammers, soldering irons, planes, shovels, etc., are remarkably like those used today; but certainly the most marvelous instruments found are the surgical instruments, beautifully executed, and of designs exactly similar to some recently patented and reinvented. Incredible as it may appear, yet it is a fact, that the Pompeiians had wire ropes of perfect construction.

Newton's theory of gravitation was much closer to the modern than that of some of his commentators. He denied that one body can act on another at a distance through a vacuum and called that theory an absurd error.

AN INGENIOUS WATER CYCLE.

Transmission of Human Thought.

Julius Emmner, jr., claims to have perfected an invention by means of which the thoughts of any person may be accurately recorded, and intelligibly transmitted to any other person, without the intervening medium of speech. In other words, the invention is one as will enable man to pry into the brain recesses of his fellow man, and discover what ideas are harbored there. Heretofore, the thoughts of man have been sacred and inviolate, so far as he may have chosen to conceal them, except as a general idea as to their nature, whether pleasant or disagreeable, may have been gathered from the facial expression. That there could exist any natural or scientific means by which thought could be stripped of the cloak of secrecy which mantles it, and the silent and mysterious workings of the mind be laid bare to the world, through any scientific device, has been held among the impossibilities.

Emmner is the inventor of the long distance telephone, and has been at work for three years upon what he designates his "thought machine." The principle involved is similar to that which is responsible for Edison's phonograph. The latter, as is well known, is an accurate mechanical register of sound. Science teaches that sound is a succession of atmospheric waves, set in motion by vibration. These waves are borne to the ear, and analyzed by what might be termed an aural keyboard; each key receives the corresponding sound wave, the impression is conveyed to the brain, and there is experienced a sensation, sometimes musical, sometimes discordant, sometimes harsh, sometimes soft, sometimes clear, sometimes dull or muffled—depending upon the character of the sound waves which have produced the impression. Science has been enabled, through this knowledge, to register these waves upon a super-sensitive cylindrical surface, and to create, at will, the same sensations of sound which were received. If thought, therefore, were an actual and physical effort, involving external manifestations and disturbances, similar to those occasioned in the production of sound, why was there not a scientific possibility of registering and reproducing these disturbances? That thought manifested itself in some tangible and material way, was argued from the fact that occasional persons of peculiar and highly sensitive organisms have been able, apparently, to receive intelligible impressions of the mental processes of others through simple physical contact. These persons have been known as "mind readers," and elements of the supernatural have entered into them in no slight degree in the minds of no small number. If these discernments of the thoughts of others were genuine, was there not some rational and natural law by which they were accomplished? Such were the arguments in favor of the possibility of a "thought machine," and the premises upon which Emmner claims to have perfected his invention.

The invention is similar to the phonograph. There is a delicately sensitized surface for the registration of impressions. An electric current aids in the transmission to, and indication upon, this surface, of the disturbances which thought is claimed to produce. The machine is attached, by a certain connecting wire, to the temples. Impressions are borne along this connection to the registry surface, and there transcribed by the machine. The receiver is removed, and a transmitter applied. The machine is applied to the temples of another person. The same impressions formerly received, are borne back along the connection, to be taken up by the brain, analyzed by that organ, and resolved into conscious thought. And there you are, your brain teeming with the same thoughts and ideas which were registered upon the sensitive surface by the first comer.

Impossible, you say. Emmner says actual, and besides, recent scientific developments have gone far toward rendering the word impossible obsolete. Of course the right to skepticism is reserved to every one until there shall be an actual and physical demonstration of the workings of the machine. But if it should prove a reality and not a myth, it would be in nowise startling. And if it should prove a myth,

it is not for any one to say that the future will not dissipate the myth and develop the reality. For science is a region of which we are yet barely across the threshold, and that which is but the simple scientific wisdom of today was the supernatural of yesterday, and what is the supernatural of today is the science of tomorrow.

The possibilities of a thought machine as above described would be marvelous. In criminalogy it would be unerring and infallible in the detection of guilt or innocence. It would do away with the necessity of criminal trials, and preclude the possibility of escape of a criminal from justice. In medicine, the physician could receive the actual impressions of the patient, and be placed upon an equal footing with disease, in combating its advances. Insanity and its vagaries could be understood and intelligently treated. Plots, conspiracies, nefarious designs and unlawful schemes of suspected persons could be laid bare and frustrated. The liability of man to a test at any time would, through fear, bring about a minimum of evil. In fact, it would revolutionize the whole social fabric, and inaugurate a new era.

The American Boiler Stoker.

The stoker consists of a hopper, capable of holding one hundred and fifty pounds of coal, placed between the ash-pit doors, occupying but twenty-four inches of space in front of the boiler and allowing the doors free swing. Directly beneath the hopper is an incased gear, driving a "worm," which enters the furnace by an opening but ten inches in diameter in the boiler front. Within the front is a "V" shaped coal-bed, three feet wide at the top, two feet deep and extending the length of the grate bars. Along the top of the coal bed, on each side is placed a row of heavy iron blocks called "tuyere blocks," through which the air is discharged into the coal. These tuyere blocks are about the size of a brick, weigh 12 pounds each and are very easily replaced.

The coal fed into the hopper is conveyed into the coal retort by a steel conveyor, and when there is evenly distributed and raised in a body to the level of the "tuyeres."

As the coal slowly approaches the fire above, it becomes hot, the volatile gases are released and mixing with a supply of air from the tuyeres are exploded into a flame. This air is supplied proportionately to the amount of coal fed, and at a mild pressure.

The gases having been drawn from the coal, we have coke, and this being pushed up by the continuous feeding from beneath, spreads over the entire grate surface, thoroughly charged by the air and intensely hot.

It can be readily understood how impossible it is for the smoke producing gases ever to pass through this coke bed and not be consumed. The combustion of the fuel is complete. The non-combustible in the coal is removed in the shape of small vitrified clinkers. We have no ashes. The action of the coal under continuous feeding is very beneficial. The coal is always in action—breathing—and the air penetrates it thoroughly. Large clinkers are an impossibility, as the movement in the coal prevents their formation. The rate of feeding the coal is

regulated by the lever in front of the coal hopper. Six different speeds are obtainable, and coal may be fed at a rate of from 25 to 1,500 pounds per hour.

The power to actuate the stoker and drive a small blower is derived from a small upright engine, which with the blower forms a part of each equipment. The ability to force a boiler is unlimited, as there is no dependence on natural draft.

In applying the stoker there is no tearing down of boiler fronts or interior. Under ordinary circumstances, any boiler can be equipped and put in running order within twelve hours.

None of the mechanical parts of the stoker are subjected to heat, thus obviating the difficulties caused by expansion and contraction. The tuyere blocks are the only parts coming in contact with the fire and they are constantly filling with air, and experience has proven that they will not burn out in nine months severe duty.

Numbers of them have been placed in Toledo, Detroit, Cleveland and Cincinnati, and the company has now completed its arrangements by which it will have a capacity of ten stokers per day and will enter the entire field. Very interesting experiments have been made, looking to the adaptation of the stoker

SECTIONAL VIEW OF THE "AMERICAN" STOKER.

to puddling and heating furnaces, and tests exist, showing that by use of the stoker, the out put of iron can be increased from 30 to 40 per cent, the fuel being bituminous slack coal.

The stoker is essentially a gas producer and a gas burner. By the use of the cheaper grades of coal, a large saving is made in the difference in price of coal, and the regularity of the heat and the draft, enable the operator to steam a boiler far beyond its rated capacity. In an installation of ten boilers, 60 x 14, in a large steel works in Detroit, the stokers are developing 129 H. P. under each boiler. The absence of soot in the boiler flues is also another marked feature.

The fact that the poorer grades of coal can be burned without smoke and practically no ash is itself proof of the fact that in the conversion of the cheaper coals into gas, all of the volatile elements in the coal are removed and consumed.

As there can be no loss by dropping through grates, and the fact that the flue temperatures are very low and the evaporation results very high, is conclusive evidence that the stoker is furnishing as complete combustion as is possible. The stoker is especially adapted for use on ship board, and experiments are about to be made by a prominent railroad company, looking to its use on locomotives. The fact that hand firing can be resorted to any time, is a very valuable feature.

The deepest well on our Atlantic coast is that at the silk works near Northampton, Mass., depth, 3,700 feet.

What they say of the "Inventive Age."

Lockport, N. J.— * * * Please never stop my paper, for on of the last number I have received one dollar's worth of benefit.
                                                               C. D. O.
                          * * *

Howell, Mich.— * * * I am well pleased with your paper; am an inventor and have had some experience with patent sharks. Of course my experience cost me $20.00 but I'll know better next time. Had I been taking your paper when negotiations were being made with these sharks I would have known better.
                                                               L. C. P.

SUPPLEMENT.--Tips to Inventors.

The Infancy of Invention.

As capital is constantly being invested and expended to protect and preserve capital previously expended and invested in various enterprises all over the land, so will inventions continue—their variety and multiplicity being demanded to further the usefulness and perfection of inventions previously originated.

It was Edison who, replying to the question, "Do you think that the inventions of the next fifty years will be equal to those of the last?" said: "I see no reason why they should not. It seems to me that we are at the beginning of inventions." The truth of this prediction is illustrated in the many useful and wonderful achievements of Mr. Edison's own laboratory since giving utterance to this statement only a short time ago.

Profits from Invention.

The value of an invention is determined by no fixed rule. Fabulous sums have been made from simple and novel, as well as complex and useful, inventions. It is a fact that four-fifths of the business of the United States is transacted by the use of inventions. The benefits to mankind because of inventions, are so manifest and so common we are apt to look upon them in a matter-of-fact sort of way and fail to give the inventor the credit due him. In the majority of cases, however, the failure of an inventor to reap a reward is attributable to his own negligence, lack of forethought and indiscretion.

Nearly every human being is an inventor, but only a few obtain a *monopoly*—a patent—on the product of their brain. There are thousands of really useful articles, appliances and discoveries, in use every day by millions in all walks of life, that might have been patented had the inventor possessed the business sagacity that has actuated his more fortunate neighbor. Take for instance the open slot necessarily used in all conduit electric, or cable street railway systems. The inventor failed to get a patent on the idea and a fortune missed him.

There is money in inventions, but not always for the inventor.

The only way to make money out of an invention is through the protection afforded by a patent; not a patent in name only, but a *good patent*—one that is intelligently drawn, with claims commensurate with the scope and importance of the invention.

The profits arising from inventions in the electric field during the past twelve years have been simply astounding. In railway appliances, bicycles, typewriters, telephones, cash registers, slot machines and farm machinery, the field has been equally remunerative. And just think of that simple toy "Pigs in the Clover"—it netted the inventor, whose friends laughed at him for obtaining a patent on so simple a toy, over $150,000. The inventor of the metal plates to be attached to the worn heels of shoes (for sale in all cities) realized a fortune out of it amounting, it is said, to nearly $1,000,000. Perforated wooden seats for chairs and rubber tips for lead pencils brought the inventors big results. Howe made a million dollars from his sewing machine attachments, and the inventor of that simple lamp attachment, the inverted glass bell, to be suspended over lamps to protect the ceilings from being blackened, made the inventor rich. The "Darning Weaver," a device for repairing stockings, is a useful invention and is netting the inventor a handsome revenue on royalties. The wire nail and gimlet-pointed screw are fortune makers, and wire nails caused the invention of automatic machinery that manufacturers them so cheaply it does not now pay the carpenter to spend his time in picking a nail up when it drops, if it requires ten seconds to do so. The inventor of the well-known "safety pin" lived in luxury all his life, after discovering a means of concealing the point of a pin in such manner as to prevent scratching. The inventor of roller skates made nearly a million and the inventor of the needle-threader for a long time made $10,000 a year.

Relation of Capital to Invention.

Mr. Edward P. Thompson, one of the most entertaining writers on the subject of invention, says that "every invention, before the introduction into practical use, passes through two stages; namely, mental and physical"—mental when in the mind of the inventor only, and physical when the mental invention is put into bodily form by hand, or by hand with the assistance of a convenient tool. "A mental invention," says the writer, "sometimes does not become a physical invention because the inventor lacks money, technical knowledge or diligence. Such a mental invention often becomes a physical invention by the assistance of a capitalist, an educated person, or diligent companion." This being true the *mental* inventor, the person who, for lack of means possibly, would fail to make his invention a physical reality—such a person should take into his confidence a friend or companion to share the prospective benefits of his invention. Thousands of meritorious mental inventions are never worked out because of the over-timidity of the inventor, his exagerated greed for *all* the benefits to accrue instead of half the loaf, which in many instances is, or would have been, ample reward. Mr. Thompson truly says: "Inventors and capitalists should be more willing to co-operate. It is too often the case that the former must pay for his own experiments and all patent costs before a capitalist will even take the trouble to look into the merits of the alleged invention. On the other hand it is too often true that the capitalist seeks to join with the inventor, but the latter wants too high a price at the beginning."

Who Can Apply for Patents.

Patents are issued to any person who has invented or discovered any new and useful art, machine, manufacture, or composition of matter, or new and useful improvement thereof, not known or used by others in this country, and not patented or described in any printed publication in this or any foreign country, before his invention or discovery thereof, and not in public use or on sale for more than two years prior to his application, unless the same is proved to have been abandoned; and by any person who, by his own industry, genius, efforts and expense has invented and produced any new and original design for a manufacture; any new and original design for the printing of fabrics; any new and original impression, ornament, pattern, print or picture to be printed, painted, cast, or otherwise placed on or marked into any article of manufacture, or any new, useful and original shape or configuration of any article of manufacture, the same not having been known or used by others before his invention or production thereof, nor patented or described in any printed publication, upon payment of the fees required by law and other due proceedings had.

If it appears that the inventor, at the time of making his application, believed himself to be the first inventor or discoverer, a patent will not be refused on account of the invention or discovery, or any part thereof, having been known or used in any foreign country before his invention or discovery thereof, if it had not been before patented or described in any printed publication.

Joint inventors are entitled to a joint patent; neither can claim one separately. Independent inventors of distinct and independent improvements in the same machine can not obtain a joint patent for their separate inventions; nor does the fact that one furnishes the capital and another makes the invention entitle them to make application as joint inventors, but in such case they may become joint patentees. The receipt of letters patent from a foreign government will not prevent the inventor from obtaining a patent in the United States, unless the invention shall have been introduced into public use in the United States more than two years prior to the application. But every patent granted for an invention which has been previously patented by the same inventor in a foreign country will be so limited as to expire at the same time with the foreign patent, or, if there be more than one, at the same time with the one having the shortest unexpired term, but in no case will it be in force more than seventeen years.

Protection to Inventors.

What is a patent? It is a monopoly or grant, in the United States, for a term of seventeen years, to the patentee, his heirs or assigns, of the exclusive right to make, use and vend the discovery throughout the United States, as the inventor's rights may appear in the specifications and patent granted.

This means a great deal to the inventor who has secured a *valid* patent containing all the claims so worded as to prevent infringement and loss in contest. Thousands of inventors, obtaining patents through unreliable and inefficient attorneys or agents, find themselves possessed of patents *in name only*, and of no value when combatted by infringers with capital and the aid of able legal talent. A good patent costs no more than a weak and worthless one. Therefore how shortsighted are those inventors who employ cheap attorneys, saving $5 or $10 in fees, only to find themselves loosers of *all* they have paid when the contest comes.

The Need of Reliable Attorneys.

The Revised Statutes of the United States provide that "before any inventor shall receive a patent for his invention, he shall make application therefor in writing to the Commissioner of Patents, and shall file in the Patent Office a written description * * * of the same in such full, clear, concise and exact terms, as to enable any person skilled in the art or science to which it appertains, or with which it is most nearly connected, to make, construct and use the same; and in case of a machine, he shall explain the principle thereof and the best mode in which he has contemplated applying that principle, so as to distinguish it from other inventions."

To carry out these provisions it is necessary for the inventor to first make a clear, concise and complete drawing, or a working model of his invention or discovery, and send it to THE INVENTIVE AGE, or some thoroughly reliable attorney, who, before making application for the patent, should make a thorough and rigid examination of the Patent Office to determine upon its novelty or patentability. If the invention has been anticipated by some one else, or if it lacks novelty, or if for any reason a patent can not be granted, or, if granted, would be of no worth or value, then the inventor does not want to incur the expense of making application and paying attorney's fees and government fees. For making this thorough examination THE INVENTIVE AGE and all reliable attorneys charge $5, which fee is, under some circumstances, however, taken out of the additional fees paid by the inventor in case letters patent are applied for. The fees of patent attorneys vary somewhat, but the average fees for obtaining a United States patent are about $65—the government fees being $15 on filing the application and $20 on issuing a patent—the balance being the fees for preparing specifications, making searches, etc. The inventor is sometimes favored in terms given for payment of the fees, more detailed information regarding which can be obtained by enclosing a 2-cent stamp with enquiry to THE INVENTIVE AGE, Washington, D. C. The reason why the inventor should have a preliminary examination of the Patent Office made before applying for a patent lies in the fact that if the case is rejected the fees paid to the government and the attorney are lost

All patents obtained through us will receive special mention in THE INVENTIVE AGE and in cases of unusual merit inventions will be illustrated free of charge to our clients.

This publication, reaching capitalists, manufacturers and business men throughout the world, is of value in assisting to bring an invention before the public in case its promotion or sale is desired by the patentee.

INVENTIVE AGE

Patent Department.

PATENTS obtained in all Countries.

LEGAL, TECHNICAL, and SCIENTIFIC ADVICE given by personal interview or letter.

SPECIFICATIONS settled by Counsel.

WRITTEN OPINIONS furnished.

SEARCHES carefully made.

LEGAL DOCUMENTS prepared and, if desired, settled by Counsel.

ASSIGNMENTS, AGREEMENTS, LICENCES, Ftc., Registered.

DESIGNS Registered.

TRADE MARKS specially Designed and Registered.

☞ All communications relating to Legal or Patent matters, and checks, express orders and postal notes should be made payable and addressed to THE INVENTIVE AGE, Washington, D. C.

Advertisements inserted in this column for 20 cents a line (about 7 words) each insertion. Every new subscriber sending $1.00 to THE INVENTIVE AGE will be entitled to the AGE one year and to five lines three times FREE. Additional lines or insertions at regular rates.

FOR SALE.—One-half or whole interest in my American (or N. S.) patent, the simplest and most feasible railway frog ever invented. This is a practical invention endorsed by promient railway people. Here's a chance for some enterprising citizen of the United States. For full particulars, address James Baird, Chignecto Mines, Nova Scotia. 1-3

FOR SALE.—I have invented a perfect Nut Lock. Who will pay for the patent for three or four states? This is a practical invention. Address INVENTIVE AGE, or W. H. Dillon, Virginia, Ill. 1-3.

FOR SALE.—Territory west of the Mississippi, of Letters Patent No. 536,870 issued April 2nd, 1895. This fruit and food preserving jar is meeting with great favor from the public, and its construction and superiority are described in the July and December issue's of the INVENTIVE AGE. Address R. A. Gilchrist, 39 Hanover Street, Wilkes Barre, Pa. 1-3

WILL DIVIDE.—Any person who will invest just a little money about $65.00 to assist an inventor in getting valuable patent will get half of the patent assigned to him. Write for particulars to Thos. C. Davis, Ridgely, Md. 1-3

WANTED.—Automatic Pin and Link Car Coupler. Parties without working drawings need not apply. J. H. Snow & Co. Indianapolis, Ind. 1-2.

FOR SALE.—A combined square, level and triangle; also a Display Apparatus for store windows, using life size dancing figures; for sale outright or by states. No patent selling agent need apply. For particulars address F. W. G. Boetticher, West Duluth, Minn. 1-3

FOR SALE.—A two-thirds interest of a valuable Canadian Patent, No. 38,240, for an improvement in Thill-Coupling. It is superior to any in use and can be cheaply constructed. Address J. S. McClellan, Cambridge, N. Y. 12-2

FOR SALE.—Patent Hook and Eye. Can be inserted without sewing on the garment. Strong, neat and economical. For particulars, address, Ellen Donnelly, Hempstead, N.Y. 11-1 '96.

FOR SALE.—Patent 544,943. Stove Pipe Joint Cover. Great Seller. Staple Article. No Stove Pipe should be put up without it. Will make royalty offer. Liberal terms. Write to James Woodside, Hawarden, Ia. 12-2

FOR SALE.—A complete set of Patent Office Reports and Official Gazette of the U. S. Patent Office from 1790 to 1894. H. B. Dickinson, Ravenna, O., Adm'r of Estate of Bradford Howland. tf.

FOR SALE.—My Patent, 540,179, issued May 28th, 1895. Improved Combined Step Ladder, Clothes Rack and Ironing Table. Will sell entire interests or State rights. Address the Inventor Josiah C. Mifflin, Du Quoin, Ill. 11-1 '96.

FOR SALE.—All the necessary material for making a small dynamo, finished ready to wind, only $3.50. Send for circulars. W. I. Beecroft, Bangor, Maine. 11-1 '96.

THE INVENTIVE AGE can recommend the "Climax" watch, advertised in another column, as being, undoubtedly, the best stem-winder watch for the price in the market. It is a good time keeper, and either a plain or imitation engraved cases can be had. This watch is fully timed and regulated and fully guaranteed for one year, the same as Waltham or Elgin.

BUSINESS SPECIALS.

Advertisements under this heading 20 cents a line each insertion—seven words to the line. Parties desiring to purchase valuable patents or wanting to manufacture patented articles will find this a valuable advertising medium.

WANTED.—To Manufacture on Royalty. Or, For Sale Patents Nos. 546,882 and 546,883. Price $6,000. Dr. Gunther, Schorenksville, Pa. 11-11.

WANTED.—The Perfection Sack Tie. The best known agents wanted everywhere. Circulars and prices free, would contract on royalty or sell whole U. S. Patents. Address, J. R. Eichenberger, Patentee, Burton City, Ohio. 9-12

Want a Fountain Pen?

One of the very best in the market, a standard article, warranted, will be sent as a premium with THE INVENTIVE AGE. The retail price of the pen is $2.75. We will send the AGE one year and the pen for $2.75.

Aftermath.

NEW YORK policemen are now experimenting with bicycles.

The Brpol Company's bid to light the streets of New York City was 40 cents a light for 720 lamps and 45 cents for 92 lights.

THE Deleware Iron Works, at New Castle, Del., has received an order for 600 iron telegraph poles to be used at Cairo, Egypt.

Plans have been perfected for the holding of an electrical exposition in New York City next May, under the auspices of the National Electric Association.

THE second electric motor has been received and tested by the B. & O. railroad company. It is somewhat different in construction than No. 1, possessing some improved features, and the test was entirely satisfactory.

IT is reported that a syndicate of American capitalists have secured a concession from the Chinese Government to build a railroad 200 miles long from tidewater to Pekin, tapping a valuable coal mine district on the route.

MANAGER Bernstein, of the Hallstead, Pa. Textile mill, has, after much experimenting, invented a new device for oiling high speed machinery which gives promise of being entirely successful. Spindles that revolve with the enormous rapidity of ten thousand revolutions per minute formerly required constant watching and oiling, together with an enormous waste of lubricants, but the new device which is largely automatic, requires attention not oftner than once a week, giving better results with less expense.—Hallstead Herald.

THE census of Yucatan is now being taken, and it is estimated will show about 500,000 people. This is one of the most advanced states of Mexico in the matter of education and the people are quite industrious.

The last consul reoprt shows that laborers in the cities are paid by piecework in the industrial pursuits, and average about 50 cents (gold) per day; drivers and conductors on the tramways, about the same; locomotive engineers and stevedores, 75 cents to $1 (gold); and railway brakemen, 50 cents. The average work day is eight hours.

Recent Australian Patents to Americans.

The following list of applications has been specially prepared for the Inventive Age by Mr. George G. Turri, Certificate Patent Agent, Melbourne, Australia:

J. M. Rademaker, San Francisco. Amalgamator.

F. E. Heinig, Louisville, Machine used in heading sheet metal pails.

O. B. Peck, Chicago, Centrifugal ore separators.

J. V. Ruynbeke & W. F. Jobbins, New York, Process of and apparatus for the recovery of glycerine.

W. W. White, New York. Sewing Machines.

C. L. Kline, New York, Key washing machinery.

E. Chaquette, Bridgeport, Air compressors.

E. W. Cornell & H. H. Knapp, Adrian, Can labelling machine.

Cantwell & Company, Patent Agents, Calcutta, India.

We have received from this well-known and long established firm in the east their handy little phamphlet which deals fully with the laws and regulations appertaining to the taking out of Patents and registering of Trade Marks in India and other Eastern countries. It will be found a very useful adjunct to the file of Patent Agents in this country and more so to those in this city, the headquarters of the fraternity. The manual contains a list of charges which we think are very moderate, and, which will we feel sure fill a long felt want in that direction.

A junior member and representative of the firm, Mr. Harry Cantwell, is at present in this city which he proposes to make a permanent residence.

All communications to him with reference to any information regarding Patent Practice in the East may be directed in care of the Inventive Age office, Washington. We may add that the above firm are the sole agents for the "Inventive Age" in India, British Burmah and Ceylon.

An Extraordinary Offer.

By special arrangement we are enabled to offer that great success in illustrated magazinedom, "Cosmopolitan," and the INVENTIVE AGE one year for $1.85. In clubs of three or more $1.75.

During 1896 it is promised that both Cosmopolitan and Inventive Age will be greater magazines than ever. Cosmopolitan is a magazine of about 1400 pages and 1000 illustrations during the year. No magazine excels it. Old as well as new subscribers can take advantage of this offer.

"BUBIER'S POPULAR ELECTRICIAN" is the name of a monthly publication which contains a vast amount of valuable information on all electrical subjects. Its department of "Questions and Answers" will be appreciated by students and amateurs desiring information or instruction on any problem that may arise. THE INVENTIVE AGE has made special arrangement whereby we can supply that popular dollar journal and THE INVENTIVE AGE—both publications one year—for $1.50.

William H. Rau, of Philadelphia, Pa., announces that the half-tone engraving business so long conducted by him has been made the basis for the formation of a stock company entitled the International Engraving and Illustrating Company. He will retain an active interest in the new concern, and all of its work will be done under his personal supervision. The photographic and lantern slide business will be conducted by himself as heretofore at 1324 Chestnut Street.

STANDARD WORKS.

"The Electrical World." An illustrated weekly review of current progress in electricity and its practical applications. Annual subscription........ $3.00

"Electric Railway Gazette." An illustrated weekly record of electric railway practice and development. Annual subscription........ $3.00

"Johnston's Electrical and Street Railway Directory." Published annually........ $5.00

Dictionary of Electrical Words, Terms and Phrases. By Edwin J. Houston, Ph. D........ $5.00

The Electric Motor and its Applications. By T. C. Martin and Jos. Wetzler........ $3.00

The Electric Railway in Theory and Practice. The first systematic treatise on the electric railway. By O. T. Crosby and Dr. Louis Bell........ $2.50

Alternating Currents. By Frederick Bedell, Ph.D., and Albert C. Crehore, Ph.D........ $2.50

Gerard's Electricity. Translated under the direction of Dr. Louis Duncan........ $2.50

Theory and Calculation of Alternating-Current Phenomena. By Charles Proteus Steinmetz........ $2.50

Practical Calculation of Dynamo Electric Machines. By A. E. Wiener........ $2.50

Central Station Bookkeeping. By H. A. Foster........ $2.50

Continuous Current Dynamos and Motors. By Frank P. Cox, B.S........ $2.00

Electricity at the Paris Exposition of 1889. By Carl Hering........ $2.00

The Elements of Static Electricity. By Philip Atkinson, Ph.D........ $1.50

Electricity and Magnetism. By Edwin J. Houston, Ph.D........ $1.00

Electric Measurements and other Advanced Primers of Electricity. By Edwin J. Houston, Ph.D........ $1.00

The Electrical Transmission of Intelligence and other Advanced Primers of Electricity. By Edwin J. Houston, Ph.D........ $1.00

Electricity One Hundred Years Ago and To-day. By Edwin J. Houston, Ph.D........ $1.00

Alternating Electric Currents. By Edwin J. Houston, Ph.D. and A. E. Kennelly, D.Sc. $1.00

Electric Heating. By E. J. Houston Ph.D and A. E. Kennelly, D.Sc........ $1.00

Electromagnetism. By E. J. Houston, Ph.D and A. E. Kennelly, D.Sc........ $1.00

Electricity in Electro-Therapeutics. By E. J. Houston, Ph.D., and A. E. Kennelly, D.Sc........ $1.00

Electric Arc Lighting. By E. J. Houston, Ph. D., and A. E. Kennelly, D. Sc........ $1.00

Electric Incandescent Lighting. By E. J. Houston, Ph.D. A. E. Kennelly, D.Sc........ $1.00

Electric Motors. By E. J. Houston, Ph.D., and A. E. Kennelly, D.Sc........ $1.00

Electric Street Railways. By E. J. Houston, Ph.D., and A. E. Kennelly, D.Sc........ $1.00

Electric Telephony. By E. J. Houston, Ph.D., and A. E. Kennelty, D.Sc........ $1.00

Electric Telegraphy. By E. J. Houston, Ph.D and A. E. Kennelly, D.Sc........ $1.00

Alternating Currents of Electricity. By Gisbert Kapp........ $1.00

Recent Progress in Electric Railways. By Carl Hering........ $1.00

Original Papers on Dynamo Machinery and Allied Subjects. By John Hopkinson, F.R.S........ $1.00

Davis' Standard Tables for Electric Wiremen. Revised by W. D. Weaver........ $1.00

Universal Wiring Computer. By Carl Hering........ $1.00

Experiments with Alternating Currents of High Potential and High Frequency. By Nikola Tesla........ $1.00

Lectures on the Electro-Magnet. By Prof. Silvanus P. Thompson........ $1.00

Dynamo and Motor Buildings for Amateurs. By Lieutenant C. D. Parkhurst........ $1.00

Reference Book of Tables and Formulæ for Electric Street Railway Engineers. By E. A. Merrill........ $1.00

Electric Railway Motors. By Nelson W. Perry........ $1.00

Practical Information for Telephonists. By T. D. Lockwood........ $1.00

Wheeler's Chart of Wire Guages........ $1.00

Practical Treatise on Lightning Conductors. By H. W. Spang........ .75

Tables of Equivalents of Units of Measurement. By Carl Hering........ .50

Copies of any of the above books or of any other electrical book published will be sent by mail, postage prepaid, to any address in the world on receipt of price.

THE INVENTIVE AGE,
Washington, D. C.

$ THINK OF THE MONEY MADE from inventions—novelties, or simple useful labor-saving devices. If you have made a discovery, or worked out a mechanical problem we'll give reliable advice as to patentability. The "Inventive Age," illustrated magazine, 6th year, in interest of inventors guarantees work of its "Patent Department," and illustrates and describes useful inventions free. Complete, valid, strong and comprehensive patents; best terms; advice free. Address THE INVENTIVE AGE, Washington, D. C.

INVENTIVE AGE BUILDING.

PREMIUMS TO SUBSCRIBERS.

Read the following offers to new subscribers:

OUR $1 OFFER.

THE INVENTIVE AGE one year and two copies of any patent desired, or one copy of any two patents........ $1 00

THE INVENTIVE AGE one year and a list of 50 firms who manufacture and sell patented articles........ 1 00

THE INVENTIVE AGE one year and Altograph map of the City of Washington 1 00

THE INVENTIVE AGE one year and a five line (35 words), advertisement in our "Patents For Sale," or "Want" column, three times........ 1 00

OUR $1.50 OFFER.

THE INVENTIVE AGE one year and Robt. Grimshaw's famous book "Tips to Inventors"........ $1 50

Address all communications to
THE INVENTIVE AGE,
Washington, D. C.

A Great Novelty.

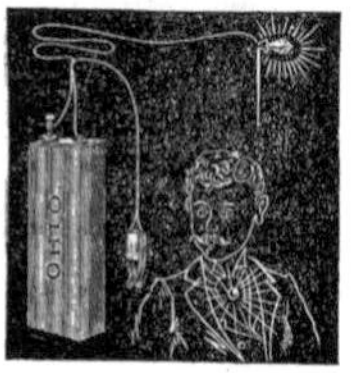

ELECTRIC LIGHT
FOR THE NECKTIE.
$1.50.

Complete with Powerful Pocket Battery and all accessories, postpaid.
The Lamp an Efficient Beauty.
Battery 3 1-2 by 3-4 by 1 3-4 inches.

This elegant outfit is practical, brilliant and satisfactory, and every part is guaranteed.

Electricians who want a piece of jewelry that is emblematical, or anybody who appreciates a real novelty, can, by ordering now, secure a complete outfit for $1.50. Order at once of

THE INVENTIVE AGE,
Washington, D. C.

Given FREE for four new subscribers.

HOW TO MAKE A DYNAMO.

By ALFRED CROFTS.

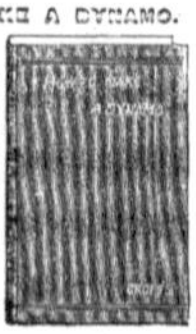

A practical work for Amateurs and Electricians, containing numerous illustrations and detailed instructions for constructing Dynamos of all sizes, to produce the Electric Light, containing 96 pages of genuine information which will enable anyone to construct a Dynamo either for pleasure or profit.

Large 12 mo,
Cloth, 75 Cts.

The Inventive Age

AND INDUSTRIAL REVIEW

A JOURNAL OF MANUFACTURING INDUSTRY AND SCIENTIFIC PROGRESS

Seventh Year. }
No. 2. }

WASHINGTON, D. C., FEBRUARY, 1896.

} Single Copies 10 Cents.
} $1 Per Year.

AN AMERICAN LINER AGROUND.

How the Great Steamship St. Paul Appeared on the Beach at Long Branch.

How ferocious and formidable in appearance the modern locomotive, whistling around abrupt curves, snorting and puffing on "up grades," flying over level prairies and through deep canons, suspended in midair on dizzy bridges over rushing waters—how impressive this triumph in mechanical skill and human ingenuity—while on the track; but how useless, how utterly helpless, clumsy and desolate when off the track—when lying in the ditch a shapeless, irregular jumble of wheels, eccentrics, rods, valves and flues! How majestic and queenly, the modern trans-oceanic steamship, how proudly and swiftly she plows through the waves of the mighty deep, her load of precious freight scarce giving the matter of danger a single thought; but how crestfallen, how weak and helpless this floating palace when cast ashore by nature's wrath or "hard aground" through frailty of human calculation!

The illustration herewith presented is an excellent view of the new American line steamship St. Paul as she appeared on the day she went ashore near Long Branch on the New Jersey coast in a dense fog. Her commander, Captain Jamison, attributed the accident to an error of the leadsman, who, he says, reported seventeen fathoms of water when there were only seven. The St. Paul went ashore about half-past one o'clock a. m. She struck about four hundred yards off the shore, almost in a direct line with Seaview avenue, East Long Branch, and right in front of the Grand View House. The vessel had been running on dead reckoning, and at first no one knew where she had struck. The life saving station men at Nos.3, 4 and 5 flashed their signals and by and by Captain Jamison and the pilot found

they were in the vicinity of Long Branch. It was high tide when she struck and although held as in a vice—powerless to move in either direction, entirely at the mercy of the elements—her passengers knew nothing of their predicament until daybreak, when the early riser found after the tide had gone out, he could thrown a stone onto dry land—the famous beach of Long Branch where in season thousands of pleasure seekers stroll upon the sands or take a "header in the brine."

THE STRANDED STEAMSHIP ST. PAUL ON THE LONG BRANCH BEACH.

In addition to her large passenger list the St. Paul carried 200 bags of mail, $1,300,000 in gold and 11,000 tons of miscellaneous freight. After repeated unsuccessful efforts of numerous tugs to pull the stranded steamer into deep water the work of transferring her freight to lighters began and the wrecking companies inaugurated a systematic scheme of floating the monster vessel which was finally accomplished on the 4th ult., just 11 days after striking the sand in which she embedded herself to a depth of ten feet or more..

Three Chapman and three Merritt wrecking boats

tugged at the vessel during the early hours. There were fifty powerful tugs opposite the St. Paul ready to give assistance, but they were not utilized. Six immense kedge anchors were planted in the sea about 1000 yards from the stranded vessel. The kedge anchor is a huge mud hook, which, deeply imbedded, is pulled upon by the two forces in opposite directions. The cable that connected the kedge anchors with the steamship was made fast to the drum of the St. Paul's heaviest hoisting engine. This permitted the St. Paul to wind up, or draw in, the cable, while the tugs, with opposing strength, gave a purchasing power. Little by little she moved, until about a third of her length in distance was traversed, and then she came to a final stop, and not another inch could she be moved. The strain on the hawsers was not relaxed, in order to keep the vessel from moving inshore.

A local account of the scene at Long Branch while the great steamship was ashore says that "the news that a big steamer was ashore had spread rapidly. Soon after daybreak hundreds were on the beach. In the afternoon there were 5000 persons viewing the stranded vessel and the fleet of tugs bobbing at a respectful distance from her. These spectators came by train from all the neighboring towns and villages, in carriages, in buggies, on bicycles and in nondescript vehicles.

One could almost throw a stone from shore to steamer. There was a graceful swinging motion to her bow as the mighty St. Paul defied the waves. And the stately vessel—for stately she looked even in her helplessness—appeared to treat with scorn the curious thousands who came to make a holiday out of her misfortune. With the principal features of the St. Paul readers of the INVENTIVE AGE are familiar. She is one of the staunchest and fastest vessels—as one jocular spectator remarked—on land or sea.

The Inventive Age

Established 1889.

INVENTIVE AGE PUBLISHING CO.,

8th and H Sts., Washington, D. C.

ALEX. S. CAPEHART. MARSHALL H. JEWELL.

The INVENTIVE AGE is sent, postage prepaid, to any address in the United States, Canada or Mexico for $1 a year; to any other country, postage prepaid, $1.50.

Correspondence with inventors, mechanics, manufacturers, scientists and others is invited. The columns of this journal are open for the discussion of such subjects as are of general interest to its readers.

Technical matter is particularly desired. We want practical information from practical men.

Nothing will be published in the editorial columns for pay.

The INVENTIVE AGE is thoroughly independent.

Advertising rates made known on application. Special facilities for furnishing cuts of any patented article together with descriptive article. Business specials 25 cents a line each insertion, 7 words to the line. No advertisement less than 50 cents.

Address all communications to THE INVENTIVE AGE, Washington, D. C.

Entered at the Postoffice in Washington as second-class matter.

WASHINGTON, D. C., FEBRUARY, 1896.

EDWIN A. HILL, of Connecticut, has been appointed confidential clerk to Commissioner of Patents Seymour, vice John W. Street, resigned.

An English trade paper says: "The transfer from England to the United States of the control of the world's iron trade means an acquisition that is only the first of a long series yet to come."

THE article on curious inventions, from the pen of W. Harvey Muzzy, of this city, will be found interesting and is fairly illustrative of the impractical vein of inventive genius which the Patent Office has to deal with.

IN the case of the City of Carlsbad vs. Kulnow, United States Circuit Court of Appeals, second circuit, the decision sustains the title of the City of Carlsbad to the exclusive use of its name as a trade-mark, for medicinal preparations.

IT is not unlikely that following the discovery of acetylene gas will come a means of utilizing it for motive power. The possibility of using it in connection with bicycles and "horseless" vehicles is a suggestion to the inventor already being acted upon in more than one direction.

THE wording of a recent patent granted to Mr. J. Swinburne, indicates the ridiculousness of the English patent system under which the inventor so often receives a patent in name only: "Improvements connected with electric telegraphs, organs, condensers, deposition of nickel and cobalt, and with packing eggs and other fragile articles, and coating eyelets and such objects, and making bleaching powder."

JOHN A. COCKERILL, spcial correspondent of the New York Herald, denies the report that American industries are likely to suffer by Japanese piracy and cheap labor. Some time ago it was reported that a Japanse bicycle, equal to the American make, was being imported to this country and sold at retail in San Francisco at $12. Mr. Cockerill says that the Japanese wheel is not durable and can in no way compete with the American machine.

MANY readers of the INVENTIVE AGE will be interested in the proceedings of the recent meeting of the American Association of Inventors and Manufacturers, published in this issue. Many of the papers read and synopsis of discussions had at this meeting will appear in subsequent issues of the INVENTIVE AGE. Desirable changes in the patent laws and the question of additional facilities in the Patent Office were matters receiving earnest attention. The meeting was dignified by the presence of Commissioner of Patents Seymour, who addressed the meeting on the urgent necessities of the Patent Office and changes in the patent laws calculated to better the service, and secure greater benefits to inventors and manufacturers whose products are made largely by patented devices.

BY the reports of the daily papers, Mr. Edison must have another fit of indigestion, says Bubier's Popular Electrician. He is reported as saying some more sour things against our present patent laws. For a man who has made several fortunes out of his patents and still continues patenting his inventions, it is at least inconsistent to make the remarks he is accredited with having made. He is reported to have said: "Our present patent laws, which as interpreted by the courts, encourage perjury and put a premium on fraud, are worse than a farce." We have no doubt but there are imperfections in our present patent laws, but would it not be better to try and suggest a remedy, than to vent one's spleen on the government?

It is expected that the new electric line from Washington to Alexandria—the extention of the Mount Vernon line to this city—will be completed by April 1st. The General Electric Company have completed six motors for the line and it is said these motors have a guaranteed speed capacity of sixty miles an hour and that when outside the District limits where no obstructions are encountered this rate of speed will be approximated and that the trip from the corner of Fourteenth street and Pennsylvania avenue in Washington to Mount Vernon, a distance of seventeen miles, in thirty-five minutes, will be accomplished. The new coaches are nearly sixty feet long, and on account of the high speed the trains are to be run the cars are nearly as heavy as ordinary railway coaches.

NEXT to the discovery of a means of perfect transmission from one mind to another of thought-waves—mind reading by purely mechanical means—the experiments of Prof. Routgen in photography without light and of Prof. Klupathy in the photographing of concealed subjects—the penetration of opaque substances—are most bewildering, and afford the widest range of speculation as to the possibilities of the future. If we can believe in the success of these discoveries—if the inner recesses of a man's mind can be reached, his thoughts silently recorded and faithfully and silently transmitted to another's brain; if by means of a vacuum tube light it is possible to penetrate walls of stone and brick, and reveal the interior of buildings and faithfully and indellibly photograph the scenes therein; if light is not necessary and in darkness, through the means and influence of electric radiation solely, objects and scenes can be transferred to plate and paper—if these things can be accomplished who can contemplate the possibilities of the future? Who knows but what is considered superhuman and supernatural today, may be found but the immutable working of natural laws tomorrow. Even the seat of life itself and the prolongation of its existence in human form may yet become a principle of medical and physical science?

A Wonderful Automatic Machine.

An automatic basket machine in which R. G. DuBois, of Washington, is interested was on exhibition in this city some time ago, says the Washington correspondent of the Hallstead (Pa.) Herald, and I had the pleasure of taking Secretary Carlisle and Senator Harris to see the novel exhibit. After watching the marvelous operation of the machine as it took the little bands, the sides and bottoms and placed them with great precision and nailed them in position with great accuracy, and with a speed that produced three thousand of finished baskets a day, Senator Harris turned to the inventor and asked: "How did you think of this? The inventor replied: "I live up in Central New York where for the last twenty years I have seen these baskets made by hand, and seven years ago I came to the conclusion that they ought to be made by machinery, and this is the result of my labors." "Well," said Senator Harris with a smile, "My friend, while you were about it why didn't you make the machinery talk? It seems to do everything else."

The greatest distance the human voice has been heard is said to be eighteen miles—the record being held by the Grand Canon of the Yellowstone.

NOTES.

Largest Steam Ferry in the World.—The largest steam ferry in the world, to ply between Manitowoc and Ludington, Lake Michigan, is now being built. The steamer will cost $300,000, will have three screws, and will carry thirty cars. The length of the steamer between perpendiculars will be 331 feet, and her overall length 350 feet.

* * *

A Substitute For Gold.—A French journal describes a new and promising substitute for gold. It is produced by alloying ninety-four parts of copper with six of antimony, the copper being first melted and the antimony afterward added; to this a quantity of magnesium carbonate is added to increase its specific gravity. The alloy is capable of being drawn out, wrought, and soldered just as gold is and is said to take and retain as fine a polish as gold. Its cost is a shilling a pound.

* * *

Cable Line to Iceland.—A great new scheme has been in active progress in London for some time past. It is nothing less than to lay a submarine telegraph line from Shetland, the furthest northern outpost of the British telegraph system in Europe, to Iceland under 500 miles of untraveled sea. The funds necessary for this great undertaking have already been secured, and Great Britain, Denmark, and Iceland will jointly guarantee those who advance the money and interest of 6 per cent, for a number of years.

* * *

Weighing Common Air.—The weight of air has often been tested by compressing it in receptacles by the air pump. That it really has weight when so compressed is shown by the fact that the weight of the vessels is increased slightly by filling them with compressed air, and that such vessels become specifically "lighter" as soon as the air contained in them is exhausted. Many elaborate experiments on the weight of air have proven that one cubic foot weighs 536 grains, or something less than 1¼ ounces. The experiment of the weight of air is supposed to be made at the surface of earth with the temperature at 50 degrees Fahrenheit. Heated air, or air at high elevations, is much lighter.

* * *

A Bridge of Concrete.—A concrete bridge, having a clear span of 164 feet and 26 feet wide, has been constructed over the river Danube at Munderkingen, Austria. It is stated that, while stone is scarce and costly there, good Portland cement is produced in large quantities. In building this bridge the centring was covered with oiled paper, on which was laid the concrete, the latter consisting of one part cement, 2½ parts sand, and 5 parts of broken stone, thoroughly mixed. Blocks of this concrete have shown a resistance of 187 tons per square foot in seven days, 235 tons in twenty-eight days, and 308 tons in five months. The concrete was applied in layers twelve inches thick, starting at the abutments and working toward the crown where it is 3¼ feet thick, and midway to the crown it is 4½ feet thick.

* * *

Steel Engraving Effects in Photography.—An invention which marks a great step forward in the art of photography has just been patented in this country and abroad by Herman Mendelssohn, photographer of Worcester Mass. It is a process of producing a photograph so much like a steel engraving that no one but an expert can tell the difference. Photographs by this new process contain the lines dots or stipples always found in steel engravings, and have the general softness, roundness, and high lights of the engraver's work, and all without impairing the likeness in the least. The process has so far been tried only in photography, but the inventor is confident that it can be used with equal success in lithography and photo-engraving.

Old World Wages.

German tanners earn $187 a year.
A German potter earns $157 a year.
Sailmakers in Ireland make $6.96 a week.
Shipwrights in Belfast make $8.14 a week.
Upholsterers in Dublin make $8.26 a week.
Lacemakers in India receive 25 cents a day.
An Italian mason make $3 to $3.60 a week.
Stevedores in Italy make about $7.44 a week.
A carpenter in Bremen makes $5.20 a week.
An English painter averages about $8 a week.
Chiness tea packers are content with $1 a week.
An iron worker in Syria can make $3 a week.
A weaver in Jurusalem earns 50 cents a day.
A stonecutter in Cairo makes 40 cents a day.
A fez maker in Turkey can earn 70 cents a day.
Liverpool boiler-makers are paid $3.63 a week.
Joiners in England receive about $8.51 a week.
A turbin maker in Teheran earns $2.60 a week.
A Russian teamster receives about 40 cents a day

Propelling Vessels by Wave Motion.

A Providence (R. I.) dispatch says: "A new invention, which promises to revolutionize the coastwise freight carrying trade of the world has been successfully tested here. The invention is nothing more nor less than the utilization of the forces of the ocean to obtain therefrom a means of motive power for craft at sea, by an ingenious device of using a swinging cargo attached to air compressers in such a manner that every motion of the vessel however light, whether pitching or oscillating, acts as a means to compress the air, which, being conveyed to an ordinary upright boiler, quickly attains the necessary amount of pressure, and, let into the engine, starts it in motion. In a vessel of 3000 tons noly one-third of the space would be used, and in this space there would be a large steel compartment which would be hung on trunnions in such a manner

as to meet every motion of the waves. In this 1000 tons of cargo would be housed.

A Couple of Patent Appeal Cases.

On the 8th of last month the patent appeal case of John Milton against Albert F. Kingsley was decided by the Court of Appeals in favor of the plaintiff. Milton it appeared had taken out patents on cognate devices. Being without means to prosecute applications pending, he entered into a contract with Kingsley, October 8, 1892, reciting that Milton was the inventor of the smoke-consuming device, and that Kingsley agreed to furnish the funds for prosecuting it and for experiments for introducing the invention into general use. In consideration of devoting all his time to it, Kingsley was to have three-fifths interest. After a test of the invention by a railroad company the iron of the conduits in

the device could not stand the heat and fire clay was substituted. As to whose idea this was the parties could not agree. Rival applications were made. Milton appealed from the commissioner of patents' decision, and the Court of Appeals decided that if Kingsley made the improvement he did so as the agent and partner of Milton, and hence the latter should have letters patent awarded him.

Another case, that of John J. Carty against Milo G. Kellogg, was also decided at the same time. It was claimed that Carty conceived the idea of a telephone switchboard and filed an application June 10, 1885. Though notified of certain defects in it he did not resume the case until November 17, 1887. July 30, 1887, Kellogg filed an original application and pushed it to a conclusion July 31, 1888. The reasons given by Carty for failing to prosecute his patent within the legal limit of time were not satisfactory to the commissioner of patents or the Court of Ap-

peals, and the decision in favor of Kellogg was affirmed.

In the official test of the pneumatic guns erected to guard the entrace to "Golden Gate" the three and one-half inch pneumatic guns were tested for capacity, rapidity and distance. Four rounds of shells, each containing 100 pounds of dynamite, were thrown 5,000 yards and five rounds of projectiles, each weighing 1,140 pounds, were loaded, and fired in eight minutes twenty-three seconds.

According to Industries and Iron it is rumored that the inventor of the Linde refrigerating machines proposes to utilize liquid air in the construction of a refrigerating system that will out-distance all others in economy and efficiency. The method is not elaborated, but it seems that the effort to do something along new lines is being made. Ingenuity never needs a night cap.

The Detroit Boiler Explosion.

Last month we reproduced from Power, of New York, an illustration showing the condition in which the terrible boiler explosion in Detroit left the Journal building. In this issue we reproduce, from the same source, a view showing the position of the exploded boilers when the debris was cleared away. It appears that the tubes of the exploded boiler pulled out of the back sheet instead of breaking squarely across as was at first reported, and went with the front head being bent into the shape of a Z. The back portion broke off just back of the manhole the sides turning down nearly flat. As we understand, it was the rivets not the sheet that parted both on the round-about seam near the manhole and the back tube sheet.

It has long been the custom to regard America as the land of invention, and Americans as a specially

endowed inventive race. A nation, seven-tenths of whose manufacturing capital (twelve hundred millions sterling) is based on patents, and of whose people almost every second man, to say nothing of women, may almost be said to be an inventor, or interested in patents, must be thought to have exceptional claim to inventive distinction far beyond the rivalry of this country. And yet intelligent and impartial American citizens who visit this land find our inventive productions in no way inferior to their own, and, our people endowed with the inventive faculty as abundantly as theirs are. But either from lack of incentive, or other causes, we do not, to anything like the same extent, turn our inventive powers to the practical and remunerative account that the born American does.—*Invention*.

Over half a million bicycles were sold last year and that record will doubtless be eclipsed this season.

CELLAR OF JOURNAL BUILDING, DETROIT, SHOWING THE POSITION OF THE EXPLODED BOILERS WHEN DEBRIS WAS CLEARED AWAY.

INVENTORS AND MANUFACTURERS.

Proceedings of the Recent Meeting of the Association in Washington.

The fifth Annual Meeting of American Associaton of Inventors and Manufacturers was held at board of trade rooms, in this city on the 21st of January.

There was an unusually good attendance of members and the proceedings were full of interest. The report of the Secretary and Treasurer detailed the work of the Association during the past year and showed that a very great deal of effective and useful work had been done at very small expense. The Association has no salaried officers and the legislative committee, under the lead of its very able chairman, Arthur Steuart, of Baltimore, has given much time and attention to the duties devolving upon them entirely at the expense of the individual members. Letters were read from J. C. Anderson of Chicago; A. J. Moxham, President of the Johnstown Iron Works; Judge R. S. Taylor, of Indiana; and H. M. Boies of Scranton, who were prevented by important business from attending the meeting, but who strongly commended the objects of the Association and evinced a very enthusiastic support.

President Gatling read his annual address as follows:

GENTLEMEN OF THE ASSOCIATION:—The United States, although young, has an interesting and remarkable history. It is the most progressive nation in the world. It is in the van of nations in wealth and the general prosperity of its citizens. It will be interesting and instructive to consider and point out some of the causes and events that have brought about these marvelous results. In studying the history of this country we find its early settlers were men and women of sterling worth. They were imbued with love of liberty and the highest religious sentiment. They endured many hardships, fought Indians and subdued the forest, and finally, when oppressed by the Mother country, they rose up in arms and carried on a seven years war with England, and after the colonies (now states) had gained their independence they had the good sense to form the National Government, with powers defined by the Federal constitution, which, in many respects, is the wisest document of its kind ever formed by man. A kind Providence favored the colonist in having such wise men as Washington, Jefferson, Franklin and others to guide and direct their public affairs. It was fortunate that the people at the close of the revolution were poor and helpless, as these conditions gave them a realizing sense of the necessity of forming a more perfect union. Had not such a union been formed there would have been thirteen poor and feeble nationalities in the place of our present populous and prosperous nation.

A number of events have occurred since the formation of our government which have contributed to the extension of the area and prosperity of the country. Among the number may be enumerated the following: Jefferson when President in 1803, bought of France what is known as the Louisiana purchase, which gave us the mouth of the Mississippi River and embraced not only the State of Louisiana, but the country now occupied by the States of Arkansas, Missouri, Iowa, Minnesota, and the most of that vast country lyiug west of the states mentioned and the Rocky Mountains, and which, with Texas, embrace the best and most fertile agricultural region in the world.

In 1845 Texas, an empire in itself, was annexed to the United States. Its admission brought on the war with Mexico, at the close of which Mexico ceded to the United States by treaty New Mexico and California. It is needless to say that the acquisition of these extensive regions added immensely to the domain and to the agricultural and mineral resources of the United States. There is no estimating the value of such possessions, which give us access to the Pacific Ocean. Florida and Alaska have been added to the Union, but are of little value compared to the area of country before mentioned. The interior water communications of the United States surpass those of any other country. The Mississippi River and its tributaries furnish some ten thousand miles of steamboat navigation. The great lakes also furnish extensive navigation. It is eleven hundred miles from Buffalo to the head of lake Superior, and there is more commerce on the lakes alone, than is possessed by most nations.

There are other causes and influences that have contributed to the nation's greatness: The influence of home life in the United States, which is of the best type, has done much to improve society. Who can estimate the imfluence for good the mother has over her children? Our free school system lays the foundation of intelligent political action, and the press is also a great educator of the people. The printing presses of today, which print thousands of copies per hour, were unknown to our forefathers; so were railroads and steamships, the telegraph and telephone. The pulpit has an immense moral and religious influence in teaching the thoughtless self-sacrifice and making the masses better citizens. The laws made in accordance with the terms of the constitution have been a boon to the people. Our tariff laws have incidentally encouraged the growth of manufactures, which now produce most of the products and articles that are consumed by the people of this country and which contribute to their prosperity and independence. The patent laws have also done much to promote the country's progress; such laws have encouraged new inventions and have brought wealth and comforts to the people. The interest of the manufacturers and inventors are co-ordinate. In other words, manufacturing industries are promoted by invention, which in turn is fostered by the manufacturing industries. It may be truly said these co-ordinate interests go hand in hand and to a great extent dominate the progress of the world. The Patent Office business is the only department of this government that pays its own expenses. The total cost of the Patent Bureau last year was $1,195,557. The excess of receipts over expenditures was over $157,000, and the total receipts over expenditures to the credit of the Patent Office in the Treasury of the United States amount to over four and a half millions of dollars. The Patent Office is worthy of the highest consideration. The very best facilities should be provided for its maintenance. The office of the Commissioner of Patents should not be of a political character, but should be made continuous as that of the Supreme Court Justices, and the salary of the Commissioner should be equal to that of a member of the cabinet. Such increase of salary would be the means of securing the best talent and would be

PRESIDENT R. J. GATLING.

an inducement for the Commissioner to continue in office. This would be desirable, as now, after learning their duties they wish to resign and go into other business which yields them a better income. The number of examiners and their salaries should be increased and their salaries paid out of the patent fund. Such changes would enable the work of the Patent Office to be conducted efficiently and without delay.

The government profits to the extent of over $150,000 a year by reason of the income it derives from the Patent Office. Justice should be done to inventors. It is not right that they should pay so largely to the patent fund and also be required to pay their proportion of other taxes assessed by the government. Inasmuch as inventors have contributed so largely to the erection of the Patent Office it would seem to be but right and proper for that building to be dedicated and used exclusively for patent business, and a Museum of Arts. It would be the part of wisdom for congress to pass an act separating the Patent Office from the Interior Department and make the Patent Bureau a department of its own, with a Cabinet officer at its head; such a bureau should be required not only to attend to patent matters, but look after the manufacturing industries of the country. Such a department would no doubt be of even more service than that of the agricultural department. Our patent laws, although not perfect, are undoubtedly better than those of any other nation, and they have greatly promoted and encouraged invention. The United States have in use more labor saving machines than is possessed by any other country, and such machinery has materially aided in producing wealth and giving comforts to the people. It is a well known fact that such labor saving tools and machinery have been the means of enabling the laboring classes in this country to get higher wages than are paid in any other country. Such in brief are some of the good effects flowing from our patent laws, and it should be the duty of congress and all good and intelligent citizens to shield and protect our patent laws from the attacks of all enemies.

Foreign patent laws seem to have been made solely to bring in revenue to the sovereign without any regard to the rights or the interest of the inventor. No features of such laws should be engrafted or incorporated into our patent system. The patent laws of most foreign countries require that the patentee must pay a periodical tax in order to keep his patent alive, and also require the invention to be worked within a certain period of time. It is needless to say these requirements are onerous and discourage invention. It often happens that some of the most valuable inventions are made by poor men who have not the means to pay such taxes or to work their inventions in a given time. It frequently occurs that new inventions are made before there is any demand for the same, and often years may elapse before there is any benefit derived from new inventions, owing to the time it usually takes to perfect them. If a patentee should die before he has fully perfected his invention, his widow and children might not be able to meet the expenses required to keep the patent alive, and in such case there should be some redress.

The great questions to be decided are:

First. Do new inventions benefit mankind? and

Second. Do the patent laws encourage the people to make new inventions?

If the foregoing questions are decided in the affirmative it should prove that no impediment should be put in the way to retard invention, but that all reasonable efforts should be made to encourage inventions. As previously stated, our patent laws although needing improvement in many respects, are better and give more encouragement to inventors than any foreign patent laws, and have been the means of the issue of more patents in this country than in all the world besides. Invention and discoveries should be allowed to go on. What a blessing it would be to the human race if some man could invent something that would take the place of alcohol and that would be harmless in its use. And would it not also be a comfort and a blessing if some lover of humanity could invent some cleanly and harmless product that would take the place of tobacco? Who can say but what such discoveries may yet be made? Men who lived in the past century never thought or dreamed of the steamship, railroad, telegraph, telephone, steam printing presses, suspension bridges, reapers, mowers, sewing machines, iron plows, friction matches, and the thousands other inventions now in common use. Notwithstanding this is a mechanical and inventive age, some men are to be found who seem to be asleep, and who take but little notice, or give any thought of the world's progress. Such persons through their ignorance, would, if they had the power, abolish all our patent laws and be content to go back to the old stagecoach and the sickle. Among the more modern inventions that deserve special attention are the bicycle, the electric car and the horseless carriage, all of which have come to stay, and which will cheapen and facilitate travel, and, to a great extent, supersede the need of horses. It may be stated as a trueism, that whatever saves labor, time and money, will, in the end, result in good to the human race.

I cannot close without saying a few words in praise of our form of government. All sovereign power in this country resides in the people. In other words, all rights begin with and go from the people, and it should be the duty of all patriotic citizens to educate their children to believe that the mission of the United States is to teach the world the great benefits and blessings that flow from free government. A man that does not appreciate our form of free government is unworthy of American citizenship. The people of this great nation should not allow anger, or petty issues, to disturb the common good; they should love liberty and be mindful of the prosperity and welfare of all, and they should not lose hope that the nation's flag will be loved and revered as long as time endures.

PROPOSED CHANGES IN THE PATENT LAWS.

The report of the legislative committee, presented by the chairman, is a very valuable and interesting paper, to which we call especial attention. It should be noted by all our readers that the Association does not propose or encourage any attempt to make radical changes in the patent laws, but that the aim of the Association is to act in a conservative manner in maintaining the present patent system, which is believed to be the best in the world, and to gradually effect such modifications as the solid sense of all parties concerned agree as being necessary. The report was as follows:

Many proposed amendments of the patent law were introduced into congress during the last session, some of them of a radical character. These were much discussed, especially before the Patent committee of the House, but none were put upon their final passage.

They proposed changes concerning which differences of opinion exist, and, while many of them meet with general approval, it would be a task of great difficulty to undertake anything like an extensive revision of the existing statutes. Your committee are of the opinion that it would be unwise at present to assume the advocacy of any sweeping measures of intended reform.

There are, however, some particular defects in the present laws which can be cured by simple amendments, as to the expediency of which it is believed there is no serious difference of opinion, and which your committee recommend as practicable and useful.

These amendments relate to the following points:

1. Under the present statutes, an applicant for a patent has two years after any action by the office, of which he shall have received notice, before he is obliged to take any further steps in prosecuting his application. This provision renders abuses possible that have been frequently called to the attention of the public by several Commissioners of Patents, and there is practical unanimity of opinion that this period should be shortened. As the law now stands, it is possible for an applicant for a patent to keep his application pending in the office for an almost unlimited time without forfeiting his rights against the public, and finally to take out a patent for an invention which has been in public use perhaps for years. Your committee therefore propose to change the time, within which action must be taken by the applicant at such stage of the proceedings in prosecuting an application, from two years to six months. This, it is believed, will give ample time for applicants residing at the greatest distance from Washington to communicate with the office, and will promote diligence and prevent the mischief above noted.

2. It is also proposed to amend the law so that the patenting or publication of an invention in any foreign country, if more than two years prior to the application in this country, shall be a bar to a patent. As the law now stands, an invention may be published and patented abroad, and become by these means familiar to the world, and, years after the foreign inventor, who may have had no real appeciation of what he had invented, or an inventor in this country, who can prove that he made his invention before such foreign publication or patenting, can obtain a patent here. It thus happens that an invention familiar to the literature of the art, but which may not have gone into actual public use in this country, may be patented here years after it has been known. The same reasons that compel the applicant for a patent to apply within two years after the invention has gone into public use in this country, make it reasonable that he should apply for his patent within two years after it has been patented or published abroad.

3. Another amendment, as to the expediency of which there is a general concurrence of opinion, relates to the limitation of actions. By the late decision of the Supreme Court in the case of *Campbell vs. Haverhill* (153 U. S., 610) the law has been settled that the state statutes of limitations relating to actions on the case, and which of course, are different in the different states, apply to actions for the infringement of patents brought upon the law side of the court. It has seemed to your committee eminently desirable that there should be a uniform statute of limitations throughout the United States, applying both to cases at law and in equity, and they therefore propose that there shall be no recovery of profits or damages for any infringement committed more than six years before the commencement of the suit.

4. The law requires that assignments of patents shall be in writing, but there is no provision whereby an acknowledgment may afford *prima facie* proof of such instruments.

It is consequently necessary, wherever proof of execution is required, to offer direct evidence thereof, which often gives rise to great delay, inconvenience, and expense. To remedy this your committee propose that a certificate of acknowledgment of these instruments before a proper officer shall be *prima facie* evidence of execution.

5. The subject of the limitation of the term of a United States patent by that of a previously granted foreign patent to the same inventor or his assigns under the present law has been much discussed in the courts, and it was finally determined by the Supreme Court, in the case of *Bate Refrigerating Company vs. Sulzberger* (157 U. S., I.), that, under the proper construction of that statute, a foreign patent granted to the same inventor after the application in this country, but before the issuing of a patent to him in this country, limited the term of the United States patent so that it should expire at the same time with the foreign patent. The provision thus interpreted has been the subject of animadversion, both in judicial opinions and by the profession, as inflicting a hardship upon American inventors, and the present statute is regarded as unsatisfactory. It provides that every patent granted for an invention which has been previously patented in a foreign country shall be so limited as to expire at the same time with the foreign patent having the shortest term. Under this law it is often difficult to ascertain when a patent expires, and the limitation imposed does not seem to be adapted to accomplish effectively any of the objects which have been suggested as reasons for the enactment of the limiting provision. Your committee propose, as an amendment that the granting of a foreign patent to the same inventor, or his assigns, shall not affect the term of the United States patent, unless the application for said foreign patent was filed more than seven months prior to the filing of the application in this country, in which case no patent shall be granted here. A similar provision exists in the laws or treaties of most European countries, except that the term there is six months instead of seven. The additional month is suggested for this country because of its greater distance from the foreign countries in which patents are granted, and in this regard is in harmony with the provisions of the International Convention, of which the United States is a member. This provision, it is believed, will accomplish the object which the congress had in view in framing the present law, and will obviate all of its inconveniences; and, so far as your committee is aware, is a change which meets with general assent among those who have investigated the subject.

Your committee therefore recommend that the following changes are desirable in the law relating to patents for inventions:

(The amendments are indicated by italics, or inclosed in parenthesis.)

Amend Section 4894 of the Revised Statutes of the United States so that it shall read as follows:

"SEC. 4894. All applications for patents shall be completed and prepared for examination within *six months* after the filing of the application, and in default thereof, or upon failure of the applicant to prosecute the same within *six months* after any action therein, of which notice shall have been given to the applicant, they shall be regarded as abandoned by the parties thereto, unless it be shown to the satisfaction of the Commissioner of Patents that such delay was unavoidable."

Amend Section 4886 of the Revised Statutes so that it shall read as follows:

"SEC. 4886. Any person who has invented or discovered any new and useful art, machine, manufacture or composition of matter, or any new and useful improvement thereof, not known or used by others in this country (before his invention or discovery thereof), and not patented or described in any printed publication in this or any foreign county, before his invention or discovery thereof, (or more than two years prior to his application), and not in public use or on sale in this country for more than two years prior to his application, unless the same is proved to have been abandoned, may, upon payment of the fees required by law, and other due proceedings had, obtain a patent therefor."

And, also, the third clause of Section 4920 of the Revised Statutes, so that it shall read as follows:

"THIRD. That it had been patented or described in some printed publication prior to his supposed invention or discovery thereof, (or more than two years prior to his application for a patent therefor.)".

Amend Section 4921 of the Revised Statutes by adding thereto the following:

("But in any suit or action brought for the infringement of any patent, there shall be no recovery of profits or damages for any infringement committed more than six years before the filing of the bill of complaint or the issuing of the writ in such suit or action, and this provision shall apply to existing causes of action.)"

Amend Section 4898 of the Revised Statutes by adding thereto the following:

("If any such assignment, grant, or conveyance of any patent shall be acknowledged before any notary public of the several states or territories or the District of Columbia, or any Commissioner of the United States Circuit Court, or before any secretary of legation or consular officer authorized to administer oaths or perform notarial acts under Section Seventeen hundred and fifty of the Revised Statutes, the certificate of such acknowledgment under the hand and official seal of such notary or other officer shall be prima facie evidence of the execution of such assignment, grant, or conveyance.)"

Amend Section 4887 of the Revised Statutes so that it shall read as follows:

("SEC. 4887. No person otherwise entitled thereto shall be debarred from receiving a patent for his invention or discovery, nor shall any patent be declared invalid, by reason of its having been first patented or caused to be patented by the inventor or his legal representatives or assigns in a foreign country, unless the application for said foreign patent was filed more than seven months prior to the filing of the application in this country, in which case no patent shall be granted in this country. This section, as hereby amended, shall not apply to any patent in this country granted prior to the passage of this act, nor to any applications for a patent in this country then pending, or to any patent granted on such a pending application.)"

Following the reading of his report Mr. Steuart stated that: "There is pending now before congress another bill with the formation and drawing of which I had somthing to do, which was filed in the last congress, for the enactment of a new trademark law. It is of the greatest importance to us in this country that we should have a comprehensive trade-mark law, providing for the registration of trade-marks which are used in interstate commerce. At present we have no provision in our law permitting the registration of trade-marks used in interstate commerce. Trade-marks can only be used in foreign commerce or commerce with an Indian tribe. It is also of importance that we should have a general national law covering the whole country, providing a criminal remedy for the infringement of trade-marks. It is entirely within the province of congress to pass such a provision, and this bill has been carefully prepared for the purpose of covering the disputed and difficult questions which caused the overthrow of the act of 1876 by the Supreme Court. I want the authority of the association, as chairman of the legislative committee, to ask that that bill be passed in the interest of the manufacturers of this association. I consider it to be of the highest importance to you all and to those of us who are patent lawyers, and to the general interests of the country. I think it would be a great benefit to throw the weight of this association and its members in favor of having that bill passed.

Hon. John S. Seymour, Commisioner of Patents who was present by invitation of the association delivered the following address:

MR. PRESIDENT AND GENTLEMEN: I esteem it a great honor to be invited to address this body of inventors and men who are interested in inventions, and I should come before you with remarks carefully prepared in writing were it not for the press of time that forbids me to do anything with that deliberation which I would like, and which the dignity of this subject and of the occasion demands in this instance.

I would not, however, come before you without preparation of another sort; for I can say that I have very carefully considered the subject of classification in the Patent Office, not merely recently, but for a long time past, as it has indeed been pressed upon my attention in a way that cannot be avoided.

I need not say to you what the proceeding is when an application comes before the Patent Office, or how the examiner goes to that place where is classified all applications and patents in the same art; but I want to say with deliberation that there is no classification in the Patent Office that is complete or reasonably satisfactory or philosophical. All issued patents go in one drawer or another; that is, in one sub-class or another, in one or another of the 5,880 different sub-classes in the Patent Office, and there the first search is made. But it is not true that in any one place is contained all that is to found in print or even all that is to be found in granted patents upon any one subject; and while the classification there is as particular as the circumstances have permitted it to be, the enormous work of making it or keeping it up, and of having cross referrence made with such accuracy as to cover the whole field, has made it practically impossible to examine in the time limits allowed in the Patent Office an application with such certainty as to be able to say that it is novel.

The labor involved in classifying the 562,000 American patents, and now, 950,000 foreign patents that we have in the Patent Office is enormous. It is not merely to classify them as the librarian classifies his books, so that he can go and put his hand upon any one when called for or when he desires to examine it; but the minute contents of that paper must be so examined that the officer in charge of an application may say with reasonable certainty—and it ought to be so classified—that he has before him everything in print upon that sub-class. When, in such an art as printing, there are thousands of patents granted, and when in that art upon one element in the machine, the webb in the printing press, there are perhaps five or six hundred patents granted, and when that webb movement is applied to a great many other arts in such a way as that the courts pronouce the movement in that other art an anticipation of the one in the printing art, then it is necessary, certainly, to have cross references in the Patent Office extending through all that cognate matter that is called the analogous art.

I am not much disturbed with the thought that the Patent Office will ever become strict in the matter of granting patents. It always inclines toward the inventor. Its desire always is to issue a patent rather than withhold it. It has been the settled policy of the office apparently from its foundation to grant patents with great liberality, rather than to scrutinize them so closely as that there is danger of harm being done to an inventor because he does not secure his patent; but the real harm is in having a patent go out without having had cited to the inventor an anticipating reference which he might have avoided if he had known of it, and which he did not avoid, by amendment, because it was not cited to him.

Some time ago I had a number of cases examined where courts had set aside patents, and in more than half of those cases, it was because the diligence of counsel had unearthed references that apparently—I do not say really—had not been cited to the applicant when the patent was before the Patent Office, probably because in many cases they were not found.

I do not want to have my remarks construed into any serions criticism upon the Patent Office; but it is obvious that if a classification is made, and it is somewhat inadequate, it is really a dangerous machine to work with, and more than anything else needs that revision which begins at the bottom and goes over the classification from beginning to end. It is my belief, and I believe it is the belief of those who have carefully examined the subject, whatever their position may be with reference to the Patent Office, that it is perfectly possible to have in the hands of an examiner in every sub-class in the Patent Office, everything that is in print upon that subject, and have it so classified that both examiner and the public can readily obtain access to the whole art, whatever the application presented may be.

I know that is a great undertaking. We have undertaken to count the cost of it. It is believed in the Patent Office that it would cost as much as the salary of an examiner in every division in the Patent Office—of an extra examiner and of a primary examiner—and four or five assistant examiners who shall be in some one place, and that there shall be attendants, messengers, and so on, and the expense of some slight printing, in all amounting to as much as $64,500 a year, and that it would take, to carefully revise all the matter which we have there as many as five years. But I believe that it is a work that is so indispensable that it may be said with truth that the condition of granting valid patents is such a classification and division.

Mr. Draper has presented to the House of Representatives a bill upon this subject. The Committee on Patents to whom that bill was referred, had a full meeting, the only full meeting that they had during this session, and I am informed it was very carefully considered. They examined into the matter very carefully. The committee instructed the chairman to report that bill, the one which you now have in your hand, Mr. President, to the House of Representatives unanimously. Mr. Draper has made a report upon the subject in which he has gone over the grounds carefully. It will be upon the calendar of the House shortly, if it is not now there, and it will be considered some day at this session to be fixed by the committee on rules. There has been no opposition to it upon the merits, so far as I know up to this time. There is a disposition to consider that no increased appropriation can be allowed to any of the departments upon any subject.

THE PRESIDENT. Can it not be paid for out of the Patent Office fund?

MR. SEYMOUR. They make no appropriations directly, Mr. Chairman, out of any fund. The appropriations are always out of funds not otherwise appropriated, and on that point there has been no patent fund maintained separately and apart from other moneys in the treasury for some years. It is figured up by the reports of the commissioners, and it is stated always that the patent fund amounts to more than four and a half millions above every dollar that has been expended for the patent system since it was established. It is a fact that since 1861 every year there have been thousands of dollars turned into the treasury over and above all the expenses of the Patent Office, derived from fees received at the office for the granting of patents or for the granting of certified copies, the sale of copies, and other matters. Last year—the figures have not yet been published—the net excess of receipts over expenditures was $160.000. The year before that, when perhaps the Patent Office felt the hard times as much as at any time in its recent history the net excess was $89,000; and so for every single year since 1861 there has been an excess, sometimes as much as $300,000, and sometimes considerably less. From 1836 when the Patent Office was established upon its present basis, in every year but eight, there has been an excess of fees at the Patent Office over and above all expenditures. It seems to me that it might be justly observed that the Patent Office, if it requires this money to do its work carefully, is entitled to an additional expenditure of as much as the ordinary excess of its fees over and above its present expenditures.

There are men who are sincerely opposed to it upon the ground of expense. I think the chairman of the committee on appropriations, whom I would not mention by name except for the purpose of clearly presenting this matter here, Mr. Cannon, of Illinois, is said to be thinking about this matter and whether it may not be too expensive, after all; whether it does not after a time imply a large amount of printing, whether on the whole it is not better to postpone this, in view of the state of the treasury; and such is the opinion of Mr. Dockery, of Missouri, one of the members of the sub-committee on appropriations, to whom the original bill is to be referred. But with those exceptions I do not know of any who are opposed to it; and I do not know of any person familiar with the Patent Office who has examined the subject who has expressed himself as opposed to it, since the thing was first proposed.

I thank the association for permission to speak of this matter to you.

The views expressed by Mr. Seymour were approved by the meeting, which subsequently adopted the following resolution:

WHEREAS, a classification of patents in such manner as to provide the means for rapily making an accurate search through the prior art is indispensable to the grant of valid patents; and

WHEREAS, a bill introduced by the Hon. William F. Draper has been unanimously reported by the Committee on Patents of the House of Representatives, and is now pending in the House, No. H. R. 3455;

Therefore Resolved, that having examined the proposed legislation we hereby give the same our hearty approval, and request the favorable action of congress thereon.

One of the most interesting addresses delivered during the session was that of Mr. F. H. Richards. of Hartford, who commented upon the questions contained in the association schedule, as follows:

MR. PRESIDENT AND GENTLEMAN: I have here the rough draft of some remarks which were taken down during a discussion of some points on the circular of the legislative committee. I will not, however, follow the text very closely.

Respecting questions numbered 1, 2 and 3:

In my opinion, the principal value of patents for inventions consists in the complete freedom of the inventor to realize from his invention and patent during any part, especially the latter part, of the term of the patent.

Mark, the idea is there that the value of the right consists in the freedom that he has under it.

It is well known that in a great many cases—probably the majority of the cases—useful inventions are discovered and made know before there is a market for them, and that sometimes the period required for creating a demand uses up the greater part of the life of the patent.

I am quite sure if you go through New England and investigate the history of the more prominent classes of automatic machinery used there, you will find it is eminently true that it takes from five to twenty years to develop a class or new line of machinery.

Furthermore, it happens that a large number of inventors of meritorious inventions are totally unable, by reason of financial or other disabilities, to obtain remuneration until the latter portion of the term of the patent, and then only by the assistance of others, either directly or indirectly given.

On that point I wish to emphasize those other difficulties, the principal one of which is time and effort under that co-operation which seems in our civilization to be indispensable to success. I submit to you that it is an almost unheard of thing for an inventor to succeed commercially, proceeding single-handed. Perhaps that never occurred to you in that light; but it is an almost unheard of thing, and it takes time to secure the confidence of other people in his invention whereby he can secure that indispensable co-operation.

It is also understood that meritorious inventions are quite as likely to be made by one class of individuals as another, some of the best of them being produced by inventors who have not only little or no experience in patent practice but who are too poor to develop their inventions, and especially to defend the same, or keep their patents alive if a periodical tax is required. In view of these circumstances, it becomes the more evident that a United States patent should be and remain free from compulsory working on the part of the patentee, and from the payment of any taxes other then the original charge for the patent; it being one of the objects of a United States patent to benefit the inventor, and thereby benefit the public indirectly and not merely to raise revenue for the sovereign as in the case of European patents. Our experience in patent litigation should be convincing as to the impractibility of establishing whether or not a patent is "worked" or what should constitute a working of a patent; also what degree of acquiescence in the rights of the patentee would be necessary.

The object of the United States patent laws is to reward the inventor and thereby indirectly benefit the public. If in any case the public do not wish to use the invention, the inventor should not have his right abridged thereby; and if the public prefer to wait one year, or fifteen years, before wishing to adopt or benefit by the use of the invention, that is a matter that concerns the public, and should not be charged against the inventor.

In other words, if a certain invention be not produced, the public are still as well off as they were before.

It is not the object of the United States patent laws to compel any particular individual whether an inventor or not, to enter upon the manufacture of his own invention or the invention of any other party. When the inventor has produced his invention and made known the same with the particularity required by the laws he has fairly earned the reward offered, which then becomes a personal right. If this were not so, it is safe to say that one of the greatest inducements to the exercise of invention

F. H. RICHARDS.

would be taken away; and if the right should be abridged in the way suggested—either by a periodical tax, or the requirement of working, or the restriction of the right because of sale—the progress of invention would be immediately and very largely reduced.

On that point I can speak from some experience, having had the advantage of having been a mechanic, or, as we say in short phrase, having gone through the shop, I had the fortune in the early seventies to invent what proved to be the basic patent of the air-cushion door-check, now almost universally used. At the time that invention was made, efforts were made to get manufacturers to put it on the market. It was not only rejected, but it was spoken of by every one to whom we applied as being something which would not sell, for which there was no demand, which people would not have on their doors if it was given to them, etc. It was seven years before we heard of anyone that wanted that patent. It was one of the attorneys for the Golden Door-Check Company of Boston who came and bought it up. You are all familiar with the device.

I know by a life-long experience among that very class of men who are making these improvements that this certainty of the duration of their right and privilege is the strongest motive that causes the exercise of the inventive faculty. No other one consideration compares for a moment with the fact of that inducement. I am afraid that it is a fact that is not appreciated by our public men generally. I doubt if anyone can appreciate it except they have been intimate for a long period of time with that class of people, a class of people to whom this country owes a great deal for its progress.

Following this are some remarks regarding furnishing copies, and so on—a matter of comparatively little importance. Of course it is desirable that the copies should be had as freely as is consistent with cost to the government of producing them. Some one suggested three cents a copy. From what I have been told from time to time, I think that is too low. I think possibly five cents a copy might be desirable, but the cost is not a serious objection.

MR. SEYMOUR. Three cents by classes would be sufficient.

MR. DODGE. And five for single copies.

MR. SEYMOUR. Ten cents for single copies, possibly, but three cents for classes. I figured it out to my satisfaction once and I think it could be done. When yon see how large the classes are in certain lines, it becomes serious to pay ten cents a copy for them.

MR. RICHARDS. In regard to the time for making amendments, of course that is a matter of some importance. I am inclined to think that in view of the changes which have come about in the way of doing business, and the greater celerity with which information can be obtained and disseminated, and especially in view of the more general understanding of patent practice, it would seem to me reasonable that the statutes should be changed to limit the time within which to make response to an official action to one year instead of two years. I think in many cases six months is a little too short a time. In many cases it would result in hardship from the present practice, although personally I am not averse to some change in that respect.

If I had time and circumstances would warrant it, I should be glad to say something upon what we may consider the liberating power of modern machinery. It is through the productiveness of machinery taking the place of the slave of antiquity and even of later times that the facilities and resources of society have been so extended as to liberate a part of the people from what is called necessary work. If you think of it a moment, and think that we had to perform all our work in the old way, it would take the entire labor of every man, woman and child in the country to furnish just the necessary things that we have without leaving any leisure time, either by individuals or by classes, for intellectual pursuits, for the acquirement of luxuries, or for any of these occupations which take up one-half of the time of the majority of city people.

I have seen it stated as long ago as 1856 that if at that time the people of London were to have brought all the water they then used, by the methods which were in use a hundred years before, it would have taken the entire labor of the whole population of London to simply procure the water they were actually using, and they would have had no time for any other purpose.

The manufacture of articles by machinery has so liberated the workmen of the world that they have been able to turn their attention to intellectual pursuits, for the production of luxuries, to a great many arts and sciences that otherwise could never exist; so that indirectly machinery has a very much wider utility than merely making money for the owner of it.

In reply to a question as to whether it would be possible to relax the patent law giving to inventors the exclusive use of the patent in cases where the inventor does not make or use his invention, Mr. Richards said:

"That is one of the most important questions that could possibly be asked, and one to which I referred somewhat forcibly, perhaps, and without going into the matter fully. I believe from all my experience of thirty years in this kind of work, if any such laws or practices are established, they will take away at once the strongest inducement that now exists, for the exercise of the inventive faculty. The possibility that a man can create this property and then have it taken, as it were, without his consent, will destroy the activity of a large number of inventors; for invention is an industry, and when a man operates a factory and produces wrenches or hammers, the wrench or hammer he makes takes a piece of personal property which he can do with as he pleases; and because he has a thousand hammers stacked up in his store, and the carpenters happen each to want a hammer, they still have no right even under our present law to come and insist that they shall be sold hammers at any particular price which they may set themselves or through their agents.

I have indeed given this whole subject a great deal of consideration. I have talked with some others who were in the inventive business, so to speak, and as I say I have been familiar with mechanics who have been producing these improvements for many years, and I am perfectly satisfied in my own mind that that change would undermine the development and prosperity of our American patent system, which is one of the crowning glories of the United States of America.

On the same subject, Mr. W. C. Dodge made the following remarks:

"I would like to ask Mr. Richards and the other gentlemen present if they ever knew, saw or heard of an inventor who did not want to get his invention into use. I have had some thirty odd years experience in connection with manufacturers and inventors, and I never knew of such an instance. Unless an inventor can sell his patent or put it into use himself, it is worth no more than a blank piece of paper to him.

I can conceive of the suggestion—and I know it has been practiced. That was first done in the sewing machine business. Those companies, after they had formed their combination, bought up subsequent inventions in sewing machines simply for the purpose of preventing the public from having them. That brings me to the first point touched upon by Mr. Richards, of getting inventions into use. People who have not had experience in that line have no conception of the difficulty that an inventor has to get an invention adopted, unless he is a man of sufficient capital himself to establish the industry. As was testified by Mr. Bessemer before the parliamentary committee in England, it takes from eight to ten years to get the invention introduced generally.

There is another reason which makes it difficult to get them introduced. Suppose I make an imprvement in a gun, which is better than the gun in use now, better than the one being manufactured by the Winchester Arms Company, or Mr. Gatling here, if you please. I go to him and want him to buy my patent. "Ah," he says, "but I have invested in my machinery and plant and factory today half a million dollars. All my dies, my tools, everything, is adapted to the manufacture of this gun, and it would cost me an enormous sum to make a new set to manufacture your gun." Therefore the manufacturer says: "No, I will not do it."

That is the greatest difficulty the inventor has when he attempts to introduce an invention.

I do not suppose any man in his senses who has any appreciation of what invention does for this country, and what our patent system has done for it, would propose to put any obstacles in the way of the inventor to the free and exclusive right to control his invention; and as I said in the beginning I cannot conceive of an instance where the inventor is not anxious from the very moment he has made his invention to get it introduced and before the public just as soon as he can, because until he does that, it is of no earthly use to him—either to sell it

W. C. DODGE.

to somebody else to manufacture, or manufature it himself. In the great majority of cases, he is obliged to sell to somebody else, because they have not the means, as a rule, of doing it themselves.

There should be no obstacle in the way of the inventor to enjoy his patent, free and unlimited, just as he would any other piece of property. In fact, I do not think congress has the power to place any such condition in the law, because the constitution gives to congress power to give the inventor only the exclusive right during the life of that patent. They have not the right to give him any other than that. They have no power to legislate under that clause of the constitution, to give him any other than the exclusive right. They may attach, of course, all the conditions they please to the grant of the patent in the first instance; but when the patent is granted, as the Supreme Court has said more than once, it is property, the same as any other franchise is, during its existence, and congress cannot legislate to interfere with it.

I would do nothing to interfere with the patent system in any way, shape or manner, or with the rights of the inventor to control his invention after the patent has been issued. On the contrary, what we want to do is to make it easier and surer that when he does get his patent, he has a valid patent, one that is safe for the capitalist and others to invest their money in and build up a new industry.

Mr. Bessemer testified that he spent $20,000 in gold before he produced one pound of steel. He says he was not a manufacturer of steel, but he knew if he could succeed in perfecting that invention he had a big thing. He said: "I would not have spent one moment of time nor invested one pound of that money had it not been for the hope of recouping myself by my patent, because after I had spent all that time and all that money, see how I would have stood as compared with others. My competitors without having invested any money and without having spent that time could come in and take the benefit of what I had done, and thus they would be that much better off than I was."

In response to the circular sent out by the Association, which was printed in the INVENTIVE AGE of December and January a number of interesting replies were received, some of which we will publish next month.

The following resolutions were adopted by the Association:

Resolved, that it is the sentiment of this Association that a national trade-mark law should be passed which will provide for the registration in the United States Patent Office of trade-marks in interstate commerce, and also to give the Circuit Courts of the United States jurisdiction of infringement cases envolving said registered trade-marks, and providing a criminal remedy for the infringement of registered trade-marks.

Resolved, that this Association believes that the Patent Office is entitled to all the facilities required for carrying on its work. That its present crowded condition is a serious hindrance to its efficiency that should be remedied promptly.

The constitution was amended in several important particulars which experience has shown were desirable to put the organization in better practical working shape.

The objects of the Association are now defined to be: "The just and adequate protection of the American patent system as authorized by the Constitution of the United States."

The clause fixing the annual dues was made to read: "The regular annual dues of five dollars shall be paid by members admitted during the first half of any calendar year; those admitted during the second half of any calendar year shall pay two dollars and fifty cents as their dues up to the beginning of the next following year.

A proposition was made and discussed to hold semi-annual meetings of the Association in other cities than Washington, and the Executive Council was given authority to call such meetings if found desirable.

The old board of officers and directors was elected with the addition of Mr. F. H. Richards of Hartford, as a member of the Council, in place of Col. F. H. Seely, whose death occurred during the last year.

In next month's issue and subsequent issues other papers and proceedings of this meeting of Inventors and Manufacturers will be published.

The Late Albert F. Andrews.

Since our last issue another prominent inventor and manufacturer, Albert F. Andrews, of Avon, Conn., has passed away. Mr. Andrews in earlier years of his young manhood was a lecturer on phrenology and physiology. In 1852 he secured a patent for an endless safety fuse, which he manufactured for a few years Then selling the business to his brother he removed to New Haven to engage in the manufacture of malleable cast iron and steel cast-

ings. In 1880 he returned to his native town and resumed the manufacture of fuses. In 1883 his factory was burned but he at once rebuilt on a much larger scale. The Climax Fuse company, a joint stock company, was organized and Mr. Andrews was chosen its president, and he succeeded in establishing the company on a solid basis, remaining at its head until he sold its interest, since which time he had not been in active business. Mr. Andrews was also a chemist of no mean ability, and back in 1884 took enough interest in politics to run for congress and later for governor on the People's ticket. His death occurred on the 12th ult. He was born in Avon in 1824 and was highly respected by all who knew him.

Abrasion of Horse Hair.

Of course every one knows that the hair used in making a violin bow is taken from the tail of a horse, but just why some bows give so much sweeter chords than others has, to a greater or less extent, always remained a trade secret. In passing the finger over a horse hair in one direction a slight roughness is observed very similiar in character to the barb of a fine fish-hook. In order to bring out the proper tone from a violin and that the up stroke of the bow will give the same tone as the down stroke, these millions of infinitesimal points are arranged to point in opposite directions. In other words, the hairs are placed in the bow end for end, so as to cause the same amount of friction whichever way the bow is drawn. Artists appearing in public who are noted for the superior quality of tone of their instruments, are very particular upon this point; in fact many of them "re-hair" their bows themselves, and will allow no one to do it for them. When this roughness of the hair wears off a most decided difference of tone is at once noticed and the hair cast aside. It is said that some of the most noted violinists will "re-hair" their bows for every concert, not daring to run the risk of losing their reputation by using a bow that does not contain entirely new hair.—*Digest of Physical Tests.*

Over 400 diamonds are known to have been recovered from the ruins of Babylon. Many are uncut, but most are polished on one or two sides.

BOOKS AND MAGAZINES.

Popular Science News in its new and improved from, publishes 24 pages of intensely interesting and valuable reading matter covering the whole field of science, invention, botany, chemistry, archœology, electricity, etc. It is edited by Benj. Lillard, 19 Liberty street, New York: $1 per annum.

* * *

A publisher's announcement in the issue of Electrical Engineer of January 8th reads as follows: "The present issue of The Electrical Engineer is No. 401. The first number of this paper appeared January, 1882, when there was no other electrical paper in existence in this country. In entering upon its fourteenth year, the Engineer looks back upon a career of useful existence and of steady growth, through periods of public prosperity and adversity, and it believes itself to be today in a position of stability and influence that will enable it to do even better work in the future than it has done in the past. As to actual facts, the Engineer would like to point out that while its first monthly issue contained only 16 pages, all told, the present weekly issue contains no fewer than 88 pages."

* * *

With the January number of Volume X, the "Social Economist" of New York is continued and enlarged under the new title of "Gunton's Magazine of American Economics and Political Science." The magazine will continue the work of the "Social Economist" in promoting a strictly American school of social Economics in the United States, as distinguished from the English (Manchester) and other foreign schools. In its new form it will discuss a wider range of public questions bearing on American industrial and social progress.

* * *

Popular Science has donned a new dress and added new departments on invention and electricity. The size of the magazine has been increased and in every department the completeness is only equalled by the excellence of the matter published A "Gigantic Stegosaurus," "Yellowstone's Death Gulch." "Photography in Colors," are interesting features of the January issue.

* * *

"Digest of Physical Tests," is the name given to a quarterly publication by Frederick A. Riehle, 1424 N 9th street, Philadelphia, which will be found of great interest to engineers and others interested in the mechanic art of physical testing. The magazine is edited by H. M. Norris and L. R. Shellenberger and is a resume of practical tests made in the laboratories of the world—in fact the only publication of the kind published. Among its regular contributors are some of the most distinguished civil and mechanical engineers in the country.

* * *

The anniversary number of the Iron Trade Review was a hummer. It contained an extensive review of the iron and machinery trade of the year, and interviews with leading manufacturers showing the improved condition of business.

* * *

With the January number the National Car and Locomotive Builder was consolidated with the American Engineer and Railroad Journal, with the title of American Engineer Car Builder and Railroad Journal. It will hereafter be published monthly under the editorial supervision of Mr. M. N. Forney, assisted by Mr. Waldo H. Marshall, heretofore editor of the Railway Master Mechanic, published in Chicago. The size has been increased and the excellent features of both former publications retained.

* * *

"Home Study," is the name of a new monthly paper, published under the auspices of the Colliery Engineer Company, of Scranton, Pa. It is designed as an elementary magazine for students of the industrial sciences and readers of the technical press who need a better knowledge of arithmetic, geomertry, trigonometry and the principles of physics and drawing to enable them to derive the best results from their reading.

The $1 Edison Motor.

The cut shown herewith represents one of the most satisfactory and perfect of all small dynamos, capable of running toys, fans, light models and for innumerable uses. To the youth scientifically inclined, it offers great satisfaction and benefit.

It is a marvel in price, and if not made in large quantities would cost from $5 to $10. We can fur-

nish the motor for $1 and will send one free to any person forwarding us three new subscribers.

Twenty-four years ago, electricity, as a mechanical power, was unknown. Now $900,000,000 is invested in various kinds of electrical machinery.

Some Queer Inventions That Have Been Patented.

While we are constantly reading with interest the newspaper and journalistic accounts of the great inventions of the age, let us glance for a moment at some of the unique inventions that have been patented and which may at first blush excite our ridicule; but be not too quick with your condemnation for it is a well known fact that nearly all great inventions were met with sneers and ridicule at the time of their introduction.

THE TAPEWORM TRAP.

We are indebted to a member of the medical profession for a very unique device for entrapping the unwary taenia mediocanellata, better known as tapeworm. This, as described by the inventor, consists of a trap somewhat resembling a quinine capsule,

and which is baited, attached to a string and swallowed by the patient after a fast of suitable duration to make the worm hungry and ready to snap at the bait. The patient retains the string in his hand and when he feels a nibble at the trap he gives a quick jerk and safely lands Mr. Taenia Mediocanellata. The trap is then again baited and swallowed and the operation repeated.

A VERY PECULIAR OVEN.

We are not prepared, in the present enlightened age of electricity to sniff with contempt at some very unusual electrical phenomenon, presented to us, although at first blush it may strike us as very ridiculous or absurd.

A patent was recently issued on an electrical oven composed of a metallic box-like structure wound about with wires; asbestos being interposed between the oven and the first layer of wires and between each succesive layer of wires. The whole is surrounded by a protecting casing—a suitable opening being left for the introduction of articles of food

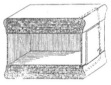

into the interior metal box. The inventor claims that if a turkey, or any other article of food, were placed in the oven and the electric current turned into the wires, the turkey would cook from the CENTRE outward. In other words, the cooking would commence within the body of the turkey and proceed gradually outward until the skin would brown. If the turkey were to be removed from the oven before the cooking was completed, it would be found, upon cutting it, that it was thoroughly cooked internally while it yet remained raw externally or near the surface. The only explaination given, is, that the oven is so wound with the electric coils that the lines of the different magnetic fields all converge or meet at the centre, thereby so disturbing the normal conditions as to generate heat at this point. The heat is also generated by the usual resistance of the wire to the passage of the electric current.

This oven, if it accomplishes what is claimed for it, would certainly be of great assistance to the housekeeper. No more sticking of straw into cakes and the like to ascertain if they are cooked internally.

A COMBINED PLOW AND CANNON.

The inventors says:

"This plow is constructed in the usual manner with the exception that the beam which is of metal is bored and formed into a cannon. As a piece of light ordnance its capacity may vary from a projectile of one to three pounds in weight. Its utility of the two fold capacity described is unquestionable, especially when used in border localities. As a means of defense in repelling surprises and skirmishing attacks on those engaged in peaceful avocations, it is unrivaled, as it can be immediately

brought into action by disengaging the team, and in times of danger may be used in the field, ready charged with its deadly missiles of ball or grape."

A further use the inventor seem to have overlooked might be here remarked, viz. that the invention could be aptly employed for the encouragement of contrary minded mules hitched to the plow, by loading with a light charge of buckshot and red pepper.

If this inventor was wide awake now he would take advantage of the present unsettled condition of the country to introduce his invention to the attention of the farmers and residents along the seaboard and frontier.

AN ECCENTRIC BICYCLE.

The principal feature of this bicycle is the rear wheel. This (to gain leverage the inventor claims) is journaled eccentrically, that is, with its axle out of the centre of the wheel. The consequence of this peculiar construction will be that the rider will be rising and falling as he rides along, with a sweep-

ing, gliding motion very similar to the ocean's swell. If he tires of the long steady swell he can get a short choppy sea by simply increasing his speed. It is thought though that the rider's appearance would be anything but graceful, as the up and down motion, especially with speed, would give a rather jumping-jack effect.

A DOG BICYCLE.

In construction the bicycle is slightly larger than is usual and the rider's seat is placed over the rear wheel. The front wheel is constructed of sufficient size to accommodate two good sized dogs therein and is built on the order of a tread-mill and very much resembles the wheel of a squirrel's cage. The dogs are chained within this wheel and in their en-

deavors to run forward turn the wheel, at least so the inventor explains. It is thought though that there would be some difficulty in starting and stop-

ping the dogs but no doubt the inventor contemplated opening a canine training school where the special dogs to be used with this wonderful invention could be purchased by the dozen ready to be slipped into the wheel and started and stopped by the signal of the rider.

A MUSICAL CHAIR.

We can expect, in the future, upon entering a friend's parlor and taking a seat, to be greeted with the strains of some popular opera. At least, this is made possible by a recently patented invention for a musical chair. This chair has the appearance of being an ordinary one and, to all intents and purposes is, with the exception, that under the seat is

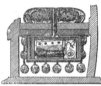

concealed a music box. The seat proper is flexible so that, when sat upon, it will be slightly depressed. This depression of the seat operates a catch and releases the drum of the music box. Different chairs or articles of furniture may be loaded with different operas. At receptions the hostess will of course take great care that her guests are seated in the chair containing their favorite opera.

A STEAM MAN.

This man is an intricate piece of machinery. His legs are operated by suitable levers, cranks and steam cylinders so that when hitched to a carriage he has every appearance of an ordinary man drawing a coach behind him. He is supported in an up-

right position by the shafts of a coach. A pipe protrudes from his mouth and the smoke and steam are discharged through the same thus making it appear that he is enjoying a quiet smoke while dragging his heavy burden along the street. The levers, cranks and other parts are of course hidden by suitable clothing or livery.

A LUMINIOUS CAT.

Those who now tolerate mice or rats in or about the house certainly must be blind to the fact that a luminious cat, which costs very little to the fact that a luminious cat, which costs very little to the fact that a luminious cat, which costs very little and can be placed in any dark corner or nook and effectually scares

away all such pests. This cat is stick, or stamped from sheet metal or other like material so as to represent in appearance the exact counterpart of its animated feline sister. It is painted over with phosphorous so that it shines in the dark like a cat of flame. After being used for about a week the place is forever free of either mice or rats.

AN AERIAL RAILWAY.

This invention consists of a cable anchored at each end and extending from one point to another, say from Philadelphia to New York. A balloon fully

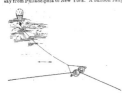

charged with gas is provided with a roller or grooved wheel pendent therefrom by suitable ropes or rods. This roller or wheel is adapted to fit upon and roll along the under side of the cable as the balloon passes forward by the action of the wind or propelling devices.

As the balloon passes forward the cable is lifted, at the point of engagement with the wheel, to some distance above the ground. A car for carrying passengers, parcels, baggage or the like is suspended from the said wheel and is of course always kept free of the ground by the balloon and carried forward by the wheel pendent therefrom.

A TOY COW THAT CAN BE MILKED.

The body of this animal is formed of either wood or metal and is an exact representation of an ordinary Jersey cow as she stands while being milked.

A suitable tank is arranged within the body of the cow and can be filled with real milk from the outside. This tank is provided with suitable nipples. These nipples are operated in exactly the same manner as in milking the real every day article. This movement of milking the Jersey also transmits motion to her pivoted jaws and it thus appears that she is quietly chewing her cud while being milked.

A HOTCH-POT OF PATENTED INVENTIONS.

A cannon motor, to the mind of one of our inventors, is the ideal method of generating power. This motor consists of an endless chain carrying teeth or projections; said chain being supported on suitable wheels for transmitting the power imparted to it. This power is derived from two cannons arranged on opposite sides of the machine and firing alternately against devices that engage the projections on the chain—one cannon being reloaded while the other is being discharged.

According to another of our inventors the proper buffer for adoption by our large railways, is a track rising from the main track at an angle of about 45°. An engine striking this buffer would soon be traveling skyward and the increasing gravity would readily bring it to a standstill. Yes.

An automatic gallows is the subject of a recent patent. With this device no special person can be pointed out as the man who sprung the trap sending the unfortunate convict to his death. The device is set and the trap sprung by a time catch.

Boiler explosions and consequential injury to life and property, in the future, will be accountable for only by those who wilfully ignore the great invention of one of our citizens. This comprises one boiler surrounded by another. When the inner boiler bursts the outer one catches the flying iron, water and steam and any injury is thus averted, at least so the aforesaid citizen is opinioned.

W. HARVEY MUZZY.

Monstrous Fighting Machines.

The two new United States battleships, Kentucky and Kearsarge, now building at Newport News, Va., are not ships at all; they will be monstrous floating fighting engines, provided with the most powerful guns, and the most effective devices known to modern naval warfare for the destruction of an enemy's fleet and the protection of its own batteries. These vessels will embody all the best principles of the best battleships afloat and will excel all others of this or any other nation.

There is no battleship afloat today and hardly any fleet that could strike a single blow so terrific as the Kentucky or the Kearsarge. The reason for this is that each carries on the forward and after deck a double turret, each turret carrying two 13-inch guns. No European power has placed guns larger than 12-inch on any new vessel. The Kentucky, therefore, can concentrate the fire of two 13-inch guns and two 8-inch guns simultaneously upon any antagonist. This is a blow which all ordnance authorities agree in stating no ship afloat could resist. A single blow of this kind would disable, if it did not instantly sink, the greatest war ship on the seas. No armor now made for naval purposes could protect a ship against these four guns simultaneously fired, the impact of whose shots would, it is thought, be even greater than a direct blow from the heaviest ram. Either from bow or stern the Kentucky or the Kearsarge could deliver this irresistible force which nothing could withstand.

The length of these vessels will be 368 feet; extreme beam, 72 feet 2 inches; mean draught, 23 feet 6 inches, and displacement 11,500 tons.

The Kentucky and the Kearsarge will carry their full coal supply of 1,210 tons with the greatest of ease, their bunker room being so ample that the coal can simply be dumped into it without the labor of trimming. This quantity will be ample for all ordinary contingencies of cruising and for the service in time of war along the coast, as at cruising speed of ten knots it will enable them to steam 6,000 miles, and at thirteen knots nearly 4,000, or at full speed to cross from New York to Queenstown. The cost of these ships will be about $2,250,000 each.

General R. G. Dyrenforth, the rain maker, has a scheme to dispel the famous London fog. He has been in correspondence with leading officials of that city, and, it is said, a fund of £50,000 will be raised with which to conduct experiments. His plan is to establish fog stations below the city and begin his compaign against it as the fog rolls in from the sea. His bombardment of the skies would produce rain, he says, and when that was started the fog would be dispelled.

When water freezes it expands with a force estimated at 30,000 pounds per square inch. No material has been found which can withstand this preasure.

The new star for the State of Utah, making forty-five in all, will be added to our national ensign July 4th.

What Constitutes an Invention.

INVENT :—"To find out, as a new means, instrument, or method, contrive by ingenuity."—Standard.

"To find out by original study or contrivance."—Century.

"An invention in mechanics consists not in the discovery of new principles but in new combination of old principles. The principles of mechanics are few, simple and well understood; their combination are various and inexhaustible."—Tyler vs. Devel.

It appears in these days that there are many who think to serve themselves or their country by belittling and tearing to pieces those to whom the country is indebted for that measure of material prosperity which it enjoys.

Under the above caption, a discoverer lets himself loose in Milling. He has discovered that there are no inventors, ergo, all who represent themselves as such and receive patent are perjurers and frauds. Possibly he thinks this sort of teaching will benefit mankind. If so, he must think it would benefit mankind to stop just where we are now and have no more new contrivances to lighten the weary steps of labor, or add pleasure to hours of rest. Because it is beyond question if the men who devote themselves to the evolution and production of the little sweeteners of life would cease to do so, if they were deprived of the patent and title of inventor.

He starts out by declaring "Inventions are not nearly so numerous as most people imagine." "Not one in a thousand patents granted are for invention." And then he goes on to illustrate : "Roller mills are certainly not in the list, for we copied the idea of the roll itself, of corrugations and differentials, and the main principle of adjustments, from Europe." If this means anything, it means that having been brought from Europe, it was never invented. "The Smith purifier was merely the application of a traveling brush to a European device." and so he goes on to establish a principle which would invalidate every patent which was ever issued in this country and close up the patent office instantly, because no patent has ever been issued which could not be taken to pieces and its materials shown to be old. As the court said in Tyler vs. Devel. "the principles of mechanics are few, simple, and well understood, (and it might have been added, and have been for centuries), but their combinations are various and inexhaustible."

"The question then naturally arises, what is an invention and who is entitled to the name inventor?" These questions can best be answered by an illustration. Columbus discovered America; thousands of men since then have spent the best part of their lives exploring the continent Columbus discovered . . . So it is with the inventor. . . . Usually his creations are crude and results imperfect, but however imperfect or fruitless the results, they are the necessary prelude to the triumphant harmony which subsequent brains and hands evolve therefrom." So, the first man who perceived he had a body, was a discoverer, and all those patient investigators who have separated and classified its organs and hunted the deadly bacilli to their hiding places are merely explorers devoid of creditable right to be called discoverers.

Old Hero utilized the power of steam and was an inventor. Watt, Stevenson, and the thousands who have contributed of their brains to make the steam engine more and more useful to man are only hangers on to the tail of the original kite, but are not entitled to be called inventors, they are only explorers.

When a man sets out to reform the language, it were well to know whether the game will be worth the candle. "A rose by any other name would smell as sweet," and it would make little difference to the ingenious man whether he is called inventor or explorer, provided he gets his patent; but this discoverer does not propose he shall have any patent because he has not made any invention. Fortunately for the ingenious man, the dictionary, the law, the courts and common sense says that the man who improves is truly an inventor, and entitled to all the joys and profits to come out of that happy condition.

The American Hotchkiss Arm Company chartered at Norfolk, Va., is capitalized at $1,000,000 for the purposes of securing American rights for the patents owned by the Hotchkiss Company of England. The officers are Alfred Debuys, of New York City, President ; John B. Summerfield, of Brooklyn, Vice President ; Charles H. Gulick, of Washington, D. C., Secretay and Treasurer. James H. Wilcox and R. A. Dodie, of Norfolk. with the above named officers, constitute the board of directors.

As a result of the competitive tests of machine guns firing the ammunition made for the new naval small arm, which have been in progress for many mouths, an order has been placed with the Colts Company for fifty of their automatic guns.

PHOTOGRAPHING IN THE DARK.

Experiments Proposed for Determining the True Nature of the Photographic Agency.

[BY EDWARD P. THOMPSON, M. E.]

When the electro magnet was first discovered, it was considered by scientists as well as others, as little more than a toy; but when further experiments were performed, additional knowledge concerning the magnet was disclosed and after many years inventors made numerous applications resulting in the present wonderful and almost magical telephone, electric car, electric telegraph, and, in fact, nearly all the remarkable electrical inventions which embody the magnet as the leading feature.

Now that it is reported that Professor Routgen has discovered a radiant force adapted to pass through opaque objects and produce a photograph of an intervening object, it would be well for those who have the proper laboratory to carry out systematic investigations in whatever direction may seem to suggest new results. In suggesting these experiments, we should not be too careful to be conservative as to whether or not they are worth trying.

The first experiment is based upon the intimation that the Professor finds that the rays from Crooke's tube, when properly excited, will not penetrate clear glass, but will, to a certain extent, pass through ground glass. It is stated further that aluminum affords a far better transmitter of the rays than glass. Now it is common to construct Crooke's tube of glass. The question arises, why not construct them of aluminum and thereby save a great deal of the penet-ating power which certainly is overcome by the glass of which the tubes are made? Perhaps the Professor has tried this. Furthermore the tube should be constructed of other materials and the result noticed. Physicists will understand of course that the tubes must be evacuated and hermetically sealed. There is no reason why an aluminum tube could not have all the air pumped out of it and sealed by fusion or any other way as, for example, by a suitable stopper.

If these rays possess such a wonderful power as to pass through aluminum and some other opaque materials, and to produce chemical action upon a photographic plate, and if it is doubted as to whether the rays are light rays or not, experiments should be performed to determine their similarity to light rays. The mental selenium is a very wonderful substance in that it will not condcut an electric current in the dark, but it will if exposed to light rays. It should be remembered that light rays are made in reality of three sets of rays characterized by their distinctive properties. There is one set of low vibration and invisible producing the greatest amount of heat. Another set of a higher rate of vibration producing the greatest amount of luminosity and another set of still more rapid vibration and invisible, and called actinic rays producing the greatest amount of chemical action. There may be another set of still higher vibration that can be produced by the Crooke's tubes, and that have the new property not only of producing chemical action, but of being able to pass through certain opaque objects. How is selenium effected?

An experiment can be performed of passing these rays, after going through aluminum, through a prism and determining whether any peculiar differences of power are noticed in different parts of the spectrum.

One of the properties of light and especially of the actinic rays is to blacken what is know as transparent phosphorous. Will the Routgen rays produce a similar effect upon phosphorous?

There is a wonderful substance known as luminous paint and consisting of one or more of the chlorides of barrium, strontium, and calcium, made by a peculiar process sometimes from shells. When exposed to light the same become stored by the luminous paint which will continue to radiate the light for several hours when taken into the dark. The Routgen rays after passing through the opaque substance should be directed upon this material and too points should be noticed: First, whether any light will be radiated in the dark; and, secondly, whether there are any invisible rays radiated in the dark that would produce a chemical effect upon a photographic plate.

The actinic rays of light also have the property of igniting a mixture of chlorine and hydrogen gases. This mixture should be exposed to the rays of Professor Routgen after passing through aluminum and the effect studied.

Inventors should consider the possibility of the following results: The passing of the Routgen rays through an opaque object and their conversion into luminous rays. If, by any means, a device can be invented for reducing the rate of vibration of actinic rays or of the Routgen rays which, upon the above outlined theory, have a very high rate of vibration, then there will be obtained luminous rays. It seems strange if no one was able to produce this reduction of vibration. It is in accordance with the laws of the conservation of energy. If a pendulum is vibrating at high rate, it is know how to make the pendulum vibrate at a lower rate.

Some one has suggested the idea of using the Routgen rays to photograph objects in the dark. It should be remembered that the object must be between the source of the rays and the photographic plate and it might not be convenient for a detective, for example, to induce his prey to assume that convenient position. In photographing by light the object may be in any position while the photographic plate accompanied by the lenses may be in any other position.

Inventors will immediately begin to imagine many uses of a force of a chemical nature that will pass through an opaque material. Not any one inventor will think of useful applications but some out of many are likely to do so, and, therefore, the more experiments and suggestive ideas that are published, the more rapid will be the useful results, if any, to be obtained from this very interesting discovery.

Lyle's Improved Screen Door.

The accompanying illustrations are those of an inventor and an ingenius and useful invention. The patentee of the combination wire screen and steel panel door is Mr. W. R. Lyle, of Ripon, Wis. It is

claimed for this invention that it excels all other screen doors for summer use and that it combines with the adjustible steel panels a storm door for cold climates and in a mild climate no other door is necessary The door is never taken down, is solidly constructed and under rights granted by the patentee can be manufactured and sold at such a low figure as to place it at no disadvantage in competition with even the inferior makes of screen doors. The inventor of the door is also the inventor of an automatic staple driver to nail wire cloth and moulding on screen door frames. The doors are made of 1⅛ in. pine, mortised and tennoned, covered with the best wire cloth, pointed and stained. The steel panels cost but 40 cents per door. Any information regarding agencies or manufacturing on royalty can be had by addressing Mr. W. R. Lyle, Ripon, Wis.

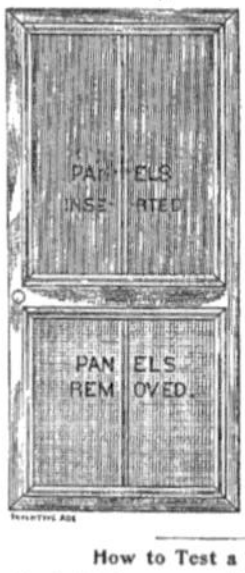

How to Test a Thermometer.

Prof. Sofshu says: "Before purchasing a thermometer invert the instrument; the mecury should fall to the end in a solid 'stick.' If it separates into several small columns, the tube contains air and will not register accurately. Nine persons out of ten think the mercurial column is round; but this is not the case; it is flat, and the opening in the tube is as small as the finest thread."

The lower tunnel on the Standard mine, Cœur d'Alenes, has been driven 900 feet. When completed it will be 3,300 feet long, and wide enough for a double track for electric ore cars. For five months of the past year the Standard paid $442,000 in dividends: August, $71,000; September, $78,000; October, $93,000; November, $95,000; and December, $105,000.—*Spokane Miner*.

Association of Manufacturers.

An organization known as the National Association of Manufacturers has, after considerable effort on the part of a few earnest enthusiasts, become a fixed fact—the first really important session of the body being held in Chicago on the 21st ult. The objects of this association vary somewhat from those engrossing the attention of the American Association of Inventors and Manufacturers, in that the former association will have to deal more particularly with business—the extension of trade in this and foreign countries, and incidentally legislation in that direction. The proceedings of the recent meeting were characterized by great earnestness and something of the wonderful development of manufacturing interests in this country can be gathered from the statistics appearing in the address of Thos. Dolan, of Philadelphia, president of the association. In 1860 the value of the manufactured product of the United States was but $1,900,000,000—this representing the total gain from the foundation of the government. Between 1860 and 1890 the gain was nearly $7,500,000,000, or about 300 per cent in 30 years. It appears from the tenor of the addresses made but one sentiment—wise protective traiff laws—actuated the assembly, and it may be reasonably expected that the closer and more sympathetic union of the manufacturers of the country will result in popularizing to a greater degree than now exists the American idea of industrial development and commercial reciprocity.

Bill to Cancel Certain Patents.

The following is the full text of Congressman Lacey's bill, No. 281, entitled "a bill to enable the United States to terminate and cancel letters patent for inventions in cases of general public importance." The object of this measure is to provide suitable recompense to the inventor for any device or discovery desired by the government and to prevent extortion on the part of the inventor where public policy justifies or demands governmental control of the invention:

SECTION 1. That any letters patent of the United States that may be hereafter issued to inventors may be terminated at any time by the enactment of a special Act or special Acts for that purpose, and from the date of the taking effect of such special enactment or enactments all rights under such letters patent shall cease and determine.

SECTION 2. That no letters patent shall be terminated under the provisions of section one of this Act unless a grant of not less than twenty-five thousand dollars nor more than one hundred thousand dollars shall be made in such special Act to compensate the inventor or his assignee or assignees for the loss of such patent: Provided, That Congress may further grant an allowance to any inventor, in the case of thus terminating a patent, for a sum equal to the amount expended by such inventor in perfecting his said invention.

Two Thousand Words A Minute.

Patrick B. DeLaney, in his lecture before the New York Electrical Society at Columbia College a few days ago, successfully demonstrated that his new telegraph instrument could deliver a message at the rate of 2,000 words a minute.

This means that 32,000 messages of 70 words each can be sent every day from New York to Chicago. Mr. Delaney has figured it out that at five cents for each message of seventy words there would be a satisfactory profit to the telegraph companies. The two cents which the government receives for forwarding a letter is almost equally divided between the cost of railway conveyance and local delivery of the letter. The cost of transmitting the words of letters at the rate of 2,000 words a minute, if the government should do the telegraphing, would be insignificant, and therefore the outlay of the government would not be limited to one cent for local delivery. This being so, the possibility of a 70 word letter delivered 1,000 miles away in one hour for one cent at once becomes attractive.

To Obviate Puncture of Bicycle Tires.

A gentleman in London writes: "A bicycle tire here has been constructed in this way: A rubber tube is stretched over a mandrel twice the diameter of the tube in its normal state, then a thin gum tube is drawn over this and solutionised. I might add that the tube when enlarged to twice its diameter is also stretched to twice its length. Then a thin pure rubber tube is placed on the outside and solutionised, after which it is removed from the mandrel and turned wrong side out. In the event of a puncture occurring this causes a contraction in every direction absolutely closing the puncture."

SUPPLEMENT.--Tips to Inventors.

The Infancy of Invention.

As capital is constantly being invested and expended to protect and preserve capital previously expended and invested in various enterprises all over the land, so will inventions continue—their variety and multiplicity being demanded to further the usefulness and perfection of inventions previously originated.

It was Edison who, replying to the question, "Do you think that the inventions of the next fifty years will be equal to those of the last?" said: "I see no reason why they should not. It seems to me that we are at the beginning of inventions." The truth of this prediction is illustrated in the many useful and wonderful achievements of Mr. Edison's own laboratory since giving utterance to this statement only a short time ago.

Profits from Invention.

The value of an invention is determined by no fixed rule. Fabulous sums have been made from simple and novel, as well as complex and useful, inventions. It is a fact that four-fifths of the business of the United States is transacted by the use of inventions. The benefits to mankind because of inventions, are so manifest and so common we are apt to look upon them in a matter-of-fact sort of way and fail to give the inventor the credit due him. In the majority of cases, however, the failure of an inventor to reap a reward is attributable to his own negligence, lack of forethought and indiscretion.

Nearly every human being is an inventor, but only a few obtain a *monopoly*—a patent—on the product of their brain. There are thousands of really useful articles, appliances and discoveries, in use every day by millions in all walks of life, that might have been patented had the inventor possessed the business sagacity that has actuated his more fortunate neighbor. Take for instance the open slot necessarily used in all conduit electric, or cable street railway systems. The inventor failed to get a patent on the idea and a fortune missed him.

There is money in inventions, but not always for the inventor.

The only way to make money out of an invention is through the protection afforded by a patent; not a patent in name only, but a *good patent*—one that is intelligently drawn, with claims commensurate with the scope and importance of the invention.

The profits arising from inventions in the electric field during the past twelve years have been simply astounding. In railway appliances, bicycles, typewriters, telephones, cash registers, slot machines and farm machinery, the field has been equally remunerative. And just think of that simple toy "Pigs in the Clover"—it netted the inventor, whose friends laughed at him for obtaining a patent on so simple a toy, over $150,000. The inventor of the metal plates to be attached to the worn heels of shoes (for sale in all cities) realized a fortune out of it amounting, it is said, to nearly $1,000,000. Perforated wooden seats for chairs and rubber tips for lead pencils brought the inventors big results. Howe made a million dollars from his sewing machine attachments, and the inventor of that simple lamp attachment, the inverted glass bell, to be suspended over lamps to protect the ceilings from being blackened, made the inventor rich. The "Darning Weaver," a device for repairing stockings, is a useful invention and is netting the inventor a handsome revenue on royalties. The wire nail and gimlet-pointed screw are fortune makers, and wire nails caused the invention of automatic machinery that manufacturers them so cheaply it does not now pay the carpenter to spend his time in picking a nail up when it drops, if it requires ten seconds to do so. The inventor of the well-known "safety pin" lived in luxury all his life, after discovering a means of concealing the point of a pin in such manner as to prevent scratching. The inventor of roller skates made nearly a million and the inventor of the needle-threader for a long time made $10,000 a year.

Relation of Capital to Invention.

Mr. Edward P. Thompson, one of the most entertaining writers on the subject of invention, says that "every invention, before the introduction into practical use, passes through two stages; namely, mental and physical"—mental when in the mind of the inventor only, and physical when the mental invention is put into bodily form by hand, or by hand with the assistance of a convenient tool. "A mental invention," says the writer, "sometimes does not become a physical invention because the inventor lacks money, technical knowledge or diligence. Such a mental invention often becomes a physical invention by the assistance of a capitalist, an educated person, or diligent companion." This being true the *mental* inventor, the person who, for lack of means possibly, would fail to make his invention a physical reality—such a person should take into his confidence a friend or companion to share the prospective benefits of his invention. Thousands of meritorious mental inventions are never worked out because of the over-timidity of the inventor, his exagerated greed for *all* the benefits to accrue instead of half the loaf, which in many instances is, or would have been, ample reward. Mr. Thompson truly says: "Inventors and capitalists should be more willing to co-operate. It is too often the case that the former must pay for his own experiments and all patent costs before a capitalist will even take the trouble to look into the merits of the alleged invention. On the other hand it is too often true that the capitalist seeks to join with the inventor, but the latter wants too high a price at the beginning."

Who Can Apply for Patents.

Patents are issued to any person who has invented or discovered any new and useful art, machine, manufacture, or composition of matter, or new and useful improvement thereof, not known or used by others in this country, and not patented or described in any printed publication in this or any foreign country, before his invention or discovery thereof, and not in public use or on sale for more than two years prior to his application, unless the same is proved to have been abandoned; and by any person who, by his own industry, genius, efforts and expense has invented and produced any new and original design for a manufacture; any new and original design for the printing of fabrics; any new and original impression, ornament, pattern, print or picture to be printed, painted, cast, or otherwise placed on or marked into any article of manufacture, or any new, useful and original shape or configuration of any article of manufacture, the same not having been known or used by others before his invention or production thereof, nor patented or described in any printed publication, upon payment of the fees required by law and other due proceedings had.

If it appears that the inventor, at the time of making his application, believed himself to be the first inventor or discoverer, a patent will not be refused on account of the invention or discovery, or any part thereof, having been known or used in any foreign country before his invention or discovery thereof, if it had not been before patented or described in any printed publication.

Joint inventors are entitled to a joint patent; neither can claim one separately. Independent inventors of distinct and independent improvements in the same machine can not obtain a joint patent for their separate inventions; nor does the fact that one furnishes the capital and another makes the invention entitle them to make application as joint inventors, but in such case they may become joint patentees. The receipt of letters patent from a foreign government will not prevent the inventor from obtaining a patent in the United States, unless the invention shall have been introduced into public use in the United States more than two years prior to the application. But every patent granted for an invention which has been previously patented by the same inventor in a foreign country will be so limited as to expire at the same time with the foreign patent, or, if there be more than one, at the same time with the one having the shortest unexpired term, but in no case will it be in force more than seventeen years.

Protection to Inventors.

What is a patent? It is a monopoly or grant, in the United States, for a term of seventeen years, to the patentee, his heirs or assigns, of the exclusive right to make, use and vend the discovery throughout the United States, as the inventor's rights may appear in the specifications and patent granted.

This means a great deal to the inventor who has secured a *valid* patent containing all the claims so worded as to prevent infringement and loss in contest. Thousands of inventors, obtaining patents through unreliable and inefficient attorneys or agents, find themselves possessed of patents *in name only*, and of no value when combatted by infringers with capital and the aid of able legal talent. A good patent costs no more than a weak and worthless one. Therefore how shortsighted are those inventors who employ cheap attorneys, saving $5 or $10 in fees, only to find themselves loosers of *all* they have paid when the contest comes.

The Need of Reliable Attorneys.

The Revised Statutes of the United States provide that "before any inventor shall receive a patent for his invention, he shall make application therefor in writing to the Commissioner of Patents, and shall file in the Patent Office a written description * * * of the same in such full, clear, concise and exact terms, as to enable any person skilled in the art or science to which it appertains, or with which it is most nearly connected, to make, construct and use the same; and in case of a machine, he shall explain the principle thereof and the best mode in which he has contemplated applying that principle, so as to distinguish it from other inventions."

To carry out these provisions it is necessary for the inventor to first make a clear, concise and complete drawing, or a working model of his invention or discovery, and send it to THE INVENTIVE AGE, or some thoroughly reliable attorney, who, before making application for the patent, should make a thorough and rigid examination of the Patent Office to determine upon its novelty or patentability. If the invention has been anticipated by some one else, or if it lacks novelty, or if for any reason a patent can not be granted, or, if granted, would be of no worth or value, then the inventor does not want to incur the expense of making application and paying attorney's fees and government fees. For making this thorough examination THE INVENTIVE AGE and all reliable attorneys charge $5, which fee is, under some circumstances, however, taken out of the additional fees paid by the inventor in case letters patent are applied for. The fees of patent attorneys vary somewhat, but the average fees for obtaining a United States patent are about $65—the government fees being $15 on filing the application and $20 on issuing a patent—the balance being the fees for preparing specifications, making searches, etc. The inventor is sometimes favored in terms given for payment of the fees, more detailed information regarding which can be obtained by enclosing a 2-cent stamp with enquiry to THE INVENTIVE AGE, Washington, D. C. The reason why the inventor should have a preliminary examination of the Patent Office made before applying for a patent lies in the fact that if the case is rejected the fees paid to the government and the attorney are lost

All patents obtained through us will receive special mention in THE INVENTIVE AGE and in cases of unusual merit inventions will be illustrated free of charge to our clients.

This publication, reaching capitalists, manufacturers and business men throughout the world, is of value in assisting to bring an invention before the public in case its promotion or sale is desired by the patentee.

INVENTIVE AGE

Patent Department.

PATENTS obtained in all Countries.

LEGAL, TECHNICAL, and SCIENTIFIC ADVICE given by personal interview or letter.

SPECIFICATIONS settled by Counsel.

WRITTEN OPINIONS furnished.

SEARCHES carefully made.

LEGAL DOCUMENTS prepared and, if desired, settled by Counsel.

ASSIGNMENTS, AGREEMENTS, LICENCES, Ftc., Registered.

DESIGNS Registered.

TRADE MARKS specially Designed and Registered.

☞ All communications relating to Legal or Patent matters, and checks, express orders and postal notes should be made payable and addressed to THE INVENTIVE AGE, Washington, D. C.

BUSINESS SPECIALS.

Advertisements under this heading 20 cents a line each insertion—seven words to the line. Parties desiring to purchase valuable patents or wanting to manufacture patented articles will find this a valuable advertising medium.

WANTED.—Publisher to publish and introduce Walsh's Perpetual Calendar and Almanac no royalty; or I will sell the copyright. A complete calendar, absolutely perpetual. No complicated rules or computations. James A. Walsh, Helena Mont. 3

Want a Fountain Pen?

One of the very best in the market, a standard article, warranted, will be sent as a premium with THE INVENTIVE AGE. The detail price is $2.75. We will send the AGE one year and the pen for $2.75.

Aftermath.

A 20-ton bell will be built for the new city hall in Milwaukee by Campbell and Sons, the famous bell founders.

Newell Narth, said to be the inventor of the gimlet pointed wood screw, died lately at the country infirmary in Akron, Ohio, in a demented condition.

It is reported that about 200 railway carriages are now lighted by electricity in Sweden, and in Denmark the same system is also in use on the better trains.

The *Cosmopolitan Magazine* has offered prizes amounting to $3,000 for horseless carriages making the best records and showing the most good qualities in a trip between its New York city Office and its publishing plant at Irvington-on-Hudson.

Congressmann William A Stone has introduced a bill in the house to incorporate the Columbia Company of Washington and authorities him to supply gas and electric light to the citizens of Washington, D. C. The incorporators are all Pittsburgers, as follows: John W. Chalfant, J. N. Pew, W. T. Marshall, W. B. Miller, W. L. Pierce, T. S. Bigelow and S. G. Pew.

Escaping gas from a leak in the main of the Boston Gas Company on East Canton street, Boston, on January 15, filled a whole block of houses on each side of the street, and but for the prompt discovery 300 lives would have been sacrificed. As it was, one person died, three others were in a precarious condition and 21 o be s were under treatment the following day.

Lemuel W. Serrell, the well known patent lawyer of New York, has associated his son Hafeld with himself in business, and hereafter the firm will be known as L. W. Serrell and Son. Mr. L. W. Serrell is one of the most efficient practitioners before the patent bar and one of most earnest champions of the American patent both system and the rights of the inventor.

The increase of the capital of the Newport News Shipbuilding and Dry Dock Co. from $3,000,000 to $6,000,000 is another striking illustration of the wonderful growth of this gigantic plant, which is fast becoming one of the greatest industrial institutions of the world. With a capital of $6,000,000, and $3,600,000, unpressed by bonds, this ship-yard now ranks with the greatest ship-yards of the world.

Any person sending us the names and addresses of five inventors who have not yet applied for patents will receive the INVENTIVE AGE one year free.

Quick Work.

The Jewell Belting Co. of Hartford, Conn. possesses many friends in the electrical field, not alone for the excellence of its product, but also for the promptness and ability in filling "rush" orders. Recently a break-down occurred in the station of the Albany Street Railway Company, and a belt was telephoned for to the Jewell company Sunday afternoon. Mr. Charles E. Newton, the secretary and manager of the company, realized the urgency of the the case, and, putting men to work on the order immediately, was able to ship the belt so it was in Albany on Monday. The belt was a 48-inch one, and the promptness of the delivery saved the Albany company much annoyance.

Another Great Combination.

Popular Science is one of the best magazines published. It is 20 year's old and covers all the sciences in a popular and instructive way. We can furnish Popular Science and the INVENTIVE AGE one year together each for $1.50. This combination rate can be offered but a short time as the subscription price of Popular Science is soon to be increased. Address the Inventive Age, Washington, D. C. If Kell's Cyclopedia is also wanted—the three for $2.75.

THE INVENTIVE AGE wants live agents in every part of the country. Readers of this notice will confer a great favor by sending us names of persons who would likely take an interest in the matter. Our cyclopedia premium offer is a great winner. The book alone is worth double the amount, $5, charged for the two.

An Extraordinary Offer.

By special arrangement we are enabled to offer that great success in illustrated magazinedom, "Cosmopolitan," and the Inventive Age one year for $1.85. In clubs of three or more $1.75.

During 1896 it is promised that both Cosmopolitan and Inventive Age will be greater magazines than ever. Cosmopolitan is a magazine of about 1600 pages and 1000 illustrations during the year. No magazine excels it. Old as well as new subscribers can take advantage of this offer.

" BUBIR'S POPULAR ELECTRICIAN" is the name of a monthly publication which contains a vast amount of valuable i..formation on all electrical subjects. Its department of "Questions and Answers" will be appreciated by students and amateurs desiring information or instruction on any problem that may arise. THE INVENTIVE AGE has made special arrangement whereby we can supply that popular dollar journal and THE INVENTIVE AGE—both publications one year—for $1.50.

STANDARD WORKS.

"The Electrical World." An illustrated weekly review of current progress in electricity and its practical applications. Annual subscription $3.00

"Electric Railway Gazette." An illustrated weekly record of electric railway practice and development. Annual subscription $3.00

" Johnston's Electrical and Street Railway Directory." Published annually $5.00

Dictionary of Electrical Words, Terms and Phrases. By Edwin J. Houston, Ph.D. ...$5.00

The Electric Motor and its Applications. By T. C. Martin and Jos. Wetzler $3.00

The Electric Railway in Theory and Practice. The first systematic Treatise on the electric railway. By O. T. Crosby and Dr. Louis Bell $2.50

Alternating Currents. By Frederick Bedell, Ph.D., and Albert C. Crehore, Ph.D......$2.50

Gerard's Electricity, Translated under the direction of Dr. Louis Duncan..........$2.50

Theory and Calculation of Alternating-Current Phenomena. By Charles Proteus Steinmetz. $2.50

Practical Calculation of Dynamo Electric Machines. By A. E. Wiener...........$2.50

Central Station Bookkeeping. By H. A. Foster $2.50

Continuous Current Dynamos and Motors. By Frank P. Cox, B.S..................$2.00

Electricity at the Paris Exposition of 1889. By Carl Hering........................$2.00

The Elements of Static Electricity. By Philip Atkinson, Ph.D...................$1.50

Electricity and Magnetism. By Edwin J. Houston, Ph.D.........................$1.00

Electric Measurements and other Advanced Primer's of Electricity. By Edwin J. Houston, Ph.D...........................$1.00

The Electrical Transmission of Intelligence and other Advanced Primer's of Electricity. By Edwin J. Houston, Ph.D............$1.00

Electricity One Hundred Years Ago and To-day. By Edwin J. Houston, Ph.D....$1.00

Alternating Electric Currents. By Edwin J. Houston, Ph.D. and A. E. Kennelly, D.Sc. $1.00

Electric Heating. By E. J. Houston Ph.D and A. E. Kennelly, D.Sc................$1.00

Electricity in Electro-Therapeutics. By E. J. Houston, Ph.D., and A. E. Kennelly, D.Sc. $1.00

Electric Arc Lighting. By E. J. Houston, Ph.D., and A. E. Kennelly, D. Sc.......$1.00

Electric Incandescent Lighting. By E. J. Houston, Ph.D. A. E. Kennelly, D.Sc...$1.00

Electric Motors. By E. J. Houston, Ph.D., and A. E. Kennelly, D.Sc..............$1.00

Electric Street Railways. By E. J. Houston, Ph.D., and A. E. Kennelly, D.Sc.......$1.00

Electric Telephony. By E. J. Houston, Ph.D. and A. E. Kennelly, D.Sc..............$1.00

Electric Telegraphy. By E. J. Houston, Ph.D and A. E. Kennelly, D.Sc..............$1.00

Alternating Currents of Electricity: By Gisbert Kapp............................$1.00

Recent Progress in Electric Railways. By Carl Hering...........................$1.00

Original Papers on Dynamo Machinery and Allied Subjects. By John Hopkinson, F.R.S.............................$1.00

Davis' Standard Tables for Electric Wiremen. Revised by W D. Weaver........$1.00

Universal Wiring Computer. By Carl Hering................................$1.00

Experiments with Alternating Currents of High Potential and High Frequency. By Nikola Tesla..........................$1.00

Lectures on the Electro-Magnet. By Prof. Silvanus P. Thompson................$1.00

Dynamo and Motor Building for Amateurs. By Lieutenant C. D. Parkhurst........$1.00

Reference Book of Tables and Formulæ for Electric Street Railway Engineers. By E. A. Merrill...............................$1.75

Electric Railway Motors. By Nelson W. Perry $2.00

Practical Information for Telephonists. By T. D. Lockwood.......................$1.00

Wheeler's Chart of Wire Gauges........$1.00

Practical Treatise on Lightning Conductors. By M. W. Spang....................$.75

Tables of Equivalents of Units of Measurement. By Carl Hering...............$.50

Copies of any of the above books or of any other electrical book published will be sent by mail, postage prepaid, to any address in the world on receipt of price.

THE INVENTIVE AGE,
Washington, D. C.

PREMIUMS TO SUBSCRIBERS.

Read the following offers to new subscribers:

OUR $1 OFFER.

THE INVENTIVE AGE one year and two copies of any patent desired, or one copy of any two patents.............$1 00

THE INVENTIVE AGE one year and a list of 50 firms who manufacture and sell patented articles.................1 00

THE INVENTIVE AGE one year and Autograph map of the City of Washington 1 00

THE INVENTIVE AGE one year and a five line (35 words) advertisement in our "Patents For Sale," or "Want" column, three times........................1 00

OUR $1.50 OFFER.

THE INVENTIVE AGE one year and Robt. Grimshaw's famous book "Tips to Inventors" $1 50

Address all communications to
THE INVENTIVE AGE,
Washington, D. C.

GREAT PREMIUM OFFER! = =

 Valuable
Sensible
Economical

Zell's Condensed Cyclopedia
Always Sold at $6.50 Complete.

Complete in One Volume. A Compendium of Universal Information, Embracing Agriculture, Anatomy, Architecture, Archeology, Astronomy, Banking, Biblical Science, Biography, Botany, Chemistry, Commerce, Conchology, Cosmography, Ethics, The Fine Arts, Geography, Geology, Grammar, Heraldry, History, Hydraulics, Hygiene, Jurisprudence, Legislation, Literature, Logic, Mathematics, Mechanical Arts, Metallurgy, Metaphysics, Mineralogy, Military Science, Mining, Medicine, Natural History, Philosophy, Navigation and Nautical Affairs, Physics, Physiology, Political Economy, Rhetoric, Theology, Zoology, &c.

Brought down to the Year 1890. Substantially Bound in Cloth with the Correct Pronunciation of every Term and Proper name, by L. Colange, LL.D. Handy Book for Quick Reference. Invaluable to the Professional Man, Student, Farmer, Working-Man, Merchant, Mechanic. Indispensable to all.

In this work all subjects are arranged and treated just as words simply are treated in a dictionary--alphabetically, and all capable of subdivision are treated under separate heads; so that instead of one long and wearisome article on a subject being given, it is divided up under various proper heads. Therefore, if you desire any particular point of a subject only, you can turn to it at once, without a long, vexatious, or profitless search. It is a great triumph of literary effort, embracing, in a small compass and for little money, more (we are confident) than was ever before given to the public, and will prove itself an exhaustless source of information, and one of the most appreciated works in every family.

It cannot in any sense--as is the case with most books--be regarded as a luxury, but will be in daily use wherever it is, by old and young, as a means of instruction and increase of useful knowledge.

Thousands of leading men offer the highest testimonials. It is not a cheap work; it is standard.

Never before, even in this age of cheap books, was such a splendid offer made, viz:

$2 For Cyclopedia, Postage Paid.
For Inventive Age, One Year. **$2**

An Extra Copy of Cyclopedia to any person sending us Six New Subscribers. The Same Offer is made to old Subscribers renewing for another term.

Address THE INVENTIVE AGE, Washington, D. C.

Inventive Age

INDUSTRIAL REVIEW

JOURNAL OF MANUFACTURING INDUSTRY AND SCIENTIFIC PROGRESS

WASHINGTON, D. C., MARCH, 1896.

Single Copies 10 Cents.
$1 Per Year.

UMPH.

the Great
npletion.

etter illus-
aving ma-
:ago Drain-
·ise—by all
iece of en-
ng machin-
c of nearly

Much has
rk but it is
:iated after
lustrations
ecent num-

height of five feet above low water level (1847) of Lake Michigan. Where the excavation is wholly through earth, the banks of the canal have a two-to-one slope, no masonry being used. The grade in the rock sections is one foot in 20,000 and the canal is designed for an ultimate flow of 600,000 cubic feet of water per minute, providing for a future population of 3,000,000 people. The work began September 3rd 1892, and contracts for completion expire April 30th 1896.

According to present estimate, it will cost $27,303,-216 to complete the work, which, when finished, will be 35 miles in length, and will necessitate the removal of 39,972,762 cubic yards of material. In the "rock section" of the canal, which is 160 feet wide and 35 feet deep, 12,071,668 cubic yards of material

the bottom, with the sides sloping outward. The wide portion constitutes twenty of the twenty-eight miles. The canal for the remaining eight miles traverses soft earth and can be dredged out later to ship canal width; at present the width will be 110 feet at the bottom, with sloping sides. The minimum depth of the canal will be twenty-two feet.

One of the most conspicuous features of progress on the canal has been the application of compressed air for power purposes. Examples are notable in operation of extensive plants on the line of the canal, where powerful compressors force air at working pressure through main pipes one or two miles long on the canal bank, and distribute it to the rock drills. In several instances advantage has been taken of its pressure to supply power to operate

BROWN CANTILEVERS, FIRST EVER USED, CHICAGO DRAINAGE CANAL.

· the more will be channeled, drilled, and blasted out of solid   drainage pumps either in the place of or in addition

The Inventive Age

Established 1889.

INVENTIVE AGE PUBLISHING CO.,

8th and H Sts., Washington, D. C.

ALEX. S. CAPEHART. MARSHALL H. JEWELL.

The INVENTIVE AGE is sent, postage prepaid, to any address in the United States, Canada or Mexico for $1 a year: to any other country, postage prepaid, $1.50.

Correspondence with inventors, mechanics, manufacturers, scientists and others is invited. The columns of this journal are open for the discussion of such subjects as are of general interest to its readers.

Technical matter is particularly desired. We want practical information from practical men.

Nothing will be published in the editorial columns for pay.

The INVENTIVE AGE is thoroughly independent.

Advertising rates made known on application. Special facilities for furnishing cuts of any patented article together with descriptive article. Business specials 25 cents a line each insertion, 7 words to the line. No advertisement less than 50 cents.

Address all communications to THE INVENTIVE AGE, Washington, D. C.

Entered at the Postoffice in Washington as second-class matter.

WASHINGTON, D. C., MARCH, 1896.

THAT the Willson patents for making calcium carbide are not of a fundamental nature and are, in fact worthless, seems to have been pretty well established by the investigations of "Electricity."

THOSE who desire to read all the discussions had at the recent meeting of the American Association of Inventors and Manufacturers, should begin their subscriptions with the February issue. New subscribers received this month will receive the February issue extra—the $1 for subscription paying for the issues of thirteen months.

THAT acetylene gas is not yet a commercial success is due largely to the excessive cost of producing calcium carbide. The cheapest it has been sold for in Paris is said to be $180 a ton. The claims made that calcium carbide can be made in North Carolina or at Niagara Falls for $25 to $40 a ton have not been verified.

LAST month was the banner month for Zell's Cyclopedia. From every state in the union and even from European countries the orders came in for the combination offer of the INVENTIVE AGE and this great book of reference, in elegant and substantial binding and excellent print for $2. Readers of the AGE, delighted with the book and the magazine have advised their neighbors and an unprecedented increase in our subscription list has resulted.

AT a recent meeting of the "Present Day Club," in Dayton, Ohio, an intensely interesting paper was read by Mr. John H. Patterson, president of the National Cash Register Company, on "The Relations of Employers to Employees." Mr. Patterson owes his success in business in a great measure, he believes, to the liberality and consideration shown his employees. His ideas of profit-sharing, of promotion from the ranks, of premiums for meritorious work, of the cleanliness, sobriety and integrity of employees, are set forth in this paper, which but for its length would find a worthy place in these columns.

THE Manufacturers' Record, of Baltimore has suggested that congress be asked to create a new department of national government, to be termed a Department of Manufactures and Commerce. The suggestion met with the approval of the meeting of the National Association of Manufacturers in Chicago, and many congressmen have expressed their advocacy of the proposition. The enormous value of the manufactured product of the United States, representing nearly $10,000,000,000 in 1890—a gain of over 300 per cent in 30 years—certainly dignifies and intensifies the argument. The stimulating of the manufacturing interests and the giving of employment to labor is of even greater importance than the bureau of agriculture, the success of which must largely depend on the success of the former. There are more substantial reasons for a government department of manufactures than of agriculture.

WE CAN see no good whatever in the bill providing for the termination of letters patent by act of Congress upon an appropriation to the inventor of "not less than $25,000 nor more than $100,000." It would be dangerous legislation. In fact any legislation is dangerous and unjust to the inventor that contemplates for one moment a controversion of the theory that a patent is property and the exclusive property of the inventor for the period and under the provisions of the constitution and laws of Congress. It follows then that Congress has no power or right to fix either a mimimum or maximun value on an invention. The relations of the inventor and the government are in the nature of a contract that cannot be abridged or violated without the consent of both parties.

What are Cathode Rays?

In commenting on the speculations appearing in the February issue of the INVENTIVE AGE regarding the wonderful advancement in science—the discovery of a means of transmitting human thought and penetrating opaque substances and photographing hidden objects without light—the McConnellsville (Ohio) Herald says:

The transmission of thought-waves from the brain to a plate like a photograph plate, and from this same plate days after, to the mind of another, has been accomplished. An electrical engineer, high up, has done it. And he has done it by rejecting the "continuous ether" theory. We have been asked a hundred times what is meant by cathode rays. Now everybody knows what the poles of a common galvanic battery are; and that one of them is "positive" and the other "negative." Now if the direction of the electrical current is noted carefully, it seems to go in at the one and out at the other. The one it goes out of is the negative, and is called a cathode, the other the anode. Now if the cathode be held, so to speak, against the end of a vacuum tube, violent motion is set up among the remaining molecules, fluorescence appears and the X rays are there. The X rays are wherever electricity is, whether in mineral or in animal matter, but good effects will never be obtained without vacuum conditions.

We have photographed with an incandescent light; did it at Chandler's the night it was done in London, also with a magnet, and aver that photographic effects can be done with any object that can be electrified, whether it be the brain, body or simply mineral matter. In short, matter has now found its master, matter nor space in the future can have any secrets. And this we affirm, for all planetary matters—mineral, vegetable, animal. The lights at spiritual seances will fool people no longer, nor spirit photography, and all that—the old ends; the new begins:

An interesting address on this subject was delivered before the Manufacturer's Club in Philadelphia but it remained for a gentleman in the audience to present the most plausible theory. The Roentgen rays as they are termed, cannot be seen by the human eye; but it is practically beyond dispute that they are a part of the light emitted by the sun. Now why can they not be seen? Sound is vibration. Light is vibration. Sound is audible to human beings when the vibrations are of such a kind that the ear-drum can respond to them or repeat them. It is a known fact, however, that some vibrations are so rapid, or intense, that the ear-drum is unable to respond to them, and so, although sound is there, it is not heard by the ear. The human eye responds to the vibrations which produce light. But the visible rays of light differ in the intensity of their vibrations: and thus we have different colors. Now, as there are sound-vibations which the ear cannot hear, there may be (almost certainly there are) light-vibrations which the eye cannot see, and, not seeing them, the space occupied by them appears to be darkness. But, as a tunningfork can take up the inaudible sound-vibrations, so a sensitive photographic plate can take up and record the invisible light-vibrations. This is the reasonable theory of the Roentgen rays which the lecturer omitted to refer to, until one of his hearers presented it as altogether the most interesting and important part of the subject; quite comprehensible by anybody. The penetrating power of the invisible rays is a strange fact of the phenomenon; but there is a fair probability that it may be attributed to the very intensity, and therefore high force, of the rays.

PHOTOGRAPHY WITHOUT LIGHT.

James M. Rusk Suceeds in Making a Photographic Print by Static Electricity.

A print has been made, or a photograph taken, by means of static electricity without intervening medium, without a Crooke's tube and without an electric light. The manner in which this print was made is explained by Mr. James M. Rusk—who by the way is a nephew of "Uncle Jerry," formerly Secretary of Agriculture—as follows:

I took a common "No. 27 X" photographer's plate; on the back of this plate I placed a piece of tinned iron; on the front of the plate I placed a common needle; this was all done in a dark room; they were then tied up in two thicknesses of photographer's paper through which neither the light of the sun, nor of an incandescent light, nor of an arc light, could penetrate. The package was put in a heavy paper box, then box and all were put in a second box; the package was then taken to an electric light plant and the sensitive side placed within six inches of a rubber belt—the machinery of course running, and the belt statically charged. The plate was then taken, left over night in the photographer's office and developed in the dark the next morning.

The same night that a London photographer used an incandescent light bulb for photographing hidden objects, Mr. Chandler and I did the same thing here. Anybody can do it. I have also succeeded in photographing with a magnet. How I have done this I do not propose to state until futher experiments have been made, and they cannot be made here at Malta or McConnelsville.

I make this general statement, that wherever static electricity can be obtained, there photographic effects can be produced. The experiments of Ingles Rogers of looking long at a postage stamp then staring long at a sensitive plate and getting an image, also the other experiment detailed in the Amateur Photographer of this same individual thinking of a postage stamp and staring at a sensitive plate and producing an effect that looked like a postage stamp, has been taken by some of my experiments out of the region of the impossible, placing them in the region of the highly probable.

There is much that I might say and back it up with actual experiments, but I am crowded for time for everything. I wish to make the following assertions which, though apparently dogmatic, have an abundance of proof behind them which will come out later on:—

The Roentgen ray is longitudinal; it is the highest cosmic energy possessed by the atom; such a thing as matter can not exist without the Roentgen ray existing; with the proper stresses placed upon the object to be photographed, no existent matter can resist its penetrating power. It can be refracted; it is the crest of the negative electrical ray; it is the cloudland above the high peaks on which Tesla has been standing. In short it is the dividing line between Tesla and John Uri Lloyd. I repeat, that it is a force present wherever atomic matter exists. Its discovery will, in all probability, enable us to tell what magnetism is. My own experiments tentative with the crudest of materials warrant my saying that natural and animal magnetism are identical. What this means, anyone, read in modern so-called psychical phenomena, will readily grasp.

Mr. Rusk is now pursuing his investigations further and claims to have made some other discoveries that will astonish the world when given out.

SPECIAL OFFER TO PHOTOGRAPHERS.

THE INVENTIVE AGE makes a feature of illustrating new inventions and new triumphs in engineering and mechanics.

Under this heading may be classed the building of canals and waterways, modern vessels and war ships, modern buildings, interior views of model machine shops and factories, railroad bridges, views of engineering achievements of every nature, natural wonders and discoveries, new machines, engines, motors and developments in electrical science, novelties, labor saving devices, etc.

We desire the assistance and co-operation of amateur and professional photographers everywhere. Every photographer has in his collection, or can obtain, one or more views that can be used in the AGE. We also want photos of prominent inventors as well as their inventions.

Not only will we give the artist credit for any view used but in addition we will forward the AGE free one year to his address. In instances of special merit and views of extraordinary achievements of genius and labor, cash prizes will be awarded.

Readers of THE INVENTIVE AGE in all parts of the world will confer a favor by advising local photographers of our request.

It is also desired that accompanying each view, there also be sent a complete description of the subject or enterprise, or that the address of some person be given from whom complete information can be obtained.

Among the deep coal mines in Europe is one at Lambert, Belgium. Depth, 3,490 feet.

NOTES.

Over 900,000 Miles of Wire.—It has been figured out that the total length of telegraph lines in the world is 904,701 miles, and the total miles of wire used on the same, 2,682,583, or enough to go around the globe at the equator over 107 times. The total miles of line in the United States January 1, 1895, was 190,393, with total miles of wire, 790,792.

* * *

Photography Under Water.—A French photographer has arranged an alcohol lamp so that while it is immersed he can throw powdered magnesium into the flame and thus secure a very brilliant light under water. In this manner he has been able to obtain some clear and beautiful photographs of the bed of the Mediteranean. Oxygen is carried down in the apparatus to promote combustion.

* * *

Horseless Vehicles Have Come to Stay.—The autocar is coming, and coming to stay. Two exhibitions have been arranged for, for horseless carriages on a large scale. The first will take place at the Imperial Institute. The Prince of Wales has consented to open it about the middle of May. The directors of the Crystal Palace have arranged for an exhibition in that building during the months of May and June; it is intended to be on a most extensive scale. The latter will not be confined to autocars alone, but it is to include all new inventions connected with carriages, whether drawn by horses or motor.s—*Invention.*

* * *

Products from a Ton of Coal.—From a ton of ordinary gas or bituminous coal may be produced 140 lb. of coal tar, in addition to 1,500 lb. of coke, and 20 gallons of ammonia water. By destructive distillation the tar will yield 69·6 lb. pitch, 17 lb. creosote, 14 lb. heavy oils, 9·5 lb. naphtha yellow, 6·3 lb. naphthalene, 4·75 lb. naphtol, 2·25 lb. alizarin, 5·4 lb. solvent naphtha, 1·5 lb. phenol, 1·2 lb. aurine, 1·1 lb. benzine, 1·1 lb. aniline, 0·77 lb toluidine, 0·46 lb. anthracene, and 0·9 lb. toluene. From the latter is obtained saccharine, which is a substance 230 times sweeter than the best cane-sugar, one part of it giving a very sweet taste to 1,000 parts of water.—*Textile Manufacturer.*

* * *

Something About Ivory.—African ivory is now conceded to be the finest. The first quality of this comes from near the equator, is closer in the grain and has less tendency to become yellow by exposure than Indian ivory. The finest transparent African ivory is collected along the west coast between latitude 10 degrees north and 10 degrees south. The whitest ivory comes from the east coast. It is considered to be in the best condition when recently cut; it has then a mellow, warm, transparent tint, as if soaked in oil, and very little appearance of grain or texture. Indian ivory has an opaque, dead white color, and a tendency to become discolored. Of the Asian varieties Siam is considered to be the finest, being much superior in appearance and density.

* * *

Stone Forests of Arizona.—The regions of the Little Colorado river in Arizona abound in wonderful vegetable petrifications—whole forests being found in some places which are hard as flint, but which look as if but recently stripped of their foliage. Some of these stone trees are standing just as natural as life, while others are piled across each other just like the fallen monarchs of a real wood forest. Geologists say that these stone trees were once covered to the depth of 1,000 feet with marl, which transformed them from wood to solid rock. The marl, after the lapse of ages, washsd out, leaving some of the trees standing in an upright position. The majority of them, however, are piled helterskelter in all directions, thousands of cords being sometimes piled uy on an acre of ground.

* * *

Dilute Air for Electric Light Filaments.—The remarkable genius of Nikola Tesla is again emphasized by his succeseful experiments looking to the substitution of dilute air in electric bulbs for the filaments now used. He sends currents of high tension through space, without any conductor, at a voltage many times greater than that employed in electrocution. He receives in his person currents vibrating a million times a second, of two hundred times greater voltage than needed to produce death. He surrounds himself with a halo of electric light, and calls purple streams from the soil. His experiments are of the utmost promise to the industrial world. His aim is to hook man's machinery directly to nature's, pressing the ether waves directly into our service without the intervention or the generation of heat, in which such an enormous proportion of the energy goes to waste—90 per cent in arc lighting, 94 in incandescent.

* * *

Double-Acting Piston Pump.—Considerable attention has been given in the scientific papers to the double-acting piston pump, in which the discharge is made approximately constantly throughout the stroke by a simple arrangement of the valves of the two pump cylinders, and the pistons are driven from a common shaft with cranks at 120° apart. The cylinders are exactly alike and side by side, being cast in one piece, and are double acting; one cyclinder draws water from the suction pipe, which has a branch to each end of the cylinder, and discharges into the second cylinder, there being direct communication between the cylinders, at both ends, while the second cylinder, drawing water from the first, pumps it into the discharge pipe, all the six valves necessary for these operations opening toward the discharge. If a complete stroke be divided into twelve parts equal, the delivery in each of these periods is the same. The delivery at each end of the pump separately varies from a maxim to nothing but the joint delivery of the two ends is nearly constant. The delivery of the pump for one complete revolution is three times the capacity of one of the cylinders.

The Japanese Incursion.

The possible incursion of the Japanese into labor circles in the United States, and the still greater probability of their early competition with us in competitive manufacture, has called considerable attention of late to the rates of wages that prevail in Japan. It must be admitted that they are wonderfully ingenious people, and that their manufactured articles may very readily come in competition with us in our home market. One of the Japanese Commissioners at the Cotton Centennial Exposition, at New Orleans, was asked how it was possible for them to make as handsome an exhibit as they did there at that time, and especially in the higher branches of manufacture. His answer was that, in many things, they began their work where we left off. In other wordr, they had the advantage of all that we had ever learned, and adding to this accumulated knowledge, their own ingenuity, patience, and cheap labor, they were capable of doing wonderful things. The rates of wages current in Japan have been furnished to our government by Consul-General McIvor and have been published. Counting the Japanese *yen* as worth 68 cents in American money, Consul-General McIvor reports the rates for farm laborers, males, $1.44 per month; females, $1.20. Common laborers 19 cents per day; carpenters, plasterers, stone cutters, sawyers, roofers, etc., from 24 to 31 cents per day.

The free trade policy so urgently advanced by a number of the extremists in the last Congress would soon produce very queer conditions in the Uuited States, if such policy be permitted to develop itself as fully as its advocates desire. It may be that we should consider all mankind a common brotherhood, but the first duty of every good citizen is to look after the interests of his own country in preference to any other country, and of every good man to look after the interests of his own household in preference to those of others. To many persons competition in our markets by the Japanese seems a very remote thing, but they may be assured that under anything like a free trade policy it is much nearer now than has ever previously seemed possible. The Japanese are ingenious, enterprising, industrious, and aggressive, and are bound to carry their wares to the best market. Had not we better preserve America for the Americans?—*New Orleans Planter.*

Work on the Panama Canal.

It is perhaps not generally understood that work on the Panama Canal has again started up. The Panama *Star and Herald* says that the new Panama Canal Company is doing practical work on the canal. The number of laborers employed at the different sections last month, including Panama, Colon, and La Boca, was 4,362. The work of deepening the channel on the Pacific side of the canal has just been resumed, and, as a whole, more considerable activity is apparent in connection with the work than has been seen for a long period.

Electric Coat for Motormen.

Major A. H. Swanson, of Texas, where they are supposed not to need such things, suggests that motormen can be kept warm in winter by hitching the trolley circuit on to their overcoats, which have a resistant network of wire imbedded in them, and thus warm up. This is something like the idea that has been practically carried out in the "Electrotherm."

The Double-Wedge.

EDITOR INVENTIVE AGE:—It is an old saying that the inventor starves while the capitalist and improver get rich out of his invention. In many cases this is sadly true, but a notable exception to the rule is to be found in the case of Jacob S. Schuckers the inventor of the double-wedge in the linotype machine. Some of you may not know what a linotype machine is, so I will tell you briefly. It is for automatically casting a line of printing type to save the labor of setting up the individual type by hand. Ten years ago there was no such word as 'linotype'" nor such a machine. Now fully two thousand of these machines click out the type for several hundred of our leading daily newspapers such as the New York Tribune, Herald, Times, and World. The capital of the company controlling the patents is $5,000,000; its stock is selling on the market at two hundred and twenty-six. The machine is the creation of a little German named Otto Mergenthaler, who, a few years ago, was as poor as a church mouse. Today he draws royalties amounting to $60,000 a year, and his machine is regarded as one of the mechanical marvels of the age. But you want to know something about Schuckers. I will tell you. When Mergenthaler had completed the main part of his machine he met with great difficulty in devising means for justifying the lines of type so that they would not look like blank verse when put in a column. This was one of the greatest obstacles he had to overcome, and without it his machine would not be practical for general work. At this critical period in the developement of his machine, Schuckers hit upon the idea of a peculiar kind of double-wedge which did the work of spacing the words automatically and perfectly, It was the key to the whole situation. Mergenthaler claimed to have hit upon the same idea at the same time, and filed an application in the Patent Office. Schuckers deposited his application also. An interference suit was thereupon instituted. Mergenthaler was backed by millions. Schuckers was poor, and his intimate friends warned him against the folly of fighting such a mighty antagonist. But paying no attention to their advice he went to work with a heart full of courage, and organized a stock company of a million dollars.

Those who heard his story bought stock in his company, and the money was used to fight the all-powerful Merganthaler Company.

It was a long hard fight which covered several years. In the meantime the Mergenthaler people were building and renting hundreds of machines all of which contained the indispensable double-wedge.

The Mergenthaler Company employed hundreds of skilled machinists to try to avoid the claims of Schuckers. They bought up every conceivable variation of the double-wedge, and frequently maintained that they had other devices which would circumvent Schuckers. Regardless of these pretensions which were doubtless transparent to the Schucker's side, the original double-wedge appeared in all the machines going out of the Mergenthaler factory. Finally, after years of anxious waiting, Schuckers was declared by the Patent Office to be the prior inventor, and a patent was issued to him accordingly.

Upon receiving his patent Schuckers immediately brought suit against the millionare Mergenthaler Company. More years of waiting and expense ensued, as the two inventors and their backers fought bitterly on. But Schuckers' prospects grew brighter while Mergenthaler's grew darker, until finally, fearing defeat, the Mergenthaler company tendered Schuckers an offer of $416,000. The Schuckers folks called a meeting to consider the matter, and many were strongly opposed to taking less than a million, but the majority agreed to accept. The Mergenthaler company then gave a check for this handsome sum. Schuckers was made independent. His backers quadrupled their investment, and the Mergenthaler Company had a new and valuable patent to prolong and strenghthen their monopoly.

RHESA G. DUBOIS.

A new optical instrument, which Dr. Fraser Harris has proposed showing to the Glasgow Philosphical Society, is known as the stereo-photo-chromoscope. Its purpose is to photograph an object in such a way that the "positive" of a picture, viewed as a transparency, will present the object not only in natural colors, but also with stereoscopic effects.

THE publishers feel that every reader, having an opportunity to do so, will please a friend by explaining to him the opportunity he has to obtain Zell's Cyclopedia and the INVENTIVE AGE for one year for $2.

MARVELOUS ENGINEERING TRIUMPH.

(Continued from first page.)

were being talked of the Chicago Drainage Canal was completed, and the rock drill was one of the most potent factors to expedite the work.

The application of the channeling process for canal construction was first introduced on this work. Deep channels are cut on the boundary lines. These

REMOVING TOP LIFT BY INCLINES.

channels are about two inches in width, and are made as deep as the channeling machine will admit, that is, from 10 to 14 feet. The channeler runs ahead of the work of excavation, the cuts being made first, after which the intervening rock is broken by blasting in the usual way. As the depth of the canal cut is about thirty-five feet, it was found necessary to make three distinct channels, one above the other, and each independent of the other, the offset being only about eight inches. The advantage which the channeling process in canal construction gives, is a solid bank maintained free from irregularities and weakness produced by blasting. Were it not for the channeler, it would be necessary in many places to build up the bank of the canal by masonry or otherwise. The smooth wall made by the channeling machine facilitates the flow of water, lessening the friction. The excavation requires a slightly smaller amount of explosive because of the release lines made by the channel cut, and there is also an advantage in preventing the claims for extra work which might otherwise be made by the contractor, the rock on the canal being paid for between the channels. It is usual to put in a row of about eighteen holes between the channels. These holes are about two inches in diameter at the bottom and twelve feet deep, and are charged with 40 per cent dynamite, and blasted simultaneously, throwing down the entire bench, and breaking it up into pieces varying in size from a brickbat to pieces of several tons weight.

The immensity of the work can scarcely be conceived except by a trip along the entire length of the canal. The excavated earth, glacial drift, rock, etc., are heaped up on each side of the great channel to a height of nearly a hundred feet, making even more striking the impression of great depth upon looking down into the ditch. What is to become of this enormous deposit is a problem yet to be decided. Now the mass of grayish slaty looking soil, clay, and gravel lies undisturbed and will remain indefinitely, as will the tens of thousands tons of rock blasted out.

In the construction of this tremendous work machinery commensurate to the task has been used.

parallel with the channel, and can be moved bodily by steam power as easily as a bucket is carried up the incline to be dumped at the further extremity of the bridge, as it is called. This apparatus works rapidly, and it is so successful that at the present time eleven of these cantilevers are in use on the various sections of the canal.

A unique and imposing mechanical device is in operation at Willow Springs, about midway between Chicago and Joliet. It is an enormous piece of ma-

chinery and can be seen four miles. It is in the shape of an immense piece of iron bridge work formed like a great chair rocker. Its length is 750 feet and it is more than 100 feet high. It spans the channel, with its horns, or ends, running up over each of the banks of excavated earth, and it is moved on tracks running parallel with the channel

towers are connected by a by means of which great excavated earth are rapid ing places.

Over twenty firms of co in hand, employing betw

An important circumsta building of the canal is power near the City of Jo that part of Niagara whic ized. This power will be works governing the outl river—the Desplaines—thr Where the channel begins tom is twenty-four and s This datum is the low-wat in 1847 and is 579.61 feet i at Sandy Hook. When th southern end of the chann siderabe incline into the c Canal basin at Joliet, kn The fall from datum at the level of the upper poo feet in four and one-third a gratuitous water forc power—sufficient to furnis of canal boats the whole l Assistant Chief Enginee belief that it would not on cable to utilize the power City of Chicago by electri

A "Glowworm" Electr

A New Jersey electrici claims to have discovered t other words, that he can m accordance with a new pri tion. He proposes to emul stead of having the presen in the ordinary incandesce whole surface of the gla illumination. Mr. Moore' directed along the lines o tricity, which he claims son why we cannot have there is why we cannot hav without striking all the music. He claims to be at divisions of energy, and e ing elements. He employ 110 volts, and from this he very favorably with sunsb good negative is concerne a one-volt current is enoug held close to the eyeball, drew out the steel, leaving the eye without serious injury.

The Great Unknown.

In a geneal way the Psychical Societies or the Psychological Associations all over the world are doing much to advance the cause of true science. There are certain phenomena which heretofore have

§ 14, CHICAGO DRAINAGE CANAL.

been described as psychic or superphysical in their nature, but which in reality belong directly to the domain of the exact sciences. Heretofore all these classes of phenomena, such as hynoptism; magnetism, mesmerism, clairvoyance, clairaudience, *id omnia genera*, have been thrust aside by the schoolmen as quite unworthy of scientific research. Thanks to the general advance of scientific knowledge amongst all classes and the widening of the

and recognized among the schoolmen as factors having a direct bearing on our physical life. True science has no line of delimitation. The universe of nature, infinite in its manifestations, is the true field for scientific investigation. The more scientific one's mind, the more interesting any apparently insoluble problem becomes. The subtle influences that govern that complex machine, our brain, are as much a legitimate object of study and research as the composition of the air we breathe or the interrelations of the molecules towards each other in the laboratory of the chemist. The difficulties presented in formulating a basis for investigation have perhaps had as much to do with the disinclination of scholars to pursue their studies in this direction as anything else. Now however, that the existence of superphysical matter, or matter beyond the reach of chemical reaction, is recognized, and its properties (in a small measure to be sure) are beginning to be understood; now that it is foreshadowed that there is a grand harmony of ethereal impulses which controls all forms of physical matter, scientific research will be devoted to the ascertaining on the edge. This great annulus swings slowly and continuously round and round in the water at the foot of the falls.

With the rapid broadening of our scope of enquiry the recognition of ethereal influences translated into force by impinging on physical matter, all psychical research is beginning to find a firm foundation upon which to stand. Before long these superphysical phenomena, which all fair-minded persons must recognize as a fact, will be as much legitimate objects of scientific investigation as are the constituents of the more material elements composing this planet. Far from sneering at these earnest investigators, all scientific minds should lend their aid toward the elucidation of the many problems which have been regarded as without the legitimate scope of scientific investigation.—*Invention.*

Two Full Moons in a Month.

Probably not many people are aware that last December was different from any other since the beginning of the Christian era. December, 1895, enjoyed the unique privilege of having two full moons, which is a phenomenon that has not occurred in any December in 1,896 years. The coincidence of the last event of this kind happening in the same year as the birth of Christ was not widely noted, however, or it would probably have been looked upon by many as significant. The occurrence was a purely astronomical one.—*N. Y. World.*

Successful Heating by Electricity.

The experiment of heating a theatre by electricity was tried recently at the London Vaudeville, and with complete success. Storage batteries were employed, connected with radiators formed of non-conducting materials, and hence becoming heated

Gum and Microbe.

There are two places in Washington, D. C., particularly attractive to the student and not without considerable interest to the average visitor at the Capital city. One is the Medical Museum of the U. S. Army; the other is the Naval Museum of Hygiene. In the former, "man's aggressive inhumanity to man," is seen in portions of human anatomy pierced with bullets, cut with swords and broken by shot and shell. Rows of grim skeletons that once stepped lively to martial music, here mutely attest the "individual" failure of war, illustrating the fact that "the path of glory leads but to the grave." Here disease, in its most horrible forms points a moral in awful specimens that make one wonder at the possibility of human suffering.

The museum contains more than 3,000 anatomical specimens, besides 11,000 pathological and 3,000 campartive anatomy exhibits. In this place death is defied in the preservation of bone and tissue. A part of the bony structure of Guiteau the assassin, is here, and a section of John Wilkes Booth's backbone and spinal cord add their quota to this collection of the "dregs of life." Egypt also has contributed to the gruesome exhibit, with those things so emblematic of her antiquity, the mysterious mummies. Among these is an excellently preserved head of one of Pharoah's male subjects. The classical features and contour of face and head of this specimen are almost perfect, and the hair, a reddish brown, curls about the well-shaped cranium, as naturally as it did two thousand years ago when its possessor was a citizen of Memphis.

While in the museum so much attention is paid to defunct man, the health of the living is also looked after there. The pysical troubles of U. S. soldiers are made a study, and disease germs, which are more potent than rifles and cannon, are isolated, classified, cultivated, and even corked up in bottles in the laboratory, so that their habits and effects can be marked with certainty. If the doctors of the museum should uncork their "vials of germy wrath," there is no telling how many varieties of disease the people of Washington would have to contend with.

If the Medical Museum is a "den of horrors," the Museum of Hygiene is a storehouse for scientific curiosities. As in the former place, there is a laboratory here, where the analytical work for the Navy Medical Department is done. But the main feature of this place is in its two-thousand models, apparatuses, plans and diagrams, illustrating sanitary subjects. If one desires to build a house he can find instruction here necessary for its sanitary equipment; even a model bed, washstand, bath, etc., warranted microbe proof, will be shown him. If he wishes to set up a hospital, the most improved methods of taking care of the sick can be seen, from a stretcher to a perfect hospital ward. And in connection with such an enterprise, the searcher for information can if desired, find in the museum models of firstclass furnaces for cremating; or, if preferred, he has a burgular-proof vault at hand for inspection.

The latter is made of boiler iron riveted together, and weighs four-hundred pounds. Besides these cheerful reminders of sic"k" transit, there is a model of the Tower of Silence at Bombay, India, which offers that "distributive" method of sepulture. All about the exhibit rooms are inventions for comfort and for circumventing the deadly germ and other micro-organisms—the living problem has precedence. Heating, ventillation, plumbing, filtering and the minor details of sanitation are here set forth. The process of fumigating and disinfecting ships is also shown, with their sick-bags, beds and hammocks.

The work of the Naval Museum is devoted particularly to sanitation on shipboard, and much of its scientific investigation is in the interest of the "power afloat."

When a warship is equipped for a cruise she is a floating drugstore as well as as a fighting machine. A cruiser, such as the Cincinnati, carries in her medical department nearly three-hundred different kinds of drugs and medicines, besides food extracts and other dietary provisions for the sick. The amount of drugs carried by a ship of this class, would lead one to believe that life on the ocean is often disturbed by something more than sea-sickness. The Cincinnati's pharmacopœia supply would run something in this way: Antipyrin, 150 gm.; ammonii iodium, 25 gm., ammonii benzoas, 50 gm.; belladonna plasters, 2 boxes: chloroform, 3,500 gm.; cinchonae tinct. 1,500 gm.; glycerine 2,000 c. c.: iodoform, 400 gm.; quinine 250 gm.; bromide potash, 600 gm.; and so on through the long list. Of surgical instruments and appliances this vessel has a collection formidable enough to make a Jack Tar forswear war. These cutters, twisters, gougers, binders, etc., number more than one hundred.

In time of peace looking after the sick on board of a man-of-war is not more difficult than the labors of a country doctor, but in the reign of shot and shell, the ship's physician has his hands full.

Before a fight begins everything is in readiness for rendering aid to the wounded. Knives, saws, tourniquets, forceps, splints, bandages, anaesthetics, antiseptics, etc., are at hand, and though—when the fight begins—the ship rolls to the sea and trembles with the discharge of her mighty guns, while the iron hail pours on and around her, the knife cuts through the shattered limb of the unfortunate sailor and the needle pierces his torn flesh, in its painful but necessary office.

Although our warships are well equipped with drugs and medical apparatus, there is a lack of space on most of them for taking proper care of the sick. The Cincinnati, for instance, is 300 feet long 42 feet beam, of 3,200 tons displacement, with engines of 10,000 horse-power; has 11 guns in her main battery and carries 317 persons, men and officers. Yet her sick bay has less than 200 square feet of floor space, affording room for but two cots and three hammocks, besides the small amount of furniture contained.

This lack of accomodation on ship board would result in serious difficulty if there were no other means provided for taking care of the wounded after a battle at sea. But an emergency of this nature will be met in future marine war, by the hospital ship, which will hover, like an angel of mercy, on the outskirts of the battle, respected as neutral by the combatants, and when the slaughter is over steam up and get her suffering frieght. Up from the sick bay, wardroom, or other part of the warship, which has been turned into a temporary operating room, the wounded will be brought on the backs and in the hands of sailors and marines—as this is said to be the best method of transporting the injured.

There are a number of inventions for hoisting wounded men from the holds of ships and conveying them to other vessels. One way is to secure the patient with straps to a stretcher, which, after being gotten on deck is allowed to slide gently down inclined skids to the receiving vessel. The patient is then lowered to the sick ward (if it is a hospital ship to which he is taken) where physicians, skilled nurses and all modern hospital science await him.

How different is the sanitary status of the modern warship, from the old time craft. In the latter health and comfort were less thought of than the ability to house a great number of men and cannon. We read of an English ship carrying over a hundred cannon and nearly a thousand men, when in fighting condition, while the air in her lower holds was so foul that men were often suffocated by it. Today the warship is scrupulously clean, comfortable and healthy. Ventillation is obtained by means of pipes and blowers which drive the air into all parts of the ship, and these blowers are supplemented with windsails and air-scoops, so that at all times there will be plenty of pure air.

The Cincinnati's blowers make 425 revolutions per minute, and in this amount of time drive 10,000 cubic feet of air through her ventillating tubes. The need of much cool ozone on ships of great steam power, is apparent when we learn that the temperature in the engine room of the modern cruiser mentioned here, sometimes reaches 135°; in the fireroom 168°, and in the dynamo room 110°.

Besides other conveniences the Cincinnati has on board porcelained bath tubs, shower and douche baths. In the officers' bath rooms the floors are laid in patterns with red, white and black marbles. An ice-making machine, and of course, an evaporator of 3,000 gallons daily capacity, add their valuable services to the ship's complete outfit. J. E. P.

One hundred and twenty years ago, in 1775, the Paris Academy of Science withdrew its standing reward of 500,000 francs which had been offered for a "perpetual motion machine." It was plainly stipulated in the offer that the machine should "be self active," so much so, at least, that when once set in motion it shall continue to move without the aid of external forces and without the loss of momentum until its parts are worn out. During the years that the above reward was the standing offer thousands of men became insane over the problem. At last, at the of the date given in the opening, the impossibility of constructing such a machine having been demonstrated, the offer was formally withdrawn. No government or society of standing now offers a reward for such a machine.

The research laboratory just opened in Paris seems to mark a new stage in scientific progress. For a nominal monthly sum, paid to the Municipal Council, any person may use the laboratory, and will be provided with materials for his experiments.

Duluth, Minn., is the first to start the new industry of evaporating potatoes. They have demonstrated the fact that the tubers can be thus treated and preserved the same as apples.

Proposed Changes of the Patent Laws.

At the last meeting of the Association of Inventors and Manufacturers of patented inventions, held in this city, many valuable suggestions were made by its members, with respect to our patent system, and it was gratifying to see present the Hon. John S. Seymour, the present Commissioner of Patents, and to hear his timely words as to the necessity for a full classification of patented inventions of this and foreign countries.

In the Washington Post of the 3rd inst. appeared the following as a proposed change in the patent laws:

> Senator Burrows introduced a bill in the Senate yesterday to make it illegal for the Commissioner of Patents to reject an application for patent on the ground that the several features which it combines have been anticipated in several different patents. It requires that all rejections shall be based upon a single invention, embracing a complete anticipation of the device applied for.

The question involved in this proposed change of the patent laws is one with which it is very difficult to deal, and congress should be very careful in its treatment of it.

It is well understood among competent patent attorneys, Patent Office examiners, and the Commissioner of Patents, that a new combination of old devices is patentable, notwithstanding the fact that one or two of the devices may have been combined before for a like purpose and that the function of each, separately, may be old. It, however, is difficult to determine when patentable invention is developed by the uniting or combining of several old devices so as to form a new organization or machine. The courts have often held that the mere "aggregation" of old parts of several different machines in a single machine, is not a patentable invention; also that the constituent elements of a mere combination patent, has no equivalents in the broad sense of the word as used in connection with an invention which is new broadly. The test ought to be, in our opinion, in the determination of the question of patentability, has the invention improved or advanced the condition of the arts, or has it added, in the smallest degree even, to the benefit of mankind; and if by this test the affirmative answer can be given, a patent ought to be granted.

Inventors, and congress also, must bear in mind that the public has a full right to all that is found in expired patents; and that, often, it is not invention to pick out of three expired patented inventions or fully organized machines, certain parts of such machines, and put the same into one whole. However, such a combination may have required invention to bring the three parts into the one machine—that is in co-active relation, and, if this be the case, it may be found to be a patentable combination instead of a mere aggregation. Questions of this sort require the nicest discriminating minds to dispose of them, and usually an examiner in the U. S. Patent Office, assisted by argument of competent patent counselors, is able to arrive at a correct decision in regard thereto, and it would probably be useless, if not dangerous, to pass such a law as is proposed. It is to be hoped that there will be very little tinkering with out present admirable system of patent laws; but where matters like those advocated by the Commissioner of Patents as to classification, in his recent address before the Association of Inventors and Manufacturers; and the proposed amendment, as suggested by the patent committee of said association, to the abominable 25th section found in the law enacted in July, 1870—which curtails the term of United States patents three years, because foreign patents were first secured for the same inventions—are involved, it is safe, and but right to amend the law at once in such particulars, as both inventors or patentees and the public, would be greatly benefited thereby.

ROBERT W. FENWICK.

Valuable Bicycle Attachments.

Two recent inventions will especially interest bicycle manufacturers and riders. Raymond P. Sholl of Bartonville, Ill., is the inventor of a new and improved bicycle crank, by means of which "deadcenter" is overcome and lost motion is prevented. With 6½ inch throw of cranks it is claimed this invention gives 9 inch leverage and a gain of over 20 per cent in power over straight cranks.

N. J. Prichard of Shenandoah, Va., is the inventor of a valuable speed gear for bicycles which seems to possess much merit. Its advantages are admitted by experts who have investigated the invention.

The gear gives two wheels in one and in ascending grades the wheel can be geared 20 inches less that when the rider reaches the top; in descending grades the rider can allow his feet to remain on the pedals as they do not revolve. The attachment can also be used as a brake without contact with tire. It can be attached to any machine and adds but one pound to the weight. This is many times overcome by the advantages gained. The attachment can either be placed on the front sprocket or rear wheel. The inventor's enthusiam seems justified from the fact that all who have used a wheel equipped with it pronounces it a very desirable attachment.

Power Transmitting Machinery.

The Dodge Manufacturing Company of Mishawaka, Indiana, had a very complete exhibit of their patent "rope transmission of power" system at the Atlanta Exposition. Of it an Atlanta publication, Dixie, says:

Their celebrated Independence Wood Split Pulleys are known and used all over the world, this company being the originators and inventors of the system of standard bored split pulleys with a standardized bushing system to facilitate carrying pulleys in stock, as well as enabling manufacturers to easily change pulleys to shafts of different diameters. All their pulleys are made separable and in halves, and are of the highest degree of perfection, in workmanship and material. As evidence of the popularity of these pulleys: twelve years ago pulleys were not carried in stock, while now every trade center has numerous stocks and requirements are filled promptly. Over 1,250,000 Dodge Independence pulleys are now in use and they are sold all over the globe wherever machinery is in use. The bushing system makes a stock of 1,000 Independence pulleys equal to 40,000 iron pulleys. The dynamo pulley is exhibited and meeting with much favor

become the most popular mode of distributing and transmitting power.

The Field is Still Large.

We venture to state that no matter what your occupation, you cannot do a day's work without the use of from one to a dozen patents—in fact, cannot even dress yourself without contributing to "royalty." Let's see. First you arise from a patent spring bed; if a laborer, your overalls are patented; you put on shoes made on a patent last, eat breakfast cooked on a patent stove, and, if a farmer, you hitch to a patent plow or reaper, as the season may be. If a mechanic, nine out of ten of your tools are patented; you drive a patent nail with a patent hammer; bore a hole in a corner with a patent ratchet brace bit, and so on all along the line. Still the market is in no danger of being overstocked with useful inventions. This is a big country, continuously developing wants. It won't be crowded in your life-time. It will take five hundred millions of people before its resources will begin to be tested to the straining point. Your invention may be just what the people want, and, if so, after it is introduced, some dealer somewhere will be selling it

Engineering Science in Industrial Progress.

Engineeing science has been an indispensable factor in industrial progress. It has been the brain of enterprise. The best means, and the most economical methods of developing our resources, have been outlined for the practical man on the chart prepared by the engineer. His foot print is everywhere. The bridge that spans the river, the tunnel that perforates the mountain; the canal that intersects the broader streams; the mine that has its chambers in buried coal and ore; the industrial plant that hums with the spindle, throbs with the machine and glows with the furnace, are what they are in economy of space and stability of structure by reason of engineering skill. Its efforts and its achievements are largely the history of material progress, and what may yet be in its future story will be but a circle around the same axis. To tally its services would be impossible, and it is only by following it along its special lines, that we can reach an approximate idea of its services.

At a recent meeting of the Engineers' Club of Philadelphia, a synopsis of some recent engineering achievements in the Great Lake region was given by John Birkinbine. Among other examples per-

DISPLAY OF DODGE MANUFACTURING COMPANY AT ATLANTA.

amongst electrical machinery manufacturers. Their capillary oiling system seems to be all that could possibly be desired for journal bearings, giving a constant cleanly flow of oil without gumming or stoppage and without undue commotion in the reservoir. The entire line shafting of Machinery Hall is operated in this type of bearing and is most satisfactory. Bearings will run from three to six months with but once filling the oil chamber. The most important part their exhibit is the model power plant of the Independence Power Co., built expressly for this Exposition. It consists of a main jack shaft upon which are mounted grooved laboring sheaves carrying ropes in all conceivable manners and angles to model buildings, and is a thoroughly representative show of the possibilities with fibrous rope driving. The system is patented by the Dodge Mfg. Co., and known as the Dodge American System of Rope Transmission. The distinctive features are the continuous system of winding, making an endless rope, and the use of a traveling automatic slack take-up to preserve an even tension in the rope under all laboring and atmospheric conditions. This system is applicable for heavy power and for both exposed and indoor service, and destined to

every hour in the day. Then you can decide whether to purchase a yacht or endow a college; we trust it will be the latter. Lamentable as it may seem, in one sense, no matter how many people there may be, there will always be enough ambitionless plodders who wait for the "other fellow" to think of it, to support the active, observant, cute-witted inventor who gets in on the ground floor.

Cable Road in a Volcano's Crater.

A Mexican dispatch says that engineers have completed the survey of the volcano Popocatepetl for the purpose of determining the best location for an aerial cable railway to the summit. It has been determined to start the line from the ranch of Timacas, on the northwest, and tourists will be able to make the ascent to the summit, nearly 18,000 feet above the sea, with entire ease, and also descend into the crater, where the work of extracting sulphur is going on.

The development of China has begun. A first-class double track railway is to be built from Pekin to Tien Tsin.

tinent to the subject we note those referring to the mining of iron ore. The shovel system of mining on the Mesaba Range, radically changing the cost and labor of production, is a noteworthy adaption of appliances to environments. In dock facilities for shipping ore, time and labor have been wonderfully economized. In some instances by means of pockets and adjustable spouts, 2500 tons of iron ore are loaded in a freight-boat within forty-five minutes. Ore has been unloaded for less than a cent per ton. In the canal system, with its locks and grades, the transportation of ore and other freight is prompt and economical. As showing the advantages of this system and its importances in the distribution of products, the shipments through the St. Mary ship canal are estimated at about 17,000,000 tons in 1895. In other related matters, the same striking features of improvement in appliances and methods are strongly presented. In all and each of these, we have evidences of the incessant and progressive work being done by the engineer, in our industrial develpment. As is in the past so in the future; the same service will be indispensable to the material progress of all nations, and will undertake its several tasks with the same assurance of success —*Age of Steel.*

TALKS ON THE PATENT LAWS.

Opinions of Berliner, Shaw, Billings, Davis, Fraser and Other Prominent Men.

Letters from patent attorneys, inventors, manufacturers and others indicate wide-spread interest in the discussion on needed and proposed changes in the patent laws appearing in the February number of the INVENTIVE AGE. The January meeting of the American Association of Inventors and Manufacturers was indeed an interesting one and many members unable to attend sent letters of regret, some of them containing valuable suggestions worthy of publication. The papers presented were from able and representative Inventors and Manufacturers and much space in our February issue was given up to the regular proceedings, the verbatim address of the Commissioner of Patents, who honored the association with his presence, and the report of the Legislative Committee recommending important changes in the patent laws. In this and subsequent issues, space will be freely devoted to the publication of papers presented, discussions had, and responses received to the list of subjects sent out by the Executive Council of the Association to ascertain the true sentiment of the inventive and manufacturing world on topics of vital interest to the great army of inventors throughout the country. The list of subjects to which many responses were received was as follows:

1. The propriety of a Federal Statute of Limitations in Patent Causes.
2. The importance of having American Patents granted for a definite fixed term without regard to the date of expiration of corresponding foreign patents.
3. Should not a patent become void by failure to manufacture and use the invention in such manner that it may be brought to the attention of the public within a fixed period from the date of the grant?
4. Would it not be beneficial to our Patent System to so amend the Patent laws that innocent users of devices and processes covered by old and unworked patents, of doubtful validity, may not be compelled to pay tribute to purchasers of such patents?
5. Should not the Commissioner of Patents be clothed with authority to appoint, on application, Special Examiners, for taking testimony in Interference Proceedings the same as Special Examiners are appointed by the Courts in Equity Cases?
6. Are the New Patent Office Rules abridging the time for response to official action desirable and sustainable under the statute?
7. Should or should not the United States Patent Office be made a separate and distinct department and be provided with a building of its own?
8. In view of the necessity, already arisen, of storing a part of the models elsewhere than in the Patent Office should not models be placed, all together, in a suitable building specially designed for that purpose?
9. How can the rules of procedure in the Patent Office be amended or changed to facilitate and improve the character of the work which it is intended to accomplish?

ARTHUR C. FRASER.

Arthur C. Fraser, of New York, submitted the following extended seriatim reply:

1. A federal statue of limitations in patent cases seems to me desirable, as tending to remove uncertainty as to the rights of litigants, and as promoting uniformity of practice throughout the entire country. I fail to see any valid objection to the enactment of a reasonable limit, say five years, within which an action for infringement must be brought.

2. In the patent professions, the opinion is I believe unanimous that American patents should be granted for a definite fixed term independent of the date of expiration of foreign patents for the same inventions. The operation of Section 4887 of the existing statutes is a hardship to domestic inventors, who are often compelled to sacrifice either the right of protecting their inventions abroad or a part of the term of their domestic protection. In effect, the existing law tends to discourage an American inventor for protecting his inventions in foreign countries, so that its effect is in many cases to leave the field of manufacture unrestrictedly open to foreigners, while confining it in this country to the inventor and those deriving rights from him. A serious objection to the existing law is the uncertainty which it induces as to when a patent will expire, since to determine the date of expiration it is necessary to ascertain what rights have been granted in foreign countries, and to interpret the laws of those countries in order thereby to determine when the foreign patents shall expire. A simple provision that every patent granted should expire at the end of 17 years from the date of application for the first foreign patent for the same invention, and in no case should remain in force more than 17 years, would seem to me a desirable substitute for Section 4887.

3. Theoretically it would appear desirable that patents should become void in case the inventions are not brought into use within a reasonable time. But practically any such provision would be extremely difficult of just enforcement. The provisions of the laws of foreign countries regarding working of the invention which are intended to accomplish this end, are ineffective because easily evaded, and onerous because they impose a serious burden of expense upon patentees of limited means. The expedient of requiring the payment of taxes at stated intervals to keep a patent alive is effective to terminate a patent when its owner has concluded to abandon it, but is oppresive in enacting an unjust tax upon patentees who are giving the public the benefit of their inventions. If such a tax is made so light that it ceases practically to be a burden to patentees of small means, it will be paid by those who are sleeping on their rights rather than suffer their patents to lapse so that the scheme is found to be effective only in proportion as it is burdensome. Many inventions are not immediately available because of being in advance of their time, of which the pneumatic wheel tire is a conspicuous example, and in such cases any system which terminates the patent because it is not brought into use, deprives the inventor of his only chance of reimbursement for his genius and skill, whereas under the present system he has an opportunity even in the 17th year of his patent to derive some benefit from it. And yet it would seem just in the interest of the public at large that when a patentee has failed or neglected to bring his invention into use, his patent should not wholly stand in the way of its being devoloped and introduced by others. To this end it might be provided by law that when a patent has been in existence for five years without the invention having been introduced upon the market, any person desiring to introduce the invention may apply to the courts for permission to use the invention, and upon a hearing, the courts, if it finds the patentee to have failed through his own neglect to bring the invention into public use, may direct the patentee to grant the petitioner a license upon such terms as the court may impose, in default of which the patent shall be declared void. Such a law properly safeguarded, would tend to stimulate patentees to greater diligence in putting their inventions upon the market, and would insure that after five years the public, if it were not supplied with the invention by the patentee, should not be wholly cut off from use thereof in case anyone else were to come forward desirous of manufacturing the invention. A further provision to the same end that might be feasible, is to require each patentee to file annually with the Commissioner of Patents a sworn statement of the extent to which his invention has been introduced into use, if used at all, and make the failure to file such statement work the lapse of the patent. This would provide record evidence of the amount of diligence and effort claimed in behalf of the patentee, and would be free from objection of hardship attaching to the payment of an annual tax.

4. The question of the protection of innocent infringers of patents has been much discussed. The difficulty with any provision of law which relieves one part of the public from the penalty of infringement to which another part of the public is liable, is that it necessarily violates the provision of the constitution under which all patents are granted, namely, that "Congress shall have power . . . to promote the progress of science and useful arts by securing . . . to inventors the exclusive rights to their . . . discoveries." Congress has no power under the constitution to confer upon an inventor any right which is not an exclusive right for a limited time. If so called innocent infringers be excepted from the exclusiveness of the right conferred, the right ceases to be exclusive. The only means of protection against "old and unworked patents of doubtful validity" that seems to me feasible, lies in the direction of limiting the duration of the exclusive right under such patents, rather than in rendering the right non-exclusive. I am well aware of the hardship sometimes resulting to innocent manufacturers and users from the defense of suits based upon patents which have been long disused, and have been searched out and purchased by business competitors for the sole purpose of harassing and crushing out the competition of the defendants. If it were possible to avoid it, our patent system should never be prostituted to oppressions of this kind. But of the many plans which have been suggested for avoiding this abuse, none seems to me feasible in view of the constitutional limitations upon our patent system, except those in the direction which I have already indicated.

C. E. BILLINGS.

Hon. C. E. Billings, of Hartford, Conn., wrote as follows:

I believe in limiting the time on patents granted for inventors to manufacture and bring same to the notice of the public and if not worked within the time, same should become void.

The patent office should be a separate department and be provided with a building of its own, and separate building should also be provided for the storage of models, especially designed for the purpose and of fire proof construction.

THOMAS SHAW.

Thomas Shaw, of Philadelphia, wrote:

I wish to present one feature in regard to the meetings which is in imitation of other organizations that meet annually, selecting a different location each time. I believe our Association would be greatly benefitted and prospered by meetings in New York, Philadelphia and other prominent cities. The local papers would advertise the Association and the proceedings, and those interested would add to the organization from each separate locality. * * * I am of the firm conviction that all patents granted to citizens of the United States should stand upon an independent basis, unaffected by any foreign patents, and that no law of foreign countries that confiscates a Foreign patent of an American citizen for such trifling cause as non-payment of tax, etc., should step in and hold sway over the United States law and compel a like confiscation in our U. S. Courts with no other possible result than the carrying the results of an unjust law across the seas to work a great injury to an American citizen by pursuing the outrage to the doors of our very capital of freedom.

I hold that failure to manufacture in specified time should not render the patent void as a matter of liberal policy in favor of a poor inventor who is unable to procure funds to manufacture. If such a feature was engrafted in the law, capitalists would hold off from engaging in the manufacture, hoping thereby to render the patent void and thus escape payment to the inventor as is now the case in most foreign countries.

EMILE BERLINER.

The following excellent suggestions from E. Berliner, of Washington, D. C.

I submit the question whether in the curriculum of High schools there should not be emdodied a course on Patents and Patent laws—elementary of course. Nearly every citizen of this country becomes once in a while interested—financially or otherwise—in a patent, and it is certainly deplorable to see the utter ignorance of most of such people in even the common sense principles of patents, their meaning, scope, liability of infringing, together with the general course of Patent law, etc. A great deal of harm results from such ignorance. Many hundreds of thousands of dollars are lost yearly by the gullibility of investors, by disappointments of embryo inventors, by companies formed for exploiting inventions, etc., and worst of all, every victim becomes a new recruit for the army of malcontents who decry the whole patent system and would want to kill the goose which has laid so many golden eggs of progress for these United States.

A course such as I have indicated based on a proper text-book would, in the course of a few years, remedy the trouble as must be evident to every thinker on the social fabric of our society and civilization.

The question of the establishment in the city of Washington and other large cites of Scientific Museums on the order of the "Urania" in Berlin. In this great establishment are exhibited complete sets of expensive scientific apparatus of every department, like spectral analysis, electricity, sound, light, microscopy, Crooke's tubes, Hertzian resonators and in fact all such instruments as can only be collected by richly endowed scientific institutes.

But the principal addition to this is a trained set of scientific assistants whose duty it is to operate for any visitor any of the various apparatus he may be interested in. The educational value of such a museum particularly to the great mass of inventors who have not had the advantage of college training and college facilities would be very great and the museum can be made self supporting by the charge of a moderate admission fee.

In the Urania in Berlin there is also a lecture hall in which popular scientific lectures are given by competent professors at stated intervals; for this a moderate additional fee is also charged.

A. G. DAVIS.

A. G. Davis, of Baltimore, Md., says:

I would suggest, if it were possible, that when a basic patent is issued by the Patent Office all other patents bearing upon that basic, or improvements upon any portion of that basic patent, ought to be so stated in issuing a new patent—that is, it ought to state, that this patent is an improvement upon the basic. As the law now stands, patents are granted on improvements, and those who take them out fancy they have a valid patent, when in truth, they are of no earthly use to the inventors without making arrangements with the patentees of the basic to use the improvement.

It is a very difficult matter for manufacturers to go back years in the Patent Office and investigate the validity of some patent which has been running a number of years with no effort on the part of the owner to push his invention; and we find that several years, almost until the patent expires, it has laid dormant, and at that late day a claim is made of infringing, when the party who did infringe may be perfectly innocent of the act."

F. H. RICHARDS.

Discussing questions 1, 2 and 3, (Feb. issue of INVENTIVE AGE) Mr. Richards took strong grounds in opposition to any proposition that contemplates in any way an abridgement of the inventor's rights, or rather that will operate to deprive him of any exclusive results of his inventive genius at any period of the present term. He believes that in the majority of cases inventions are discovered and made known before there is a market for them, and the period required for creating a demand often uses up the greater portion of the life of the patent. On the question of opposition here and there to the patent systen, Mr. Richards, in the course of his remarks at the January meeting of the Association Mr. Richards said: "I wish to add a few words on one point which has been brought up. The one thing which the rural population seem to antagonize most strenuously is the practice of massing patents under one ownership. For instance, the inventor of a reaper may after a while get a small market for his machines and earn a little more money than he has spent perhaps during ten years. He then commences to make further improvements. What for? For the most simple reason in the world. That his efforts have so far educated the public up to the point where they demand something better than he first supplied. If he does not make improvements, somebody else will do it in competition and he will lose his trade. It happens that during all this time he has been busy himself manufacturing reapers. He has ceased to be an inventor for the time being and has become a business man. He has assistants, he has larger business; and in course of time he employs inventors; and he masses improvements upon improvements until it becomes a great business. In the case of the Singer Sewing Machine, it is said that the company own one hundred patents. This is necessary in order to interest capital, in order to make it safe for capital to go into the business and in order to give the opportunity for making a quicker success of that business. It is necessary to have more protection than can be afforded by any one patent at that stage of the art.

That is a very important matter today, in a great many industries. I happen to be interested in two or three of that class at the present time. I own one, in particular, in which patents have been issued in this country since 1850, up to 1895. Over 450 patents have been issued to different men, and never until 1884 did any one man get enough protection to warrant him in putting a thousand dollars into the business of making that particular machine. The parties who are interested now in it have got invested at the present time over one hundred thousand dollars. They have been at it several years and have a good deal of experience back of it. They have got that money there and they expect to get it out. They expect to make a successful business, confer a lasting benefit on the public, and when they get through and the patents have expired, the public will then have an established successful business, and a facility for carrying on business which it had not before. They will have free from that time forever an article which was not in use before but for which there was a great need.

I submit that our country is full of just such examples as that. When an art is partially developed and some of its fundamental patents have expired, it is impossible to get sufficient protection by any one patent or even a few patents to warrant capital going into the development of the business. If it is necessary in order that the public shall finally get the benefit of this established and successful business, that these patents shall be massed together and one protect the other, I say it is a public benefit to have them so massed, and our laws should not stand in the way of it. The life of a patent is short anyway, and I know in a great many cases the public can attain similar results by no other means. They are monopolies for the time being, but without an occasional monopoly, how are you going to obtain a general benefit all along the line.

If our readers desire any book not mentioned in our advertising columns we can furnish it. Books mentioned in our Review column can be obtained through us.

Will Science Disprove Man's Accountability?

The wonderful achievements in the field of science, the rapidity with which the great secrets in nature's laws are being revealed, the possibilities today admitted, but yesterday declared ridiculous, the discoveries of magnetic, electric, physicological and other agencies controlling and directing the animate as well inanimate, are calculated to relieve the odium of delusion or crankiness to a great extent now attached to many earnest individuals, advanced thinkers and seekers of truths long hidden and still in the mist of obscurity. The application of the wonderful illumminating or electrical wave agent in the field of photography, the partial success of the wizard Edison and other electricians in photographing the brain and other secreted parts of the human anotomy are matters just now receiving the attention of the greatest scientists and mystifying the most learned in all the sciences. That crime and sin may be mainly the result of physical degenerations and imperfections is a theory that seems now in a fair way to be demonstrated a fact.

During the past thirty years Mr James M. Rusk, now editor of the McConnellsville (Ohio) Herald, has been an ardent student of the occult sciences, and his physical, psychical, theosophical and metophysical researches during this period have heen carried on with a view of discovering if possible how much the so-called "vice" of habitual or periodical drunkenness was due to heredity environment, to disordered function, to defective organism, one and all, or any one of them, or any combination. His conclusions will be readily understood from the following article from the Chicago Tribune called out by the seven fold murder of a Chicago anarchist:

The scientific examination of the head of Klaettke, which proved him to be a degenerate and unaccountable for the murder of his family, and the discoveries of Roentgen, who has photopraphed the interior of the human body, are calculated to start in any reflecting mind exciting and even painful reflections concerning the moral accountability of man. The parellelism between a man's body and his character has been noticed ever since the race began. Not only can one form a tolerable judgement of another man's character from seeing him at a distance, but young children and even dogs can do the same thing. It would be just as irrational and dangerous to ignore the different aspects of different men as to ignore the the different appearances of men and wild beasts.

These facts, though the logical inference from them is unmistakable, have never had any effect on the universal belief in human accountability. But the discoveries of science for the last fifty years have all served to emphasize and corroborate the parallelism between a man's bodily peculiarities and his traits of character. Guiteau, Klaettke, Prendergast and Holmes have all been declared irresponsible degenerates. A German proffessor had his brain photographed by Roentgen, and was so terrified by the peculiarities it exhibited he could not sleep for weeks. Not little by little, but by gigantic strides science seems to be demonstrating a physical basis for every human trait of character and for every human act or even thought.

The theological mind of course resents this process. It claims the degenerate, like the insane, are exceptional and that we cannot reason from them to sane and normal human beings. But science answers, these unfortunates differ from other men only in degree. Moreover, if the degenerate deserves no blame, how can the normal man deserve any praise? Theology says a man can, by his own will, overcome his mental peculiarities and a corrupting environment. But science replies, some men can and some men can not, according to the character impressed on them at the start. So the tendency of everything is to spread the mantle of charity and forgiveness over poor human nature.

This assault on human accountability is based not only on constitutional characteristics, but on disease as well. A painful illustration of this is afforded in the case of a noble and brilliant Chicago woman who recently passed away. This woman seemed to be two different persons in one. At one moment she evidently possessed a charity so broad, and a philanthropy so sweet and pure she might be considered angelic. At another moment she seemed the exact reverse of this. A Christian minister might have felt it his duty to tell her "this kind goeth not out but by prayer and fasting," even if he did not point her to the fearful retributions of eternity. But in this woman's last illness the surgeon's knife laid bare a vast internal cancer, which in anything short of an angelic character would have produced absolute diabolism; and, like a dying swan, the sweetest strains of poetry fell from her pen in the agonies of dissolution.

No statistics are preserved from which we can learn how many millions of wretched men and women have gone down to their graves hated by all who knew them, and consigned to an everlasting hell by furious ecclesiastics, but who by one puncture of a lancet wisely directed would have been transformed into philanthropists and saints.

X Ray Spectacles.

A London inventor, claims to have discoverd a device which enables the human eye to penetrate opaque objects. The instrument is described as consisting of a cylinder of card-board, the inner surface of which is coated with a material that becomes fluorescent under the action of the Roentgen rays. The lens is at one end of the cylinder. The object to be examined with its coverings is placed between a Crooke's tube and the cylinder; on looking into the tube through the lens, the observer sees the outline or shadow of the concealed object, which is thrown on the fluorescent interior.

When a ray of light comes to the eye from an object the effect is not essentially different from that produced when a similar ray from the same object strikes the sensitive plate in a photographer's camera. If the plate is sufficiently sensitive to be affected by light of that quality a photograph is made. If our eyes are adapted to the same light we see the objects from which it is refracted.

Hence while the invention of "Roentgen spectacles" is improbable, and perhaps impossible, it is not absurd. Any invention which will enable the eye to use "X" rays as they are now utilized for photography will certainly enable us to "see through" our friends as completely as we now think we do. There is no doubt of it whatever.

Firing the Indiana's Big Guns.

On a recent run of the battleship Indiana from Newport to Hampton Roads, all the guns of her battery, including the 13 inch rifles, were fired. It was the first time they were fired on board ship, and the test showed that the gun mounts and their installation were entirely successful. Such a result was to be expected; so the test was not so important in this respect as it was in another, namely, the effect of the blast of the heavy guns.

The recent test showed no damage to woodwork or glass, nor any serious injury to the officers and men engaged in the firing, but it was conclusively demonstrated that in certain positions of the turret the man in the sighting hood would be exposed to serious discomfort and sometimes actual injury when the 13 inch gun was fired.

The heavy guns of the monitor Amphitrite were fired in the same manner, to ascertain the effect on her structure and on living animals placed under the decks over which the guns were fired. It was found in her case that no injury resulted to the ship with 12 inch guns, nor to the animals, so far as observed.—*Scientific American.*

Electro-Magnets as Lifting Agents.

Electro-magnets as lifting agents in connection with cranes came into use several years ago, though, on the whole, their employment in this way has been comparatively restricted. In one of the large English foundries—at Sandycroft—however, they are now again applied to that purpose, and in sizes which permit of readily lifting, by their means, weights of as much as two tons. The magnets are attached to a crane, and take a current of about 5 1-2 amperes, at 110 volts, the current supply being controlled by a switch. Some measure of the service gained from these magnets may be obtained from the statement that with one of them three men can do in about fifteen minutes the work which previously occupied twice as many men for about an hour and a half.—*Cassier's Magazine.*

Important Invention In Weaving.

One of those inventions which will inevitably displace a large amount of labor because of its increased economy, has recently found its way into textile manufacturing. It is a loom which automatically feeds bobbins into the shuttle. This work hitherto has been one of the chief duties of the weaver. Now five persons can take care of 80 of these looms, whereas, at present, one person looks after six. The new process will run for an hour or two after every one has left the factory at night and through the noon hour. It is said that it will save one-half the labor cost of manufacturing.—*Buffalo Express.*

In New York City there is said to be the most populous single block in the world. It contains over 5,000 human beings, packed like sardines, in vile tenement houses.

A Few Clever Inventions.

An inwardly-turned seam joint for tin cans to facilitate labeling.

A method of heating and regulating the heat of fowl or chicken brooder by steam heat.

A new method of using dilute hpdrobromic acid as a solvent for extracting the alkaloids of cincho na bark.

An improved plow which can be swung to or from land without stopping the team or leaving the handles.

A method to prevent soap cakes from drying or evaporation by dipping them in a solution of gelatine or parafine.

An eye cup which consists of an electric battery, for application to the eyes in electrotherapy, as a palliative or curative.

A hollow auger, with an opening for carrying away the chips, the cutter being held in a recess above the opening by a plate secured by screws.

An attachment for the pockets of ladies' dresses, consisting of a spiral spring along the edge of the mouth of the pocket to render it self-closing.

In the manufacture of boots and shoes, a heel counter made of aluminum and strengthened by steel springs equidistant from the centre of the outside.

A small-flanged third wheel with outreaching arms, to be attached to an ordinary safety bicycle by means of which the rider can travel on a railroad track.

A fishing apparatus in which the line is mounted on the end of a spring, which, in the event of a fish taking the bait, closes an electric circuit and rings a bell.

A mucilage bottle having an auxiliary inner cork that is depressible and which causes the brush to recede after using, and holding it up free from the contents of the bottle.

A waterproof fabric to prevent the rotting of the covers of umbrellas by the rusting of the framework, the fabric being placed between the cover and the top-notch joints.

A roof for metallurgical and other reducing furnaces, made of a series of tubes through which water constantly flows, the interstices between the tubes being filled with fire-brick.

A telegraph indicator for steamships, the apparatus being provided with bell indicators on the bridge for sounding an alarm if the engineer fails to follow the captain's directions.

A system of facilitating the removal of roofing tiles, the holes which receive the nails being made in the form of key-hole slots. Nails can be driven placing the tile and the tile hooked on.

A method of permitting filaments of incandescent lamps to be replaced when deteriorated, whereby a second filament is combined with the first, the leading-in wire projecting so that it can be bent round to make contact with the metal plate when required.

The Earth to be Eléctricuted.

According to a writer in the Philadelphia Times, "the earth for three feet is alive with the invisible power (electricity) and forms the secret of vegetable life. Waves of electricity are constantly passing through the soil in unseen billows, thus keeping the soil from souring." This fact is simply demonstrated, the writer says, his directions being as follows:

Go to some rock-bound pool, dip out a small quantity of the polluted water, place it in a bottle, cork and set aside in a warm place for a short time. Then take the bottle into a dark room, shake the bottle, draw out the cork, and you will see tiny forks of blue lightning shoot out from the bottle, and if you keep perfectly quiet, you will hear faint mutterings like thunder. This comes from the flint-like rocks preventing the unbroken flow of electricity through the soil and from the air becoming charged and emptying itself into the water.

Having thus established the correctness of his theory that electricity "is fire—the fire of friction," this prophetic scientist thus foretells the awful fate that awaits this world of ours:

In the ages to come the charge of electricity will keep on accumulating, until some commotion of the earth will cause it to ignite, when, in the twinkling of an eye, our world, with all it contains, will be enwrapt and consumed by a conflagration that will startle if not frighten the inhabitants of other planets as they look down upon the flaming mass and see it burn up one of the greatest works of the Almighty's creation.

Nobody quite knows who invented billiards. One account says that the game was first played in Italy and another that it first saw the light in Spain. It is also affirmed that it was first played in England in the middle ages. It is a historical fact that the Knights Templar brought it back with them to that country on their return from the second crusade. There is also good reason to believe that the game was played in the monasteries of France in the sixteenth century.

Copying American Machinery.

While abroad last year Robt. Grimshaw, the noted author and inventor, wrote the following article to Power and Transmission:

I was much interested in an article in a recent issue of Power and Transmission, by the American Consul at Chemnitz. in reference to the habit of the Germans and others of copying American machines as exactly as possible, without authority of the designers. But the consul is not altogether right, nor altogether fair. I do not at all defend the practice of getting an exclusive agency for an American machine, and then instead of selling the original, selling copies of it, either as American machines or as German copies of it. Such a practice is dishonest. But it could very readily be prevented by inserting in the agency contract a clause calling for the purchase of or payment of license on a minimum number of machines (or articles) the first year, a certain greater number the second, and so on, under penalty of losing the agency. And the matter could be kept in control as against surreptitiously made machines or articles, by numbered license plates furnished the agent from the main office of the patentee, in case the machines were to be built in Europe; so that every machine not bearing a license plate either affixed by or furnished by the American builder or patentee, should be known to be spurious.

Our consul forgets, too, one very important factor in the exploitation of an article. There are such things as patents in Europe, but they differ from ours, in most European countries. In some of them the patent is issued to any one demanding it, and in most of them it must be "exploited" within a given length of time, and the annual taxes paid thereon, or the patent falls into the public domain. Now if the American manufacturer fails to patent his invention in Germany or other European country, he has no legal rights (I say nothing of moral rights) which the European manufacturer is bound to respect. They can go ahead and make the machine or article just the same as for many decades American publishers could go ahead and print English or other European books—without the proprietor's consent or knowledge, and without payment of royalty or license. There is no more unfairness in an European manufacturer making in Europe an unpatented American invention than in an American publisher printing in America an uncopyrighted European publication—in fact not so much; for for many years Europeans could not obtain copyright in America, whereas the European patent offices were open, all along, to American inventors.

Another most important factor to be considered, in the case of inventions patented in Europe, is that law calling for the thing to be made in the countries in which it is patented. This is a law intended for the protection of European industries, just as our tariff is intended for the protection of American industries. I don't say anything just here, about the wisdom of it; but as to the moral side it is all right. It is a purely business matter, which each country has a right to regulate for itself, according to its best judgment or even according to whim. We compel European authors who get copy right in the United States to have their works printed from plates made in the United States from type set in the United States; and that is a similar law to the European ones of compelling patented articles to be made in the countries in which they are patented. If the American inventor wants to realize from his invention without competition, in Europe, he should patent it in the country where he wants to do business; and then he should either sell his patents or have them worked on royalties. Otherwise he has no "rights" (no *legal* rights) that any one is bound to respect or that he can enforce. He is just in the position of an inventor in the United States, who fails to patent his invention there. If no patents were obtainable, the case would be different; but they are to be had, and in most countries but Great Britain, ridiculously cheap.

The fact that if the American inventor does not patent his invention in Europe, the first man who comes along and can properly describe it, can do so (in most countries,) should be an additional incentive to the inventor to protect himself early, and have something to sell, instead of only something to regret.

I am not to be considered as altogether admiring all of the European patent laws. They are very various and in some cases absurd. Just now I am particularly disgusted with the French laws. I am arranging for the manufacture in France, of an article of German invention, and requiring for its manufacture not only special machinery costing $20,000 or so, but special composite sheets of steel and iron combined requiring special skill and machinery to make it. In order to arrange for a company to manufacture this article, and in order to teach workmen how to make it, it is necessary to have a few dozen perfect articles in France as samples. But not only will the introduction of a single one of these German-made articles in France break the patent, but unless I do have some made in France within a given time, the patent falls to the ground! I can introduce the Belgian-made or the English-made article into France, without breaking the patent, because those countries are in the patent union with France; but Germany wanted to stay out of the union, and now in many respects German inventors are out (in the cold.) It has been necessary for me, after a delay of months, to have the article made in France on a small, inperfect scale at great expense and to start the manufacture in Belgium before we were really ready, so as to enable us to introduce into France enough of the articles to show capitalists whom I wish to interest, and manufacturers whom I wished to bid on the article to be made in quantities on license, in case we decided not to build a factory.

The law as to exploitation within a certain time in the country of patents is particularly hard on the inventors of such articles, as cannons, dry-docks, etc., orders for which must come from the government, or from very large concerns, and for large sums even for a single one. The inventor of a cannon said in one country to the government, in vainly endeavoring to get an extension of the time for exploitation, "give me an order for a cannon and I'll be only too glad to make one as soon as metal can be poured, cooled and turned. But can I invest $20,000 to $75,000 in producing a cannon which, even if you approved, you might decide was the wrong size?"

The best way for Americans to realize on inventions in Europe is to arrange for selling the patent or having them worked on royalty. With honorable firms the matter can be arranged before patenting.

New Members A. A. I. M.

During the last year, a large number of very prominent inventors, manufacturers and others have joined the American Association of Inventors and Manufacturers. (Don't confuse this association with a Philadelphia patent brokerage concern.)

Among the new members of the Association are:

Robert P. Linderman, President Bethlehem Iron Co., South Bethlehem, Pa.

Fred Meriam Wheeler, Builder of Steam Pumping Machinery, New York City.

C. F. Wieland, Manufacturer of Machinery, St. Louis, Mo.

C. M. Spencer, Manufacturer of Machine Screws, Windsor, Conn.

James Lyall, Manufacturer of Cotton Machinery, New York City.

Charles Hall, Manufacturer of Bolts and Wrenches, New York City.

Nicholson File Co., Providence, R. I.

W. G. Nourse, Marine Iron Works, Chicago, Ill.

S. Frisbie, Treasurer Upson Nut Co., Unionville, Conn.

Robert Dawes, Mechanical Engineer and Inventor, Philadelhpia, Pa.

Herman Dock, Mechanical Engineer, Philadelphia, Pa.

Frederick Augustine Drake, Capitalist, Windsor, Conn.

John G. Battelle, Manufacturer Iron and Steel Sheets, Piqua, Ohio.

D. Wwight Wiman, Manufacturer Malleable Iron, Moline, Ill.

James B. Williams, Manufacturer of Soaps, Glastonbury, Conn.

R. D. Wood, Manufacturer of Machinery, Philadelphia, Pa.

William Quinby, Lawyer, Boston, Mass.

Francis Meisel, Kidder Press Manufacturing Co., Boston, Mass.

John L. Dalgleigh, Treasurer Barbour Silver Co., Hartford, Conn.

C. C. Kimball, President Smyth M'f'g. Co., Hartford, Conn.

Arthur E. Hobson, Supt. Barbour Silver Co., Hartford, Conn.

Geo. Best, Manufacturer of Carriages, Hartford, Conn.

Herbert C. Warren, Hartford, Conn.

Billiardelle.

This is the name of a minature billiard table apparatus invented by Mr. Arthur Munch of Saint Paul, Minn., and patented by him through Mr. R. G. DuBois a few days ago. The old popular and scientific billiards are arranged in a compact form for home use. The cue is a novel little device adapted to be held in one hand when striking the ball as seen in the accompanying illustration. The balls are knocked about by a little button on the end of a spring-arm projecting out from the side of the body portion of the cue. The torce of the blow is regulated beforehand by a notched trigger which may be released by pressing the button at the top end with the fore-finger. By this means the player can shoot the ball with a light or heavy stroke in any direction from a sitting position without changing seats, during the entire game, thus avoiding the clearing away of furniture and walking around the table thereby making the game less tiresome, and more sociable.

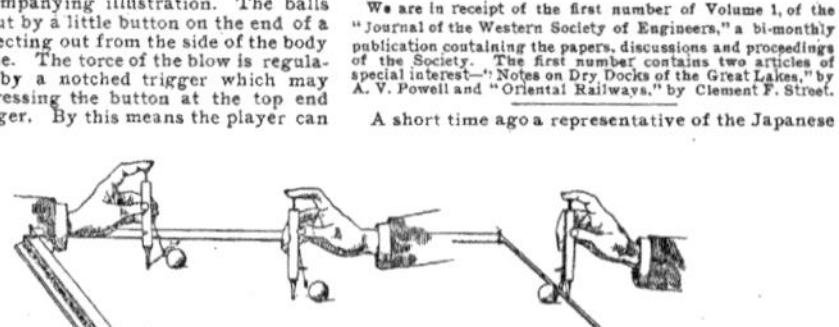

BOOKS AND MAGAZINES.

The "Southern Trade Review" is the name of a new monthly journal at Nashville, Tenn., covering the field indicated by its name. The first issue contains many special articles written by men experienced in the subjects of which they treat. There also appears in this number an illustrated article on the coming Tennessee Centennial to be held in Nashville from the pen of Leland Rankin.

* * *

A novel and extremely useful little cyclopedia is now being published in serial form known as "Alden's Living Topics Cyclopedia." It is a record of recent events—up-to-date occurences. This work will supplement all high-class cyclopedias, by the adeqaute treatment of new topics, and by bringing the treatment of older topics up to the latest possible date. Topics of the day—political, social, scientific, etc—are treated in alphabetical order, and as often as the alphabet is covered, a new series begins and the same course is resumed. Volume I is now ready; cloth 50 cents. New York: John B. Alden, Publisher.

* * *

One of the most ingenius, usefel and simple perpetual calendars ever issued is that of which James A. Walsh, a leading attorney of Helena, Montana is author. It is absolutely a perpetual calendar and almanac for all time—past and future. It contains the various phases of the moon from A. D. 1800 to A. D. 2100 and by simple computation before and after these centuries as well. It furnishes a complete calendar for each year, both according to the Gregorian and Julian systems or reckoning. It is of especial value to lawyers as by it can be learned at a glance on what days of the week, years ago, any contract, vote, deed, will or other instrument was executed. It is a small volume of only 56 pages and can be found at Bretanos' and other book stores.

* * *

Prof. Chas. H. Saunders, of Yale, is the author of a little work containing a fund of useful information for mechanics and engineers—as its title says, a "Hand Book of Practical Mechanics," for use in shop and draughting room. It contains tables, rules and formulas and solutions of practical problems by simple and quick methods. It is published by the Student Publishing Co., Hartford, Conn.

* * *

Charles Henry Cockrane, M. E., is the author of a very interesting and usefsl work just published by Lippincott Co., of Philadelphia. Its title is "The Wonders of Modern Mechanism," and is a resume of recent progress in mechanical, physical and engineering science. It is an up-to-date work, especially in the portion devoted to electrical science. The horseless carriage, pneumatics, conduit electric systems of street car propulsion, the experiments of Nikola Tesla, acetylene gas, the "chaining of the Niagara Falls," etc., also are intelligently and popularly treated. In the language used in this work the author seems to have struck the happy medium between the over technical and the unscientific field.

* * *

No work on subjects of universal interest to mechanics has come to our notice that seems so full of meat as that recently issued by Norman W. Henley & Co., of New York, entitled "Shop Kinks and Machine Shop Chat." The author is that well-known mechanical engineer, author and inventor, Robert Grimshaw. The book consists of a series of over 500 pactical paragraphs in familiar language, showing special ways of doing work better, more cheaply and more rapidly than usual. There over 200 engravings. In the preparation of this little work the author, in addition to the notes taken during thirty years of his own active life, weaves in, with due credit, the interesting and useful experiences of many engineers, machinists and scientists with whom he has been associated. It is a work, the price of which, $2.50, no mechanic will regret.

* * *

The Photographic Times of New York will just now prove particularly interesting to all who are interested in the new discoveries in photo-electric science. The March issue contains handsomely illustrated articles on "Roentgen Rays and Skotography" and kindred subjects.

* * *

The "New Power Developments at Niagara Falls" is a subject graphically treated by Orrin E. Dunlap in the March number of Cassier's magazine.

* * *

"Gleanings from the Patent Laws of all Countries," is the tittle of a little work issued in two parts, received with the compliments of the author, Mr. W. Lloyd Wise, J.P., F.R.G.S., from the publishers Cassell & Co., London. The text of the book is very neatly arranged and in addition to explanations of the patent laws information as to points of practice and area, population, chief productions and other statistics properly classified are given for each country. The first portion treats of twenty-two countries.

* * *

We are in receipt of the first number of Volume 1, of the "Journal of the Western Society of Engineers," a bi-monthly publication containing the papers, discussions and proceedings of the Society. The first number contains two articles of special interest—"Notes on Dry Docks of the Great Lakes," by A. V. Powell and "Oriental Railways," by Clement F. Street.

A short time ago a representative of the Japanese government secured in Chicago an electric light plant for a palace in Japan. There now exist half a dozen *fac-similes* of the same plant in actual use, perfect models of the original sample, made in Japan by Japanese workmen, without outside aid.

SUPPLEMENT.--Tips to Inventors.

The Infancy of Invention.

As capital is constantly being invested and expended to protect and preserve capital previously expended and invested in various enterprises all over the land, so will inventions continue—their variety and multiplicity being demanded to further the usefulness and perfection of inventions previously originated.

It was Edison who, replying to the question, "Do you think that the inventions of the next fifty years will be equal to those of the last?" said: "I see no reason why they should not. It seems to me that we are at the beginning of inventions." The truth of this prediction is illustrated in the many useful and wonderful achievements of Mr. Edison's own laboratory since giving utterance to this statement only a short time ago.

Profits from Invention.

The value of an invention is determined by no fixed rule. Fabulous sums have been made from simple and novel, as well as complex and useful, inventions. It is a fact that four-fifths of the business of the United States is transacted by the use of inventions. The benefits to mankind because of inventions, are so manifest and so common we are apt to look upon them in a matter-of-fact sort of way and fail to give the inventor the credit due him. In the majority of cases, however, the failure of an inventor to reap a reward is attributable to his own negligence, lack of forethought and indiscretion.

Nearly every human being is an inventor, but only a few obtain a *monopoly*—a patent—on the product of their brain. There are thousands of really useful articles, appliances and discoveries, in use every day by millions in all walks of life, that might have been patented had the inventor possessed the business sagacity that has actuated his more fortunate neighbor. Take for instance the open slot necessarily used in all conduit electric, or cable street railway systems. The inventor failed to get a patent on the idea and a fortune missed him.

There is money in inventions, but not always for the inventor.

The only way to make money out of an invention is through the protection afforded by a patent; not a patent in name only, but a *good patent*—one that is intelligently drawn, with claims commensurate with the scope and importance of the invention.

The profits arising from inventions in the electric field during the past twelve years have been simply astounding. In railway appliances, bicycles, typewriters, telephones, cash registers, slot machines and farm machinery, the field has been equally remunerative. And just think of that simple toy "Pigs in the Clover"—it netted the inventor, whose friends laughed at him for obtaining a patent on so simple a toy, over $150,000. The inventor of the metal plates to be attached to the worn heels of shoes (for sale in all cities) realized a fortune out of it amounting, it is said, to nearly $1,000,000. Perforated wooden seats for chairs and rubber tips for lead pencils brought the inventors big results. Howe made a million dollars from his sewing machine attachments, and the inventor of that simple lamp attachment, the inverted glass bell, to be suspended over lamps to protect the ceilings from being blackened, made the inventor rich. The "Darning Weaver," a device for repairing stockings, is a useful invention and is netting the inventor a handsome revenue on royalties. The wire nail and gimlet-pointed screw are fortune makers, and wire nails caused the invention of automatic machinery that manufacturers them so cheaply it does not now pay the carpenter to spend his time in picking a nail up when it drops, if it requires ten seconds to do so. The inventor of the well-known "safety pin" lived in luxury all his life, after discovering a means of concealing the point of a pin in such manner as to prevent scratching. The inventor of roller skates made nearly a million and the inventor of the needle-threader for a long time made $10,000 a year.

Relation of Capital to Invention.

Mr. Edward P. Thompson, one of the most entertaining writers on the subject of invention, says that "every invention, before the introduction into practical use, passes through two stages; namely, mental and physical"—mental when in the mind of the inventor only, and physical when the mental invention is put into bodily form by hand, or by hand with the assistance of a convenient tool. "A mental invention," says the writer, "sometimes does not become a physical invention because the inventor lacks money, technical knowledge or diligence. Such a mental invention often becomes a physical invention by the assistance of a capitalist, an educated person, or diligent companion." This being true the *mental* inventor, the person who, for lack of means possibly, would fail to make his invention a physical reality—such a pe,aon, should take into his confidence a friend or companion to share the prospective benefits of his invention. Thousands of meritorious mental inventions are never worked out because of the over-timidity of the inventor, his exaggerated greed for *all* the benefits to accrue instead of half the loaf, which in many instances is, or would have been, ample reward. Mr. Thompson truly says: "Inventors and capitalists should be more willing to co-operate. It is too often the case that the former must pay for his own experiments and all patent costs before a capitalist will even take the trouble to look into the merits of the alleged invention. On the other hand it is too often true that the capitalist seeks to join with the inventor, but the latter wants too high a price at the beginning."

Who Can Apply for Patents.

Patents are issued to any person who has invented or discovered any new and useful art, machine, manufacture, or composition of matter, or new and useful improvement thereof, not known or used by others in this country, and not patented or described in any printed publication in this or any foreign country, before his invention or discovery thereof, and not in public use or on sale for more than two years prior to his application, unless the same is proved to have been abandoned; and by any person who, by his own industry, genius, efforts and expense has invented and produced any new and original design for a manufacture; any new and original design for the printing of fabrics; any new and original impression, ornament, pattern, print or picture to be printed, painted, cast, or otherwise placed on or marked into any article of manufacture, or any new, useful and original shape or configuration of any article of manufacture, the same not having been known, or used by others before his invention or production thereof, nor patented or described in any printed publication, upon payment of the fees required by law and other due proceedings had.

If it appears that the inventor, at the time of making his application, believed himself to be the first inventor or discoverer, a patent will not be refused on account of the invention or discovery, or any part thereof, having been known or used in any foreign country before his invention or discovery thereof, if it had not been before patented or described in any printed publication.

Joint inventors are entitled to a joint patent; neither can claim one separately. Independent inventors of distinct and independent improvements in the same machine can not obtain a joint patent for their separate inventions; nor does the fact that one furnishes the capital and another makes the invention entitle them to make application as joint inventors, but in such case they may become joint patentees. The receipt of letters patent from a foreign government will not prevent the inventor from obtaining a patent in the United States, unless the invention shall have been introduced into public use in the United States more than two years prior to the application. But every patent granted for an invention which has been previously patented by the same inventor in a foreign country will be so limited as to expire at the same time with the foreign patent, or, if there be more than one, at the same time with the one having the shortest unexpired term, but in no case will it be in force more than seventeen years.

Protection to Inventors.

What is a patent? It is a monopoly or grant, in the United States, for a term of seventeen years, to the patentee, his heirs or assigns, of the exclusive right to make, use and vend the discovery throughout the United States, as the inventor's rights may appear in the specifications and patent granted.

This means a great deal to the inventor who has secured a *valid* patent containing all the claims so worded as to prevent infringement and loss in contest. Thousands of inventors, obtaining patents through unreliable and inefficient attorneys or agents, find themselves possessed of patents *in name only*, and of no value when combatted by infringers with capital and the aid of able legal talent. A good patent costs no more than a weak and worthless one. Therefore how shortsighted are those inventors who employ cheap attorneys, saving $5 or $10 in fees, only to find themselves loosers of *all* they have paid when the contest comes.

The Need of Reliable Attorneys.

The Revised Statutes of the United States provide that "before any inventor shall receive a patent for his invention, he shall make application therefor in writing to the Commissioner of Patents, and shall file in the Patent Office a written description * * * of the same in such full, clear, concise and exact terms, as to enable any person skilled in the art or science to which it appertains, or with which it is most nearly connected, to make, construct and use the same; and in case of a machine, he shall explain the principle thereof and the best mode in which he has contemplated applying that principle, so as to distinguish it from other inventions."

To carry out these provisions it is necessary for the inventor to first make a clear, concise and complete drawing, or a working model of his invention or discovery, and send it to THE INVENTIVE AGE, or some thoroughly reliable attorney, who, before making application for the patent, should make a thorough and rigid examination of the Patent Office to determine upon its novelty or patentability. If the invention has been anticipated by some one else, or if it lacks novelty, or if for any reason a patent can not be granted, or, if granted, would be of no worth or value, then the inventor does not want to incur the expense of making application and paying attorney's fees and government fees. For making this thorough examination THE INVENTIVE AGE and all reliable attorneys charge $5, which fee is, under some circumstances, however, taken out of the additional fees paid by the inventor in case letters patent are applied for. The fees of patent attorneys vary somewhat, but the average fees for obtaining a United States patent are about $65—the government fees being $15 on filing the application and $20 on issuing a patent—the balance being the fees for preparing specifications, making searches, etc. The inventor is sometimes favored in terms given for payment of the fees. more detailed information regarding which can be obtained by enclosing a 2-cent stamp with enquiry to THE INVENTIVE AGE, Washington, D. C. The reason why the inventor should have a preliminary examination of the Patent Office made before applying for a patent lies in the fact that if the case is rejected the fees paid to the government and the attorney are lost.

All patents obtained through us will receive special mention in THE INVENTIVE AGE and in cases of unusual merit inventions will be illustrated free of charge to our clients.

This publication, reaching capitalists, manufacturers and business men throughout the world, is of value in assisting to bring an invention before the public in case its promotion or sale is desired by the patentee.

The Late Eckley B. Coxe.

By the recent death of Mr. Eckley B. Coxe, the country lost one of its greatest inventors, and the commercial world one of its most active spirits. Although Mr. Coxe had figured less prominently as a patentee than many other American inventors, this was largely because of his extensive business interests, which, for many years, called for his unceasing attention. During that period, however, he very greatly improved the machinery for preparing anthracite coal, introducing the same into the extensive collieries of Messrs. Coxe Bros. & Co., and also furnished, during recent years, some of the more important machines to other collieries. One of these machines was for the screening of coal, separating the same into the required successive sizes and delivering it in the best shape for marketing. This screen is of the gyratory type and consists of a series of screens, one above the other, and so arranged that the fresh coal are balanced so that the machine, although of great weight and power, operates quietly and without any appreciable jar to the buildings. These important machines have not only been put into use in this country, but has been extensively adopted in Europe, where Mr. Coxe secured patents on the leading features of the invention. At time of his decease Mr. Coxe had only just completed, in fact had hardly completed, the development of his variable-blast traveling grate furnaces for burning the smaller sizes of anthracite coal. In this important invention Mr. Coxe introduced a new principle, that of subjecting the fuel to ignition fusing it to a high state of combustion, and gradually reducing the pressure of the air blast with the corresponding reduction in the percentage of carbon in the burning fuel. By means of this process the subject of a separate patent Mr. Coxe made an important advance in the art, and proceeded with his usual vigor to put the invention into practice on a large scale, installing such than twenty of the machines at his own works during a period of two years.

Mr. Coxe had himself obtained more than fifty United States patents, and owned more than a hundred additional patents, taken out in his interest by his employee or associates. Some of these patents related to adaptations of his machines to the requirements of steamships and locomotives, for which purposes, it is believed, his invention will finally prove of great importance.

Aftermath.

"Electrography" is the name given to the new discovery of a means of photographing through opaque substances.

THE EXCELSIOR Manufacturing Company, of St. Louis, has assigned, with ample assets it is slated, but unavailable. An effort will be made to reorganize.

THE scientific world awaits with much interest, the confirmation of the report that Dr. Nansen, the Norwegian explorer, has succeeded in reaching the north pole by "drifting" in the currents of the Polar Sea.

DYDOCK expansation of the steamship St. Paul, illustrated in the last issue of the INVENTIVE AGE, revealed the fact that she sustained no serious injuries from her 10-day sojourn on the New Jersey coast sand bar. On the 29th the liner New York also "run aground" in a fog.

THE first needle factory in the United States, according to a Chicago dispatch, is soon to be established in that city. The needles are to be made by a machine invented by Eugene Fontaine, of Detroit, which, it is claimed, can turn out 2500 needles an hour at a materially lower cost than the present price of imported needles.

THE fact that many regular readers of the INVENTIVE AGE are frequently sending for back numbers "to complete their files for binding"--copies that have been mislaid and lost--as an argument in favor of securing one of those magazine files advertised elsewhere in this magazine. It is simple, inexpensive and just the thing for the AGE or any other magazine.

An Extraordinary Offer.

By special arrangement we are enabled to offer that great success in illustrated magazindom, "Cosmopolitan," and the INVENTIVE AGE one year for $1.25. In clubs of three or more $1.15.

During 1896 it is promised that both Cosmopolitan and Inventive Age will be greater magazines than ever. Cosmopolitan is a magazine of about 1400 pages and 1000 illustrations during the year. No magazine excels it. Old as well as new subscribers can take advantage of this offer.

Another Great Combination.

Popular Science is one of the best magazines published. It is 39 years old and covers all the sciences in a popular and instructive way. We can furnish Popular Science and the INVENTIVE AGE one year postage paid for $1.50. This combination rate can be offered but a short time as the subscription price of Popular Science is soon to be increased. Address the Inventive Age, Washington, D. C. If Self's Cyclopedia is also wanted--the three for $2.75.

THE INVENTIVE AGE wants live agents in every part of the country. Readers of this notice will confer a great favor by sending us names of persons who would likely take an interest in the matter. Our cyclopedia premium offer is a great winner. The book alone is worth double the amount, $2, charged for this offer.

ANY person sending us the names and addresses of five inventors who have not yet applied for patents will receive the INVENTIVE AGE one year free.

STANDARD WORKS.

"The Electrical World." An illustrated weekly review of current progress in electricity and its practical applications. Annual subscription........................$3.00
"Electric Railway Gazette." An illustrated weekly record of electric railway practice and development. Annual subscription........$3.00
"Johnston's Electrical and Street Railway Directory." Published annually........$5.00
Dictionary of Electrical Words, Terms and Phrases. By Edwin J. Houston, Ph. D....$5.00
The Electric Motor and its Applications. By T. C. Martin and Jos. Wetzler........$3.00
The Electric Railway in Theory and Practice. The first systematic treatise on the electric railway. By O. T. Crosby and Dr. Louis Bell.........................$2.50
Alternating Currents. By Frederick Bedell, Ph.D., and Albert C. Crehore, Ph.D........$2.50
Gerard's Electricity. Translated under the direction of Dr. Louis Duncan........$2.50
Theory and Calculation of Alternating-Current Phenomena. By Charles Proteus Steinmetz..........................$2.50
Practical Calculation of Dynamo Electric Machines. By A. E. Wiener..........$2.50
Central Station Bookkeeping. By H. A. Foster.........................$2.50
Continuous Current Dynamos and Motors. By Frank P. Cox, B.S..........$2.00
Electricity at the Paris Exposition of 1889. By Carl Hering.........................$2.00
The Elements of Static Electricity. By Philip Atkinson, Ph.D..........$1.50
Electricity and Magnetism. By Edwin J. Houston, Ph. D..........$1.00
Electric Measurements and other Advanced Primers of Electricity. By Edwin J. Houston, Ph.D.........................$1.00
The Electrical Transmission of Intelligence and other Advanced Primers of Electricity. By Edwin J. Houston, Ph.D..........$1.00
Electricity One Hundred Years Ago and Today. By Edwin J. Houston, Ph.D........$1.00
Alternating Electric Currents. By Edwin J. Houston, Ph.D. and A. E. Kennelly, D.Sc. $1.00
Electric Heating. By E. J. Houston Ph.D and A. E. Kennelly, D.Sc..........$1.00
Electro-magnetism. By E. J. Houston, Ph.D and A. E. Kennelly, D.Sc..........$1.00
Electricity in Electro-Therapeutics. By E. J. Houston, Ph.D., and A. E. Kennelly, D.Sc......................$1.00
Electric Arc Lighting. By E. J. Houston, Ph. D., and A. E. Kennelly, D. Sc....$1.00
Electric Incandescent Lighting. By E. J. Houston, Ph.D. A. E. Kennelly, D.Sc....$1.00
Electric Motors. By E. J. Houston, Ph.D., and A. E. Kennelly, D.Sc..........$1.00
Electric Street Railways. By E. J. Houston, Ph.D. and A. E. Kennelly, D.Sc........$1.00
Electric Telephony. By E. J. Houston, Ph.D. and A. E. Kennelly, D.Sc..........$1.00
Electric Telegraphy. By E. J. Houston, Ph.D and A. E. Kennelly, D.Sc..........$1.00
Alternating Currents of Electricity. By Gisbert Kapp..........$1.00
Recent Progress in Electric Railways. By Carl Hering.........................$1.00
Original Papers on Dynamo Machinery and Allied Subjects. By John Hopkinson, F.R.S.........................$1.00
Davis' Standard Tables for Electric Wiremen. Revised by W. D. Weaver........$1.00
Universal Wiring Computar. By Carl Hering.........................$1.00
Experiments with Alternating Currents of High Potential and High Frequency. By Nikola Tesla..........$1.00
Lectures on the Electro-Magnet. By Prof. Silvanus P. Thompson..........$1.00
Dynamo and Motor Buildings for Amateurs. By Lieut. Commander C. D. Parkhurst....$1.00
Reference Book of Tables and Formulæ for Electric Street Railway Engineers. By E. A. Merrill.........................$1.00
Electric Railway Motors. By Nelson W. Perry.........................$1.00
Practical Information for Telephonists. By T. D. Lockwood.........................$1.00
Wheeler's Chart of Wire Gauges..........$1.00
Practical Treatise on Lightning Conductors. By H. W. Spang.........................$.75
Tables of Equivalents of Units of Measurement. By Carl Hering..........$.50
Copies of any of the above books or of any other electrical book published will be sent by mail, postage prepaid, to any address in the world on receipt of price.

THE INVENTIVE AGE,
Washington, D. C.

Australian Patents to Americans.

The following list of applications has been specially prepared by Mr. George G. Turri, Certificated Patent Agents, Melbourne, W. B. Kid.-ler, Boston, Typewriting Machines; H. Belfield, Attorney of the foreign Electric Traction Co., New York, Electric Railways. C. H. Reed, New York, fastening devices for envelopes, boxes etc. A. H. Brisnell, Toronto, Electric propulsion of cars. B. Phillips, attorney of the The Moore Electrical Co., New York. Phosphorescent electrical illumination.

INVENTIVE AGE BUILDING.

GREAT PREMIUM OFFER! = =

Valuable
Sensible
Economical

Zell's Condensed Cyclopedia

Always Sold at $6.50 Complete.

Complete in One Volume. A Compendium of Universal Information, Embracing Agriculture, Anatomy, Architecture, Archeology, Astronomy, Banking, Biblical Science, Biography, Botany, Chemistry, Commerce, Conchology, Cosmography Ethics, The Fine Arts, Geography, Geology, Grammar, Heraldry, History, Hydraulics, Hygiene, Jurisprudence, Legislation, Literature, Logic, Mathematics Mechanical Arts, Metallurgy, Metaphysics, Mineralogy, Military Science, Mining, Medicine, Natural History, Philosophy, Navigation and Nautical Affairs, Physics, Physiology, Political Economy, Rhetoric, Theology, Zoology, &c.

Brought down to the Year 1890. Substantially Bound in Cloth with the Correct Pronunciation of every Term and Proper name, by L. Colange, LL.D. Handy Book for Quick Reference. Invaluable to the Professional Man, Student, Farmer, Working-Man, Merchant, Mechanic. Indispensable to all.

In this work all subjects are arranged and treated just as words simply are treated in a dictionary--alphabetically, and all capable of subdivision are treated under separate heads; so that instead of one long and wearisome article on a subject being given, it is divided up under various proper heads. Therefore, if you desire any particular point of a subject only, you can turn to it at once, without a long, vexatious, or profitless search. It is a great triumph of literary effort, embracing, in a small compass and for little money, more (we are confident) than was ever before given to the public, and will prove itself an exhaustless source of information, and one of the most appreciated works in every family.

It cannot in any sense--as is the case with most books--be regarded as a luxury, but will be in daily use wherever it is, by old and young, as a means of instruction and increase of useful knowledge.

Thousands of leading men offer the highest testimonials. It is not a cheap work; it is standard.

Never before, even in this age of cheap books, was such a splendid offer made, viz:

$2 For Cyclopedia, Postage Paid.
For Inventive Age, One Year. **$2**

An Extra Copy of Cyclopedia to any person sending us Six New Subscribers. The Same Offer is made to old Subscribers renewing for another term.

Address THE INVENTIVE AGE, Washington, D. C.

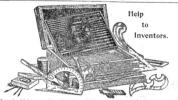

The Inventive Age
AND INDUSTRIAL REVIEW
A JOURNAL OF MANUFACTURING INDUSTRY AND SCIENTIFIC PROGRESS

Seventh Year. No. 4.   WASHINGTON, D. C., APRIL, 1896.   Single Copies 10 Cents. $1 Per Year.

Westinghouse-Baldwin Electric Locomotive.

The electric locomotive illustrated herewith, is the 1,000 horse power size of the series of electric locomotives which the Baldwin and Westinghouse combination have arranged for standards. The sizes range from 100 horse-power to 1,600 horse-power, the smaller sizes being used for mines and light tramways, and the larger for tunnels.

These locomotives are built under patents owned and controlled by the Baldwin Locomotive Works and the Westinghouse Electric and M'f'g. Co.

The principal feature of the new system of construction is the use of geared motors. This permits the construction of locomotives that are easy to repair, reasonable in cost and simple in construction. Before the subject was taken up by the Baldwin-

will not be sent out on the line unless an entire train crew be put on the engine. The reason is that in case of wreck, the firemen and engineer are often killed and the locomotive becomes an obstruction in the track with no one to flag it.

Most people are accustomed to a steam locomotive appearance and naturally expect that the electric locomotive will look something like it; but there is no reason why this should be, and if the electric had been built first, the steam locomotive with its numerous projections, stacks and protuberances, would have seemed to be a very queer looking machine. Those who have built electric locomotives with the peculiarly shaped superstructures have found that such are not acceptable to railroad companies and hereafter a plain car-like body will be

hence there is not the limitation to long runs with electric locomotives that there is in the case of steam locomotives where the grates become clogged after a short run. A large steam locomotive can often generate a thousand horse power for fifteen or twenty minutes, and then time must be provided for cleaning the fires. In the case of the electric locomotive, 1,000 horse power can be developed as it is only dependent upon the central station.

The speed of the electric locomotive is dependent upon the gearing; that is, the velocity of the armature which revolves is more than sufficient to run the locomotive at more than 120 miles an hour and has to be geared down, so that the wheels of the locomotive will revolve slower. By making the reduction of the gearing suitable, the speed may be

WESTINGHOUSE-BALDWIN ELECTRIC LOCOMOTIVE FOR TRUNK LINE WORK.

Westinghouse combination electric locomotives had been built for mine work and for tunnels; but the attempt had been made to give them the appearance of steam locomotives, with the result that the locomotives were neither handsome in appearance nor handy and serviceable for the railroad company. All electric locomotives necessarily have the driving gear underneath the floor, as there are no boilers, smoke stacks, steam domes, etc., as is usual in steam locomotives; hence the superstructure of the electric locomotives may be made in any form, but it has been found upon investigation and enquiry among railroads, that they would prefer that the superstructure be made ample in size to receive railroad freight; that is, freight belonging to the road itself, which is often shipped from one point to another on the road, and receive baggage, mail and wrecking apparatus, replacing frogs, chains and jack screws, crow bars, etc. Furthermore, an order has been recently issued in some roads that a light locomotive

given in case of duplicate orders. In other words, those who have built electric locomotives, attempting to retain something like the appearance of a steam locomotive, have found they have made a mistake. In some cases the attept has been made to give them the appearance of Ericson's Monitor. This also is not only found to be very expensive, but serves no useful purpose, inasmuch as the sloping ends simply interfere with the repair of the machinery and do not give useful space.

The power of this locomotive to travel at high speed is somewhat greater than that of the standard engines on the Pennsylvania Railroad which haul the limited trains. When moving at a freight train speed, it will haul more than the heaviest consolidation engines of the Pennsylvania road. For the reason that the power of the electric locomotive is not determined by the power of the locomotive itself, but by the power of the central station, it is evident that there is practically an unlimited supply,

made almost nothing. This locomotive with a gearing of 1-2 to 1 is perfectly adapted to run from 70 to 100 miles an hour, and with a gearing of 4 to 1, is adapted to pull heavy loads at a freight train speed.

This locomotive is about 38 feet over the pilots, about 9 feet 6 inches wide and 13 feet 6 inches high. The weight is 150,000 lbs., cost about $16,000.

On a steam locomotive, an engineer has to operate about twenty handles, keep watch of the fireman, the height of the water in the boiler, the steam pressure, besides looking on ahead; while on this electric locomotive there are but three handles; the reversing switch, the controller and the air brake handle, so that the attention of the engineer can be confined to the track ahead.

The air brakes are operated in the usual way by a pump which is driven by an electric motor. The machine is practically noiseless, and is placed under the floor of the locomotive. The head lights are illuminated either by electricity or oil.

⁂Inventive Age

Established 1889.

INVENTIVE AGE PUBLISHING CO.,

8th and H Sts., Washington, D. C.

ALEX. S. CAPEHART.   MARSHALL H. JEWELL.

The INVENTIVE AGE is sent, postage prepaid, to any address
in the United States, Canada or Mexico for $1 a year; to any
other country, postage prepaid, $1.50.
Correspondence with inventors, mechanics, manufacturers,
scientists and others is invited. The columns of this journal are
open for the discussion of such subjects as are of general interest
to its readers.
Technical matter is particularly desired. We want practical
information from practical men.
Nothing will be published in the editorial columns for pay.
The INVENTIVE AGE is thoroughly independent.
Advertising rates made known on application. Special facilities for furnishing cuts of any patented article together with
descriptive article. Business specials 25 cents a line each insertion, 7 words to the line. No advertisement less than 50 cents.
Address all communications to THE INVENTIVE AGE, Washington, D. C.

Entered at the Postoffice in Washington as second-class matter.

WASHINGTON, D. C., APRIL, 1896.

THE platform of business men, without regard to
politics, is fair trade, not free trade.

"RADIOGRAPH," is the name suggested by Electrical Review for the new means of photography by
Roentgen rays. A very good coinage. Pass it
along.

THE new Directory of the iron and steel works of
the United States, shows an enormous capacity for
the production of steel in this country. From being
7,740,900 gross tons in January, 1894, our converting
capacity today is 9,472,350 gross tons. an increase
n two years of 22 per cent.

THE investigation of copyright laws which has
been carried on for several weeks by the house committee on patents, will probably lead to a more or
less comprehensive revision of the copyright system.
One of the principal changes likely to result will
be the establishment of a bureau of copyrights in
connection with the congressional library. which
now has charge of all the copyright business, but
which not a sufficiently large clerical force to handle
this work properly.

WORK on the electric line connecting Washington
and Baltimore is progressing rapidly and contracts
for the power houses and much of the equipment
have been let. In planning the power stations the
requirements provide for the operation of express
trains at the rate of over sixty miles an hour between Baltimore and Washington. Provision in
the design of the machinery is made for running
such trains on half-hour schedules, local through
trains on similar schedules, and trains between
Washington and Laurel and Baltimore and Ellicott
City on twenty-minute schedules. With the exception of the Baltimore portion of the line, the grades
are slight and the construction of the road throughout is planned for speed.

IN his opening address before the February meeting of Mechanical Engineers in Pittsburg, President Joseph D. Weeks reviewed the evidence supporting the claim that Wm. Kelley, of Pittsburg,
was the inventor of the Bessemer process. He
showed that in 1847, seven years before Bessemer's
discovery, Kelley, then operating the Suwanee furnace near Eddyville, Ky., used the process at his
forge. In concluding the speaker said : "Do not
these experiments cover all the points of Holley's
definition of the essential features of the Bessemer
process : 'The decarbonising of crude cast iron by
the air blast in a vessel independent from the blast
furnace or furnace in which it was melted, and without the application of external heat?'" And if Kelly
did this in 1847, while Bessemer did not conceive the
idea until 1854, at the earliest, is not Kelly the original inventor, and in calling it the Bessemer process
has not Kelly been unfairly deprived of the credit

due him, as was Columbus, when this continent was
named America ? While the mechanical appliances
that made possible the rapid production of pneumatic steel were Bessemer's, and the idea of using
spiegeleisen to remove oxygen and to recarbonize
the metal was Mushet's, of Cheltenham, the original
idea of decarbonisation by blasts of air was William
Kelly's, of Pittsburg."

ENGLAND'S LETHAL SLEEP.

Under the title "Are We the Chinese of the Nineteenth Century," we have constantly referred in
these columns, to the loss which threatens England
in her commercial supremacy. During the past
few weeks another example of the strong foreign
competition with which our nation has to compete
has come before our eyes. We refer to the much-
talked-of rail order of 10,000 tons for Japan, which
has been placed with the Illinois Steel Company,
and the order for 12,000 tons of steel rails for Chili,
which has been undertaken by a German firm.
English houses were in both cases invited to tender,
but the prices quoted in this country being higher
than the quotations in the case of the successful
competitors, the order was lost. We do not for an instant suppose that these contracts have been necessarily lost by the demerits of our home-made stuffs.
We have ever been cognisant that Germany has
quoted prices for foreign exports below the cost at
which she can supply home purchasers, and this sort
of competition on the part of Germany has long been
a known quantity in the English market, but it is a
new thing in our experience to find American quotations running against us in somewhat the same
manner.

This is perhaps the first time we have had seriously to face this question in a large way on the
part of the United States, and this must open our
eyes to the fact that American manufacturers are
not likely to stop at the Japanese orders secured by
the Illinois Steel Company. American resources are
vast, and though they are reckoned to have only half
a dozen or so large steel rail manufacturing firms
in the United States, it is possible that without
any great inconvenience, they may produce twice or
three times as large a tonnage of rails as they have
done during the past few years. The most hopeful
British subject must see the serious consequences for
British trade, and these consequences do not stop
here. * * * Our fathers gained a goodly reputation
for themselves, a reputation upon which we are now
inclined to batten. Meanwhile our very daily bread
is being filched from under our noses, whilst we
drink more and more deeply of the lethal stream,
and hug ourselves in our smug complaisancy, crying Peace when there is no Peace. It is time for
every Englishman to be up and doing.

If we had only ourselves to contend against, then
we might say that competitions would ensure the
maximum wages to the laboring classes, but in the
face of the overwhelming foreign industries which
pour into this country, we are bound to take steps
to protect ourselves and our workmen against its evil
effects. The very admission of foreign prison made
goods has undermined one branch of British industry, and at the present day we see that that branch
stands in danger of total extinction. We cannot
produce goods of the required standard at anything
like that price at which the foreign authorities
can ship them over here. It would be a better day
for Old England, once called Merrie England, were
producers and consumers in every department to
make a stand to stamp out this fatal indifference,
and revitalise our English trade. If this does not
happen soon there must inevitably follow a rude
awakening.

Nicola Tesla, the well known electrician, says that
he is satisfied that he has a machine which when
perfected, will enable him to make practical experiments in distributing electric waves about the earth,
so that messages may be conducted to all parts of
the globe simultaneously. He believes that electric
waves may be propagated through the atmosphere,
and even the ether beyond, a disturbance of the
waves at any point being instantly felt at every
other point along them. He declares that he believes that the transmission of news about the earth
by electric waves in the place of wires is no longer a
dream.

The value of the invention of a machine for making oval cigarettes, by Bernhard Baron, of New
York, under which it is understood the National
Cigarette and Tobacco Company will operate in the
manufacture of oval cigarettes, may be understood
from the fact that heretofore oval cigarettes have
been made by hand at the rate of 600 a day, whereas
the improved patented mechanism will make the
machine-made oval cigarettes at the rate of between
400,000 and 500,000 per day.

NOTES.

X Rays Focussed Downwards.—Once again
our American cousins are to the fore. At Toronto
University Herr Rontgen's rays are focussed downwards by means of a glass bell over the Crookes'
tube. In this way instaneous photographs were
taken.

* * *

Sword Making a Lost Art.—In a lecture on
Japanese swords, before the Franklin Institute of
Philadelphia, by Benjamin Smith Lyman, November 8, 1895, the lecturer said : "The most famous
Japanese sword-maker was Masamune (about A. D.
1290); and, next, his pupil, Muramasa (about 1340);
then Yoshimitsu (about 1275); and Munechika (about
990). Of Masamune's swords it is often said, they
are so fine they will cut a hair falling in the air, or
cut in two the very hardest suzuki bean as it falls ;
or, if held in a stream will cut in two a sheet of
paper floating down. The swords of Muramasa are
said to be so finely tempered as to cut hard iron like
a melon."

* * *

Another Illusion Dispelled.—It is commonly
supposed that tattooing is impossible of eradication.
From Paris however, comes a method by which
these marks may be removed if desired. First paint
over the tatoo marks with a concentrated solution
of tannin ; afterwards, by means of fine needles,
make a series of pickings over the tattooed design.
Over the surface thus pricked pass a stick of nitrate
of silver. In a few minutes we see detached the
black prickings previously made, and we know that
the superficial layers of the skin contain a tannate
of silver. To ensure success this surface must be
powdered with tannin two or three days. An in-
flammatory action is set up lasting for several days.
The pricked parts turn black, forming a thin scab,
very adherent to the deeper skin, but painless. At
the end of from 14 to 18 days, the scab falls off, and
in its place a superficial red mark is seen which
gradually fades away, until at the end of a few
months all signs of discoloration disappear. An
tiseptic precautions should be observed in performing this operation. The old tattoo needle may be
used to remove all the tattoo marks.

* * *

An Electric Chaise for Her Majesty.—We
hear that an electric chaise was ordered by the late
Prince Henry of Battenberg for Her Majesty the
Queen. So far the vehicle is incomplete, only the
skeleton, as it were, to be seen. It is built upon a
double frame-work of tubes with a head tube for
steering much as a bicycle has, the steering handle
being something like that used in bath chairs. The
body of the carriage is composed of aluminium, and
it runs upon three wheels, fitted with michelin tires ;
the propelling force is a dynamo driven with Ful-
men accumulators. From the description given
of this electric chaise, it should be a very graceful
example of what an autocar can be made.—
Invention.

The opening of the new Pacific Short Line bridge
between the Iowa and Nebraska banks of the Missouri river at Sioux City, recently, was an occasion
that was greeted by residents and users with due
approbation, and by contractors and makers of supplies as the completion of an enterprise that has
taxed large capital, skill and labor. The bridge is
2,200 feet in length, four spanned, cost nearly
$1,300,000, and over 7,500,000 pounds of iron and
steel entered into its construction. J. A. Waddell,
of Kansas City, was the engineer, employed under
the direction of the Credits Construction Company.

SPECIAL OFFER TO PHOTOGRAPHERS.

Eloquent Champion of American Shipping.

That Hon. Joseph B. Foraker, Senator-elect from Ohio, is an earnest champion of the true American principle and awake to the necessities of the hours was shown by the recent utterances of that gentleman in Cincinnati.

"I want to see the Monroe doctrine; recently so much talked about, upheld and enforced against all the world. And I shall stand by the administration that stands for America, be that administration republican or democratic.

"I want to see our merchant marine restored. There was a time when our merchant marine was the pride of every American. It is today but a mortification to us all. We once carried ninety per cent of our foreign trade in American bottoms, under the American flag. We now carry less than

HON. JOSEPH B. FORAKER, OF OHIO.

thirteen per cent. We are paying out annually more than $150,000,000 in gold to foreign ships for transportation of freights and passengers. The time has come to remedy that. The way to remedy it is not with subsidies and bounties, but by going back to the first thing practised by George Washington and the founders of this Republic when they applied the principles of protection to the water as well as to the land.

"I want to see the Congress of the United States provide that the fifty per cent or more of imports that come into our country free of duty, shall come in free, provided that they come in American bottoms and under the American flag. I want to see it provided that the dutiable goods brought in American ships shall be allowed a rebate on that account.

"And when we make these new reciprocity treaties, which we hope to make soon in the future, I want to see incorporated in every one of them a provision that the goods mentioned in the reciprocity treaty shall have the benefits of that, provided they are carried in the ships of the reciprocity country. When that shall be done, as done it can and should be, there will be no longer an elbowing by Great Britian of the American marine off the waters of the globe.

"Shipbuilding will revive, and once again the flag of the United States will be seen floating in all the channels of trade and commerce. And then after that will follow easily and naturally what we should have had ere this, an American navy able to protect us, let come what may. When Mr. Cleveland sent to Congress his Venezuelan message it had more good results than one. One of its good results was to impress the American people with our defenceless situation. We should realize that the great wars of the future, if there be any at all with which we are to be concerned, are far more likely to be on the water than on the land. We should order accordingly. It is a patriotic duty to do it."

Although there are many inventions in that line, the pefect mucilage holder—clean, economical and handy for office use—is yet in the mirage of the future.

A canopy, not unlike those used on baby carriages, is one of the latest novelties in connection with bicycles.

What Cuban Annexation Would Mean.

The Annexation of Cuba to the United States would put its entire commerce with the rest of the Union in the hands of the American shipowners. The act of 1792, still in force, prevents foreign ships from engaging in the American coastwise trade, so that the trade with Cuba would be just as completely protected as the trade between New York and Virginia now is, or as fully as it is between any of the States of the Union.

Cuban sugar and Cuban tobacco, manufactured and unmanufactured, would then come to the Union free of any duty either export or import. What an impetus that would give to the developement of Cuba. It would cause a remarkable developement in all of the great staple products of Cuba. It would cause a very large emigration from the United States to Cuba. It would give stability to trade between the rest of the United States and Cuba, impossible now under Spanish domination.

Cuba would prosper and bloom as never before. The United States would enormously increase its domestic commerce. German beet sugar would be under an additional disadvantage, and Cuban sugar would find its market wholly in the rest of the Union.

Cuba, if a part of the United States, would obtain all of the goods from the rest of the States that she now purchases in foreign countries, because the same tariff that now excludes foreign imports from the United States would exclude them equally from Cuba, at the same time permitting perfect free trade between the products of the United States and Cuba.

As a part of the Union it would be everything for Cuba to gain and nothing to lose. The nation that has absorbed Spanish Florida and French Louisiana and Mexican Texas, need fear nothing from the absorption of Cuba. Every three or four years the United States absorb a foreign population the equal of Cuba's. Cuba's manifest destiny ought to point in the direction of Cuba's best interests, and, unquestionably, her interests are all in the Union.— *Seaboard.*

X Rays From Heat Radiation.

Prof. H. S. Greene, of Harrodsburg, Ky., Academy and S. F. Spillmann, photographer, have produced X-rays from ordinary heat radiation, without any Rhoumkodff coil or Crookes tubes. It is said they have succeeded in producing some excellent skiagraphs of various objects. The mysterious rays are found to exist in ordinary heat radiation, and have been made to penetrate several thicknesses of opaque substances. They pass readily through metals, vulcanite, cardboard, etc. Photographs taken by the agency of the newly discovered rays on exhibition at Mr. Spillmann's studio, where they have been viewed by many people.

A Novel Typewriter.

Mr. R. J. Fisher, cashier of the First National Bank, of Athens, has perfected a typewriter which will write on the pages of a bound book as well as on ordinary letter paper. It weighs only thirteen pounds, yet makes a maximum line of eleven inches, sufficient for the largest records. Ink is used which cannot be washed off the surface of the pages by acid, and which is practically indelible. The typewritten lines are nine inches long and accurately spaced, while all the characters are as clear and legible and of uniform shade as can be made by the best typewriters now on the market.

Deer Park, on the Crest of the Alleghenies.

To those contemplating a trip to the mountains in search of health or pleasure, Deer Park, on the crest of the Allegheny Mountains, 3,000 feet above the sea level, offers such varied attractions as a delightful atmosphere during both day and night, pure water, smooth, winding roads through the mountains and valleys, and the most picturesque scenery in the Allegheny range. The hotel is equipped with all adjuncts conducive to the entertainment, pleasure and comfort of its guests.

There are also a number of furnished cottages with facilities for housekeeping.

The houses and grounds are supplied with absolutely pure water, piped from the celebrated "Boiling Spring," and are lighted with electricity. Deer Park is on the main line of the Baltimore and Ohio Railroad, and has the advantages of its splendid Vestibuled Limited Express trains between the East and West. Season excursion tickets, good for return passage until October 31st, will be placed on sale at greatly reduced rates at all principal ticket offices throughout the country.

The season at Deer Park commences June 22d, 1896.

For full information as to rates, rooms, etc., address George D. DeShields, Manager, Deer Park, Garrett County, Maryland.

Valuable Invention for Mariners.

The brain of the inventor has not been idle in matters concerning the protection of life and property. When unusual hazards are incurred as in railway or water travel the ingenuity of man has been tireless in limiting these risks as far as possible. Among the latest of these we note what is known as the eophone, the invention of a Baltimore scientist. This instrument is designed to be used on ship-board in case of fog or darkness, and it is claimed that by its means the direction of any sound can be detected with absolute accuracy. It consists of two bell-mouthed sound receivers separated by a central diaphragm. The tubes are carried within the chart house, and the instrument can be turned from below to any direction. It has taken years to perfect this instrument, and in its now perfected condition it is being generally accepted as a success. It is likely to be as indispensable a part of marine equipment as the compass. Science has once again rendered a great service to the practical side of every day life.— *Age of Steel.*

'Seeing by Electricity.

The discoveries in electrical science promise to come fast and marvelous during the coming decade and it has been suggested that seeing by telegraph, with a telephone attachment at the same time, may soon place individuals on opposite sides of the continent in all but personal contact with each other. It is announced from London that a Mr. J. G. Vine has been able to photograph at one end of a wire objects exposed between two vacuums at the other end of the wire. The same experimenter declares that he will soon be able to photograph objects at any distance by means of the X-rays.

Everybody Can Now be Polite.

Mr. James C. Boyle, of Spokane, Washington, is the inventor of a novel saluting device—an attachment to the head-gear that will premit of the most graceful sort of salutations without effort and without the use of the hands in any manner. The mechanism consists of a combination of gear-wheels,

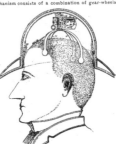

HOW TO BE POLITE.

spring fingers, levers, weight block, etc., so assembled that by the force of gravity when the wearer of the hat having the novel mechanism within it, desires to salute another person he simply tips his head forward and the hat is automatically raised in front, effecting a graceful salutation. When the person making a salutation resumes an erect position, the hat drops back to its normal condition. The inventor believes it an excellent device for troops of soldiery on dress parade or there may be a sign or placard placed on the hat having the improvements within it, and the saluting device be used to attract attention of the public on a crowded thoroughfare to the advertisement on the hat, the novelty of its apparant self-movement calling attention to the hat and its placard.

One of the great industries of Marsailies is the importation of over 200,000,000 pounds of peanuts annually which are converted into "olive" oil (imitation) and exported to the United States as the genuine article. Prof. F. E. Weatherby, of Pittsburg has invented a process or machine for the treatment of the peanuts and the turning out of various grades of oils for food or lubricating purposes.

An Ore Hoist and Conveyor.

The accompanying illustrations are of interest to dock companies, railroads and others operating and connected with handling ore and coal from vessels to docks and cars or vice versa. They give a general idea of a plant recently designed and built under patents and rights granted the King Bridge Co., of Cleveland, on the docks of the P. Y. & A. Ry. Co. at Ashtabula, O., operated by M. A. Hanna & Co. The King Bridge Co. has built a number of machines of this character, the first being as far back as 1882, probably the first plant of the kind built for handling ore and coal in this way. The machine shown in the illustration is superior to the ealier type built by this company, in the improvements that experience has shown to be of benefit in dispatch and economical maintenance. These two points mean much when the shortness of the navigation season upon the lakes is considered.

The hoists are built entirely of wrought and cast steel which has been subjected to thorough inspection throughout its manufacture. Care has been taken that the stucture should be well designed and detailed from an engineering standpoint, which is of the first importance when the very hard usage that is required of it is considered.

t   As shown in the illustration, the hoists are arranged in pairs; that is, two bridges have one support in common at the rear, designated as the rear

KING BRIDGE CO.'S ORE AND CONVEYOR.

tower, while, there is a separate support or leg at the front for each bridge. All of the towers are supported on wheels placed on substantial tracks, consisting of two rails at the rear and one in front. The moving of the plant from one point to another is done by steam or electric power at the rear, and at the front by means of hand gearing.

The principal dimensions of the machine are 180' between towers, with 88' cantilever and 34' apron on the front tower, which can be raised and lowered by power at will so as not to interfere with the masts of vessels when they are taking position in front of the plant. There is a clear space under the bridges from the top of the dock of about 30', which gives ample room for storage purposes, providing it is not convenient to unload a vessel directly into the cars. The bridges are on a grade to increase the storage capacity of the dock.

The plant which the King Bridge Co. has just built at Ashtabula has thus far given entire satisfaction. The dock company has since ordered an additional set of two legs, which is now being constructed ready for operation next spring. The dock will then have a plant of 10 legs, which is now considered no more than enough to handle boats of the latest atyle, which are designed some 432 feet in length over all, with 12 hatches, and a capacity of 6,000 tons of ore on 19 feet draft.

Geo. B. Raser, the manager of the P. Y. & A. dock at Ashtabula, O., selecting at random from his records, reports that on Sept. 26, 1895, the steamer Roman, with 2,500 tons of ore commenced unloading in cars at 6:30 A. M. and finished at 5:30 P. M. On Sept. 28, 1895, the steamer German, with 2,540 tons of ore, commenced unloading in pile at 7:00 A. M.

and finished at 5:30 P. M. An allowance in each case should be made for a shut-down of one hour at noon, and in neither of these cases was any special effort made to obtain a record

The general arrangement of machinery, mode of operation and important details and various safety devices are all covered by the Rasch and other patents controlled by the King Bridge Co. It is their intention to operate this branch of their business more extensively here after. They are also engaged in perfecting a coal handling device for unloading cars directly into boats.

Nikola Tesla's Interesting Experiments.

Nikola Tesla contributes to the Electrical Review on account of some of his recent experiments in "shadowgraphy" or radiography, from which the following extract is taken :

I am producing strong shadows at distances of 40 feet. I repeat, 40 feet and even more. Nor is this all. So strong are the actions on the film that provision must be made to guard the plates in my photographic department, located on the floor above, a distance of fully 60 feet, from being spoiled by long exposure to the stray rays. Though during my investigations I have performed many experiments which seemed extraordinary, I am deeply astonished observing these unexpected manifestations, and still more so, as even now I see before me the possi-

bility, not to say certitude, of augmenting the effects with my apparatus at least tenfold! What'may we than expect? We have to deal here, evidently, with a radiation of astonishing power, and the inquiry into its nature becomes more and more interesting and important. Here is an unlooked-for result of an action which, though wonderful in itself, seemed feeble and entirely incapable of such expansion, and affords a good example of the fruitfulness of original discovery. These effects upon the sensitive plate at so great a distance I attribute to the employment of a bulb with a single terminal, which permits the use of practically any desired potential and the attainment of extraordinary speeds of the projected particles. With such a bulb it is also evident that the action upon a fluorescent screen is proportionately greater than when the usual kind of tube is employed, and I have already observed enough to feel sure that great differences are to be looked for in this direction. I consider Roentgen's discovery, of enabling us to see, by the use of a fluorescent screen, through an opaque substance, even a more beautiful one than the recording upon the plate.

Since my previous communication to you I have made considerable progress, and can presently announce one more result of importance. I have lately obtained shadows by *reflected rays only*. thus demonstrating beyond doubt that the Roentgen rays possess this property. One of the experiments may be cited here. A thick copper tube, about a foot long, was taken and one if its ends tightly closed by the plate-holder containing a sensitive plate, protected by a fiber cover as usual. Near the open end

of the copper tube was placed thick plate of glass at an angle of 45 degrees to the axis of the tube. A single terminal bulb was then suspended above the glass plate at a distance of about eight inches, so that the bundle of rays fell upon the latter at an angle of 45 degrees, and the supposedly reflected rays passed along the axis of the copper tube. An exposure of forty-minutes gave a clear and sharp shadow of a metallic object. This shadow was produced by the reflected rays, as the direct action was obsolutely excluded, it having been demonstrated that even under the severest tests with much stronger actions no impression whatever could be produced upon the film through a thickness of copper equal to that of the tube. Concluding from the intensity of the action by comparison with an equivalent effect due to the direct rays, I find that approximately two per cent of the latter were reflected from the glass plate in this experiment. I hope to be able to report shortly and more fully on this and other subjects.

In my attempts to contribute my humble share to the knowledge of the Roentgen phenomena, I am finding more and more evidence in support of the theory of moving material particles. It is not my intention, however, to advance at present any view as to the bearing of such a fact upon the present theory of light, but I merely seek to establish the fact of the existence of such material streams in so far as these isolated effects are concerned. I have already a great many indications of a bombardement occurring outside of the bulb, and I am arranging some crucial teats which, I hope, will be successful. The calculated velocities fully account for actions at distances of as much as 100 feet from the bulb, and that the projection through the glass takes place seems evident from the process of exhaustion, which I have described in my previous communication. An experiment which is illustrative in this respect, and which I intended to mention, is the following: If we attach a fairly exhausted bulb containing an electrode to the terminal of a disruptive coil, we observe small streamers breaking through the sides of the glass. Usually such a streamer will break through the seal and crack the bulb, whereupon the vacuum is impaired; but, if the seal is placed above the terminal, or if some other provision is made to prevent the streamer from passing through the glass at that point, it often occurs that the stream breaks out through the side of the bulb, producing a fine hole. Now, the extraordinary thing is that, in spite of the connection to the outer atmosphere, the air can not rush into the bulb as long as the hole is very small. The glass at the place where the rupture has occurred may grow very hot—to such a degree as to soften; but it will not collapse, but rather bulge out, showing that a pressure from the inside greater than that of the atmosphere exists. On frequent occasions I have observed that the glass bulges out and the hole, through which the streamer rushes out, becomes so large as to be perfectly discernible to the eye. As the matter is expelled from the bulb the rarefaction increases and the streamer becomes less and less intense, whereupon the glass closes again, hermetically sealing the opening. The process of rarefaction, nevertheless, continues, streamers being still visible on the heated place until the highest degree of exhaustion is reached, whereupon they may disappear. Here, than, we have a positive evidence that matter is being expelled through the walls of the glass.

When working with highly strained bulbs I frequently experience a sudden, and sometimes even painful, shock in the eye. Such shocks may occur so often that the eye gets inflamed, and one can not be considered over-cautious if he abstains from watching the bulb too closely. I see in these shocks a further evidence of larger particles being thrown off from the bulb.

The Family Doctor says that this is the secret of avoiding colds. The man or woman who comes out of an overheated room, especially late at night, and breathes through the mouth, will either catch a bad cold or irritate the lungs sufficiently to cause annoyance and unpleasantness. If people would just keep their mouths shut and breathe through their noses, this difficulty and danger would be avoided. Chilla are often the result of people talking freely while out of doors just after leaving a room full of hot air, and theater-goers who discuss and laugh over the play on their way home are inviting illness. It is, in fact, during youth that the greater number of mankind contract habits of inflammation which make their whole life a tissue of disorders.

Secretary Lamont's approval of the new plans of the East River bridge connecting New York and Brooklyn, will enable its construction to begin immediately. The lowering of the height from 145 to 135 feet will save at least a million dollars in its cost. Work, it is expected, will be well under way by June, and although the bridge will be the greatest structure of its kind in the world, it is believed that it can be completed in three or four years.

Novel Bicycle Course.

Paris furnished the greatest novelty in the way of a bicycle academy that has thus far been devised. This establishment is called the "Palais-Sport," and is remodeled from a building formerly used for panoramic scenes. The spiral pathway extends from the main floor of the academy to a point near the roof. The ascent is gradual, being about 2·5 to the 100, the total height being 36 feet. The pathway is divided into two paths by an inverted V-shaped board screen, the entire length of course, including the ascent and descent. being over a thousand yards.

The path is extended at the top into a spacious

SPIRAL PATHWAY PALAIS-SPORT IN PARIS.

platform which enables the rider to make an easy turn before taking a long coast to the main floor below. A high screen protects the wheelman from being precipitated below in case of accident. A spacious room is reserved for spectators. The outer wall of the spiral is decorated somewhat elaborately with pastoral scenes, giving the effect of the country. The bicycles are brought to the main floor from the storage room by means of elevators. The lower floor contain various waiting, reading and reception rooms, etc. In Paris the popularity of the bicycle is as marked as in this country and in the novel spiral course of the "Palais-Sport" the ascent and descent is so gradual as to make the resort even more popular than a level course.

AN INTERESTING PAPER

Invention and the Distribution of it Benefits.

Editor of Inventive Age:

The able and eloquent address on "The Inventive or Creative Age," delivered by Mr. James T. DuBois at a recent meeting of the National Statistical Association, presented in a striking manner the magnitude of the benefits conferred on the civilized world by modern inventions. In the discussion which followed, a question was raised as to how far the working classes have participated in these benefits. That they have participated to a considerable extent was not denied by any one, and a majority of those taking part in the discussion seemed to think that they have participated to a satisfactory degree. In this latter opinion I, for one, can not concur. Various estimates have been made as to the extent to which labor-saving machinery has increased the productive power of industry. An estimate which I have recently heard cited and attributed to Hon. Carroll D. Wright, Commissioner of Labor, puts the ratio of increase within the last fifty years at eleven to one. That is, a given number of workers equipped with the modern appliances will produce as much as could be produced by eleven times their number equipped with the appliances in use fifty years ago; and as regards the most advanced nations this estimate can hardly be considered extravagant.

Estimates have also been made in regard to the increase of wages, the fall of prices and the reduction in hours of labor within a corresponding period. I have not time to examine these in detail, but it would perhaps be a sufficiently liberal allowance if we should assume that, taking all these elements into consideration, the position of the working classes has been improved in the ratio of two and three-fourths to one. This woud make the

progress of the working class in respect to the amelioration of their economic condition only one-fourth as rapid as the general increase in the productive power of industry. It would, perhaps, be too much to expect that the improvement in the condition of the worker should be fully proportioned to the increase in the productivity of labor, because this increase implies the use of a very much greater amount of capital than was in use fifty years ago; and this additional capital must, of course, receive its compensation. The captain of industry, too, directing the labors of scores, hundreds, or even thousands of men, plays a much larger part than formerly and is, therefore, entitled to a larger share of the product. But making all due allowance for the proper reward of capital and industrial direction, I can not bring myself to think it a satisfactory state of things, that the economic condition of the working classes should be improving only one-fourth as fast as the general wealth producing power of the community is increasing.

It is not pretended that these figures have any claim to statistical exactness, but I am decidedly of the opinion that they understate, rather than overstate, the disparity between the economic progress of the working class and the increase in the productive power of industry. One reason I have for so thinking is the fact, that fifty years ago the working class consisted much more largely than now of persons carrying on independent industries on their own account in small establishments whose proprietors did a considerable part of their work, and sometimes the whole of it, with their own hands. To make a fair comparison between the earnings of labor then and now, these self-employing workers and the incomes which they earned should be included in computing the average for the compensation of labor at the earlier date. Great as the economic benefits flowing from the increased use of labor-saving machinery have been, we must not forget that the reduction of multitudes of these independent self-employing workers to the position of wage earners has been one of its incidental and, indeed, inevitable effects; since its economies could only have been effected by that transfer of industry from small to large establishments of which the comparative disappearance of the independent self-employing workman was a necessary concomitant.

But if the masses of the people have not enjoyed their proper share in the benefits flowing from modern inventions, in what direction shall we look for the remedy? I answer that the remedy is to be found in more invention rather than in less. In saying this, however, I have in mind, not merely the continuance of invention along mechanical lines, but also its use, and that in a largely increased measure, in the sphere of social organization. Just as an Edison or Tesla applies the principles of physics to the attainment of mechanical results through various ingenious devices, so must the statesman apply the principles of sociology to the attainment of desired results in human society trough devices which must find their embodiment in legislative enactments. So also the philanthropist must seek, through a knowledge of the same science, to apply its principles in devising schemes of social improvement. In his "Dynamic Sociology" and his "Psychic Factors of Civilization," Prof. Lester F. Ward has compared the social forces, consisting of the passions and desires of human beings, to the physical forces of nature, such as gravitation, heat, electricity, &c, which have been so wonderfully subordinated to human uses in

the field of mechanics; and he has emphasized the importance of such a thorough knowledge of these social forces and their modes of operation as will enable us to make them serve the general interest of humanity as freely and efficiently as do the physical forces named above, pointing out that to do this successfully requires just such a skilful adaptation of means to ends as enables the mechanician to control the physical forces for the purposes of industry and commerce.

It is mainly because sociology is less advanced than the physical sciences, that invention in the field of laws and institutions has lagged behind, and that the benefits resulting from physical discovery and mechanical invention have been so unequally distributed. But another reason may, perhaps, be found in the fact that invention in the field of social action has received no such encouragement as our patent laws have afforded in the field of mechanics. Encouragement on a precisely similar plan may never be practicable, but it is worthy of serious consideration whether some plan can not be devised whereby it may be afforded in the needful measure. Even if nothing adequate be done in this direction by governments, public spirited citizens possessing the requisite means may accomplish much. Some, indeed, are already doing so, while the urgency of social questions and the deep humanitarian feeling aroused in behalf of their solution are alone sufficing to enlist many of the world's ablest and noblest minds in the advancement of social science and its application to the improvement of social conditions. Is it entirely utopian to hope that the next half-century may witness in this direction a progress no less wonderful than that which the last has seen along the lines of physical science and mechanical invention.

EDWARD T PETERS.

The Development of Smokeless Powder.

Prof. Charles E. Munroe recently delivered an interesting address on smokeless powder, before the Cosmos Club in Washington. He said that notwithstanding the improvements made in the black powder, some 57 per cent of its products of combustion are eventually solid, and hence are either thrown out into the atmosphere as a cloud of smoke or remain as a residue to foul the gun.

In consequence of this, when rapid-fire guns or large guns such as the 110-ton gun, which throws over 800 pounds of solid steel at each discharge, are brought into action, the atmosphere soon becomes filled with smoke, which obscures the vision and renders all of these, accurate implements and delicate instruments unavailable against a hidden foe. In consequence of this development in ordinance, which has happened in such recent years, a smokeless powder has become essential for use in them if it is wished to realize from them their highest degree of efficiency. The chemicals which combine to form this powder were described.

In giving details of the processes of the various smokeless powders Prof. Munroe pointed out as a curious fact that the kneading machines which had been invented for the use of the baker in making bread, the mechanical cutters of the pastry makers, and the squirting machines for the making of spaghetti and macaroni had all found extended use in the manufacture of smokeless powders.

Peculiarities of the Gulf Stream.

Remarking upon some of the geographical charts now available to the student of physical science, it is pointed out by a writer that, elsewhere in the world, there is not so majestic a flow of water as the Gulf Stream, a remarkable body having its headquarters in the Gulf of Mexico, from thence flowing northeasterly along the shores of the United States to the banks of Newfoundland; then, rushing across the Atlantic Ocean to the British Isles, it is divided into two currents, one flowing northward to the Atlantic Ocean, the other southward to the Azores, and the velocity of this immense flow being also more rapid than that of the Mississippi at New Orleans or even of the Amazon at 100 miles above its mouth. Phenomenal, too, is the fact that, although its bed and bank are cold water, yet the vast stream is very warm, and so great is the absence of affinity or commingling between these waters, that their line of junction is distinctly visible to the eye. Further, the waters of this wonderful stream do not, in any part of their course, touch the bottom of the sea they are defended at the bottom and sides by what has been termed a trough of cold water, one of the best non-conductors; consequently, very little heat is lost, and the warm water is carried thousands of miles, losing only 4° of heat on the journey from the Gulf of Mexico to the British Isles.

Air Brakes in Street Railway Service.

The increasing speeds of cable and electric cars in our densely crowded city streets, and particularly the speeds which are getting to be common on the highways between cities and towns, in the competition with parallel steam railway lines, are making absolutely necessary the adoption of methods of braking which shall be, in power and quickness of application, at least proportionately as efficient as those which have been developed in steam railroading. From the universal use of air brakes in railroad service the conclusion is natural that compressed air has been found in practice to be peculiarly fitted for this special purpose, and it may readily be supposed that inventors have been busily at work in trying to solve the problems connected with the use of air brakes on street railway cars.

The cut presented herewith illustrates one of the new double truck coaches of the Washington, Alex-

& Co. to the public, generally. Certainly the manufacturer in the supposed case would have the right to embody the improvement in the mills by him made, and he might to that end invest much money and time in changing his patterns, his machinery, and in printing catalogues to advertise the changed form of his wind mill. Could the complainant, after all this expense had been incurred, by filing an application for a patent and procuring the issuance thereof, compel such manufacturer to abandon the manufacture and sale of mills embodying the Martin combination, and to submit to all the loss resulting therefrom, when the manufacture thereof had been entered upon and the cost necessary therefor, had been incurred at a time, when, by the voluntary act of the complainant the Martin combination was being sold to the public generally without being protected by a patent issued or applied for.

The defendants in this case contended that pub-

Fruit Jar Co. vs. Wright, 94 U. S. 92.

Andrews vs. Hovey, 123 U. S. 267 : 8 Supr. Ct. 101, and 124 U. S. 694—8 Sup. Ct. 676.

Finally the Judge said : In view of these rulings of the Supreme Court, and the express language of sections 4886, 4920 of the revised statutes, it is clear that, if it is made to appear that, before filing an application for a patent, the inventor had abandoned the invention to the public, the patent if issued would be invalid.

This doctrine of the Court if sound, which we very much are inclined to believe it is not, will defeat any patent secured on an application filed after the inventor has sold his invention to the public generally, even if only a short time elapses between the sale of the invention to the public generally and the date of the filing of the application ; and therefore inventors will have to file their applications for patents as soon as possible, if they want to manufacture and sell to the public generally, in other words before, they sell their inventions to the public generally.

DOUBLE TRUCK CAR EQUIPPED WITH AIR BRAKE—WASHINGTON, ALEXANDRIA & MT. VERNON RAILWAY.

andria and Mt. Vernon Railway which has recently extended its lines into the very centre of Washington. There are eighteen of these cars equipped with the brakes made by the Standard Air Brake Company, of New York. This road will operate its cars at very high speeds, up to forty-five miles per hour, on account of which, and of the motor and truck arrangement, the Standard Company is using its geared compressor type, an illustration of which was shown in the January issue of the Street Railway Journal. These compressors are not mounted on the motor axels, but on the small wheel axles of the Brill Maximum Traction Trucks. A high grade of efficiency is said to have been developed by these compressors in recent trials.

The Patent Law as Construed by Recent Decisions of the Courts.

All along until a recent period, it has been accepted generally, as law, that an inventor might make and use his invention and sell machines in accordance therewith for a period less than two years; and that such sale might be to the public generally, and not abandon his right to a patent. This view is all set aside by the recent '' knock out drop '' decision of district Judge Shiras, in the Equity case of Mast, Foos & Co. vs. Dempster Mill Manufacturing Co., rendered in Circuit Court, District Nebraska, Jan. 14, 1896, Fed. Rep. 71, p. 701-706.

In this decision the Judge said:

"It certainly must be true up to May 2, 1890, when the application was filed, every person purchasing a wind mill embodying the combination, whether from Mast, Foos & Co. or anyone else, had the right to assume that he was dealing with machines not covered by a patent. Suppose a manufacturer of wind mills had purchased one of these mills, or had seen it in operation in the hands of some one who had bought it of complainants before the application for a patent was filed, and being impressed with its advantages he had concluded he would adopt it in his establishment. An examination of the records would then have shown that the combination was not patented, and that no application for a patent had been filed, and inquiry as to the facts would have shown that these wind mills were being manufactured and sold by Mast,

lic use was permitted for a period of two years before the application was filed, by the provisions of the statutes now in force, and therefore the decision of the Court was not well founded.

The Court said in response to this, that the patent, upon its face recites a condition of facts that shows that the Patentee had abandoned the invention to the public before he made his application for his patent, and therefore that it is invalid.

The Judge it seems, made this statement simply in view of the fact that the patentee had set forth in his application record, or specification of his patent, that a considerable number of the mills embodying the improvement had been put upon the market for sale prior to the filing of the application, and that the facts recited in this record, as evidence of the value and usefulness of the combination, had been ascertained from commercial experience with it. It did not seem to be a matter of consideration by the judge whether this sale and public use, generally, had been for one, two, or three months, or two years or more.

The Judge repelled the idea that this use and sale to the public, generally, might be regarded as experimental, from the language found in the record of the patentee. In support of his decision he referred to section 4886 of the revised statutes which is as follows :

"Any person who has invented or discovered any new or useful art, machine, manufacture, or composition of matter * * * not known or used by others in this country and not patented or described in any printed publication in this or any foreign country before his invention or discovery thereof, and not in public use or on sale for more than two years prior to his application, unless the same is proved to have been abandoned, may on payment of the fees required by law, and other due proceedings had obtain a patent therefor."

The Court in discussing this question said, the defense of abandonment is separate and distinct from that of having been in public use or on sale for a period exceeding two years. The Court further said, that in section 4920 it is further enacted, that the defendant, under the general issue, upon giving thirty days notice, may prove, "Fifth, that it has been in public use and sale in this country for more than two years before his application for a patent, or had been abandoned to the public"

The Court discussing the question of abandonment, before two years public use or sale to the public, generally, had occurred, cited the following Court cases :

Elisabeth City, vs. Pavement Co., 97 U. S. 124—134.

In another case prior to this decision of Judge Shiras, it was decided by the Court of the District of Columbia, that an inventor and patentee is free from the penalty of abandonment, even though he has sold his invention to the public for trial and to be retained if found practical—the contract in that case being that the machine might be returned and not paid for if not found satisfactory. The Judge said in that case, the invention had not been abandoned notwithstanding the fact that the invention was proved to have been completed two years prior to the third year's experimental test by the public, and a conditional sale.

In concluding this article we will refer to a recent novel and interesting decision of the United States Circuit Court of Appeals, namely, 4th Circuit—Boyden Power Brake Co. et al vs. Air Brake Co. et al, 73 O. G. p 1857. This decision is to the effect that a patentee who has a claim for simply an improvement on a prior patented invention is free from the charge of infringement in view of the fact of the grant of the second patent; the Court holding, in effect, that the grant of the second patent established the fact that the invention is not an infringement.

The fourth clause of the syllabus of this case reads as follows :

" The fact that a patent was granted for a specific means for operating quick action air brakes after the grant of a pioneer patent describing and illustrating a different means is, in effect, a ruling by the Patent Office that the subsequent patent does not infringe the pioneer patent, and such a decision, while not conclusive, is entitled to great respect."

As an offset to this decision we will quote from the Supreme Court decision in the case of Smith vs. Nichols, 21 Wal. 112. " A new idea may be ingrafted upon an old invention, be distinct from the conception which preceded it and be an improvement. In such case it is patentable. The prior patentee can not use it without the consent of the improver, and the latter cannot use the original invention without the consent of the former."

ROBERT W. FENWICK.

A California inventor claims to have discovered a means of preserving milk and cream for an indefinite period in all its original purity. The milk is bottled in its natural condition and retains the same taste, appearance and properties as fresh milk and cream. The process in inexpensive.

American mechanical ingenuity has become so acute it is now claimed we can produce needles that can be sold for 50 cents a thousand against $1.20 for English and 75 cents for German make.

FUTURE OF ELECTRICITY.

An Interesting Interview.

Explanation of the Character of X-Rays.

The Roentgen rays and their bearing on electrical phenomena form a subject on which Elias E. Ries. the well-known electrical engineer and inventor, has been experimenting for some time at his laboratory , 1919 Druid Hill Avenue, Baltimore.

Mr. Ries has been prominently engaged for years in solving many of the leading questions of the day in scientific matters, and has succeeded in a score of instances in the invention or suggestion of some new idea of great value to the electrical world. After referring, in a recent interview with the representative of a leading Baltimore newspaper, to the importance of Professor Roentgen's discovery, and terming it the natural outcome of the work along the same lines by Crookes, Hertz, Lenard, Lodge and other scientists, Mr Ries says that the most singular feature of Roentgen's discovery is that the X-rays are not the cathode rays, but are indirectly produced by them and that they have resisted all efforts to deflect, reflect or refract them as in the case of ordinary light waves.

Mr. Ries thinks that the X-rays may be due to a longitudinal vibration of the luminiferous ether as distinguished from transwerse or undulatory waves, although it had been claimed by some experimentors that they are due to the projection of particles of matter at high velocities.

Since the publication of Professor Roentgen's discovery, says Mr. Ries, it has been ascertained that X-rays are present in a greater or less degree in other sources of light. And also that rays having characteristics similar to these may be produced by magnetism directly. and also by electrostatic discharges or electrical excitations of a different nature. These discoveries are likely to be of far-reaching value from a physical as well as a practical point of view, and cannot fail to cause us to materially alter our views of important natural forces. They will also greatly stimulate further scientific research and invention.

"Electricity is no longer in its infancy. While it is true that in the application of electricity to manufacturing and other arts, to intercommunication and transportation, and to business, social, domestic and other purposes, much yet remains to be done, the principles of the science and the methods of its application to these various purposes are now so well understood that this phrase can no longer be applied to it. The age of electricity is today so far in excess of its progenitor, the age of steam, as to have almost superseded it, notwithstanding the steam engine is still leaned upon to a certain extent by its successor.

"Among the important possibilities of the near future in the line of electricity are the generation of current in large quantities directly from coal without the wasteful intervention of the steam boiler and engine, and the electrical production of light without heat. The realization of the former would permit of the production of electricity at a mere fraction of its present cost and would make it so cheap and abundant as to render it universal for all purposes to which it can be applied, and especially for electric heating and power. The successful realization of the latter would enable us to save nearly 95 per cent of the electrical energy now used in incandescent lighting, and give us an absolutely perfect light without heat or combustion, in comparison with which that produced from acytelene gas or any other known illuminant is exceedingly wasteful and crude.

"The possibility of communicating with other planets depends more upon the condition of life upon the planets than upon the availability of electricity for the purpose. If it could be shown that the planets or any given one of them were habitated by beings possessing perceptive and reasoning faculties simi-lar to our own, and in the same or greater state of development, the question of communicating, or, at least, of signaling by electrical means would be solved quickly enough.

"Aerial navigation is, in my opinion, more a problem of aeronautics than electricity. The solution of mechanical flight lies largely, if not entirely, in the perfection of the aeroplane. As a source of power the storage battery is, at present entirely too cumbersome and heavy for the power developed. It is probable that, for aerial navigation the most available power will be found in the use of a compact form of gas engine using a hydrocarbon vapor or a gas generated directly from coal or calcic carbide. Aluminum will doubtless be largely used in the construction of the engine and in the framework of the aeroplane because of its lightness and strength. Considerable progress in this science has already been made, but its future is as yet uncertain, so far as its employment for general transportation is concerned.

"There is a possibility, however, that some future electrical discovery may be made, by which the attraction of the earth's gravitation may be overcome or opposed by artificial means, as by imparting to an object suspended in the air, directly or inductively, a repulsive charge. Should such a discovery ever be made, it will not only solve the problem of aerial navigation, but will be so far-reaching in its importance in other directions as to dazzle the mind beyond contemplation.

"There is no reason whatever why electricity will not sooner or later permit us to see with the eye at long distances over wires, and also to photograph in New York an object held to the camera in Washington. The electrical conditions for the fulfillment of this object are as simple as those required for transmission of speech by telephone. All that is necessary is to substitute for the transmitting diaphragm of the telephone which varies the electrical resistance of the wire connecting it with the distant receiver, according to the sound vibrations that fall upon it, a transmitting lens or disc of selenium or other substance which will vary the resistance of the transmitting surface in accordance with the degree of light or shade that falls upon the transmitting disc.

"The electrical vibration thus set up in the wire can then be easily reconverted by a somewhat similar device at the receiving station into light and shade effects, which will form an exact image of the transmitting station, which, if desired, can be readily photographed or enlarged, and thrown upon a screen. In this way, it will become possible to see the person with whom one is talking over the telephone, or to see the actors upon a distant stage, while one is listening to the opera through the telephone at his home.

"At present this process involves the use of a large number of trasmitting wires and is, therefore, impracticable for commercial purposes, but I have this invented a method by which result can now be satisfactorily produced in a single wire. I have found that Roentgen rays are also capable of varying the electrical resistance of a conducting plate upon which they fall, after passing though an interposed object. It thus becomes possible for a consulting surgeon at one end of a wire to locate a bullet in the body of another person at the other end, or for a distant physician in one city to diagnose or prescribe for a patient in another.

"In the same manner, by utilizing the principle of variation of electrical resistance in various substances that are permeable to these raps, it becomes possible to construct a simple apparatus that will permit one to see directly in broad day-light, by instantaneous action, the image or shadow produced by the invisible Roentgen rays, without the necessity of photographing them. This last effect has been accomplished recently, although imperfectly, by the "fluroscope" of Edison which is based upon the principle of the fluorescent—screen effect first observed by Roentgen and subsequently utilized by Salvioni, all of which show rather vague shadows that can be viewed only by the exclusion of all ordinary light.

"It is extremely doubtful that electricity will ever penetrate the bowels of the earth as the Roentgen rays is now penetrating solid substances, if by penetration is meant to render visible. Electricity has been successfully used in locating iron and other metallic ore located under the surface where the depth is not too great, and it will undoubtedly be used for the detection of electrical and magnetic earth currents and probably for giving warning of impending seismic disturbances.

"Electricity will never, in my opinion, enable us to travel faster on sea than we do on land, unless our present method of shipbuilding is materially changed. The resistance to a vessel moving through water increases enormously with increased speed, and the amount of fuel required by a few extra knots per hour is out of all proportion to the gain in speed. Should it become possible to covert the energy stored up in coal directly into electricity a new era will have opened up for the trans-Atlantic liners of the future. A few of the vessels will then be built of comparatively small carrying capacity, exclusively for high speed passenger and express service, the smaller surface below the water line and the greater concentration of power making a speed of about twice the maximum speed probable.

"In the matter of railroad travel I may safely say that electricity will materially reduce the best records now made with steam. The limit of speed depends largely upon the condition of the track. On a substantially straight and level track with electrically-welded rails, such as I was the first to advocate nearly ten years ago, which are now coming into use, an express speed of 120 to 150 miles an hour could be easily attained.

"At speeds higher than this the air resistance to the movement of the train becomes objectionable. There is no difficulty in obtaining that speed from the electric motors and in maintaining it. Electric brakes would guard against any danger and the train could be almost perfectly controlled.

"The supply current for the electric railroad of the near future will be what is known as the altenating current, which can be economically transmitted over long distances through very small wires. It could then be converted into a current suitable for the operation of the cars which may be of the alternating or continuous type. By having several generating stations along the line and transmitting the current at high pressure a long line of railway may be successfully operated in competition with steam, or a steam road may be converted into an electric road at comparatively small expense. This may be done without interfering with its use as a steam road for freight, provided that the conditions are such to arrange a proper schedule. There is no limit to the load that electricity can move. "The overhead trolley wire in cities will sooner or later be entirely abolished, although in the outlying surburban districts it will still be used almost exclusively. Its place will be taken within the city limiis by underground conduits, or by storage batteries on the car charged with current while the car is traveling over the suburban section of the line in contact with the trolley wire, and supplying the motors directly while on the city section. I have developed street railway systems along both these lines and have no doubt but that they will soon come into extensive use."

Mr. Ries predicts that in the near future nearly all the common occupations of life will be accomplished through the agency of electricity. Its development will be most rapid, he says, as discoveries of such great importance are being made day after day.

Mr. Ries further stated that our fields will be plowed and our crops harvested by the mysterious agency, and the products will be transported over long distances by the same subtle fluid, and that at a high rate of speed. Ferry, river and seagoing vessels will someday be propelled by the same universal force. Electric launches and pleasure boats are now to be found in abundance, and electric carriages will not be far behind.

The day is not far distant, sanguinely observes Mr. Ries, when stoves, fires and ashes will be relics of barbarism and friction matches fit objects for preservation in our museums as curios of 19th century civilization. The telephone is destined to become a fixture in every household, and all marketing, shopping and errands may be done by its agency.

Electric current for power, heating, cooking and lighting purposes will be on tap in every house, just as water is to-day. Electric elevators anu lifts will convert stair climbing into a recreation and electric ventilation, refrigeration and temperature regulation will keep the up-to-date home in an ideal condition and permit the owner to manufacture his own climate, no matter what the exterior weather conditions may be.

Electricity will play music, read books and write letters for the man of the future. It will prepare the food and bring it to the table. It will even impart nourishment and strength without the necessity of eating and drinking.

Dr. Ries is also certain that telegraphic and telephone communication, will shortly be made between points without the neccessity of wires between the places, and that the currents in the earth may have a bearing on the solution of this question.

The death by suicide of Lieut. Swift, of the Ninth Cavalry in January, says Electrical Engineer, closed one of the most remarkable careers in the United States army. From the position of telegraph operator in a small town of Virginia he became a private, corporal, sergeant and lieutenant in the army in as many days as there were promotions. When Gen. Myers organized the Signal Corps of the army, Swift had considerable local reputation as a telegrapher, and to secure his services as an instructor he was enlisted in the corps and promoted from day to day until he attained the rank of Second Lieutenant. After joining the corps and beginning his work as instructor he devoted himself to the study of electricity and telegraphy, becoming one of the most skillful operators that the world has ever known and an author on the subject. There were a lot of fancy tricks in sending and receiving messages that Swift used to do. One of them was to take two messages at the same time. He could write equally well, and beautifully, with either hand, and could take a message from one instrument, writing it out with his left hand, while he copied another message with his right. He could also send a message with one hand while he received another message, copying it with the other hand.

TIPS TO INVENTORS.

Interesting Communications From Inventors and Manufacturers.

[Under this heading will appear communications on needed changes in the Patent Laws and discussions of subjects of interest to Inventors and Manufacturers—particularly in respect to those subjects enumerated herewith, and which formed the basis for the most interesting discussions had at the January meeting of the American Association of Inventors and Manufacturers (appearing in the February and subsequent issues of the INVENTIVE AGE). All readers who desire to express their ideas from to time can have access to this column. New subscribers may, if they so indicate, receive the back issues of the INVENTIVE AGE, containing all these discussions. —ED.]

1. The propriety of a Federal Statute of Limitations in Patent Causes.
2. The importance of having American Patents granted for a definite fixed term without regard to the date of expiration of corresponding foreign patents.
3. Should not a patent become void by failure to manufacture and use the invention in such manner that it may be brought to the attention of the public within a fixed period from the date of the grant?
4. Would it not be beneficial to our Patent System to so amend the Patent laws that innocent users of devices and processes covered by old and unworked patents, of doubtful validity, may not be compelled to pay tribute to purchasers of such patents?
5. Should not the Commissioner of Patents be clothed with authority to appoint, on application, Special Examiners, for taking testimony in Interference Proceedings the same as Special Examiners are appointed by the Courts in Equity Cases?
6. Are the New Patent Office Rules abridging the time for response to official action desirable and sustainable under the statute?
7. Should or should not the United States Patent Office be made a separate and distinct department and be provided with a building of its own?
8. In view of the necessity, already arisen, of storing a part of the models elsewhere than in the Patent Office should not models be placed, all together, in a suitable building specially designed for that purpose?
9. How can the rules of procedure in the Patent Office be amended or changed to facilitate and improve the character of the work which it is intended to accomplish?

SUBJECTS OF DISCUSSION.

INVENTOR'S RIGHT SHOULD NOT BE ABRIDGED.

EDITOR INVENTIVE AGE:

I have looked over the communications of the parties who have expressed their views on various ways of amending the Patent Laws. On page 40 current issue of the INVENTIVE AGE, under the heading of "Talks on the Patent Laws," I would like to make a few comments. First in refference to Mr. Fraser's letter regarding a patent that has been in existence five years and not been worked. If some one wants to work the same I simply ask why not go to the patentee instead of the courts for a license? I will venture the assertion that in nine cases out of ten the inventor will be only too willing to have his patent worked on a reasonable royalty. I have quite a number of idle patents myself and know it would be so in my own case. Hon. C. E. Billings, of Hartford, Conn., says he believes in limiting the time on patents granted for inventors to manufacture and bring same to the notice of the public. I wonder if Mr. Billings has ever presented a newly patented invention to manufacturers. I hardly think he has; that is, being unable to introduce the thing himself, has he tried to interest capital in his device? I think if he has passed this experience he would hardly want to limit the time to get his invention before the public to one hour less than the whole life of the patent. It may be that Mr. Billing's idea is to stop the work of the professional inventor. If so he could not hit upon a better plan than the time limiting idea. I plead guilty to being one of them and so far as I am personally concerned I know that a limit of the time to get my patent before the public or loose my rights in the invention would immediately put a stop to further study on inventions—that is a limit to anything less than the life of the patent. Now as I have said I have quite a number of idle patents and I ask in what way has the public been injured by these dormant patents. Capital will not take hold of them and I cannot introduce them myself and the public as yet know nothing of them as they have not been introduced. Now suppose that these patents expire without having been put in use, has the public been defrauded thereby? I hardly think so. Now suppose that some one wants to manufacture some article covered by one or more of these patents; is it any injustice that he should pay me a royalty on the manufacture? If a person should think of something in the way of an article to manufacture and was thinking of manufacturing it without seeking patent protection, and then should find the subject matter of his ideas covered by some dormant patent there would be some ground for saying that the public had been the looser. But if the matter is sifted it will be found that he (the would be manufacturer) knows all about the patent in fact in nine cases out of ten he would never have dreamed of such a thing only for the patent showing and describing it. The patent being well along in its term of life instead of going to the owner of the patent and paying a royalty on the little that is left of the life of the patent he boldly pounces into the patented device and if he happens to get snapped up, he immediately begins to bellow and cry "injured innocence" and talks of old unused patents being a "bar to progress." When the God's truth in the matter is that the old patent gave him all the knowledge he ever had in the premises. It is not ignorance of these dormant patents that makes the trouble; it is the greed of parties that want to use them without paying the owner anything for the privilege of so doing, and if they are balked in so doing they try to make it appear that if it had not been for these patents that show and describe certain things that this same matter shown and described in the patents would be laying around loose for anyone to pick up at pleasure. In other words that the inventor did not produce anything but simply went to the Patent Office and for a fee got control of something that the public knew all about to the great detriment of all concerned. I think it is safe to say that all the live patents in any branch of industry is well known to all manufacturers in that line or at least to a vast majority of them and that where there is one innocent manufacturer who has been injured by unwittingly using some unexpired patent and have been made to suffer for infringement, there have been twenty who have knowingly stolen the patented property of some poor inventor and got off scott free, and who, if they had been caught up by some sharp patent lawyer, would have at once set up the old howl about "unused patents being a great bar to progress," etc.

W. D. SMITH.

Prophetstown, Ill., March 13, 1896.

THE BURROWS BILL.

EDITOR INVENTIVE AGE:

It appears to me that the bill introduced by Senator Burrows is merely intended to formulate into a law, what has been the practice in the Patent Office for very many years. Albeit the practice is ignored and neglected very often by examiners who are careless of their duties or undesirous of admitting a mistake. It was long ago held that "piece meal" rejections were improper and that no combination could be anticipated except by an equivalent combination. That anticipation of its elements separately is no anticipation of the combination. Probably Mr. Fenwick remembers when the Commissioner made that ruling. It has been repeatedly reiterated, and is a rule founded in law and common sense. Yet it is violated every day by examiners who somehow think it right to reject by hook or by crook, what on general principles they disapprove of. When cornered they generally resort to justifications by saying so and so in view of so and so anticipates the claim. This is no more than an admission that neither "so and so" alone, would anticipate. Examiners who as that ought to be disciplined but unfortunately neither applicants or attorneys see profit enough in such a fight. Senator Burrows' bill seems to simply declare that rejections for compound reasons so to speak, shall not be lawful, and for one with that understanding I hope it may become a law.

As to your questions I should like to say:—

1. It is well to quiet contentions and therefore there ought to be a statutory limitation.

2. There is not, and never was any sense common or otherwise in the provision of the law which refers the term of an American patent to a foreign patent. If patents are good things it is because they induce inventors to produce their inventions in the United States, and the more of them we get here the greater will be the number of valuble ones. Hence the inducements ought to be multiplied rather than diminished.

3. No. To make such a law would be simply to discourage patents and to encourage secret use. It does not need any argument to prove that if all inventions were worked in secret or held unworked in secrecy, the public would not get much chance at them in 17 or any other number of years.

4. No. The invention is the absolute property by right of creation. The people have no right nor interest in it, nor care for it. The govenment proposes a contract, while its inventor is free to accept or reject. The conditions are, on one side continued and protected ownership for 17 years, on the other side complete disclosure. The people have no rights in it until the expiration of 17 years then they assume complete ownership. If an attempt should be made to force the working of patents it would be almost impossible of execution because it would be so easy to prove working in some degree.

5. It would not lessen the expense of taking testimony but probably greatly increase it. What benefits could be expected I do not see. The testimony to go before the U. S. Courts in patent cases is almost invariably taken before some local notary acceptable to both parties and that appears to be the simplest and most convenient mode.

6. No. It is both surprising and lamentable that a commissioner should so far misjudge his duty and his authority. The thing itself may not be undesirable, yet the harm which it does is infinitesimal in comparison to the convenience to applicants who are laboring with poverty and misfortune, and who would be cut off by a rule requiring quick responses. Practically the two years term is a great boon to many and a harm to very few.

7. Emphatically, yes. But it has a building of its own from which interlopers ought to be summarily removed. There will be but little hope for great improvements in the Patent Office while it remains a subordinate bureau in a department entirely out of sympathy with it. Merely a cow to be milked.

8. Yes. And the rule in force up to 1877 requiring a model with every application ought to be restored. Commissioner Payne's order was the most gigantic error ever committed by a commissioner, and that places it pretty high.

9. By making them more in accordance with the common sense duties of a *ministeral office*. If the embrgo lawyers who some 20 years ago began to graft quasi and other bogus judicial functions upon the Patent Office practice, had died a Cornin, it would have been better for the country and better for inventors, though probably worse for the horde of small lawyers who have scraped a living out of the intricacies of a set of rules which have for years been trying to justify false steps by other false steps in a continually advancing progression. The law is simplicity. The duties of the patent office is to execute the law in simplicity, but its practice is fearful and wonderful in its false premises and false logic.

R. D. O. SMITH.

Opposition to Inventions.

[WM C. DODGE in Engineering Magazine.]

One of the most remarkable things in the history of mankind is the opposition to the introduction of inventions and improvements which has existed from the earliest times, and still exists to some extent.

From the time when the earth was believed to be flat, and Galileo was denounced and imprisoned for asserting, in accordance with the theory of Copernicus, that the sun was the center of the planetary system, and that the earth had a diurnal motion of rotation, this opposition to new ideas has existed, and been manifested in the grossest outrages upon, and persecutions of, the originators and advocates of the new ideas. This has been true of inventions and improvements in the arts and sciences, as well as in governmental and religious improvements or reforms.

Much of this was due to the existence of the prevailing doctrine of the "divine right" of rulers and the arbitrary power exercised by them, and to the claim of superior wisdom and infallibility made by the then dominant church, supplemented and rendered possible by the ignorance and helplessness of the masses.

Besides, human nature seems to be subject to the same laws as moving bodies; it moves always in a direct line when once set in motion, unless interfered with by a power sufficient to deflect it from its course. In the arts and sciences, as in politics and religion, men prefer to remain undisturbed, and naturally resent any interference with their settled beliefs and habits. They look with suspicion on new suggestions or ideas, and especially such as, in their ignorance, they think will interfere in any manner with their present interests; and hence the tendency to continue in the old ruts, and violently oppose improvements or changes, and to denounce inventors as "cranks."

History shows that the great improvements in the arts and sciences have had their development only since free governments have been established, and general education introduced; and it is where these exist in the greatest perfection that the greatest advance has taken place.

In the United States, where there is the greatest freedom in governmental and religious matters, there has been the greatest advance in inventions. The growth and development of our manufacturing and agricultural interests, which is due to inventions fostered by our patent system more than to any other cause, have been marvelous, and excite the astonishment of the world. Under the benign influence of this system, in a single century, we have grown from a cluster of scattered settlements, mostly along the Atlantic seaboard, with a population of less than 4,000,000, to a powerful and compact nation of nearly

70,000,000; have increased our national territory from 830,000 to 3,314,220 square miles; have subdued the forests and built up the whole country from ocean to ocean; and have built more miles of railroad, and established more post offices, than all other nations combined. We have grown and prospered as no other nation has, until to-day we do one-third of the world's manufacturing, one-third of its mining, one-fifth of its farming, and possess one-fifth of its wealth.

With such an illustration of the benefits of our patent system, one would suppose that opposition to inventions would long since have ceased; but, unfortunately, while it has greatly diminished with the growth of intelligence and universal education, it still exists.

As illustrative of this spirit of opposition, it may be interesting to cite a few instances. When, in 1807, Papin, of France, the inventor of the digester in universal use for paper making and many other purposes, and also of the lever safety valve, made a small steamboat and ran it down the river Fulda, the ignorant boatmen, who, like some of the laboring men of the present day, thought it would injure their business, seized and destroyed it.

So, too, when Jonathan Hulls patented his steamboat in England, in 1736, he was laughed at and ridiculed in every conceivable way.

When Jacquard invented his loom, which was so wonderful that the great Arnout, French minister of war, caused him to be brought into his presence and said to him: "Are you the man who can do what the Almighty cannot—tie a knot in a stretched string?" there was the strongest oppoition to its introduction, culminating in a mob of the silk weavers, who took it from his house into the streets, broke it up, and burned the fragments.

It was the same with Hargreaves in England, when he invented his spinning jenny in 1763. He was persecuted by his fellow workmen, who seized his machine, broke it in pieces, and drove him from his native town.

That invention, with the improvements of Arkwright and Crompton, and the invention of the cotton gin by Whitney, who was outrageously defrauded of his rights, have changed the entire art of producing woven fabrics. Indeed, so far as the cotton industry of the world is concerned, they may be said to have created the industry, which to-day gives employment to millions, and has so immensely cheapened the product that it is used the world over.

This opposition to and unbelief in the possibility of the success of inventions has not been confined to the ignorant alone, but has been shared by many educated and even great men. When it was proposed to build a railroad in the United States, Chancellor Livingston, one of the greatest men in the State of New York published a letter in which, as he thought, he demonstrated the utter impossibility of the proposed undertaking. His reasons were, first, that it would require a massive substructure of masonry the whole length of the road, and that would be so expensive that it would not pay; second, the momentum of such a moving body as a train of cars would be so great that the train could not be stopped until it got several miles past the place; and, third, no one would want to risk his life flying through the air at the rate of 12 or 15 miles an hour.

So, too, Daniel Webster expressed grave doubts as to the possibility of railroads, saying, among other things, the frost on the rails would prevent the train from moving, or from being stopped, if it did move.

When Murdoch invented or discovered a means for producing illuminating gas, no less a man than Sir Humphry Davy ridiculed the idea of using it for lighting purposes, and said if it was to be used for street lighting, they would have to use the dome of St. Paul's for a gasometer. Sir Walter Scott made clever jokes about "sending light through street pipes," and "lighting London by smoke," but subsequently had his house lighted by it. Wollaston, a scientific man, said "they might as well attempt to light London with a slice from the moon." It is but a few years since the scientists of Eurore demonstrated mathematically that the electric current could not be divided for incandescent lighting, but today the contrary is demonstrated by millions of incandescent lights, illuminating every spot where civilized man resides.

But the strangest of all things in this connection is the fact that, even in this enlightened age, there are men who still insist that inventions are injurious. It is not many years since that, in a paper published at the national capital, there was the statement that the steam engine and the sewing machine were two of the greatest curses that ever befell mankind!

It is, moreover, a matter of history that in certain sections of this enlightened land prayers were fervently offered in churches beseeching that the wickedness of the newly invented sewing machine, which, it was supposed, would rob the sewing women of their means of obtaining a living, might become apparent, and its promoters be stricken by a conviction of their wrong-doing in making it, and thus be told by heaven to desist from its manufacture.

This spirit of opposition exists to-day to a greater or less extent among the labor unions, whose members, without investigating the subject, are made to believe that labor-saving machinery deprives them of employment, or at least will lessen their wages, just as the silk weavers of Lyons thought in regard to Jacquard's loom, and as the spinners of Lancashire thought in reference to Hargreaves' spinning-jenny.

It is no doubt true that, when a new invention is introduced which revolutionizes some particular art or branch of business, it at first decreases the number of persons employed in that particular line; but that is only temporary, for in a short time the result is a cheapening of the product, a greatly increased demand for it, because of this cheapening, and then necessarily an increased demand for laborers in that line, and almost universally at increased wages.

The statistics of the country show this to be true beyond the possibility of question. The records of the Labor Bureau show that from 1860 to 1880, the most prolific period in this country of inventions, and the most intensified in all directions of their introduction, the population increased 59·51 per cent, while in the same period the number of persons employed in all occupations—manufacturing, agriculture, domestic service and everything—increased 109·87 per cent; and in the decade from 1870 to 1880 the population increased 30·08 per cent, while the number of persons employed increased 39 per cent.

As shown by the investigation of a committee of the Senate, wages have increased 61 per cent in the United States since 1860. And, as all know, during that same period the cost to the people of nearly all manufactured articles has been decreased in as great if not a greater ratio. As with manufacturers, so with farming. As a recent writer has well said, "The use of patented machinery has so changed agriculture that there is more propriety in saying that we manufacture crops than in saying that we grow them." And still another writer says: "We use implements that cheapen the cost of production, and make the labor of harvesting like the sport of the fairy books."

While most people have the idea that inventions have mainly benefited the manufacturing industries, it is susceptible of demonstration that they have benefited our agricultural industries nearly as much.

In speaking of the condition of the United States, a recent English observer says:

"America has for many years enjoyed an amazing degree of prosperity, so much so indeed, that, to use the eloquent words of Edmund Burke, 'generalities, which in all other cases are apt to heighten and raise the subject, have here a tendency to sink it. Fiction lags after truth, invention is unfruitful, and imagination cold and barren.' The United States has 65,000,000 people, who spend more on dress than any other people on the face of the earth," and, who, he might have added, enjoy more of the comforts of life in all directions than any other people on earth.

Invisible forms of Energy.

Strange as it may seem, there are forms of energy about us which until but a few years ago were completely unknown. The ultra, or dark, rays of the spectrum, the pendulous wave of etheric distubance proceeding from a Leyden jar, and the manifold vibrations set up by the rapid changes in an electro-magnet are now receiving the closest attention. The marvellous nature of these phenomena has lifted a veil from our eyes and have lent to our hands new power. The existence of a medium once postulated, now proven, has been the means of enabling to solve, with a definiteness that defies contradiction, theories of light, heat, magnetism and electricity.

The more the forms of energy are examined the more they seem to identify themselves with each other, and all because of the wonderful fact that the medium, the tenuous, yielding and infinitely elastic ether, is necessary for their appearance and transmission. What are the glories of a sunrise, with its shafts of crimson and gold, but the toned undulations of this universal ether? Each substance in this world, when permeated with it at a given rate of speed, will be endowed with glass-like properties. The visible light, as we call it, is in reality but a narrow group of ether waves, and glass responds to their touch, carrying the rays of light from other bodies to our eyes because of its simple transparency. It would seem, then, that the changes in ether, which escape our notice because of the lacking delicacy to the eye to bring them to our consciousness, are imaginably seen by eyes of a greater range of power. The word "transparent" can be applied to certain other substances which to our eyes are opaque, yet through which the dark or invisible rays can pass with perfect freedom. It then but necessary to comprehend the proper meaning of the phenomena of transparency to realize that it is but a relative expression.

That rays of light may exist and play around the eaves and cornices of a house,—rays which we have never seen and, in fact, can never see,—is but a matter of simple proof. To these waves glass is opaque, while the stone wall may be of transluce clearness. Curious as it may seem, a lens of hard black rubber may be used for the collection of certain of these stray beams, and all of the processes through which sunlight may be put will be possible by the use of this opaque lens. In the very midst of Stygian darkness, rays of light are streaming forth. Heat waves to an innumerable extent undulate through the blackness of night, and in the deepest recesses of mines the dark light, the minute vibrations, pass unnoticed before an incompetent retina. It might have been that our eyes, being sensitive only to the magenetic condition of the ether, would have left us totally unconscious of the golden sunlight and blind to all the varied tints of flowers.

Though our eyes fail and and the visual sense is useless for such investigations, the path for the investigator is open. By experiment it may be discovered which of the numerous light waves will penetrate the interior of given subtance and a photograhic plate will record these impressions and record the visible signs of an invisible light.—*Power and Transmission.*

Disinfecting the Blood.

Dr. Cyrus Edson, of New York, in an article written for the forthcoming issue of the Medical Record, announces to the medical profession his discovery of a method of disinfecting the blood by the injection of a solution of carbolic acid and thus making the blood itself a disinfecting agent as it courses through the body. In the study of disinfectants Dr. Edson learned that carbolic acid is at times to be found in the blood and that nature herself increases the amount in disease over one thousand times. He reasoned, therefore, that carbolic acid is nature's remedy, but he was confronted with the fact that the injection of any known solution of carbolic acid generally produced an abscess. The problem was to find the form in which the acid could be injected without producing this effect. This he announces that he has accomplished. The solution is a colorless liquid, smelling of carbolic acid, and it is injected under the skin. According to the results reported by Dr. Edson so far the solution is credited with the cure of about 40 per cent of cases of consumption.

Labor and Time Saving Invention.

S. H. Smith, foreman of the Norfolk & Western at Bristol, Tenn., has patented a device for making fires in locomotives that is attracting much attention. It is a small machine on wheels and is run on the ground. It is stopped near an engine and a small hose attached and run into the fire-box. A shovelful of coal is thrown in on the grates and then a little ball of waste is lighted and laid on the coal. The hose is turned on and at once the fire-box is filled with a soft white blaze. The machine causes a spray of oil and air to go out at once which completely fills the fire-box. The coal is soon ignited and a heap of red-hot coals, the hose is taken out and more coal added and the fire is made. It is stated that one gallon of crude petroleum will make four or five fires by this process at a cost of about one cent for each fire. No kindling wood whatever is needed.

Experiments in Radiography.

Mr. Edison's experiments have proceeded to a point where he claims he can see moving objects through eight inches of wood.

The first surgical operation in the Johns Hopkins Hospital at Baltimore in which the X-rays were utilized was performed on March 12. It was the extraction of a scissors blade from a womans hand, where it had been imbedded for 12 years. The position of the steel was revealed by a radiograph.

Prof. W. F. Magie, of Princeton College, spoke on March 12, before the Princeton Club of New York at its regular March meeting in the Brunswick, on the Roentgen rays and what Princeton was doing toward the development of them. Stereopticon plates were exhibited showing the bony structure of small animals and illustrating the views of the lecturer on the efficiency of different tubes. A series of plates was exhibited also illustrating the use of the method in detecting diseased parts of the bone and the presence of foreign subtances in the body and in watching the progress of recovery from operations.

New Adjustable Calipers.

Arthur Munch, St. Paul, Minn., has devised an adjustable caliper, of which we give an illustration. The calipers are of the familiar form, the heavy oentral rivet being countersunk at both ends, and, by means of notches in the back countersink, being held from turning in one of the legs, the other being movable for setting. Upon the central rivet and outside the movable leg is an "adjusting bar" movable about the rivet and clamped in any position

CALIPERS.

by tightening the large thumb nut covering the upper end of it, which is split. Upon the movable leg of the caliper there is pivoted a stud, in which is shouldered a milled headed adjusting screw, the threaded end of which works in a pivoted stud upon the adjusting bar. The adjusting bar being movable, described, a short screw is sufficient for any adjustment within the range of the calipers either inside or outside. The outside caliper shows the device used as transfer calipers.

Solidified Petroleum.

The new process for solidifying petroleum invented by Paul d'Humy, a French naval officer, is given by a writer who saw numerous experiments made with the new fuel. It is stated that heavy common oil has been converted by the inventor into a solid block, as hard as the hardest coal; it will burn slowly, give off intense heat, and it shows not the slightest sign of melting. A ton of this fuel represents, M. d'Humy said, 30 tons of coal, and the space occupied by one ton of it is about three cubic feet, as against the large space filled with the coal. Having explained the uses to which his new fuel might be put, and some of the advantages it offers, M. d'Humy proceeded to show some samples of it, and to make experiments with them. On the table were a number of cakes of solidified petroleum and low grade oils, of various sizes and shapes, and looking not altogether unlike lumps of ginger bread or tharve cake. In addition to the cakes there were samples of the same fuel in dry powder and paste. The petroleum powder and paste mixed together and pressed, form a homogeneous mass, with a great specific gravity, hard almost as stone, and when burning will give off a flame 300 times its own volume, and a heat well nigh as great as oxygen. M. d'Humy placed in the grate in his room a piece of solidified petroleum about ten ounces in weight, and on its being fired with a match it gave off a powerful, steady flame. A little later he put a small quantity of the powdered petroleum in the grate on the top of cinders, and when fired the flame filled the grate. He put a shovel of ashes on the top and the flame came through them, and all appeared to be in a blaze. Tests were applied to discover if there was either smoke or smell, and neither could be detected. There was also on view a black cake made from heavy bituminous oil, very dense and very heavy, and the flame from that burning under a boiler would, he said, cover all the space. He also showed a sample of disintegrated wood, which, when mixed with solid petroleum, made a very heavy, hard bulk, forming a splendid fire for use on board ship or under an ordinary boiler. The fuel can be pressed to any shape and used for a variety of purposes.

Those World's Fair Medals.

The Chicago Tribune, facetiously explains that the government proposes to prosecute certain firms which have claimed in advertisement that they have received World's Fair medals. Uncle Sam's indignation is so great a special bill has been introduced in the Senate providing an extremely heavy punishment for offenses of this kind in the future. It is to be hoped these cases will be pushed with all the speed and severity the law warrants. A firm that will make the preposterous statement that it has received a World,s Fair medal deserve no sympathy.

Only a few years have elapsed since the World's Fair was held, and brazenly to insinuate that the government has rushed through these medals in such unseemly haste is a gratuitous slap at our established institutions. No other motive is apparent, as the firms could not hope to deceive anybody by such a palpable and grotesque misstatement. Everybody knows about these medals. They are "nearly ready." This fact has been published at intervals for the last two years, and the public will not be deluded by anybody who asserts the medals are anything else but "nearly ready.

Some slight extenuation may be urged for the firms concerned on the ground that they are not entitled to the medals, and therefore could not be expected to be so well acquainted with he details. This should not weigh, however, against the necessity of making a precedent in these cases as a warning for thefuture. If only a light punishment is inflicted now, perhaps in after years other firms may be impelled to say they have received these medals, when in fact they will be only "nearly ready.

Science News.

Leadworking is one of the most disastrous of all trades to health.

To remove warts, Dr. Softshu says, wet them thoroughly with oil of cinnamon three times a day untill they disappear.

The largest room in the world is said to be the hall of the imperial palace in St. Petersburg. It is 160 feet long by 150 feet wide.

"Virtue increases under a weight or burden," and results increase with a comprehensive expenditure of money in good advertising mediums.

The hottest mines in the world are the Comstock. On the lower levels the heat is so great that the men cannot work over ten or fifteen minutes at a time. Every known means of mitigating the heat has been tried in vain. Ice melts before it reaches the bottom of the shafts.

It is not commonly known that the capital of China is ice-bound for five months out of the twelve, or that the stolid looking Chinese could ever be graceful skaters. The Chinese use a very inferior style of skate, of their own manufacture, a mere chunk of wood arranged to tie on the shoe, and shod with a rather broad strip of iron.

Neptune, the outermost member of the solar system yet known, is thirty times farther from the sun than the earth is, or 2,780,000,000 miles, and the tremendous line of his orbit, which incloses our comparatively small group of heavenly bodies, is so long that, although his rate of travel is three miles in a second, it takes him 165 years to complete one circuit.

A flowering plant is said to abstract from the soil 200 times its own weight in water.

Two thousand patents have been taken out in this country on the manufacture of paper alone.

In the case of Belknap vs. Schild et al., decided recently by the Supreme Court of the United States, it appeared that the defendants were the owners of a patent caisson gate used by Belknap in prosecuting government work without permission of or compensation to the owners, and they sued for an injunction and an accounting. The trial court granted the injunction and a master reported the damages at $40,000. The court held that the invention being used by an officer of the United States for the common defense and general welfare, no injunction could lie against him, and that the only damages proved being those in behalf of the United States, for which he could not beheld liable, the judgment of the lower court must be reversed with instructions to dismiss the bill, without prejudice to a suit at law against the officer for damages or against the United States in the Court of Claims.

Sir John Lubbock describes an ant which can support a weight three thousand times heavier than itself, or equal in proportion to a man holding 210 tons by his teeth.

BOOKS AND MAGAZINES.

This is probably the greatest condensation of information ever attempted in the history of book-making. It seems impossible that any considerable fraction of the enormous number of topics treated in the Britannica could be put between the covers of such a tiny volume as the book before us, and yet there are, and it is difficult to name a subject which it does not touch upon. Price one dollar. Laird & Lee, Publishers, Chicago.

* * *

An article by Cleveland Moffett in McClure's Magazine for March described the curious and important scientific uses lately made of kites, especially in the departments of meteorology, electricity and photography. It also described how to make the modern tailless kite, how to fly kites in tandem and the possibilities of the kite as a coming instrument of war. Along with other pictures, there were a number of views photographed from a kite at an elevation, is some instances of 1,500 feet.

* * *

Messrs. Frederick Warne & Co., N. Y., will issue immediately, "Sport in Ashanti; or, Melinda, the Caboceer," a tale of the Gold Coast in the days of King Coffee Kalcalli. By J. A. Skertchly. Also "The Carbuncle Clue," a Mystery. By Fergus Hume, author of "The Mystery of a Hansom Cab." The forthcoming volume in "The Public Men of To-day" series, will be "The Right Hon. Joseph Chamberlain." By S. H. Jeyes, the editor of the series. (Ready immediately). "The Shutte of Fate," a powerful story of the Lancashire Cotton Mills. By Miss C. Whitehead, with illustrations by Lancelot Speed. They have also just ready, the Fifth editon of "Electricity up to Date—For Light, Power and Traction." By John B. Verity, M. Inst. E. E. Lond. Fully Illustrated. They will also issue at once, "By Tangled Path: Stray Leaves from Nature's Byeways," a series of delightful essays on nature, animate and inanimate, for lovers of the woods and waysides. By H. Mead Briggs, arranged so as to cover the months of the year.

* * *

Captain Charles King, U. S. A., is the romancer of the Regular Ary, and of Western military life. A new story by him is always sure of a good reception by the reading public. "Bugler Fred" is a story of frontier militry experience issued in Neely's Prismatic Library, at 75 cents, with many full-page illustrations. The story is not a long one, but all the more readable because of the rapid development of circumstantial evidence against the bugler-boy, who is accused of murder.

* * *

A comprehensive account of the origin and history of the circulating medium is to be found in a volume written by J. K. Upton. The constant demand for this plain and practical treatise on the history of money in the United States has called for a new edition, and the author has revised, extended and brought up to date a book which appeals to all Americans, whatever their financial standing or financial creed. Legislation in the United States in regard to coinage, legal tender and other acts, the battle between gold and silver standards, and all matters that pertain to honest money are fully, fairly and practically treated in this book. The author is an ex-assistant treasurer of the United States, and knowns his subject practically and thoroughly. As a timely contribution to the present "silvler agitation" that is interesting the whole country, this little book will commend itself to all citizens, and should be read and studied by all. Published by the Lothrop Publishing Company, of Boston, Mass.

X Rays by Telephone.

Dr. H. L. Smith, of Boston, who has been experimenting with the Roentgen ray effects, claims to be able to obtain negatives by use of the telephone. His process is described as follows:

"The Doctor merely placed his plate-holder at the telephone transmitter, which has been so arranged as to make the bell-ringing apparatus work continuously, and left it there about half an hour. On taking out the plate it was found, after being developed, to contain distinct outlines of the small objects on the holder."

Foreign manufactures are reported as introducing a new and effective method of bleaching cotton cloth. Two horizontal rollers, one made of iron and the other of carbon, are employed in such a way as to serve as electrodes, being properly connected with a battery or dynamo, and the two lying very close to each other. The cotton cloth, before passing between them, is soaked in salt water, and the electric current, passing from one roller to the other, disintegrates the salt, and the chlorine accumulates in the goods; the other element—freed by electroysis—sodium, combines with the water instantly to produce caustic soda, which is taken up by an endless felt blanket running over the iron roller and out through a water tank. After becoming impregnated with chlorine, the cloth is rolled up and allowed to stand long enough to allow the bleaching agent to complete its work.

A contract for the reconstruction of the famous old Dismal Swamp Canal has been awarded by the Lake Drummond Canal and Water Company to Patricius McManus, of Philadelphia. The work is to begin at once and to be completed by January, 1898. The canal extends from a place on the Elizabeth, four miles from Norfolk, Va., to a place where the canal empties into the Pasquotank River, near South Mills, N. C. A feeder about three miles long is to be constructed to Lake Drummond. The total length of the canal is about twenty-two miles. Right of way 300 feet wide in Virginia and 150 feet wide through North Carolina is owned by the company.

SUPPLEMENT.--Tips to Inventors.

The Infancy of Invention.

As capital is constantly being invested and expended to protect and preserve capital previously expended and invested in various enterprises all over the land, so will inventions continue—their variety and multiplicity being demanded to further the usefulness and perfection of inventions previously originated.

It was Edison who, replying to the question, "Do you think that the inventions of the next fifty years will be equal to those of the last?" said: "I see no reason why they should not. It seems to me that we are at the beginning of inventions." The truth of this prediction is illustrated in the many useful and wonderful achievements of Mr. Edison's own laboratory since giving utterance to this statement only a short time ago.

Profits from Invention.

The value of an invention is determined by no fixed rule. Fabulous sums have been made from simple and novel, as well as complex and useful, inventions. It is a fact that four-fifths of the business of the United States is transacted by the use of inventions. The benefits to mankind because of inventions, are so manifest and so common we are apt to look upon them in a matter-of-fact sort of way and fail to give the inventor the credit due him. In the majority of cases, however, the failure of an inventor to reap a reward is attributable to his own negligence, lack of forethought and indiscretion.

Nearly every human being is an inventor, but only a few obtain a *monopoly*—a patent—on the product of their brain. There are thousands of really useful articles, appliances and discoveries, in use every day by millions in all walks of life, that might have been patented had the inventor possessed the business sagacity that has actuated his more fortunate neighbor. Take for instance the open slot necessarily used in all conduit electric, or cable street railway systems. The inventor failed to get a patent on the idea and a fortune missed him.

There is money in inventions, but not always for the inventor.

The only way to make money out of an invention is through the protection afforded by a patent; not a patent in name only, but a *good patent*—one that is intelligently drawn, with claims commensurate with the scope and importance of the invention.

The profits arising from inventions in the electric field during the past twelve years have been simply astounding. In railway appliances, bicycles, typewriters, telephones, cash registers, slot machines and farm machinery, the field has been equally remunerative. And just think of that simple toy "Pigs in the Clover"—it netted the inventor, whose friends laughed at him for obtaining a patent on so simple a toy, over $150,000. The inventor of the metal plates to be attached to the worn heels of shoes (for sale in all cities) realized a fortune out of it amounting, it is said, to nearly $1,000,000. Perforated wooden seats for chairs and rubber tips for lead pencils brought the inventors big results. Howe made a million dollars from his sewing machine attachments, and the inventor of that simple lamp attachment, the inverted glass bell, to be suspended over lamps to protect the ceilings from being blackened, made the inventor rich. The "Darning Weaver," a device for repairing stockings, is a useful invention and is netting the inventor a handsome revenue on royalties. The wire nail and gimlet-pointed screw are fortune makers, and wire nails caused the invention of automatic machinery that manufacturers them so cheaply it does not now pay the carpenter to spend his time in picking a nail up when it drops, if it requires ten seconds to do so. The inventor of the well-known "safety pin" lived in luxury all his life, after discovering a means of concealing the point of a pin in such manner as to prevent scratching. The inventor of roller skates made nearly a million and the inventor of the needle-threader for a long time made $10,000 a year.

Relation of Capital to Invention.

Mr. Edward P. Thompson, one of the most entertaining writers on the subject of invention, says that "every invention, before the introduction into practical use, passes through two stages; namely, mental and physical"—mental when in the mind of the inventor only, and physical when the mental invention is put into bodily form by hand, or by hand with the assistance of a convenient tool. "A mental invention," says the writer, "sometimes does not become a physical invention because the inventor lacks money, technical knowledge or diligence. Such a mental invention often becomes a physical invention by the assistance of a capitalist, an educated person, or diligent companion." This being true the *mental* inventor, the person who, for lack of means possibly, would fail to make his invention a physical reality—such a person should take into his confidence a friend or companion to share the prospective benefits of his invention. Thousands of meritorious mental inventions are never worked out because of the over-timidity of the inventor, his exagerated greed for *all* the benefits to accrue instead of half the loaf, which in many instances is, or would have been, ample reward. Mr. Thompson truly says: "Inventors and capitalists should be more willing to co-operate. It is too often the case that the former must pay for his own experiments and all patent costs before a capitalist will even take the trouble to look into the merits of the alleged invention. On the other hand it is too often true that the capitalist seeks to join with the inventor, but the latter wants too high a price at the beginning."

Who Can Apply for Patents.

Patents are issued to any person who has invented or discovered any new and useful art, machine, manufacture, or composition of matter, or new and useful improvement thereof, not known or used by others in this country, and not patented or described in any printed publication in this or any foreign country, before his invention or discovery thereof, and not in public use or on sale for more than two years prior to his application, unless the same is proved to have been abandoned; and by any person who, by his own industry, genius, efforts and expense has invented and produced any new and original design for a manufacture; any new and original design for the printing of fabrics; any new and original impression, ornament, pattern, print or picture to be printed, painted, cast, or otherwise placed on or marked into any article of manufacture, or any new, useful and original shape or configuration of any article of manufacture, the same not having been known or used by others before his invention or production thereof, nor patented or described in any printed publication, upon payment of the fees required by law and other due proceedings had.

If it appears that the inventor, at the time of making his application, believed himself to be the first inventor or discoverer, a patent will not be refused on account of the invention or discovery, or any part thereof, having been known or used in any foreign country before his invention or discovery thereof, if it had not been before patented or described in any printed publication.

Joint inventors are entitled to a joint patent; neither can claim one separately. Independent inventors of distinct and independent improvements in the same machine can not obtain a joint patent for their separate inventions; nor does the fact that one furnishes the capital and another makes the invention entitle them to make application as joint inventors, but in such case they may become joint patentees. The receipt of letters patent from a foreign government will not prevent the inventor from obtaining a patent in the United States, unless the invention shall have been introduced into public use in the United States more than two years prior to the application. But every patent granted for an invention which has been previously patented by the same inventor in a foreign country will be so limited as to expire at the same time with the foreign patent, or, if there be more than one, at the same time with the one having the shortest unexpired term, but in no case will it be in force more than seventeen years.

Protection to Inventors.

What is a patent? It is a monopoly or grant, in the United States, for a term of seventeen years, to the patentee, his heirs or assigns, of the exclusive right to make, use and vend the discovery throughout the United States, as the inventor's rights may appear in the specifications and patent granted.

This means a great deal to the inventor who has secured a *valid* patent containing all the claims so worded as to prevent infringement and loss in contest. Thousands of inventors, obtaining patents through unreliable and inefficient attorneys or agents, find themselves possessed of patents *in name only*, and of no value when combatted by infringers with capital and the aid of able legal talent. A good patent costs no more than a weak and worthless one. Therefore how shortsighted are those inventors who employ cheap attorneys, saving $5 or $10 in fees, only to find themselves loosers of *all* they have paid when the contest comes.

The Need of Reliable Attorneys.

The Revised Statutes of the United States provide that "before any inventor shall receive a patent for his invention, he shall make application therefor in writing to the Commissioner of Patents, and shall file in the Patent Office a written description * * * of the same in such full, clear, concise and exact terms, as to enable any person skilled in the art or science to which it appertains, or with which it is most nearly connected, to make, construct and use the same; and in case of a machine, he shall explain the principle thereof and the best mode in which he has contemplated applying that principle, so as to distinguish it from other inventions."

To carry out these provisions it is necessary for the inventor to first make a clear, concise and complete drawing, or a working model of his invention or discovery, and send it to THE INVENTIVE AGE, or some thoroughly reliable attorney, who, before making application for the patent, should make a thorough and rigid examination of the Patent Office to determine upon its novelty or patentability. If the invention has been anticipated by some one else, or if it lacks novelty, or if for any reason a patent can not be granted, or, if granted, would be of no worth or value, then the inventor does not want to incur the expense of making application and paying attorney's fees and government fees. For making this thorough examination THE INVENTIVE AGE and all reliable attorneys charge $5, which fee is, under some circumstances, however, taken out of the additional fees paid by the inventor in case letters patent are applied for. The fees of patent attorneys vary somewhat, but the average fees for obtaining a United States patent are about $65—the government fees being $15 on filing the application and $20 on issuing a patent—the balance being the fees for preparing specifications, making searches, etc. The inventor is sometimes favored in terms given for payment of the fees, more detailed information regarding which can be obtained by enclosing a 2-cent stamp with enquiry to THE INVENTIVE AGE, Washington, D. C. The reason why the inventor should have a preliminary examination of the Patent Office made before applying for a patent lies in the fact that if the case is rejected the fees paid to the government and the attorney are lost

All patents obtained through us will receive special mention in THE INVENTIVE AGE and in cases of unusual merit inventions will be illustrated free of charge to our clients.

This publication, reaching capitalists, manufacturers and business men throughout the world, is of value in assisting to bring an invention before the public in case its promotion or sale is desired by the patentee.

INVENTIVE AGE

Patent Department.

PATENTS obtained in all Countries.

LEGAL, TECHNICAL, and SCIENTIFIC ADVICE given by personal interview or letter.

SPECIFICATIONS settled by Counsel.

WRITTEN OPINIONS furnished.

SEARCHES carefully made.

LEGAL DOCUMENTS prepared and, if desired, settled by Counsel.

ASSIGNMENTS, AGREEMENTS, LICENCES, Etc., Registered.

DESIGNS Registered.

TRADE MARKS specially Designed and Registered.

☞ All communications relating to Legal or Patent matters, and checks, express orders and postal notes should be made payable and addressed to THE INVENTIVE AGE, Washington, D. C.

NEW PATENTS FOR SALE.

Aftermath.

Samuel Edison, the father of Thomas A. Edison, the great inventor, died on February 26th, at the advanced age of 93.

The public will have an opportunity to witness the new method of photographing through solids at the Electrical Exposition to be held in New York City in May in connection with the Nineteenth Convention of the National Electric Light Association.

The Elizabethton Co-operative Town Co., of Elizabethton, Tenn., has sold its entire property to Mr. George H. Towle. Mr. Towle undertakes to meet the obligations of the old company and to organize a new one, giving the old stockholder's the opportunity to come into the new organization.

A natural gas pipe line belonging to the Philadelphia Natural Gas Company, of Pittsburg, is being laid from the gas fields of West Virginia to Pittsburg. The line, when completed, will measure 101 miles in length, the longest line in the world. The company is also the largest, operating over 1,000 miles of pipes, including the above line. The latter will cost $2,000,000.

Wm. J. Mordee, the well-known inventor of railroad specialties, who died at his home in Chicago, Feb. 27th, was born in Painsville, O., 62 years ago, and began life as a boy by carrying water upon a railway construction. His latest achievement was an electrical locomotive which generates a portion of its own electricity as it travels and heats and lights the train it draws.

Messrs. John K. Cowen, recently elected president of the Baltimore & Ohio Railroad, and Oscar G. Murray, recently elected vice-president, have been appointed receivers of the road, through the default in payment of $537,000 interest, due March 1, in addition to several obligations due at the same time. The system, one of the largest in the United States, comprises over 2,100 miles.

Recent Australian Patents to Americans.

The following list of applications has been specially prepared for by George G. Turri. Certificated Patent Agent, Melbourne.

E. Waters (communicated by A. E. Woolf, New York) for manufacture of medicinal acid disinfectant and bleaching fluid.

E. N. Dickerson, New York, and J. J. Suckert, Ridgwood Bergen, for process of and apparatus for producing and liquefying acetylene gas.

P. Boyton, New York, gravity railways. G. Tarff, Agent for applicant.

J. B. Carbe, Chicago, for gas and fire engines and in the method of milking and volatilizing the gases in the same.

J. S. Bettick, Baltimore, U. S. A., improvements in machines for making all tobacco cigarettes.

E. Berottheimer, New York, for improvements in pencils.

An Extraordinary Offer.

By special arrangement we are enabled to offer that great success in illustrated magazinedom, "Cosmopolitan," and the INVENTIVE AGE one year for $1.85. In clubs of three or more $1.75.

During 1896 it is promised that both Cosmopolitan and Inventive Age will be grander magazines than ever. Cosmopolitan is a magazine of about 1400 pages and 1000 illustrations during the year. No magazine excels it. Old as well as new subscribers can take advantage of this offer.

Another Great Combination.

Popular Science is one of the best magazines published. It is 30 years old and covers all the sciences in a popular and instructive way. We can furnish Popular Science and the INVENTIVE AGE one year postage paid for $1.50. This combination price can be offered but a short time as the subscription price of Popular Science is soon to be increased. Address the Inventive Age, Washington, D. C. If Zell's Cyclopedia is also wanted—the three for $2.75.

STANDARD WORKS.

INVENTIVE AGE BUILDING.

THE INVENTIVE AGE wants live agents in every part of the country. Readers of this notice will confer a great favor by sending us names of persons who would likely take an interest in the matter. Our cyclopedia premium offer is a great winner. The book alone is worth double the amount, $3, charged for the whole.

Any person sending us the names and addresses of five inventors who have not yet applied for patents will receive the INVENTIVE AGE one year free.

ntive Age
STRIAL REVIEW
ACTURING INDUSTRY AND SCIENTIFIC PROGRESS

HINGTON, D. C., MAY, 1896.

Single Copies 10 Cents.
$1 Per Year.

urposes. There is an inland lake at
fteen miles long, varying from two to
ridth, and having a depth of between
t. This lake is separated from the
i by a narrow neck of sand and soft
uel is being cut through this neck and
e there some endless chain bucket
at work excavating the sand, and I
t about 20 cents per cubic yard to take
d of the stream and spread it over a
being made for building purposes.

uelma Railway, the Paris, Lyons &
i, and also the East Algerian, are
respects. The gauge is 4′ 8½″, the
r of wood, although steel is being used

R. Trevithick, a grandson of the Father of the
Locomotive, and the engines he has designed and is
operating would be a credit to any railway in the
world. The chief engineer of the line is Mr. M.
Nicour, a Frenchman, and while he may have had
some engineering problems to solve in constructing
the road along the Upper Nile, about Cairo the line
is prefectly straight and on a dead level. Most of
the line is laid with bull-head rails resting on cast-
iron pot sleepers, but this practice is being abandoned in favor of pressed steel sleepers. All trains
are operated under the manual block system, signals being given by semaphores at each station,
both a home and distant signal being used.

The railway station at Cairo is a new building

Inventive Age

Established 1889.

INVENTIVE AGE PUBLISHING CO.,

8th and H Sts., Washington, D. C.

ALEX. S. CAPEHART.   MARSHALL H. JEWELL.

The INVENTIVE AGE is sent, postage prepaid, to any address in the United States, Canada or Mexico for $1 a year; to any other country, postage prepaid, $1.50.
Correspondence with inventors, mechanics, manufacturers, scientists and others is invited. The columns of this journal are open for the discussion of such subjects as are of general interest to its readers.
Technical matter is particularly desired. We want practical information from practical men.
Nothing will be published in the editorial columns for pay.
The INVENTIVE AGE is thoroughly independent.
Advertising rates made known on application. Special facilities for furnishing cuts of any patented article together with descriptive article. Business specials 25 cents a line each insertion, 7 words to the line. No advertisement less than 50 cents.
Address all communications to THE INVENTIVE AGE, Washington, D. C.

Entered at the Postoffice in Washington as second-class matter.

WASHINGTON, D. C., MAY, 1896.

Among the papers in the District of Columbia devoted to inventions and patents, none has credit for so large a regular issue as is accorded to the "Inventive Age," published monthly at Washington.—*Printers Ink.*

THE official headquarters of the National Electric Light Association during the convention this month in New York city will be at the Murray Hill Hotel.

THE coming meeting of the American Society of Mechanical Engineers at St. Louis, May 19 to May 22nd promises to be an event of unusual interest. Many papers will be presented and the attendance is expected to be large.

THE following attorneys have recently been disbarred from practice before the Patent Office bureau: Thos G. Pike, Laurel, Ind.; Fred W. Seher, Sandusky, Ohio; Wm. B. Phillips, Guthrie, Okla. Ter.; Joseph Gross, New Orleans; John J. Hughes, New York.

IT is claimed for the Great Falls Electric railway, now being constructed between the city of Washington and the Great Falls of the Potomac that it is the most level railway in the world, the greater portion of the route having a fall of only one inch to the mile.

A correspondent in Age of Steel, in speaking of the defects in the Patent laws, covers the whole situation and meets the desires of inventors when he says: "The law should afford as much protection to the inventors as it does to the holder of any other kind of property."

AT last the importation of World's Fair medals has begun, but at a time so far distant from the exposition little value can be attached to the proceeding, and the majority of the diplomas will serve only as mementoes of what might have been of almost inestimable worth—in an advertising way—if issued promptly.

THE march of invention requires new and novel legislation. What would the framers of the Constitution of the United States have thought if at that time some state legislature had passed a measure to prevent the stealing of electricity? New Jersey leads off with a measure of this kind calculated to put a stop to wire tapping and cutting.

THERE is much truth in the statement of that veteran Washington correspondent, Wm. E Curtis, that: petitions to congress have become so common as to be without effect. Everyone understands how easy it is to obtain signatures to almost any kind of a petition, and congress is not in the least influenced by 100 feet or 100 yard rolls of this nature.

THE canal system, for which the bill recently introduced in Congress by Senator Hansbrough of North Dakota, provides, connects the great lakes and the St. Lawrence River with tide water in the Hudson. Among its incorporators are some of the best known men in the country. The capital stock is to be $10,000,000 and a bond issue of $2,000,000 is provided for.

ADVANCE sheets of consular reports for May contain some valuable and interesting suggestions and conclusions regarding the present and prospective trade of the United States with Japan and the effect on Japan and other leading nations of the opening up of China to the investment of capital and trade operations. It is represented that the development of both internal and foreign trade is the all-absorbing problem with every class of the people of Japan. The system of education employed in the schools is admirably adapted to the turning out of well-equipped business men, so far as a practical commercial education can accomplish such an end. Merchants, manufacturers, and, in fact, all engaged in trade actively or by investment of capital are making and will continue to make the very best use of the time intervening between the present and the coming into operation of the lately revised treaties; in borrowing, *ad libitum*, the products (in the shape of labor-saving appliances) of the inventive genius of the people of the United States, and of every other nation, for use in the work-shops of the Empire, and will return the results in merchantable goods to the people of the nations from whom they are now borrowing at prices which will make competition an exceedingly difficult problem to solve.

The raw material necessary for the production, not only of the merchandise named, but of nearly all other goods produced by the most favored of the producing nations, are found in the territory of the Empire, the material wealth and producing power of which have been enhanced in no small measure by the annexation of Formosa. The mines are rich in coal, iron, and other minerals; the soil is fertile, and, judging from the extraordinary progress being made in agricultural pursuits, every available foot of it will, in the near future, be put into a high cultivation. In many of the subdivisions of Japan, two crops are produced annually.

The daily wage paid for mining and unskilled labor necessary for the production of what may be termed raw materials will average about 10 cents (United States currency), and that paid for skilled labor, at present, will average about 18 cents.

So much for the natural conditions and prospects of Japan. As a result of the recent war China with her 400,000,000 people has been opened up as a field of investment for foreign capital and some of the most extensive iron and cotton mills in the world are now being installed by American and European capitalists. If American and Europeans fear the competition of Japan what may we expect from China of which one consul writes: "The mines, soil, and animal resources of China are as rich and productive as are those of Japan. The Chinese are noted for their business integrity and are rarely charged with failure to live up to their trade agreements. Labor of every description is cheaper, and, in nearly every other respect, equal to that of Japan, while it is in so much greater abundance that it will not be in position, for many years to come, to continually and successfully demand increased reward."

And again, in speaking of the prospects of competition by American manufacturers in the Oriental countries our minister to Tokyo, Mr. Edwin Dun, says: "The greatest competitor that, not only American, but European manufacturers as well, have to fear in Japan is the growth of home manufacturing industries here. With an almost unlimited supply of cheap, skilled labor, an abundance of coal, and magnificent water power throughout the country, there is every indication that, in the near future, the manufacturing interests of Japan will increase enormously. Inquiries are constantly being received from America and Europe in regard to the feasibility of starting manufactures of almost every kind with foreign capital and management. Existing treaties at present close Japan to foreign enterprise of this kind, but when the new treaties come into operation, there will be nothing to prevent American and European capitalists from availing themselves of the exceptional advantages that Japan will offer in almost every line of manufacture."

The American manufacturer and the American producer of raw material, will find in all the recent consul reports the most convincing arguments in favor of an American policy that will protect American industries and maintain the American standard of wages.

THE holding of the National Electrical exposition in New York this month has been forced upon the electrical world in a peculiar manner. For eighteen years the National Electrical Light Association has been holding its annual conferences in the principal cities of the country. At these gatherings the men in electrical trades have presented their wares to the members of the association, turning hotels and halls into informal exhibitions of the latest devices in applied electricity. When it was decided to hold the 1896 gathering in New York, it became evident that nothing less than an exhibition hall would satisfy the supply men who were anxious to reach the assembled electricians. And in this manner the first New York Electrical Exposition has come to pass. It will probably be repeated on a larger or smaller scale wherever the National Association holds its annual meetings.

THE number and character of the exhibitors who have applied for space in the National Electrical Exposition, New York, is a guaranty of the success of the enterprise. Nearly all the space is now under contract. The issue of Electricity of April 15th contains a full list of those exhibitors who have already obtained space and the character of a brief description of the proposed exhibit of each. The United States Patent Office is arranging to make at the Exposition a fine display of electrical models, etc., which will be peculiarly interesting and instructive as bearing upon the growth and development of electrical invention. A special representative of the Patent Office will be in charge of this exhibit, which will be under the general care of the Historical and Loan Exhibit Committee.

THE number of patents, designs, trade-marks and prints issued during the four weeks ending April 28th was 2,067. The total number of cases awaiting action in the Patent Office on that date was 9,324. Of this number 1,549 applications are in divisions 7 and 9, Hydraulics, Fire Extinguishers, etc., and Harvesters, Games, Toys, and Velocipedes, and are between two and three months old. Between one and two months in arrears are 3,728 applications and the balance are under one month. Of the total number of issues for the four weeks in question 183 were to citizens of foreign countries.

THE Court of Appeals has affirmed the decision of the Commissioner of Patents in the case of Leon Appert, of France, vs. John E. Parker, of Philadelphia. Priority of invention was the cause of the suit, Parker claiming that he had applied for a patent on a process of making wire glass, a substance combining glass with metallic netting, April 25, 1894. Appert filed an application two days later, but had received a patent in France on January 12, 1894. The Commissioner adjudged that the patent should be Appert's.

IN Cassier's Magazine for May is published a biographical sketch of Francis H. Richards, M. E. of Hartford, together with an excellent portrait. Mr. Richards is one of the most practical and successful inventors in the country and readers of the INVENTIVE AGE have had the pleasure of reading much that has emanated from his pen. He is one of the leading lights in the American Association of Inventors and Manufacturers, and at the last meeting presented an excellent paper.

NOTWITHSTANDING the fact that millions upon millions of rabbits are slain annually in Australia still these pests are on the increase. In a single year the colonial government paid a bounty on 27,000,000 rabbits' tails. Millions of acres of land practically uninhabited only because of the presence of these thrifty creatures. There is a standing offer of $125,000 for an effectual plan for subduing these pests. Here is a golden opportunity for some inventive genius.

As an evidence of the fact that our present tariff laws are operating as a wonderful stimulant to the building up of manufacturing enterprises in Japan, it is reported that a big straw matting factory has removed to that country from Milford, Conn. The entire plant has been removed to Kobe, Japan, where labor and raw material can be had for less than a third of the cost in the United States. Of course the Japanese product will be sold in this market as heretofore.

NOTES.

The Antiquity of Tin.—Tin is one of the oldest known metals. The Chinese used it in the fabrication of their brasses and bronzes from time immemorial. In the book of Numbers it is among the list of metals in which, among other things, Moses and the Israelites spoiled the people of Midian. The ancient Romans used it for coating the inside of copper and brass vessels.—*Popular Science News.*

* * *

The Sun's Rays.—Sir R. Ball, writing on the sun, says:—"For every acre on the surface of our globe, there are more than 10,000 acres on the surface of the great luminary. Every portion of this illimitable desert of flame is pouring forth torrents of heat. It has been estimated that if the heat which is incessantly flowing through any single square foot of the sun's exterior could be collected and applied to the boilers of an Atlantic liner, it would produce steam enough to sustain in continuous movements those engines of 20,000-horse power, thus enabling a large ship to break the record between Ireland and America."

* * *

Largest Arch Span in the World.—The bridge to be built across the Niagara gorge will, when finished, have the largest arch span of any bridge yet constructed. It will be among the triumphs of modern engineering, of which this age has already an unprecedented record. The Railway Gazette says "It will have a span, between centers of end pins, of 840 feet and a rise of 150 feet from the level of the pins at the skewbacks to the center of the ribs at the crown of the arch, and this latter point is about 170 feet above low water. The depth of the trusses is 26 feet; they stand 68 feet 7 inches apart center to center at the skewbacks and have a batter of 1 foot in 8 giving a width of 30 feet between centers of top chords at the crown of the arch."

* * *

Ancient Pottery.—In digging out the colossal statue of Rameses II., nine feet and four inches of Nile mud had to be removed before the platform was reached. It is known that this platform was laid in the year 1361 B. C., when Rameses was still living. Therefore three and one-half inches of accumulated Nile mud represents the lapse of a century, it being known that 3,200 years have passed since the platform was put down. Under that platform was found thirty feet more of Nile mud before the original sandy soil was reached, hence many years must have elapsed from the time of the Nile's first overflow down to the time of Rameses II. The curious part of the story is this: Pottery and fragments of the same were found on the original sandy soil thirty feet under the base of the statue, which would seem to indicate that the Egyptians understood the potter's art at a very early date.

* * *

Steel Wagon Roods.—Steel wagon roads, as advocated by Martin Dodge, State road commissioner of Ohio, are likely to have a thorough trial in several States this year. These roads consist of two rails made of steel the thickness of boiler plate, each formed in the shape of a gutter five inches wide, with a square perpendicular shoulder half an inch high, then an angle of of one inch outward, slightly raised. The gutter forms a conduit for the water, and makes it easy for the wheels to enter or leave the track. Such a double track steel road, 16 feet wide, filled in between with broken stone, macadam size, would cost about $6,000, as against $7,000 per mile for a macadam roadbed of the same width, but the cost of a rural one-track steel road would be only about $2,000 a mile. It is claimed that such a road would last much longer than stone, and that one horse will draw on a steel track 20 times as much as on a dirt road and five times as much as on macadam.

* * *

X-Ray Pictures Without Photography.—Righi describes, in "L'Electricien," and apparatus by which he obtains the Roentgen pictures without the use of the photographic plate by a method similar to that employed by Lichtenberg and Kundt. Under the Crookes tube is fixed a sheet of black cardboard, backed by a sheet of aluminum, which is connected to earth. Below this is fixed a sheet of ebonite, backed with tinfoil, which latter is connected through an air condenser to the cathode of the tube. The anode of the tube is also put to earth. If a hand is now laid on the cardboard, and subjected for a sufficient time to the action of the tube, it will be found that an electrical picture of the hand a la Rontgen has been imprinted on the sheet of ebonite. This can be made visible by the well-known mixture of sulphur and red lead, or by another mixture of talc and dioxide of magnesium, which gives an effect more closely resembling a photograph. The bones, etc., are shown in these electrographs just the same as in the Roentgen photographs.

* * *

Remarkable Telegraphy.—With a view to ascertaining the highest speed at which telegraphic characters can be legibly recorded, Mr. P. B. Delany, in some recent experiments, succeeded in transmitting by his machine system 8,000 words per minute, and obtained a plain reproduction of the signals by electrolysis on the chemically prepared receiving tape. The circuit was an artificial one of 650 ohms, 2.95 microfarads, and the electromotive force was 115 volts. This is about the equivalent of an ordinary telegraph line of 100 miles in length, or, say, New York to Philadelphia. At this speed the perforated tape upon which the messages were composed passed through the transmitting machine at the rate of 27½ feet per second, and the impulses comprising the letters traveled at the rate of 2,500 per second or 133 words, equal to six ordinary telegrams of 22 words each in a single second. At this rate the next few years must bring about great change in methods of correspondence, and, inevitably, a large portion of the ninety millions now annually expended on wheel transportation of the mails will be diverted to the telegraph. Why not?

Report of Commissioner of Patents.

The report of Commissioner of Patents John S. Seymour for the year ending December 31, 1896, has been submitted to Congress. It is perhaps the most exhaustive calendar year report in the history of the office, and embodies an interesting review of the growth of industrial arts in the United Stated during the past quarter century. The aggregate receipts from all sources during the year were $1,245,247, expenditures $1.084,496, and the total balance in the treasury to the credit of the patent fund at the beginning of this year was $4,529,886. In every year since 1861 there has been a surplus over all expenditures, and in every year but eight since the foundation of the patent office, on its present basis, in 1856, there has been a surplus.

During the year there were issued 21,996 patents and designs, exclusive of 89 re-issues, and inclusive of 2,049 issued to foreigners, 1,829 trade-marks, and 3 prints registered, and 46,899 applications filed for patents, trade-marks, labels, prints, etc., requiring investigation and action. There were also 12,345 patents which expired, and 3,428 patents forfeited for non-payment of final fees.

As an index of the comparative genius of foreigners, the four foreign countries leading in the number of patents issued to its citizens by this government are: England, 614; Germany, 539; Canada, 302, and France 202. All others fall below fifty each. In the United States the states leading in the number of patents issued to its citizens are: New York, 3,539; Pennsylvania, 2,270; Illinois, 1,876; Massachusetts, 1,793, and Ohio, 1,423, while Arizona is at the foot of the list. In proportion to population, Connecticut leads the list, with 1 to every 927 people, District of Columbia, 1 to 1,047, and Massachusetts, 1 to 1,246, while Mississippi is at the foot, with one to every 54,864, with South Carolina, on the basis of 26,076, next. The report makes the yearly recommendation for remedying crowded quarters, for increasing the scientific library appropriation; enactment to secure admission to a solicitors' bar of those entitled to practice before the office, exclusive admission of lawyers being a manifest injustice, and to allow the commissioner to fix prices of uncertified copies of patents without minimum limit. The Commissioner cites the pendency of 43,643 cases at the end of the year which had been acted on by the office and awaiting further attention by applicants, and recites the important rules made a year ago to prevent more than six months' delay by applicants in taking further action on their cases, and allowing examiners to reject claims where there have been intentional delays in cases pending more than five years before the office. He says these rules have been applied with great conservatism, yet are open to attack in the courts on the question of their validity and regarded as a matter of policy are also liable to repeal. In view of their beneficence, he recommends that they be enacted into law.

The commissioner recommends the enactment of the bill prepared by the patent committee of the American Bar Association and introduced by Chairman Draper of the House committee on patents as H. R. 3014, providing among other things for reduction from two years to six months of the period for completing and preparing applications for examination and limiting the time within which an application may be made in this country by either the foreign inventor or any other to two years after the matter has been patented or described in any printed publication in this or any foreign country. The true policy is to lay no obstacle in the way of the taking of foreign patents and encourage the course.

Concerning the publication of the Patent Office Gazette, which has in recent years involved many complications, and last year resulted in a congressional inquiry, the commissioner decries the lack of a modern process in its production, the entire work not being done by one firm, and urges, in the event of a change being inadvisable, that a contract be authorized for printing it for a longer term than one year, in view of the expensive plant required.

Reports have been received from thirty-two foreign countries, including Great Britan, Germany, Russia and France, and the total number of patents issued by those governments from the earliest period up to December 31, last, aggregate 981,961 against 562,458 so far issued by the United States.

Prior to 1870 the United States issued nearly half as many patents as all foreign nations combined. Since 1870 the United States has issued more than half as many as all foreign nations combined and in the aggregate of patents issued in the periods the United States issued more than half as many patents as all foreign countries combined. The total number of patents issued by the United States and the foreign nations prior to 1870, of which the office has a record, was 331,031; the total since 1870 is 1,215,388, and the total number of patents issued by the United States and the foreign countries herein considered is 1,544,419. Most of the civilized nations have patent systems, notable execptions being China, Holland and Greece. Many have recently been established.

The establishment of a classification division is again urged, the exigencies of the office demanding a thorough revision of the whole classification. Commenting on the suggestion that it would be better to issue all patents without examination as to novelty, leaving the sifting of records of priority to the courts, the commissioner says this would prove a most expensive system to inventors, and would entail great loss in many cases, in that errors when discovered by the courts are usually beyond correction and meritorious inventions in ignorance of previous achievements in claiming too broadly might thus be declared invalid, even though some might be found.

During the past twenty-five years twenty-five inventors have had more than 100 patents granted to them the whole number being granted to them within that period being 4,824. In these, "Wizard" Thomas A. Edison of Orange, N. J., leads, with 711; Elihu Thomson, electrician, coming next. Following are the names: Edward J. Brooks, 116; George D. Burton, 128; Luther C. Crowell, 147; Peter C. Dederick, 107; Thomas A. Edison, 711; Rudolph M. Hunter, 228; John W. Hyatt, 198; Hiram S. Maxim, 131; Arthur J. Moxham, 144; Lewis Hallock Nash, 110; Edwin Norton, 125; Freeborn F. Raymond, 146; George H. Reynolds, 101; Francis H. Richards, 343; George W. Saladen, 148; Rudolph Rickemeyer, 158; Louis Godda, 131; Walter Scott, 109; Charles S. Scribner, 248; Sydney H. Short, 111; Elihu Thomson, 394; Charles J. Van Depole, 244; George Westinghouse, jr., 217; Edward Weston, 274; William N. Whiteley, 118.

The greatest activity of the year was shown in detail inventions and necessaries to bicycles and in machines and process for making the parts, due, perhaps, in part, to the enormous accession of thinkers to the ranks of those who use them. Pneumatic tires have attracted the inventor, because of their now almost universal use on sulkies as well as bicycles. For some unknown cause there has latterly been great activity in sole-leveling machines and in the manufacture of shoes, in inventions in telephones, due supposedly to the expiration of pioneer patents, and not a little in electric locks, a new art. Others showing much activity were excavators, as the result of large enterprises just begun, notably the Chicago canal, and the carbonization of beer, based on the recent discovery that second fermentation thus can be dispensed with.

The report also comprises a summary of important typical patents granted during the last twenty-five years. Its object is to direct the attention of isolated inventors to the true field of invention instead of rediscovery.

ORIENTAL RAILWAYS.

(Continued from first page.)

...k it possible for any one who has never seen a ...dy desert to form any conception of how utterly ...ren and desolate the surface of this earth can ...ear.

...rom Suez we took a steamer to Colombo, Ceylon, ...this country is, I think I can safely say, the ...it beautiful of any we visited, as flowers and all ...ds of vegetation are probably found in such pro...ion in no other country in the world. The rail...y belongs to the government, and consists of ...ut 270 miles of 5' 6" gauge track, the main line ...rhich extends from Colombo to Bandarawella, a ...ance of about 160 miles. Between Colombo and ...tipola, a distance of 139 miles, the rise is between ...0 and 7,000 feet; the first 30 miles out of Colombo ...very nearly level, and after passing over this ...tion, which extends through a succession or rice

inches of rain had fallen in twenty-four hours, and in some parts of the island the yearly rainfall is upward of 300 inches. This rainfall makes it necessary to construct heavy retaining walls at all cuts or fills, and even with these it is sometimes impossible to hold the roadbed in proper condition. The bridges must be very substantial and are built with heavy stone abutments and central piers of wrought-iron cylinders filled with concrete, sunk to a solid foundation. The original method employed in the building of the abutments is quite interesting. The stones used are quarried in the country and worked up by the natives. They are most of them of pretty good size and elephants were used exclusively for handling and placing them. The method of working was to put an endless chain around the elephant's neck and second chain around the stone to be moved. These chains were of such a length that when the elephant dropped his head the chains could be hooked together. After this was done the elephant would lift his head and march off to the place where the stone was to be used and gently lower it in position and then place his foot on it and press it down until it was properly imbedded in the mortar. They said it was formerly the cheapest and most satis-

form to the car. The English type of carriages a... goods wagons have been used almost exclusive... until within the past three or four years, when t... American type of coach and freight car have be... coming in.

The railway system of India, which was the ne... country visited, is much more extensive than st... posed by persons who have not looked into the st... ject. At the present time there are in operation... this country something over 18,000 miles of railw... of which about 12,000 is 5' 6" gauge, 4,600 met... gauge, and the remainder 2' and 1' 6". The 5'... gauge was the first introduced, but it was so... found that many roads could not be made to p... operating expenses, and it was decided to use t... meter gauge in building some new roads and also... change the gauge of some of that built 5' 6"... order to lessen the expense of operation. T... South India Road is a notable example of the latt... as it was originally 40 to 50 miles long, and wit... 5' 6" gauge, and did not pay expenses in the har... of a private company. The India governme... agreed to guarantee interest on the bonds of t... road if the gauge was changed to one meter and t... line extended. This was done, and at the prese...

FIG. 5.—BRIDGES IN BOULAN PASS, MUSCAF BOULAN RAILWAY, BELOCHISTAN.—FIG. 6.

FIG. 7.—PUSH CAR AND COOLIES, INDIAN RAILWAY.

FIG. 8.—SPIRAL CURVE ON DARJEELING HIMALAYAN RAILWAY, INDI...

From Bombay we took a steamer to Kurrachee and there struck the Northwestern Railway, which is the largest single system in India and has 2,617 miles of line in operation. At Ruk Junction, about 250 miles north of Kurrachee, the Muskaf Boulan Railway branches off from the Northwestern and extends for a distance of about 300 miles through an exceedingly barren and desolate district across British Beloochistan to Chaman, just across the Afghanistan boundary, and forms the extreme western outpost of the British Indian government. This line was built and is maintained purely for military purposes and is probably one of the most interesting in the world. It was built under the supervision of Sir James Brown, 213 miles being completed in thirty-three months, 42,000 men being at work under the protection of a body of 5,000 troops. The line passes through a desert 50 miles wide, and during the work all food and water were carried on camels for long distances. As the line was nearing completion cholera broke out among the men and 3,000 of them, mostly natives, died within one month. From Sibi north the line was originally built through the Boulan Pass and laid on the bed of the river. For several years it was kept open with only occasional breaks of small moment, caused by heavy rains. It was found, however, that what appeared to be a more feasible route for the line from Sibi to Bostan was through Chuppa Rift, and what has since been given the name of Mud Gorge. The line was accordingly laid on this route, the distance between the two points being 133 miles. Mud Gorge, however, developed unlooked-for difficulties, as it was found that the earth was of such an unstable character that the line was unsafe. On several occasions sections of it shifted 10 or 15 feet in one night and no amount of retaining wall or piling could be made to hold it in place. While attempts were being made to overcome this difficulty, in August, 1890, a cloudburst completely wrecked the line through Boulan Pass. At just about this time it was decided that the line could not be maintained through Mud Gorge and the only possible course was that of reconstructing the line through Boulan Pass.

Chuppa Rift, which must thus unfortunately be abandoned, is a remarkable freak of nature. The hill through which it passes is an enormous ridge of rock and the rift is a split or crack, varying in width from 10 feet up to 300 or 400, and looks as though it had been caused by the earth's surface being strained beyond its elastic limit. The railway approaches the rift through a long tunnel, crosses it over a bridge shown in illustration, Fig. 4, and immediately enters another tunnel. In its course through the rift, a distance of about 3 miles, there are several short tunnels and bridges.

The reconstruction of the line through Boulan Pass was begun in November, 1891, was well under way when we were there, and it was expected that it would be opened for traffic about the 1st of January 1896. This is probably the heaviest piece of railway construction in existence and the accompanying illustrations, Figs. 5, and 6, will give some idea of the bridges. It is a double track of 5′ 6″ gauge, the distance between centers being 14 feet, the weight of the rail 100 pounds per yard, the sleepers of pressed steel weighing 160 pounds each, and the prevailing grade 1 in 24. In one five miles there are fourteen bridges upwards of 60 feet in length, and the cost of that section was $58,000 per mile.

Beyond the pass the road crosses the summit of the range through the Kojak Tunnel, and while its dimensions may be familiar to some of you, I will give a few of the most important. The length of the tunnel is 12,800 feet and the radius of the arch is 14′ 6″ above the center and 29′ below. Half of it is level and the other is on grade of 1 in 40. It passes through shale rock and about three-fourths of it had to be timbered in building. It is lined throughout in brick, usually five rings thick, although in some places there are only four, and in others six and seven. There are two tracks 5′ 6″ gauge, 12′ between centers, and the difference in the portal levels is 150 feet. The work was begun in April, 1888, and the tunnel opened for traffic January 1, 1892, 114 feet driven in one week being the best record. The work was carried on from both sides, a track of 5′ 6″ gauge being constructed over the top of the mountain for the purpose of carrying the material through to the further side. When the tunnel was built two shafts 14 feet square were put in for the purpose of ventilating, but the scheme was a failure, as the tunnel remained full of smoke. Finally the shafts were closed as an experiment and the smoke immediately cleared out, and since that time there has been no difficulty with ventilation.

One of the great features of the railways of India and Ceylon is what they call the trolley car. The natives, having no strength in their arms, are not able to pump our hand-cars, but they can get on the rails behind a car and push it at a speed of ten miles an hour on the level, one man running on each rail, as shown in the illustration Fig. 7. A good brake is put on each car, and when a grade is reached it is let loose. We went down through the Kojak Tunnel and into Chaman on one of these cars at a speed of between 35 and 40 miles an hour, and a more exciting ride I never had.

All the bridges on the Muskaf Boulan Railway and also the Kojak Tunnel have a fort at each end where guns can be mounted, and as we were winding along through the mountains it was a common thing to see sentinel towers on top of the high points commanding the railway.

The best road in India is beyond doubt the East Indian Railway. This line operates the fastest train in the country and has the finest motive power, while its roadbed is fully equal to that of any of the other lines. There are 1,843 miles in the system, of which 474 miles is double track. The main shops of the locomotive department are located at Jamalpore and employ about 5,000 men. The only rolling mill in the country is a part of these shops and in it a large amount of scrap iron is worked into merchant bar, some of which is used by the road and the remainder sold. A complete system of semaphore signals is being put in at all small stations, which consists of a home, distant and starting signal for trains going in both directions. At the large and important stations interlocking plants are being installed.

We went into Calcutta over the East Indian Railway and there found in the great docks, which have been recently completed, the greatest engineering contract we saw during our trip.

From Calcutta we went up to Darjeeling, the great summer resort in the Himalayan Mountains, and the nearest point to Calcutta where cool weather is to be found. We took the Bengal State Railway from Calcutta to the foot of the mountain range, and there we took the Darjeeling Himalayan Railway, which is one of the most interesting we visited. The gauge is 20 inches, the road is 50 miles in length, and in that 50 miles the rise is over 8,000 feet. We left the foot of the mountain in summer clothing and sweltering with heat at 8 o'clock in the morning, and at 3 o,clock in the afternoon we had to wrap our steamer rugs around us and nearly perished then with the cold. There are four spiral curves on the line similar to the one shown in Fig. 8, the radius of which is 60 feet, and also six or seven switchbacks. The rainfall in this section is heavy and some extensive retaining walls are necessary for preventing washout.

Returning to Calcutta we took a steamer for Rangoon, Burmah, and there took the Burmese State Railway and went north as far as Mandalay. The line passes through a succession of paddy fields and swamps, and while there are no engineering features of note on the line, the engineers surely performed a remarkable feat by remaining in the country long enough to build the line—that is, if the sample of heat given us was representative. At Amarapura the railway operates a very efficient ferry, for transferring goods cars across the Irrawaddy. The banks of the river are 40 to 50 feet high, and down these the tracks are laid on a gradual slope. The lower end of the tracks is connected by means of a girder to a barge which is firmly anchored out in the stream a short distance. At the upper end of the slope a winding engine is placed and the cars are attached, one at a time, to a wire rope and lowered down the track across the girder and barge to the transfer barge which is moored alongside. The transfer barge when loaded is towed across the river by a steamer which also does duty as a passenger transfer, as the passenger cars are not taken across the river.

One of the great sights of Rangoon is the elephants working in the saw-mills and lumber yards. They were formerly employed in all the large mills but at present only one continues to use them, and in this yard we saw fourteen at work at every kind of work, from sweeping the floor to piling up timbers which were said to weigh one ton.

We saw an elephant walk up to a stick of timber about 14 inches square, 25 to 30 feet long, which they said weighed one ton, get down on his knees, run his tusks under it, throw his trunk over it, get up and carry it to a pile of similar sticks and put it down with one end resting on the pile and the other end on the ground. He then walked around to the end on the ground, curled up the end of his trunk and with it pushed the timber up on top of the pile and then went to the other end and pushed it up so it lay even with the other timbers.

From Rangoon we went to Singapore, and then up to Bangkok, Siam, where the king is building a railway about 160 miles in length.

From Bangkok we went back to Singapore and took a steamer from there for Java, and the little island of Java, which looks like a mere speck on most of our maps, is about 800 miles long, 200 miles wide, and has over 23,000,000 inhabitants.

There is in operation about 1,500 miles of railway, most of which is owned and operated by the government, and is 3′ 6″ gauge. There is a continuous line from end to end of the island, with a break of gauge near the center, where there is a section of 4′ 8½″ gauge about 20 to 30 miles in length. The track is kept in excellent condition and would be a credit to any country. The ballast is gravel and broken stone. About 12 inches outside of each end of the ties a nice little rip-rap wall is built up to a height of 8 or 10 inches, which holds the ballast in place and presents a very neat appearance. The ties are completely covered with ballast, excepting one at the center of each rail, which is left open for drainage. Outside of the track on each side the sod is kept in perfect condition over the entire right of way of the railway. This sod is mown by the natives in a peculiar fashion. The implement used is a sickle about the shape, but smaller than commonly used in this country for clipping grass, but is fastened on the end of a broomstick, and the operator stands perfectly straight, and as he moves slowly along swings this implement around his head, and every time it strikes the grass it clears off a space about 6 or 8 inches in diameter. The result is a lawn mown as perfectly as though done with a modern lawn mower. The bridges are all very good, and come out from Europe, together with the capstones for the foundations, there being very little stone in the country that is of sufficiently good quality for that purpose. The piers are built of concrete, mixed with Portland cement, and whitewashed so that they make a very handsome appearance. The Dutchmen rule Java with an iron hand. While I was in the office of one of the station masters of the railway messengers would come in to deliver notes. When a messenger reached the officer's desk he would fall on his knees, place his hands before him in a very abject attitude, slip the message into the hands of the station master or on top of the desk, and then back off and go out of the room as quickly as possible.

From Java we took a steamer to West Australia and there found the railway mostly belongs to the government, is 3′ 6″ gauge, and about 300 miles in length; but as there are only about 40,000 people in the colony, you can imagine it is not a very important affair. At Freemantle a very interesting piece of engineering work is under way in the construction of a harbor and the cutting through of a ledge of rock which prevents vessels of deep draft from entering the Swan river. The river is deep and after the ledge is removed ocean steamers will be able to go all the way up to Perth, a distance of some 14 miles. The work is all done under water, the drilling being done entirely by hand from stag-

FIG. 3.—BOMBAY STATION, G. I. P. RAILWAY.

ing. The blasting is done with dynamite; and the spoil removed by English endless chain dredges. The engineers in charge state that the entire cost of drilling, blasting, dredging and dumping the spoil into the sea is less than 75 cents per cubic yard.

The new railway station at Perth is a very nice looking and well arranged building just completed.

In Japan, there is about 2,500 miles of railway in operation. of which some 500 miles belong to the government and the remainder to 13 or 14 private companies, having systems ranging from 5 miles to 600 miles in extent. The largest single system is under the supervision of European engineers, but the Nippon Railway is a notable exception to this as it was built entirely under the supervision of Japanese engineers, by Japanese capital and none but Japanese have ever been connected with its operation and management. The average cost of the railway per mile is only about $16,750, which speaks wonderfully well of its engineers. From an engineering standpoint the rack railway on the Yokogawa-Karuisawa section of the government railway is the most interesting in Japan, and is described by Mr. Francis Trevithick in a paper read before the Asiatic Society of Japan.

way up the gradient; and other buildings, 651 tsubo.

Practically all the railways in Japan are now being operated exclusively by Japanese, and when it is remembered that the first line in the country was opened in 1872, and that only a few years before that date the country was closed to foreigners, it will be seen that the progress of this country is almost magical.

I have endeavored in the foregoing remarks to give you some idea of the railways in the countries visited by the commission up to July 25, when I left it; and in closing I wish to say that each one of

COMMERCE BUILDING—391x256 FEET—TENNESSEE CENTENNIAL EXPOSITION.

CHILDREN'S BUILDING.

AUDITORIUM—300x150 FEET.

AGRICULTURAL BUILDING—300x200 FEET.

that of the Nippon railway, which comprises about 600 miles in operation and 200 under construction. The aggregate capital invested constructing 1,879 miles of railway, which was the total miles in operation in December, 1892, is given as 94,163,836 yen, which would be equivalent to a little over $25,000 per mile. During the year 1892 the number or passengers carried per mile of track was 14,300 and the tons of goods per mile of track was 1,500. The gauge of track is universally 3' 6".

Many of the lines were constructed and operated

"The construction of this line was begun in March, 1891, and opened for traffic on the 1st of April, 1893. It was therefore completed in 25 months. The principal works connected with this line were: Earthworks, cuttings, embankments, deviation of roads, etc., etc., 89,404 tsubo; (A tsubo equals 8 cubic yards); tunnels, 26 in number, with an aggregate length of 14,644 feet; bridges, 18, with an aggregate length of 1,471 feet; culverts, 20; rails laid for the main line and sidings, 8 miles 44 chains; a passing station at Kuma-no-taira, which is half

these countries has conditions peculiar to itself, which those building and operating the railway lines seem to have in every instance comprehended and met in a greater or less degree; but I am thankful it is my lot to live in a country where the conditions require a railway service which gives the traveler the most comfortable cars, the fastest and smoothest running trains, the best meals, and, in fact, luxuries which are not dreamed of in any of the many places in which it has been my pleasure to travel.

The Tennessee Exposition.

An awakening has taken place in the New South and benefitting by the experience of northern states one exposition after another is being held in that section and at last the people of the world are being made to realize the wonderful resources of this great region. The Atlanta Expositon was a success but the one now being arranged for at Nashville to celebrate the centennial of the admission of Tennessee

is being built by money given directly by the people of Tennessee, and the officers and directors in charge serve without any salary whatever. There will be more trees and flourishing grass and exquisite flowers about the buildings than any exposition has presented. No charge whatever will be made for a reasonable amount of space in any one of the edifices, but the admission of exhibits will be attended by the strictest of scrutiny in order that each may be some-

that long before that date every building will have been completed and every exhibit installed. There will be an absence of rush and incompleteness so characteristic of all previous enterprises of this nature. An immense auditorium has been completed in which to hold the inaugural ceremonies next month

In this issue we present illustrations of a number of the buildings which are now under construction, and some of them nearing completion. The greatest interest at the grounds at present, outside of the

PARTHENON.

MACHINERY HALL—526x124 FEET.

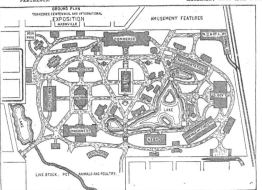

MAP OF TENNESSEE CENTENNIAL GROUNDS—200 ACRES.

TRANSPORTATION BUILDING—400x125 FEET.

WOMEN'S BUILDNG.

into the Union promises to eclipse it. It will be more elaborte and more of an international affair.

It may be said that the inception and developement of the Tennessee Centennial Exposition differs at cardinal points from any exposition ever held, in that it is more of a purely patriotic enterprise. It

thing really worth looking at. Congress has at last recognized the importance of the affair and it is likely that an appropriation of $150,000 or $200,000 will be given to assist the enterprise. The birthday festivities will commence on June 1, and last three days but the formal opening of the exposition will not take place till May 1, 1897 and will continue six months. From present indications it seems likely

building 'hemselves, centers in that portion of the park marked "Amusement Features." This portion of the plan contains about fifty acres, sloping rather sharply up a hill, in contrast to the main part of the Exposition, which is almost level. The amusement features will be placed on two broad, deep terraces near the base of the slope, and on the elevation special features will be placed.

Old English Railroad Cars.

The engravings herewith illustrate a number of cars which have until quite recently been running on an English railroad. Their origin and history are matters of some speculation, but from their construction and appearance it is evident they have been associated with a period in the history of railroads with which few men now living were acquainted. Comparison with old prints affords assurance, by reason of similarity of design, of their origin having been contemporaneous, or nearly so, with the rolling stock hauled by John Stephenson's locomotives in the early '30's.

A line of railroad, known as the Bodmin & Wadebridge, in the south of England, has recently been taken over, reconstructed and incorporated as part

Car Journal, which journal in turn acknowledges the courtesy of Mr. William Panter, Superintendent of the Carriage and Wagon Department of the London & Southwestern Railway, for the excellent photographs from which the engravings are reproduced.

Egyptian Sugar.

The British steamship "County of York," Captain Maddrell, with 3,000 tons of Egyptian sugar, the first cargo of the kind that was ever landed at Philadelphia, arrived March 3d, from Alexandria, Egypt, after a passage of thirty-two days. This cargo, which comes in bags, is of the highest grade. For years vast quantities of it have been imported to the United Kingdom and Continent. Its coming

The Keely Motor Again.

The investigation of the Keely motor by Prof. La Salle-Scott, recently sent from London to study its possibilities, has it is reported developed the fact that Keely has discovered a force hitherto unknown. The most thorough and delicate of tests have been applied and so far are in favor of the conclusion that the effects produced are not due to magnetism, electric currents or any concealed source of mechanical power. The term "dynamo" is said to be misleading, and the famous English scientist has suggested the word "vibrodyne" as most suitable to the character of the apparatus. According to the press report quoted, the Professor is of the opinion that the new force is capable of eventual economic use under certain conditions. As the investigation in progress has yet its final decision to render, it is idle to guess in a premature fashion what may or may not be the outcome. It will be the practical value of the alleged new force that will determine its status as a discovery.

The Keely motor has for some years been a subject of interest and speculation. It has had the charm of a mystery. Its possibilities have been forecast and it has also been denounced as a fake and a pretension. It will be the solving of a very wearying problem when something definite is given to the outside world as to what it is or may be, or what it is not.—*Age of Steel.*

A Water Pipe Telephone.

"I have a most remarkable telephone in my house," remarked a resident of the western addition; "I noticed that at times I could hear very distinctly the conversation in the next house. Suddenly it would be broken off short in the middle of a sentence, and I could not hear another word. It would become audible again just as suddenly. By a series of experiments I have found out that the sound is conducted by the water running through the pipes. When the water is turned on in my house I can hear all the conversation in any of the rooms next door in which there is running water. When I turn off the water, all sounds stop suddenly. I told my neighbor of it and we have put it to practical use. When I wish to speak to him, I tap on the window; he turns on the water in his house and listens while I talk to him over the water pipe in an ordinary tone of voice. When I have finished, he turns off the water in his house, and I turn it on in mine and listen. In that way we can carry on long conversations with as much ease as if he were in the room with me. Still our houses are about twenty feet apart."—*San Francisco Post.*

Largest Dry Dock in the World.

A dry dock will be built at the Erie basin in the New York harbor by the recently incorporated International Dry Dock and Construction Co., of New York large enough to admit the largest ocean vessel. The dock will be constructed on a new design,

of the London & Southwestern Railway system. The cars here illustrated constituted the entire rolling stock of this road, and the scheme of reorganization naturally included their replacement with more modern equipment. It is, very appropriately, intended that these interesting relics shall be carefully preserved.

The Bodmin & Wadebridge Railway was only six miles long and of standard gauge. There were no stations or signals on the line, and the trains were stopped wherever circumstances required; the train

to Philadelphia is due to the anticipated scarcity of the Cuban crop. The Egyptian sugar is grown and manufactured under the supervision of the Khedive, from whom this cargo was purchased by the Trust.

By reason of the insurrection in Cuba, and the burning of a number of the most productive plantations, only about half a crop has been raised, and the grinding, which has heretofore begun in the middle of December, has just been started. In all there are only four plantations that will be able to do any grinding this season, many of them having

being run each way about twice a week. It is said that the manager would start the train at the end of the line and drive over to the terminus to attend its arrival. After so many years of service these cars will be accorded a well-earned repose in the seclusion of an historical museum.

Their construction, while primitive, has evidently been enduring. Loose-chain couplings were used, and the enormous buffers are made of single blocks of timber, those only of the first-class car being supplemented with rubber springs. The third-class car is provided with a hand-brake, with a brake-shoe bearing on one wheel of each pair.

The first and second class three-compartment car is 14 ft. in length, 6 ft. 6½ in. wide and 8 ft. 5 in. high from rail. The second-class car, having two compartments, is 10 ft. long, 6 ft. 5 in. wide and the same height as the former. These two had a wheel base of 7 ft. 6 in. and 6 ft. 4 in., respectively.

We are indebted to the courtesy of the Railroad

had their machinery ruined in the course of the war. This cargo will be followed by several others from Alexandria.

The "County of York" sailed from Philadelphia in November for Greece with a cargo of case oil, and after discharging at the ancient port of Patras was chartered to go to Alexandria and load for Philadelphia.—*Traffic.*

A London exchange says that the man who devotes only a week to advertising a new thing, and then rails because he does not get a great return, is like the boy who studied with a lawyer two days and then came home and said: "The law ain't what it's cracked up to be. I'm sorry I learned it."

Willis Holbrook, of Pompey Centre, N. Y., thinks that paper would be a good substitute for metal and wood in making rims for bicycle wheels. It is light, yet tough.

and will be a decided improvement on the structures now in use. The cost of the new dock will be about $300,000, and it is expected that it will be completed within nine months. It is designed and planned so that at any time it can be extended so as to dock the largest ship afloat, or it can be disconnected at the sections so as to be used for one or more vessels. It will be the largest and best floating dry dock in the world.

To Find Length of Belt.

F. Casey, of Harrisonville, Ohio, has sent the following description of his method of finding the length of a belt required. Add the diameters of the two pulleys together, divide the result by 2 and multiply the quotient by 3 1-7, add the product to twice the distance between the centers of shafts, and you have the length required.

A New Dispensation.

[Prof. C. Coles]

(Make a Chain; for the Land is Full of Bloody Crimes, and the City is Full of Violence).—Ezekiel 7-23.

The Twentieth Century will usher in a new dispensation! Great preparations are now being made! We are living as it were, years in a single day! Science is king of the earth! Time proves all things. In 1891 when we first discovered the secret of photographing the human body, and all that lies therein, and made the following statement in the Wilkes Barre Telephone, and several other papers: The day will soon come when we shall be able to see through the human body and locate diseases," etc., we were set upon by many physicians and scientists and called a crank! A fool! To-day these same physicians and scientists, if living, know that our claims were true and right. This wonderful discovery is but the commencement of a long series of discoveries that will end in a new dispensation. Distance shall be annihilated! We shall be able to stand in Kingston and to hear, and to see, and to shake the hand of friends, though ten thousand miles away! All of our six senses will enjoy the same rights and privileges. A Thoughtophone will be constructed; that will reveal our inmost thoughts! How wonderful are the works of God, as shown through man.

We just begin to understand a little of His marvelous wisdom. God loans men wisdom, according to the age in which they live. "The thing that hath been, it is that which shall be; and that which is done is that which shall be done; and there is no new thing under the sun."

A few years ago when we read in God's Holy Book: "That every word and act was recorded in Heaven," we could not understand how that could be possible! So in order to show and to prove to us how easily our thoughts and actions could be recorded; he sent a living spirit— a living soul—down to our earth in the body of Thomas A. Edison, and he constructed a machine that he was pleased to call a Phonograph; that records our words. And he also constructed another machine that he was pleased to call a Kinetescope; that records our actions. And some day in the near future, a living soul, within some human body now on earth, will show us how easily our inmost thoughts can be told. Then comes the new dispensation! "Make a chain," for the Devil must be bound for the next two thousand years, commencing with the Twentieth Century! Oh God grant us all the power to bind the Devil, and all his subjects.

And we will praise thee; for we are fearfully and wonderfully made. Marvelous are thy works; and that our souls knoweth right well. The darkness and the light are both alike to God and he has given us the very light of darkness, as it were, in the X Rays. God is showering down the beauties of Heaven upon old Mother Earth and he or she who is too sinful to see, his merciful kindness will be cast into outer darkness.

The X Ray's will reveal all the hidden mysteries of the human body, and all the hidden mysteries of all "dark" bodies and substances. The Electric Eye will magnify these bodies (and all other matter) so as to show their now unseen or invisible forms and construction. The Vibrameter will show the true disposition of a man and record even the thought vibrations of his brain. In a few more months, if nothing happens, we shall be able we think, to have perfected a machine that we can attach to your head and read your thoughts! For the Vibrameter already shows and proves that the vibrations of thought within the brain are the same as when they escape the brain through the lips. When all these things are given us, as they certainly will be before the ending of the Nineteenth Century, or in the beginning of the Twentieth, then the evil forces will have no place to secrete themselves and the Devil will have to surrender! The world will grow better and better. Right and intelligence, will rule the nation.

The New Range Finder.

The Army Fortifications Board has made an inspection of the range-finder and relocator invented by Lieut. Rafferty. Lieut. Rafferty's invention in appearance is like a dial, with hands similar to an old fashioned telegraph signal instrument. It obviates shooting at random, and the enormous expenditure of costly ammunition will be averted. The importance of this is appreciated when one realizes that the cost of a single round fired from a modern gun is from $500 to $2,500. Every shot can now be driven where it will do the most good, or harm. Two shots fired with the Rafferty range-finder at a target over three miles distant proved a success. The first shot struck the water only 15 yards distant, while the second shot hit the barrel, destroying it. There was also used another of Lieut. Rafferty's inventions for correcting atmospheric errors, which greatly assisted in the long-distance firing.

The Next Polar Trip.

In only two more months, under present arrangements, the Andree expedition is to start on its way across the polar region; and if Capt. Nansen does not appear in public before that time, there is every assurance that scientific circles and the civilized world in general will have a good deal of interest in the daring adventure. When Herr Andree sets out on his voyage through the Polar air, every one will wish him a safe voyage.

Few Polar voyages have been taken in which success seemed to promise so much while disaster threatened such certain dangers. If the voyage is made as Herr Andree hopes to make it, the world's knowledge of Polar geography will be very noticeably advanced and the North Pole itself may be discovered by the adventurous explorer. If, on the other hand, the air ship meets with any serious mishap, the death of the aeronaut seems, unfortunately, almost certain.

Some critics call the proposed trip fool-hardy and reckless. It is not that bad, according to the views of many European scientists. The plan of the explorer has been examined by some of the best posted savants of Europe; and their verdict has been to the effect that the proposed trip is dangerous, undoubtedly, in view of the chance of a mishap; but that it may prove easily successful if no unforeseen disaster shall occur; that the plan is bold, daring, but not unsound so far as science can say.

So the work is going on near Spitzbergen, despite the news of Dr. Nansen's alleged return, and despite the dismal predictions of some of the more gloomy critics. The construction of the balloon house is almost now complete, and all of the balloon material is being subjected to rigorous tests. Some 35 tons of sulphuric acid and iron shavings are being taken to Spitzbergen, for the manufacture of the necessary hydrogen, and all the preparations are to be finished by May 25. The voyage will be a serious affair, in any event, and a short time will show the outcome.—*Boston Advertiser.*

New Use for Mica.

The uses of mica are manifold. One of its latest developments is distinctly novel. An ingenious Australian has invented and introduced a mica cartridge for sporting and military guns. The filling inside the cartridge is visible, and a further advantage is, that instead of the usual wad of felt a mica wad is used. This substance, being a non-conductor, unaffected by acids or fumes, acts as a lubricant. When smokeless powders, such as cordite or other nitro-glycerine compounds are employed, mica has a distinct advantage over every other material used in cartridge manufacture. Being transparent, any chemical change in the explosive can be at once detected. The peculiar property it has of withstanding intense heat is here utilized, the breech and barrel being kept constantly cool. The fouling of the rifle is also avoided, the wad actually cleaning the barrel.

A Steel Gun Factory in Japan.

It is stated that the Japanese Government has arranged for the establishment of a steel foundry in Japan, by Sir W. G. Armstrong and Co., Ltd., on the following terms: 1. The material shall for the present be imported from England. 2. Of the workmen to be employed 20 per cent shall be brought from England and 80 per cent shall be Japanese. 3. When a new arm is invented in England it shall be manufactured at the works in Japan. 4. For a stated number of years the Japanese Government shall give a fixed subsidy to the company. 5. On the expiration of the period during which the company receives a subsidy the works shall be sold to the Japanese Government.

Horse Fed by an Alarm Clock.

An ingenious man has invented a device for feeding his horse, and he does it with one of the ordinary alarm clocks. For instance, if the horse is to have its morning feed of grain at 5 o'clock the alarm is set, and when the morning comes the horse gets its breakfast before its owner's eyes are open. It is so arranged that the alarm pulls the slide, letting the grain run through a sluice to the manger.

The use of iron in architecture is not as new as people are accustomed to think. At Delhi is a forged iron column 60ft. high. It is 16in. in diameter at the base and 12in. at the top. Its weight is estimated at about 17 tons. From records extant it is reasonably certain that it was already in existence 900 years B. C.

Peculiarities of Ancient Pompeii.

Dr. Fletcher Horne is a very enthusiastic visitor of the partially excavated cities destroyed by Vesuvius in the year 79. In an interesting book on the subject, just published by Messrs. Hazell, Watson, and Viney, there are clever pen pictures of Pompeii and its most interesting life. The author says of the Pompeiian ladies:—"They were certainly not inconvenienced by a superfluity of clothing. A scarf crossed over the chest, a vest, tunic, and stola, with a pair of boots or sandals, constituted their entire apparel. Little though it was, they made the most of it. By raising or tightening the robe the Pompeiian beauties contrived to give the beholder a fair idea of their figures when they happened to have a good one. When they had not, they let their robe hang in majestic folds around them, and turned down their virtuous noses at the shameless damsels that were younger and better looking, who could afford to be more liberal in the display of their charms." The Doctor gives a vivid description of some of the casts of corpses made by the detritus which fell upon the city. For 1,800 years these human remains have been preserved, and the matter which enveloped them has beem filled with plaster, and made to yield an exact representation of what was once Pompeiian life. The form of a woman about 25 years of age is one of these; on her fingers were two silver rings, her garments being of fine texture. The linen head dress falling over her shoulders can still be distinguished. Overcome by the heat and suffocating gases, she had fallen on her side, one arm raised in depair, and her hands clenched convulsively; her garments are gathered up on one side, leaving exposed a limb of a very beautiful shape. The soft, yielding mud had formed such a perfect cast of the figure that one might suppose it had been taken from an exquisite Greek work of art. She had evidently been overcome in the act of flying with her little treasures—two silver cups, a few jewels, and some half dozen silver coins. Nor had she forgotten her keys, which were found by her side. What a world of tragedy there is in these representations of byegone life, enough to stock all the novelists of the age with plots for the rest of their days.

Cathode Rays Not as Bright as a Candle Light.

Now one of the first questions I have been asked in regard to these rays is this: "How did you obtain a light so intense that you could take photographs through a board an inch thick?" The answer is this: The light is not intense to the eye. It does not appear as bright as that of a fire-fly; indeed, it cannot be seen on the darkest night at a distance of three hundred feet. Yet a candle can be distinguished on a similar night at least a mile. But the rays of a candle are entirely cut off from a photographic plate by a sheet of pasteboard a sixteenth of an inch thick, or even less. The cathode rays are intense, however, to the photographic plate, which can be termed the photographic eye.—From "The New Photography by Cothode Rays," by Prof. John Trowbridge, in the April Scribner's.

Invention of the Locomotive Whistle.

When locomotives were first built, and began to trundle their small loads up and down the newly and rudely constructed railways of England, the public roads were, for the greater part, crossed at grade, and the engine driver had no way of giving warning of his approach except by blowing a tin horn. But this, as may be imagined, was far from being a sufficient warning. One day, in the year 1833, so runs a story of the origin of the locomotive whistle, a farmer of Thornton was crossing the railway track on one of the country roads with a great load of eggs and butter. Just as he came out upon the track a train approached. The engine man blew his tin horn lustily, but the farmer did not hear it. Eighty dozen of eggs and fifty pounds of butter were smashed into an indistinguishable, unpleasant mass, and mingled with the kindling wood to which the wagon was reduced. The railway company had to pay the farmer the value of his fifty pounds of butter, his 960 eggs, his horse and his wagon. It was regarded as a very serious matter, and straitway a director of the company went to Atton Grange, where George Stephenson lived, to see if he could not invent something that would give a warning more likely to be heard. Stephenson went to work, and the next day had a contrivance which, when attached to the engine boiler, and the steam turned on, gave out a shrill, discordant sound. The railway directors, greatly delighted, ordered similar contrivances to be attached to all the locomotives, and from that day to this the voice of the locomotive whistle has never been silent.—Cassier's Magazine.

New Ideas.

There are already several adding machines on the market. William Siddall, of Frontier, Mich., after years of study has perfected another.

Thomas R. Cherry of Buchannan, W. Va., is the inventor of a screen for a ladies' bicycle that will "shield from sight the form of a female"—if desired.

C. L. Coffman, of Neola, Iowa, has invented an automatic attachment to a walking plough, which is intended to correct any tendency to run at varying depths when the ground is rolling.

John S. McGranahan, of Fontanet, Vigo County, Ind., has invented a bed in which the spring is put, not in the mattress, but in the bedstead, between the end of the side rail and a special metallic support for it.

John Sheckler of Milesburg, Pa., is the inventor of a compound for exterminating thistles—the ingredients being as follows and in the proportion of salt, one barrel, kerosene oil, one gallon, vitriol, one pint, pure wood ashes, one peck.

M. C. Russell, of Larue, Marion County, Ohio, has a plan for handling a cross-cut saw without the aid of a second man, no matter how long or limber the saw is. He says it works with perfect ease. Every blacksmith can make the necessary device.

P. J. Brill, of Oostburg, Wis., has a cheap trap for catching insects that fly by night. He places it in a tree, in the right season, and in the morning finds that he has killed an astonishing number of curculios and moths. Fruit growers may be glad to consult him.

A horsewhip is so easily and frequently stolen from a carriage standing unattended on the street that many drivers carry it in their hands on alighting. This is a great nuisance. U. E. Morris, of Richmond, Ind., has devised a locking whipsocket to get around the difficulty.

Mr. Herman Huss of Princeton, N. J., is the inventor of a useful little novelty in the shape of a combined napkin ring and holder. The ring is a flexible strip, provided with hooks at intermediate points to engage the napkin and at each end with hooks to fasten into the coat waist.

J. J. Allen, of Friendship, Wis., has designed a method of connecting the sickle of a mower to the frame which will allow the former to be raised at any angle without interfering with the operation of the knives, and without pinching or cramping. This plan does away with the ordinary pitman.

At last a means has been discovered for thawing frozen pipes. It consists of a steam heated and propelled ball traveling through the pipe with flexible connection. The ball has sufficient perforations to admit of escape of enough steam to thaw the ice. Mr. Sam. D. Silver of Idaho Springs, Colo., is the inventor.

An Allegheny, Pa., man has invented an envelope so made that the mucilage is placed on the edge of the opening on the body of the envelope, where the flap laps on when sealed, instead of on the flap itself, so that when sealing an envelope one can moisten the clean paper flap and press it down with the same effect as with the old envelope.

Herr Andree's balloon, in which it is proposed to attempt the discovery of the North Pole, is partially completed. It is hoped that it may be shipped from Gothenburg, Sweden, to Spitzbergen, together with a house, which can be taken to pieces and set up again, on May 25. The exploring party sail for Spitzbergen in the stout little steamer Virgo.

In hot weather as well as in rainy weather in tropical climates the use of an umbrella is almost indispensable. Joseph Balt of Cranbury, N. J., has invented a wire support with strap or belt to fasten around the body and over left shoulder in such a manner that the umbrella can be attached and carried at any angle desired without the use of hands.

A German invented and constructed an electric mobile plow which in practical use plows eight acres to a depth of ten inches in ten hours, at a cost of $1.20 per acre. When the end of the furrow is reached, the plow is tipped over, the current is reversed and the next furrow is begun. It is a two-wheeled affair with a motor attached connecting with dynamo either on the farm or connected with a remote central power station.

A new patent governor for pumping engines was seen in operation a short time ago. The invention, which is of special interest to quarrymasters, contractors, and engineers, etc., has for its very laudable object the prevention of the accidents so frequently caused by the admission of air into the pump. By the use of this invention the pump can work at the same speed on air as on water. Whenever the water goes back there is neither shock nor knocking on the valves. A large number of experts were present at the initial experiment and expressed their satisfaction with the invention. The new patent has been taken out under the name of the Johnston-M'Adam Governor.—*Invention.*

In most grain-binders the grain is carried up over the main wheel before being bound, and a strip of canvas twelve or fifteen feet long is required as a carrier. John L. Carroll, of Creston, Iowa, has designed a binder with a low platform. This uses only three feet of canvas, and employs less power in handling the grain. Besides, the platform is open at the end, so that it will accommodate the longest grain, even corn. As its centre of gravity is low, the binder is well suited to hillsides.

A novel exhibit will be seen at the Berlin industrial exposition next summer. A Polish engineer has invented an ambulant crematory for military purposes. The object of this invention is to make use of the hygienic advantages of cremating for the disposal of the bodies of dead soldiers on battlefields, instead of burying them in numbers, thereby creating the danger of epidemics. They are mounted on low wheels, and have the appearance of the portable army baking ovens, only that they are higher and heavier.

Among the recent peculiar patents granted is one to a western inventor known as the "safety poison bottle," the object is to prevent people from getting hold of the wrong bottle in the dark. The bottle which contains the poison, whatever it is, is supplied with a contrivance, so that should a person pick it up suddenly a number of needles are projected into the hand of the one picking up the bottle. Although the bottle in some cases may be dropped the inventor believes that the safety from mistakes will more than pay for the cost.

Models of War Ships.

The models of the battle ship Oregon and the armored cruiser New York have been loaned to Messrs. Cramp & Bros. of Philadelphia for the purpose of exhibition in China and Japan as specimens of American naval architecture. Each model is an exact reproduction of the vessel it represents in every particular, and each cost the Navy Department between $7,000 and $8,000. They are about seven feet in length and give an excellent idea of the vessels as they appear in active service. Every feature of the vessel is faithfully reproduced to a carefully drawn scale. The shipment of the models to Japan was authorized by the Navy Department at the request of the Japanese minister, in order to permit of their inspection by the Japanese naval authorities. As is well known the Japanese have a high regard for the new vessels of the United States navy.

The greatest commercial drummer of the present age is the rightly-placed advertisement. It never tires, has no hotel expenses, needs no mileage tickets and finds its way everywhere. A slight charge pays for its transmission from ocean to ocean, and from the Canadas to Mexico. It travels to the outposts of civilization for the merest trifle of cost. It is a veritable globe-trotter. There is money in it. It has commercial value. It is the living seed of the future business crop. The most successful business men of to-day recognize this fact, and keep the silent drummer in perpetual motion. Those who neglect this means of soliciting trade are the losers thereby. Advertising is not a fad, nor can it be a failure, if due prudence is taken in putting the right thing in the right place.—*Age of Steel.*

One of the late developements of the X rays is in the line of detecting false diamonds. Another important application has been made surgically. By placing a subject to be examined before a screen faced with a mixture of barium, platinum, and cyanium, and allowing the rays from a tube inclosed in dark cloth to traverse the body the impression can be seen by the eye, on the surface of the screen, which is rendered fluorescent where the rays fall uninterrupted upon it, and the surgeon is consequently enabled to move the subject freely before the screen and examine the interior of the body for foreign substances or hurtful growths and distorted bones.

The Remington standard typewriter people are the owners of a patent, which broadly covers the idea of automatically moving the inking ribbon in two directions relatively to the printing point known as the "compound' or duplex feed," a great feature with all of the recent ribbon machines. It also covers broadly the tip-up platen, whereby the printing is exposed, which is also used in many machines. This patent comes as a great surprise to many typewriter manufacturers. It has been before the Patent Office fifteen years and has been the subject of much litigation, but has finally issued to the Remington people.

Studebaker, the great wagon maker, announces that he will, in self-defense, devote a portion of his extensive works to the manufacture of first-class bicycles, and sell them as low as twenty-five dollars.

Mongolian Medicines—The X Ray as a Healer.

The average American, who reads of Chinese riots and ill treatment of missionaries by the fanatical mob, who accuse the foreigners of using the eyes of Mongolian babies for medicinal purposes—is inclined to think the Chinamen are simply seeking a pretext for violence. But if one will examine the Celestial's Materia Medica, or visit the National Museum and see the curious things of which I write, the depths of a Chinaman's ignorance may be in a measure accounted for.

John Chinaman is a devout believer in the "Doctrine of Signatures—that the proper application of medicine is shown by its form or color. If the disease is in the head, the top of the ginseng root, which resembles somewhat the human body is given to the patient; for general disability, the body of the root is used, and for stomach disorders, the legs.

As a tonic birds' nest ranks next to ginseng; and this is used also for food by the higher classes. The idea of eating a bird's nest is naturally revolting; but there is no resemblance to the nests of our fethered friends in that of the Chinese swift. This peculiar structure looks like very light glue—in fact it is glue. A species of gelidium and other seaweeds are partly digested in the crop of the bird, regurgitated and moulded in shape of half a saucer upon the perpendicular face of the rock, which forms the inner bare wall of the nest.

Seed-pearls are used powdered for heart and liver trouble, and as an application for ulcers, disease of the eye, deafness and small-pox. Red roses are taken for asthma, used for flavoring medicine and eaten as preserves. Tiger bones mixed with hartshorn and the shell of the terrapin made into a jelly, is considered a remedy for disease of the bones, ague and general debility. The blistering fly, the great Chinese antidote for hydrophobia, is a large black fly about an inch long, with broad bands of yellow across its wings. It is given in powdered form, and the embryo dog is looked for in the bloody urine of the patient, caused (the blood) of course by the powerful action of the fly. A decoction of hedgehog skin is thought good for pulmonary trouble, and is taken in pills for cutaneous affections. Dried frogs are taken as a tonic, and (not on account of their hops) to bring sleep to the Celestial; while the sea-horse which never ran a race or pulled a load is used as a stimulant. The latter are worth $1.50 a pound in China. From their habit of hiding in crevices, snakes, dried and mixed with other drugs, are used as a remedy that will introduce itself into all parts of the system. The strength in the rinocerous horn is used against small-pox and fever, and dragon's teeth (the teeth of various animals known as such) are a panacea for all liver trouble. Deer's horn, antelope's horn, oyster shells, clam shells, cuttle fish bones, eel skins and locust skins, are used as pain killers by the swingers of the que, and such palate-ticklers as silk worms, salted scorpions, catapillars, maggots, and fossil crabs, are considered soverign remedies.

The Materia Medica of the Chinese comprises a vast number of drugs, and the belief that "like curses like," brings into requisition as medicines nearly everything animate and inanimate. The Pen-ts-aou-kang-mu, which corresponds somewhat to our dispensatories, published in the 16th century and still the standard Chinese medical authority, mentions about nineteen hundred drugs. Many of the disgusting animal products recorded in it have their analogues in the English pharmocopœia of less than two centuries ago, and in this day of scientific progress, we still have to swallow compounds almost as nauseating, to cure our ills. It will be an incalculable blessing to humanity and a set-back to the compounders of stomach-churners if the "soul" of electricity, the X ray, does what is claimed for it—destroys the disease germ and gives willing nature a chance to assert its healthful reign. It will indeed be an epoch in the march of science when the physician can walk into the sickroom with a radiograph in his gripsack, turn the healing light on the deadly bacteria—afflicted flesh,—and instead of prescribing drugs that destroy vitality while destroying germs, order his patient to be given only nourishment to recuperate the exhausted system.

There are nearly 1200 formulas for medicines in the pen-ts-aou, and the governing idea in the make-up is that some inexplicable power controls the virtue of medicine, without regard to physiological action. The consumption of drugs in China is enormous, and many of the natives take them while perfectly well to keep off sickness. The Chinese doctors' charges are moderate, and it is well since medicine costs so much in that country. A certain variety of ginseng is sold there for $4 an an ounce. This bifurcated root though something of a Mongolian color, is much more valuable than the Mongolian himself.

J. E. Price.

HIGHER STUDY OF MAN.

A Washington Scientist Who Devotes All His Time to It.

In a small room on the top floor of the Bureau of Education building sits a young man whose life is devoted to the study of human beings. He is not what might be called a student of human nature. His work is too practical for that. All the same he can size up a man more accurately and in less time than most men, for he has at his command a score of wonderful instruments designed for the purpose of man measurement. He can not only measure a head according to the Bertillion system, but he can to a certain extent measure its contents. This man is Dr. Arthur MacDonald, the psycho-neuralogist of the Bureau of Education.

He has long been a student of abnormal humanity—criminals, insane, inebriates, and geniuses—and it is now his purpose to extend the study to the normal classes. For this reason he has recently performed a number of his simpler, yet none the less important and interesting experiments upon the school children of Washington. He took for his subject 526 boys and 551 girls, between the ages of six and eighteen years. The first experiment was for the purpose of ascertaining the average least sensibility to distance (locality) between two points on the volar suface of wrists. This was determined by a very simple little instrument, consisting of two sharp points, the distance between which may be increased or diminished at will, on the principle of a monkey-wrench, but very much more delicate. These points are pressed gently upon the wrist, and the sensibility of the subject to distance is determined by the distance at which he is able to determine whether both points or only one are being pressed. The experiments showed that the left wrist is more sensitive than the right, and that girls are more sensitive to distance than boys.

The other experiment was for the purpose of determining the average least sensibility to heat, also on volar surface of wrists. This experiment was made with Eulenberg's thermaesthesiometer, the only one in this country, being in the possession of Dr. MacDonald. It consists of two thermometers parallel to each other, attached to each being an apparatus by which they may be heated to different temperatures. The ends of these thermometers are applied to the skin and the subject is asked to tell which is the warmer. Some subjects have been able to determine accurately up to a difference of but half a degree, but the average is less than two degrees. Again in this experiment the left wrist is more sensitive than the right, but on the other hand the boys seem to be a trifle more sensitive than the girls.

These experiments are the first ever made on the nervous system of school children. "Their practical value," said Dr. MacDonald, "is this: Any pupil 20 per cent above or below the averages for its age should be reported to the family physician. It is doubtful whether such a pupil should be allowed in school. If allowed, they should be separted from the others. There are too many bright pupils with weak bodies."

The results of these experiments on school children are as yet in the rough state, and the information which they convey so far is very general in character. It will be carefully tabulated, however, and will probably disclose some interesting facts in regard to individuals. As Dr. MacDonald says, what is true of Washington school children would probably to true of school children in general, owing to the nature of the population.

There are a large number of other experiments which Dr. MacDonald has tried on numerous occasions, for the purpose of measuring sensibility, and some of the results are extremely interesting. One of these shows the effects of mind occupation upon the breathing, which are recorded upon a delicate instrument known as a kymographion. To describe it in detail would be impossible, but nearly all the quantitative measurements of sensibility are recorded by this instrument. It consists of a clock which revolves once every minute a cylinder covered with smoked paper. On this paper a pefectly poised bamboo splinter marks the fluctuations in breathing, blood-circulation and other functions.

Dr. MacDonald not long ago invited Miss Westcott, a teacher in the Western High School, to bring some of her pupils to his laboratory for the purpose of experiments in breathing, and some valuable results were obtained regarding the effects of deep study, or concentration of the mind, arithmetical calculation, music, etc. Concentration, as is well known, lessens breathing, which is one reason why men who give their lives to study are anemic. In conducting this experiment, a belt is placed around the chest, every movement of which is recorded by upward or downward lines traced by the little bamboo pointer. The effect of laughter is to cause a great commotion on the smoked paper, showing the large increase in amount of oxygen taken into the lungs. This gives an excellent illustration of the efficacy of dining in good company. Indigestion is said to be largely the result of a decrease in breathing after eating, but when the meal is interspersed with jest and story, the supply of oxygen is increased, and dyspepsia is averted. "Laughing," said Dr. MacDonald, "is one of the best physiological things we do."

Another interesting test is that for sensibility to pain. This is ascertained by pressing to the temple or palms some sharp points on the end of a rod, with an attachment for measuring the amount of pressure, very much on the order of a spring scale.

This is an instrument of Dr. MacDonald's own invention, and he has made a number of experiments on young men and women of the wealthy classes, whom he found amusing themselves at the watering places, comparing them with similar experiments on thirty-six members of the Boston army of the unemployed. He found that members of the latter class were only half as sensitive as the young men of the wealthy classes. For the criminal classes he has found it necessary to have made an instrument, which will measure just twice the amount of pressure, as their susceptibilities are very much lower.

There are also instruments for recording the trembling of the hand and tongue, also indicated on smoked paper. Others are for measuring the sense of smell, hearing, and sight. It is maintained that in some criminals the sense of smell is very acute, and it is well known that in animals this sense is very keen, and Dr. MacDonald thinks that in time some interesting developments, will be accomplished in this direction. An ingenious instrument is that for ascertaining the electrical sensibility of the brain. It consists of two dials, one to regulate the pressure of two poles on the forehead and at the base of the brain, for accuracy in comparison, and the other to indicate the force of the current.

From the immense laboratory of Prof. Masso, in Italy, comes an ingenious instrument for testing the influence of fatigue. It consists of an arm piece in which the arm and all but the middle finger are held rigid, while to the free finger is attached a four-pound weight, which the subject is required to pull up and down until he can no longer move the finger. The member is stimulated with electricity, usually without avail, as the line on the smoked paper becomes almost straight, indicating that there is practically no movement in the finger. Then the subject is urged to make another effort to move the weight, and suddenly the lines become irregular and almost as jagged as when he first began, and the weight jumps up and down as rapidly as ever. The physical energy of the subject has been exhausted, but he gets new energy from the brain. It is a practical demonstration of the recuperative power of the muscular system. It is what is called "second wind" in the pedestrian.

The sensibility of children to suggestion is tested by a simple instrument contrived by Dr. Scripture, of Yale. It is nothing but a bare copper wire, to which a battery is attached. It gets warm when current is on, and the child is permitted to see the operation of closing the circuit. This is done several times, the child being asked to touch the wire and say whether it was getting warm or not. When this has been done several times the circuit is simply pretended to be closed and the child is again asked the question. A child very susceptible to suggestion, or a good hypnotic subject, will answer that the wire is getting warm, but the child which is governed by its own senses will reply in the negative.

Besides all these instruments, Dr. MacDonald has a complete outfit of appliances for measuring the physical man, for the purpose of comparisons. He can ascertain his weight, strength of arm, and grasp, lifting power, lung capacity, height and average stoutness. There is room for interesting studies as to the difference between quality and quantity of lungs, and various other question. He has also one of the instruments by which the Bertillion measurements are taken, and in fact when Dr. MacDonald gets through with a man, there is not much he does not know about him.

"And after all, said he, "the only way to judge a man is by his acts. It has been said that we know more about the physiology of animals than we do of men, and more about criminals and different peculiar species of men than we do of our own people.

The study of quantitative measurements of sensibility is in its infancy, and who knows what may be accomplished by it later on? People have studied rocks and plants and animals for years and years, and it is time we were studing man."

It is probable that Dr. MacDonald will read a paper on his investigations before the Anthropological Society, at which time it is expected the subject will be taken up and some more definite plans of investigation will be determined on.—*Washington Post.*

The National Electrical Exposition.

The National Electrical Exposition, which will be held in New York city from May 4 to May 30, promises to be the greatest success of the kind ever atempted. Of it the Electrical Engineer says:

Enough data with regard to the coming National Electrical Exposition in this city have appeared to show that it will possess a variety of attractions going beyond the plans of those who originally projected it. It will, indeed be fairly and generously representative of all the electrical arts, for it is hard to think of a branch that will not be included in some interesting form or other.

Although held under the auspices of the National Electric Light Association, the Exposition is very far from being restricted to the field that was covered by the Association when it was founded. As a matter of fact, the Exposition will accentuate, and direct attention to, the remarkable fact that central stations are rapidly undergoing evolution, and that every day they become less and less plants for miere lighting, and more and more reservoirs for the supply of current for a wonderfully wide range of uses, many of which are barely yet known.

All things considered, both in the scope of its displays and in the stimulas it gives to another period of prosperity in the electrical industries, the Exposition seems destined long to be memorable."

The INVENTIVE AGE has heretofore illustrated the building in which the Exposition is to be held, It is located at the corner of 43rd steet and Lexington Avenue, and is admirably arranged for an exposition of this sort. Over 150 exhibitors have already installed their exhibits and when the exhibition opens there promises to be no vacant spaces.

In connection with the opening ceremonies of the National Electrical Exhibition it is proposed to start the machinery by a circuit that has first looped in the whole continent—covering something over 6,000 miles of wire. Another feature of the Exposition will be a minature model of the plant of the Niagara Power Co., which will be operated by electricity generated at Niagara Falls transmitted from that point at night by two wires generously offered by the Western Union Telegraph Company. This working model is described as a most ingenious piece of work made by George R. Allen of Philadelphia. In the constuction of the model the scale used was one of about 100 feet to an inch and a half. Thus a comparison of this fact with the dimensions above given will give one a very fair idea of the model. All streets and buildings shown are perfect in form and direction. To one acquainted with the city of Niagara Falls, it affords pleasure to pick out well known places. The total number of buildings shown is 197, while the number of trees which grace the streets of the city in the model is 1,150. Every one of these trees Mr. Allen, the model-maker, tried to locate with precision. The model is 12 feet long by 3 feet 6 inches wide.

Still futher—and this will perhaps be considered the most wonderful feature of the entire exposition—arrangements will be made for the telephonic transmission of the actual roar made by the constant falling of thousands of tons of water at Niagara, so that the solemn booming of the distant cataract can be clearly heard by those in attendance.

Tesla, Edison, Westinghouse, Thomson, the giants of the electrical world, besides scores of others, drawn from the ranks of those who have contributed most freely to modern electrical science, are all working earnestly and harmoniously together to make the coming show the most extraordinary scientific exhibition on record.

The success of the new street railway mail service initiated by the Post Office Department in New York, Boston, Philadelphia, Baltimore, Washington, Chicago and Cincinnati has decided the Postmaster-General to introduce the system in the other large oities of the country after July 1.

Statistics of the gold and silver product of California for 1895, published in the *Mining and Scientific Press* of San Francisco, show a total gold product in the State of $15,334,317 and a total silver product of $599,779. The increase of the gold product over 1894 is $1,411,035.

The Baldwin Locomotive Works of Philadelphia have closed another contract with the Russian Government for 60 heavy freight engines, to be completed by July 1. This will make 134 locomotives built at the Baldwin works for the Russian Government since 1895.

The paper on which the World's Fair diplomas were printed, was secured from the imperial mills in Japan.

Germicidal Properties of X Rays.

The recent experiments of Prof. W. B. Pratt and Dr. Hugh Wightman of Bennett Medical College, Chicago, with X rays as disease germ and bacilli destroyers, commanded the attention of Prof. Roentgen who in an interview said:

"Yes, I have read the news with great interest, especially as I knew Prof. W. B. Pratt professionally long before he was electrotherapeutist of Bennett Medical College. I esteem him highly, and, though not being a physician, I cannot pass definitely on his discovery until the results are ocularly demonstrated in my presence, yet I think the professor incapable of promising things he cannot accomplish. "I am not acquainted with Dr. Hugh Wightman of Bennett Medical College, but as the cable mentions him as one of Prof. Pratt's associates, he is doubtless fully as trustworthy. Some great balm for man's many ailments has doubtless been found."

I am told that diphtheria was slain outright in the Chicago experiments, while no final and positive verdict is as yet given as to the effect on the bacilli of cholera, pneumonia, typhoid, and other plague germs tested. This is astonishing and partly disappoints my anticipation. I consider diphtheria and cholera the most deadly of plagues and believe positively that the bacilli of the other scourges would be the least difficult to kill. But I am confident that eventually the X ray will prove an effectual cure for all such diseases.

I will rejoice when it will be in the power of every competent physician to kill these bacilli. When once having located them the modus of annihilation will be a mere technicality.

If Profs. Pratt and Wightman have successfully completed their experiments their names should go down to posterity as benefactors of the race, since humanity is immeasurably benefited by their work."

A London dispatch on this same subject says: Scientists here are decidedly skeptical as to the germicidal properties of the X rays, through they are not prepared to deny the reported discovery in Chicago. They want details, however, before believing it.

Natural and Animal Magnetism the Same.

In speaking ef his investigations in the realm of mesmerism and magnetism Mr. James M. Rusk says: "But my studies took another turn and I have been hunting down hypnotism. I came to the conclusion more than a year ago that light explained it—light in the theosophical sense mind you. When the Roentgen ray was made known, I put my theories to the test. My experiments were based upon Maxwell's theory of light and Dana's cortex expositions.

Now if you will take up the "Arena" for December 1895, you will find a startling article by Henry Gaullier, a French writer, vouched for by Hon. Carl Schurz. This Gaullier claims that when two persons are hypnotized and one of them is made to report what he sees in regard to the other, that it is found that red and blue lights are seen issuing from the nostrils, eyes, etc., of the hypnotized person observed. Forty years or more ago, Baron Reichenbach published that sensitives placed in the dark saw red and blue lights issuing from the poles north and south respectively of a magnet.

Now was this an allusion? Did the hypnotized reporter see what he claimed or was it an illusion? There is no need of going further. A physical test can now be made. If in either case it was not an illusion, photographic or fluorescent effects can be had. I have made analogous tests and know that magnetism in nature and magnetism in man—or natural and animal magnetism are one.

An English woman, Mrs. Savage, annoyed by the trouble of having to hold on to her hat on a windy day, invented what she appropriately termed a "breeze hat grip"—a thin piece of metal with flexible projecting fingers not unlike a comb, placed on the inside of the front portion of the hat. The fingers engage the hair and hold the hat secure. It was a simple invention and a manufacturing firm, disregarding the patent obtained by Mrs. Savage began the manufacture and sale of the article. In the suit brought verdict has been rendered for the inventor and something of the importance of it may be judged by the fact that the evidence showed that 288,000 of the article had been sold in six months. This is only another exemplification of the fact that there's money in small and useful inventions.

A test run of 450 miles across the country from Fort Meade to Fort Yates and back was recently made under orders from Col. Sumner by Lieut. Cabell and two privates from the cavalry stationed at Fort Meade. The trail lay over rough roads, the broken and difficult country along the Cheyenne and stretches of prairie covered with hillocks, gopher mounds and bunch grass—conditions which made some parts of the journey arduous. Allowing one day's rest at Fort Yates, however, they were on the road but seven and a half days. A troop of cavalry would require from eight to nine days to accomplish the march, and the result tends to confirm the prediction of experts that the bicycle is destined to perform an important function in the military operations of the future.

Injurious Effects of Tobacco Using.

If statistics gathered at leading colleges have any weight it will have to be admitted that the free use of tobacco has an injurious effect upon the health of students. The best known and most thorough tests to determine this question were made at Yale and Amherst, which institutions can be taken as fairly representing the larger and the smaller colleges. In 1891 the physician at Yale published the results of his observations on the use of tobacco among undergraduates. In a class of 147 students he found that in four years the 77 who never used tobacco surpassed the 70 who did use it 10.4 per cent in gain in weight, 24 per cent in increase of height, and 26.7 per cent in growth of chest girth. But the most marked difference was in the gain in lung capacity, the non-users recording a gain of 77-5 per cent greater than the habitual chewers or smokers. This is a remarkable showing, but it is surpassed by an investigation made among the undergraduates at Amherst. It was found that during the four years of a college student's life at that school the non-users of tobacco gained 24 per cent more in weight, 37 per cent more in height, and 42 per cent more in chest girth than the tobacco users, while the increase in the lung capacity of the former was 75 per cent greater than in the latter. This larger relative increase among the non-users of tobacco at Amherst than at Yale is accounted for by the fact that the average age of the students at the former school is lower than at the latter school and hence they are more susceptible to any cause which affects them injuriously.

But the physical effects are not the only bad results of tobacco using. It harms the intellectual faculties also. According to Professor Fisk, of the Northwestern University, when a college class at Yale had been divided into four sections according to scholarship it was discovered that the highest section was composed almost entirely of non-smokers, and the lowest section almost entirely of smokers.—*Philadelphia Press.*

The $1 Edison Motor.

The cut shown herewith represents one of the most satisfactory and perfect of all small dynamos, capable of running toys, fans, light models and for innumerable uses. To the youth scientifically inclined, it offers great satisfaction and benefit.

It is a marvel in price, and if not made in large quantities would cost from $5 to $10. We can fur-

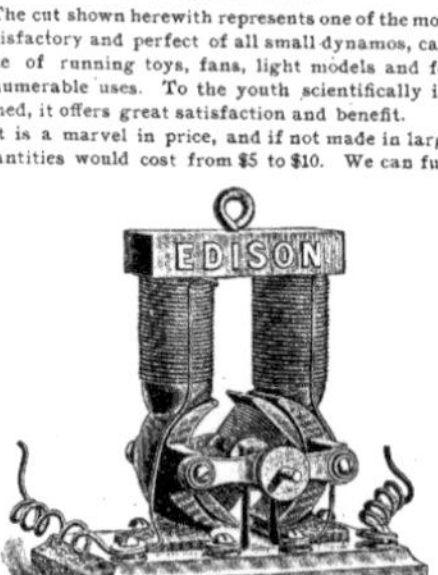

nish the motor for $1 and will send one free to any person forwarding us three new subscribers.

A New Lifeboat.

Much interest attaches to a new invention of an arrangement of floating tubes, so constructed as to prevent boats from sinking. The tubes are run along each side of the boat, and they will not only keep it afloat, even when capsized, but will support all its occupants, if need be, who can readily lay hold of them: When the boat is capsized the tubes come to the level of the water, although it is claimed by the inventor, that the mere adjustment of the tubes makes capsizing almost an impossibility.

Scientific Activity.

A year or two ago attention was called to the predictions of an eminent authority that we were entering upon a period of scientific activity that would far transcend any previous experience. The most indifferent observer cannot fail to be amazed at the manner in which his prophecy is being fulfilled. Chemists are astonished to find that the long familiar atmosphere contains a large proportion of a substance hitherto unknown—the strange and inert argon; and helmium, so long known in the spectrum of the sun, is discovered as a terrestial element. With the liquefaction of air and hydrogen we are introduced to a new chemistry of cold. The developement of the electric furnace brings great possibilities in the reduction af certain metals, and among its remarkable products yields calcium carbide, the source of acetylene, which is the first hydrocarbon to be produced artificially on a large scale, and a revolutionary achievement in chemical synthesis. Most surprising of all is the new form of radiant energy. Eager students everywhere have quickly begun experimenting with the mysterious X rays, and in a few days we are given the new art of "shadowgraphy," which promises among other marvels, that the sick can have their diseased organs brought to view, while the curious can have their skeletons photographed while they wait. The details of this new photograghy are being improved daily. Other epoch-making discoveries are almost grasped, and it is clear that, with so many roads opened to peaceful conquest, our end-of-the-century days leave no time for demoralizing wars over political boundaries.

The Inventor of Bessemer Steel.

The silly attempt that has been made by the President of the American Institute of Mining Engineers to deprive a celebrated Englishman of his just claims to a great invention will meet with little approval even in America, and with general condemnation from all well-informed persons in this country. It seems rather hard that Sir Henry Bessemer should feel constrained, in the present day, to write a long letter to some of the leading technical journals in vindication of his right to be regarded as the inventor of the Bessemer steel process. The letter, however, is an interesting one, and well worthy of perusal by all students of metallurgy. This new president of an honored society in America will doubtless have reason to regret that he has brought discredit upon himself, and, what is of far more importance, upon the society which did him the honor to elect him, by this stupid version of a story, the true value of which is well understood by the great majority of steel makers in America. Englishmen are not at all likely to take the matter seriously, or to think it worth their while to discuss it. It is well known that this Kelly, to whom Mr. Weeks would give the credit of the invention, was a maker of wrought iron in Pittsburg. Some fifty years ago he tried some experiments with molten cast iron, with the object of refining it for the after process of puddling. He did in the open what had been previously done in a "finery" furnace. He blew air down upon the metal with the view of working off the carbon and silicon. He did not succeed in producing anything better than the hard plate metal for use in the puddling furnace. He does not seem to have even dreamt of making steel at all; but when he heard of of the success of the Bessemer process, he was astute enough to rush to the American Patent office, and with affidavits and claims he succeeded in obtaining a patent for forcing air into molten cast iron, in direct imitation of Bessemer's previously published specification. Sir Henry Bessemer's claim rests upon too solid a basis to be shaken by this resurrected story which, to the honor of Americans, was buried by them many years ago.—*The Trade Journals' Review.*

An Imitation Hard Rubber.

A German patent has been taken out for the production of an imitation of hard rubber out of sawdust. It consists in mixing sawdust with chromatized glue, forming the object out of this mass by pressure between wooden or metal forms, then placing it in heated oil, varnish or tar until all the moisture is driven out. The article is then placed in a drying oven, where it is heated to between 400 and 600° F., where it soon takes on the appearance and properties of hard rubber. Sawdust of resinous woods is claimed to be particularly adapted for getting the best results with this imitation.

A Kansas man is the inventor of a means of making bricks from straw, and claims that they will be particularly suitable for paving streets. They are made of straw and wood pulp and at a cost equal to one-third the cost of making bricks from clay.

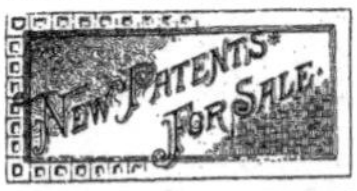

Advertisements inserted in this column for 20 cents a line (about 7 words) each insertion. Every new subscriber sending $1.00 to THE INVENTIVE AGE will be entitled to the AGE one year and to five lines three times FREE. Additional lines or insertions at regular rates.

WANTED.—Some one to join as co-partner with me to bring out Link and Pin Car Coupler No. 485,333. Rev. J. W. Poston, Holly Springs, Miss. 5-7

FOR SALE.—Good patent. Easy and cheap to manufacture. Thoroughly tested. Many in use. Recommeded by all using it. Ready seller. Price reasonable. Good investment. Pat. Jan. 9th, 1894. For references and price, terms etc., address P. O. Box, No. 1072, Tacoma, Wash. 5-7

FOR SALE.—Patent No. 501,800 issued 18th of July 1893; flash light apparatus; being compact light simple, and durable; always ready, can be used with all kinds of Cameras. Will sell the entire patent, or manufacture on royalty. A. T. Mallick, Jamestown, N. Dak. 3-5

FOR SALE.—Or manufacture on royalty United States patent 518,165, Canada patent 46,029; also several unpatented inventions; one-third to party that will furnish money to have them patented. Enclose stamp. E. H. Thalaker, Petersburg, W. Va. 3-5

WANTED.—A Partner to take an undivided one-half interest in the best 25 cent household article on earth. Is patented in both U. S. and Canada, have sold several State and County rights, also over 5,000 of the goods. I have the machinery and have reserved the right to manufacture all goods and 4 cents will cover the cost of each article. I will give the right man a snap. Address, J. E. Hyames, Gobbleville, Van Buren Co., Mich. 3-4

FOR SALE.—Patent, 492,875, Folding Hoe, entirely new principle. Will sell State or County rights cheap. Any live man can make money right in his own state or town handling the hoe. Inexpensive to manufacture, and sells on sight. For full information, address Dr. F. A. Rice, 62 Main street, Lockport, N. Y. 2-7.

FOR SALE.—Valuable Patent No. 532,324 for improvement in metal carriers or baskets for bottles and other articles. Address Patentee, H. M. Kolb, 732 Sansom street, Philadelphia, Pa. 2-4

FOR SALE.—Canadian Patent, the Highwine and Whiskey Distillers Continuous Beer Still. Manufacturers and capitalists who wish to purchase valuable patents can make very large profit. Operated in U. S. Distilleries. Address, Hermann Binz, Frankfort, Ky. 2-4

FOR SALE.—I have invented a perfect Nut Lock. Who will pay for the patent for three or four states? This is a practical invention. Address INVENTIVE AGE, or W. H. Dillon, Virginia, Ill. 1-3.

WILL DIVIDE.—Any person who will invest just a little money about $65.00 to assist an inventor in getting valuable patent will get half of the patent assigned to him. Write for particulars to Thos. C. Davis, Ridgely, Md. 1-3

FOR SALE.—Or on royalty, Patent No. 554,643; a folding suit or garment hanger; simple and cheap; big profits in it. Address, Chas. Behrend, Jr. Oconomowoc, Wis. 4-6.

FOR SALE.—U. S. Patent No. 554,735. Corn Cultivator attachment. Will be on sale April 1st, for 60 days only. For information, Address, W. S. Runyon, Lexington, Iowa. 4-7

FOR SALE.—A half interest in my "Hair Grower." I have made a discovery whereby I can grow a full head of hair on the baldest head; cure any case of dandruff or scalp disease to dead certainty. I want a partner with money to put this new discovery upon a large scale. A fortune for some one. Sample bottle sent upon receipt of $1.50. Address, Geo. W. Schoenhut, Eldora, Iowa. Box 98.

FOR SALE.—A complete set of Patent Office Reports and Official Gazette of the U. S. Patent Office from 1790 to 1894. H. B. Dickinson, Ravenna, O., Adm'r of Estate of Bradford Howland. tf.

BUSINESS SPECIALS.

Advertisements under this heading 20 cents a line each insertion—seven words to the line. Parties desiring to purchase valuable patents or wanting to manufacture patented articles will find this a valuable advertising medium.

WANTED.—Publisher to publish and introduce Walsh's Perpetual Calendar and Almanac on royalty; or I will sell the copyright. A complete calender, absolutely perpetual. No complicated rules or computations. James A. Walsh, Helena Mont. 2-4

WANTED.—Reliable manufacturer to manufacture on royalty, or will sell at a moderate figure. Caliper Adjustment. Described elsewhere. Address, Arthur Munch, 653 East 5th, st., St. Paul, Minn. 4-7

Aftermath.

CONGRESS is likely to pass the pending bankruptcy act—with the involuntary feature in it.

THE Railway Age estimates that about 40,000 miles of railroad will be constructed this year.

THE entire business center of the town of Cripple Creek, Col., was destroyed by fire on the 26th ult., entailing a loss of over $1,000,000.

THE confidence of American capital in the ultimate success of the Cuban cause is shown in the fact that the $2,000,000 loan was subscribed for in New York four times over.

IN Watertown, S. D., flax straw is being used for fuel in the electric light plant at that city at a cost of $1 a ton and two tons are considered equivalent to one ton of the best soft coal.

IT is estimated that the extension of electric traction roads in Chicago contemplated for this year will require the expenditure of over $15,000,000. The mileage is to be increased from 800 to over 1,000 miles.

SAYS the Hutchinson, Kans., *News*: "A Hutchinson man was in the City of Mexico recently and wanted to go to San Francisco. The price of a railroad ticket was $112 one day $116 another, $123 a third and $119 the next day (in silver). It was $62.50 in gold all the time.

IT is reported that a board of arbitration has been selected by the managements of the General Electric and Westinghouse Companies to undertake the solution of all problems arising in connection with the recent patent arrangement between the General Electric and Westinghouse Companies.

THE British Court of Queen's Bench has lately declared valid the English patents of the Welsbach Incandescent Light Company and has issued a permanent injunction against their infringement. The suit has been watched with much interest, as about 20,000,000 of these lights are said to be in use.

A coast paper says that three years ago this spring Peter Wilborg left his home in California for Cook's Inlet, Alaska. For nearly three years he worked unceasingly in the gold regions. Last November he came back to the United States on a visit, bringing with him $48,000 in gold dust and nuggets.

THE storage battery will shortly be given an opportunity under fair conditions to prove whether or not it is capable of handling cars practically and economically on an extensive street railway system. The Englewood & Chicago Electric Railway Company, of Chicago, has placed contracts for the construction and equipment of a large portion of its contemplated line which will embrace about sixty miles of track when completed. In Washington City it was not a success.

THE case of George E. Kirk vs. the United States, appealed from the decision of the Court of Claims, has been decided by the Supreme Court. Justice Brown delivered the opinion affirming the decision of the court below. Kirk is the assignee of Samuel Strong, who claimed to be the inventor of the letter box, in use upon the streets of cities. The court held that Strong's patent covered another device, and therefore decided against Kirk.

PREPARATIONS are being made at the Watervliet, N. Y., arsenal to build the largest gun for coast defense that has ever been constructed. The gun will be of 16-inch caliber and with the proper elevation, this gun would send a shot weighing more than 100 pounds a distance of 16 miles, using a charge of about 600 or 700 pounds of powder. This will be a costly engine of war, and will probably be mounted in New York harbor. No vessel yet constructed carries sufficient armor to resist the impact of the shell which would be hurled by this monster cannon.

Another Great Combination.

Popular Science is one of the best magazines published. It is 30 years old and covers all the sciences in a popular and instructive way. We can furnish Popular Science and the INVENTIVE AGE one year postage paid for $1.50. This combination rate can be offered but a short time as the subscription price of Popular Science is soon to be increased. Address the Inventive Age, Washington, D. C. If Zell's Cyclopedia is also wanted—the three for $2.75.

A STANDARD WORK

The American Newspaper Directory For 1895...

A BOOK OF 1397 PAGES, containing a brief description of each place where a newspaper is published, the population, railroads, county, local industries, etc. Gives a complete description of each and every newspaper published in the United States and Canada, day of issue, size, editor and publisher's names, circulation, politics or class. Also valuable and trustworthy statistics and tables. It is a vertible cyclopedia and is relied on by

The Department at Washington, Railroads, Politicians, Advertisers & all interested in newspaperdom.

Price Five Dollars.

Geo. P. Rowell & Co., Pub'r's.

10 Spruce Street, New York.

STANDARD WORKS.

"The Electrical World." An illustrated weekly review of current progress in electricity and its practical applications. Annual subscription $3.00

"Electric Railway Gazette." An illustrated weekly record of electric railway practice and development. Annual subscription $3.00

"Johnston's Electrical and Street Railway Directory." Published annually $5.00

Dictionary of Electrical Words, Terms and Phrases. By Edwin J. Houston, Ph. D $5.00

The Electric Motor and its Applications. By T. C. Martin and Jos. Wetzler $3.00

The Electric Railway in Theory and Practice. The first systematic treatise on the electric railway. By O. T. Crosby and Dr. Louis Bell $2 50

Alternating Currents. By Frederick Bedell, Ph.D., and Albert C. Crehore, Ph.D $2.50

Gerard's Electricity. Translated under the direction of Dr. Louis Duncan $2.50

Theory and Calculation of Alternating-Current Phenomena. By Charles Proteus Steinmetz $2 50

Practical Calculation of Dynamo Electric Machines. By A. E. Wiener $2.50

Central Station Bookkeeping. By H. A. Foster $2.50

Continuous Current Dynamos and Motors. By Frank P. Cox, B.S $2.00

Electricity at the Paris Exposition of 1889. By Carl Hering $2.00

The Elements of Static Electricity. By Philip Atkinson, Ph.D $1.50

Electricity and Magnetism. By Edwin J. Houston, Ph.D $1.00

Electric Measurements and other Advanced Primers of Electricity. By Edwin J. Houston, Ph.D $1.00

The Electrical Transmission of Intelligence and other Advanced Primers of Electricity. By Edwin J. Houston, Ph.D $1.00

Electricity One Hundred Years Ago and To-day. By Edwin J. Houston, Ph.D $1.00

Alternating Electric Currents. By Edwin J. Houston, Ph.D. and A. E. Kennelly, D.Sc. $1.00

Electric Heating. By E. J. Houston Ph.D and A. E. Kennelly, D.Sc $1.00

Electromagnetism. By E. J. Houston, Ph.D and A. E. Kennelly, D.Sc $1.00

Electricity in Electro-Therapeutics. By E. J. Houston, Ph.D., and A. E. Kennelly, D.Sc $1.00

Electric Arc Lighting. By E. J. Houston, Ph. D., and A. E. Kennelly, D. Sc. $1.00

Electric Incandescent Lighting. By E. J. Houston, Ph.D. A. E. Kennelly, D.Sc $1.00

Electric Motors. By E. J. Houston, Ph.D., and A. E. Kennelly, D.Sc $1.00

Electric Street Railways. By E. J. Houston, Ph.D., and A. E. Kennelly, D.Sc $1.00

Electric Telephony. By E. J. Houston, Ph.D, and A. E. Kennelly, D.Sc $1.00

Electric Telegraphy. By E. J. Houston, Ph.D and A. E. Kennelly, D.Sc $1.00

Alternating Currents of Electricity. By Gisbert Kapp $1.00

Recent Progress in Electric-Railways. By Carl Hering $1.00

Original Papers on Dynamo Machinery and Allied Subjects. By John Hopkinson, F.R.S $1.00

Davis' Standard Tables for Electric Wiremen. Revised by W. D. Weaver $1.00

Universal Wiring Computer. By Carl Hering $1.00

Experiments with Alternating Currents of High Potential and High Frequency. By Nikola Tesla $1.00

Lectures on the Electro-Magnet. By Prof. Silvanus P. Thompson $1.00

Dynamo and Motor Buildings for Amateurs. By Lieutenant C. D. Parkhurst $1.00

Reference Book of Tables and Formulæ for Electric Street Railway Engineers. By E. A. Merrill $1.00

Electric Railway Motors. By Nelson W. Perry $1.00

Practical Information for Telephonists. By T. D. Lockwood $1.00

Wheeler's Chart of Wire Guages $1.00

Practical Treatise on Lightning Conductors. By H. W. Spang .... .75

Tables of Equivalents of Units of Measurement. By Carl Hering .... .50

Copies of any of the above books or of any other electrical book published will be sent by mail, postage prepaid, to any address in the world on receipt of price.

THE INVENTIVE AGE,
Washington, D. C.

THE INVENTIVE AGE wants live agents in every part of the country. Readers of this notice will confer a great favor by sending us names of persons who would likely take an interest in the matter. Our cyclopedia premium offer is a great winner. The book alone is worth double the amount, $2, charged for the two.

ANY person sending us the names and addresses of five inventors who have not yet applied for patents will receive the INVENTIVE AGE one year free.

INVENTIVE AGE BUILDING.

PREMIUMS TO SUBSCRIBERS.

Read the following offers to new subscribers:

OUR $1 OFFER.

THE INVENTIVE AGE one year and two copies of any patent desired, or one copy of any two patents $1 00

THE INVENTIVE AGE one year and a list of 50 firms who manufacture and sell patented articles 1 00

THE INVENTIVE AGE one year and Altograph map of the City of Washington 1 00

THE INVENTIVE AGE one year and a five line (35 words), advertisement in our "Patents For Sale," or "Want" column, three times 1 00

OUR $1.50 OFFER.

THE INVENTIVE AGE one year and Robt. Grimshaw's famous book "Tips to Inventors" $1 50

Address all communications to
THE INVENTIVE AGE,
Washington, D. C.

A Great Novelty.

ELECTRIC LIGHT
FOR THE NECKTIE.
$1.50.

Complete with Powerful Pocket Battery and all accessories, postpaid.
The Lamp an Efficient Beauty.
Battery 3 1-2 by 3-4 by 1 3-4 inches.

This elegant outfit is practical, brilliant and satisfactory, and every part is guaranteed.

Electricians who want a piece of jewelry that is emblematical, or anybody who appreciates a real novelty, can, by ordering now, secure a complete outfit for $1.50. Order at once of
THE INVENTIVE AGE,
Washington, D. C.

Given FREE for four new subscribers.

HOW TO MAKE A DYNAMO.
By ALFRED CROFTS.

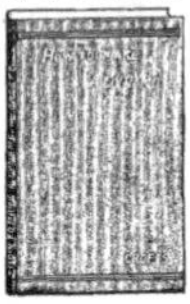

A practical work for Amateurs and Electricians, containing numerous illustrations and detailed instructions for constructing Dynamos of all sizes, to produce the Electric Light, containing 96 pages of genuine information which will enable anyone to construct a Dynamo either for pleasure or profit.

Large 12 mo,
Cloth, 75 Cts.

PATENTS.

AUSTRALIA.

List of Popular Electrical Books.

Everybody's Hand Book of Electricity.

With a glossary of electrical terms and tables for wiring. By Edward Trevert. 120 pages, 50 illustrations. Cloth, 12 mo., 50 cents. Paper, 25 cents.

Dynamos and Electric Motors.

And all about them. By Edward Trevert. Cloth 12mo. $1 cents.

Electricity And Its Recent Applications.

By Edward Trevert. A book complete on all subjects of electricity. Price. $1.00

Embracing practical hints upon power-house dynamos, motor, and line construction. By Edward Trevert. Cloth, 12mo. $2.00

Electric Motor Construction for Amateurs.

By Lieut. C. D. Parkhurst. 18mo., 115 pages, illustrated, cloth.   Price, $1.50
An excellent little book, giving working drawings.

Practical Handbook of Electro-Plating.

By Edward Trevert. 75 pages, limpo cloth, 50c.

Hand Book of Wiring Tables.

For arc, incandescent lighting and motor circuits. By A. E. Watson.   Price, 75 cents.

A Treatise on Electro-Magnetism.

By D. F. Connor, C. E. This is a very interesting book which everybody should read. Cloth.   50 cents.

How to Build Dynamo-Electric Machinery.

A complete and practical work on this subject. 380 pages, numerous illustrations. By Edward Trevert. With working drawings, $2.50.

Questions & Answers about Electricity.

A first book for students. Theory of electricity and magnetism. Edited by Edward T. Bubier, 2d. Cloth, 12mo.   50 cents.

Experimental Electricity.

By Edward Trevert. Tells how to make batteries, bells, dynamos, motors, telephones, telegraphs, electric bells, etc. A very useful book, finely illustrated. Cloth.   Price, $1.00

How to Make and Use a Telephone.

By Geo. H. Caty, A. M. A thorough and practical work on this subject. Cloth, illustrated.   Price, $1.00

Arithmetic of Magnetism & Electricity.

Up to date. The latest and best book on the subject. New ready. By John T. Morrow, M. E., and Thorburn Reid, M. E., A. M., Associate Members, American Institute of Electrical Engineers. $1.00

Electricity for Students.

By Edward Trevert. Fully illustrated. Bound in a neat cloth binding. Price, Postpaid $1.00

How to Build A Fifty Light Dynamo.

Or Four H. P. motor. By A. B. Watson Cloth.   Price, 50 cents

Address **THE INVENTIVE AGE,**
WASHINGTON, D. C.

Le Count's Cribbage.

Postage 24c.
Price 85c. Each.

It will score for three or six players, and keep tally of the games. The case is made of black walnut, nicely polished; has a solid metal top, polished and nickel plated. The slides of compartments are metal and polished. This board is superior to Boards that sell for three times the price. Full directions printed with each board.

With "Inventive Age" one year, $1.60.

GREAT PREMIUM OFFER! = =

Valuable
Sensible
Economical

Zell's Condensed Cyclopedia
Always Sold at $6.50 Complete.

Complete in One Volume. A Compendium of Universal Information, Embracing Agriculture, Anatomy, Architecture, Archeology, Astronomy, Banking, Biblical Science, Biography, Botany, Chemistry, Commerce, Conchology, Cosmography Ethics, The Fine Arts, Geography, Geology, Grammar, Heraldry, History, Hydraulics, Hygiene, Jurisprudence, Legislation, Literature, Logic, Mathematics Mechanical Arts, Metallurgy, Metaphysics, Mineralogy, Military Science, Mining, Medicine, Natural History, Philosophy, Navigation and Nautical Affairs, Physics, Physiology, Political Economy, Rhetoric, Theology, Zoology, &c.

Brought down to the Year 1890. Substantially Bound in Cloth with the Correct Pronunciation of every Term and Proper name, by L. Colange, LL.D. Handy Book for Quick Reference. Invaluable to the Professional Man, Student, Farmer, Working-Man, Merchant, Mechanic. Indispensable to all.

In this work all subjects are arranged and treated just as words simply are treated in a dictionary--alphabetically, and all capable of subdivision are treated under separate heads; so that instead of one long and wearisome article on a subject being given, it is divided up under various proper heads. Therefore, if you desire any particular point of a subject only, you can turn to it at once, without a long, vexatious, or profitless search. It is a great triumph of literary effort, embracing, in a small compass and for little money, more (we are confident) than was ever before given to the public, and will prove itself an exhaustless source of information, and one of the most appreciated works in every family.

It cannot in any sense--as is the case with most books--be regarded as a luxury, but will be in daily use wherever it is, by old and young, as a means of instruction and increase of useful knowledge.

Thousands of leading men offer the highest testimonials. It is not a cheap work it is standard.

Never before, even in this age of cheap books, was such a splendid offer made, viz:

$2 For Cyclopedia, Postage Paid. **$2**
For Inventive Age, One Year.

An Extra Copy of Cyclopedia to any person sending us Six New Subscribers. The Same Offer is made to old Subscribers renewing for another term.

Address THE INVENTIVE AGE, Washington, D. C.

The Inventive Age
AND INDUSTRIAL REVIEW
A JOURNAL OF MANUFACTURING INDUSTRY AND SCIENTIFIC PROGRESS

Seventh Year.
No. 6.

WASHINGTON, D. C., JUNE, 1896.

Single Copies 10 Cents.
$1 Per Year.

The Largest Dredge in the World.

The illustration presented herewith gives some idea of what inventive genius has done in the way of perfecting hydraulic suction dredges. Those who understand the nature of the Mississippi and Missouri rivers will realize the importance of some means of disposing of the numerous sand bars that constantly impede navigation. Experiments in dredging have been so successful that in 1895 the Mississippi River Commission decided to secure a giant in that line. The "Beta," built by Lindon W. Bates, of Chicago, the successful bidder in a list of fifteen, is the result. It far surpasses in its capacity any dredging machine ever constructed, and it is without doubt able to handle material at a lower cost than any excavating machine of any type ever

suction pipes 33½ inches in diameter. There is a battery of four 375 H P. boilers.

To convey the material the required 1,000 ft. to the rear of the dredge, a 33½-in. steel pipe is floated on the water by means of steel pontoons.

For the data and illustration herewith we are indebted to Engineering News, of New York. In its issue of April 23d was presented an exhaustive article regarding the dredge together with detail illustrations.

Prof. Langley's Steam Airship.

Prof. Langley, secretary of the Smithsonian Institution, probably the foremost student in aeronautics in the world, recently conducted the most successful experiments with mechanical flight ever recorded. Only meagre reports have been given out by the close-mouthed inventor, but it is known that

Motor Cars in England.

A bill to exempt horseless carriages from the statutory restrictions now in force, has been introduced in the House of Lords and there can be little doubt that it will pass into law very soon. A great stimulant will thereby be given to inventive talent, and we may expect to see many varieties of motor car competing for the favor of the public. Already a good many experimental machines have been made in England, most of these using an oil motor, with very low grade oil, as the propelling force. It does not yet appear that electricity has any chance of being available for this purpose, unless some new development in the formation of light storage batteries can be effected. Doubtless this is only a question of time.

THE "BETA," LARGEST DREDGE EVER CONSTRUCTED, NOW IN USE ON THE MISSISSIPPI RIVER.

built. The most powerful dredges ever built hitherto have been able to excavate in a day only about as much as this machine handles in an hour, according to its record in the recent government tests. The barge which carries the machinery constituting this dredge, is 172 feet long, and 40 feet wide. The capacity of this dredge is about 6,000 cubic yards per hour. There are six 19½-in. suction pipes, each carrying at its lower end a revolving cylindrical cutter 5 ft. in diameter. Each suction pipe is contained in a water-tight pontoon buoyed up by an air chamber, and the three pontoons on each side are braced together to form a single rigid structure. There are, therefore, two sets of pontoons with three suction pipes and three cutters each. The cutters are 5 ft. in diameter and each has 13 cutting blades to loosen the sand so that it can be drawn into the suction pipes. The discharge of these centrifugal pumps is 33 inches in diameter and the

he has succeeded by the use of steam power, in raising his aerodrome to a height of 100 feet in a course of regular spirals of about 100 yards in diameter and covered a distance of about half a mile. When the steam gave out the ship didn't tumble down, but came down gently and without injury to the ship. Prof. Langley is continuing his work and before the end of the century promises to bring forth a feasible means of navigating the air.

There are now about 150 bicycle factories in the country, having an average capital of $100,000. That makes up $15,000,000. There is one firm that has $1,000,000 capital, and several there are well up in the big thousands. Besides the exclusive bicycle factories, there are tire factories, chain factories, the concerns that manufacture tubing, pedals, saddles and other parts of the bicycle that add up toward the grand total of $30,000,000.

An Electric Current Around the Earth.

The greatest feat in telegraphy ever accomplished was the sending of a message around the world by land and submarine cable, from the New York Electrical Exposition on the 16th ult. The message dispatched by Chauncey M. Depew was sent around the world, a distance of 28,600 miles, in 50 minutes. The route of the message was via Chicago, Los Angeles, San Francisco, Vancouver, Winnipeg, Montreal and Canso to London, thence to Lisbon, Gibraltar, Malta, Alexandria, Suez, Aden, Bombay, Madras, Singapore, Shanghai, Nagasaki and Tokio, from whence, in default of passing over the Pacific Ocean, it was repeated back to New York and received in the Exposition Hall by Thomas A. Edison.

The Lake Street Elevated Railroad in Chicago has now been fully equipped with electricity.

Inventive Age

Established 1889.

INVENTIVE AGE PUBLISHING CO.,

8th and H Sts., Washington, D. C.

ALEX. S. CAPEHART.   MARSHALL H. JEWELL.

The INVENTIVE AGE is sent, postage prepaid, to any address in the United States, Canada or Mexico for $1 a year; to any other country, postage prepaid, $1.50.
Correspondence with inventors, mechanics, manufacturers, scientists and others is invited. The columns of this journal are open for the discussion of such subjects as are of general interest to its readers.
Technical matter is particularly desired. We want practical information from practical men.
Nothing will be published in the editorial columns for pay.
The INVENTIVE AGE is thoroughly independent.
Advertising rates made known on application. Special facilities for furnishing cuts of any patented article together with descriptive article. Business specials 25 cents a line each insertion, 7 words to the line. No advertisement less than 50 cents.
Address all communications to THE INVENTIVE AGE, Washington, D. C.

Entered at the Postoffice in Washington as second class matter.

WASHINGTON, D. C., JUNE, 1896.

Among the papers in the District of Columbia devoted to inventions and patents, none has credit for so large a regular issue as is accorded to the "Inventive Age," published monthly at Washington.—*Printers Ink.*

THE Iron Age says it might as well be acknowledged now as later that the real trouble with business is that this is presidential year.

THE thirty-third meeting of the American Society of Mechanical Engineers held in St. Louis, on the 19th ult., was well attended and several interesting papers were read.

AFTER a long series of experiments to determine the relative digestibility of oleomargarine and butter, conducted by Dr. Aolph Jolles, it appears that all conditions being similar, there is practically no difference either in digestibility or neutritive value.

AN ARTICLE in *Electrical World* by W. M. Stine, suggests the greatest opportunity now open to the inventor—that of devising some means of creating "cold light;" that is some means of obtaining light without heat—either in the light itself or the means used to produce it.

THE bill before Congress to amend the copyright law makes the public performance of copyrighted dramatic and musical compositions without the consent of their proprietor an offense to be punished by fine, and such a performance, when willful and for profit, becomes a misdemeanor and is made punishable by imprisonment.

A READER of the INVENTIVE AGE living in Milwaukee, Wis., having read of Prof. Langley's experiments with mechanical flight, says that he has invented a flying machine superior to any thus far devised. He says he has studied the principles employed by Langley, Maxim, Lillienthal and others and has evolved a principle entirely new.

WHEN the great Siberian railway is completed by the Russian government it will be possible to go "around the world" in a month. Some idea of the magnitude of this undertaking can be had when it is taken into consideration that the new line will require over 2,000 locomotives, over 3,000 passenger coaches and 35,000 freight cars. It is expected the work will be completed in five years.

THE management of the Rhode Island Fair Association, has sent out a circular to inventors of motor carriages, or vehicles, announcing the fact that at the State Exposition at Providence, September 7-11, a speed contest and exhibition of horseless vehicles will be held and prizes to the amount of $5,000 distributed. The "Horseless Age," New York City, will publish full instructions and particulars.

A PREMIUM of $250 is offered by the Scientific American for the best essay on "The Progress of Invention During the Past Fifty Years." It is stipulated that the paper shall not exceed in length 2,500 words. A committee of three will pass upon the merits of the essays in competition and the prize paper will be published in the special 50th anniversary number of the Scientific American of July 25.

COMMISSIONER Seymour recommends that definite steps be taken for the maintenance of a gallery, to be located in Washington, where the 158,000 valuable models of inventions now in the custody of the Patent Office can be fittingly deposited and permanently exhibited. Such a repository would be of incalculable value and interest not only to inventors but to the industrial world generally, as illustrative of the development and state of the arts.

IN COMMENTING on proposed legislation to improve the patent system, *Electricity* agrees with THE INVENTIVE AGE, *Scientific American* and other magazines in their advocacy of the measure now before Congress—the outgrowth of the meeting of the American Bar Association—but contends that an honest and capable administration of the present laws would itself prove the most efficient remedy for existing evils, and that some reforms entirely outside the Patent Office are more essential to this than any changes in the patent laws themselves. The crying evil is the admission of expert testimony in patent litigation, that is "expert" in name only—testimony that tends to obscure rather than elucidate the questions involved.

IN MANY respects the report of Commissioner Seymour, of the business of the United States Patent Office for the year ending December 31, 1895, is the best one of the kind ever issued. Its review of the growth of industrial arts within the past twenty-five years, being a recital of the most important inventions in that period, is a most valuable compilation and one that must be of great interest to all inventors. The industrial advance is shown by classes, each giving the number of patents issued, those of a foundation or pioneer character and those containing principles that have revolutionised the art or manufacture. In the next issue of the INVENTIVE AGE will appear some of the most interesting of these reviews.

IN HIS report to Congress, Commissioner Seymour makes some wise recommendations, relative to the establishing of a patent bar, as a means of removing some of the evils that inventors are now subject to. It is not argued that the bar should be composed exclusively of lawyers, because some of the most capable and honorable solicitors are not, in the full sense, lawyers. But something should be required that will in a measure serve as a protection to inventors against imposition. Congressional authority is asked to create such rules and regulations as may seem adequate to cover this matter so that an inventor may know that his representative is at least in good standing and of recognized ability.

THE Berlin Industrial Exposition which opened May 1, will continue till October 15. Consul-General, Charles de Kay, in a recent report, speaks in glowing terms of the arrangements and completeness of the enterprise. He says that although not conceived on the scale of an international Exhibition, the Berliner Gewerbe Ausstellung has been planned and carried out in such a way as to make a pleasant impression on the part of foreign visitors. The buildings are for the most part of no great height, but are relieved by picturesque turrets and pinnacles, and are scattered about a suburban park. The material is iron framework, filled in with mortar or beton. The location of the exhibition is a happy one. It lies on the left bank of the River Spree, in a park called Treptow, whose small lakes have been enlarged and connected by canals in order to furnish the necessary water views and accessories, in addition to the broader sheet of the River Spree. The chief building, in which the products of Berlin factories in metals, glass, porcelain, leather, and India rubber are to be shown, covers 60,000 square meters and is of light iron and beton work.

IT HAS been proposed to separate the tariff question from partisan politics, and to place it in the hands of business men. If such an object could practically be realized, says *Age of Steel*, we should be much nearer a bright ideal than we are at present, or are likely to be for some time to come. Until those who cast the sovereign ballot are practically unanimous on so vital a question, there can be no absolute separation of the tariff question from political interference, authority or dispute. For all that, the fact remains, that if a question so long used as a political foot ball could be made a commercial rather than a political issue, and be less a toy for the shysters of the stump, it would be a mercy deserving the gratitude of the nation. The National Commercial Tariff Convention to be held at Detroit, Mich., in the early days of June, proposes to take this matter in hand, and to inaugurate a movement that will reform if it fails to practically remove a great evil. The delegates to this convention will represent numerous important organizations. It will be in fact a parliment of business men and cannot fail as such, to gain the ear of the nation and impress its sentiments with more or less force on public attention. The plans adopted by this convention will be presented to the two great political conventions to be held in June and July respectively, and an effort be made to urge their embodiment in their respective platforms. Whatever the result of this effort may be and however near or far it may be from its announced ideal of separating a great economic question from partisan politics, one thing is practically assured that the business men of the nation are likely to be a more vigorous and forceful factor in shaping the commercial destiny of the nation.

THE question of expert testimony is one that bothers the court and the litigants in the most important patent cases. It has been estimated that in the matter of telephone suits in this country the sum of money expended for expert testimony has been in excess of that paid for legal services. Invention, of London, has a long article on the subject in which a means of avoiding "prejudicial" expert testimony is pointed out. "In the presentation of the subtle questions involved in the trial of patent cases before the judicial mind, which must finally pass judgment upon them, the counsel in the case, both solicitors and barristers, must educate themselves in the subject matter to a sufficient extent to be able to impart their recently-acquired knowledge to the court. In doing so they are obliged to draw upon the lifelong acquirements of experts in the particular branch under consideration. As only too often happens however, experts, however disinterested and able in their specialized callings, differ as widely as the opposing counsel. It has been suggested by some of the leading papers, as well as by many of the leading minds in the legal profession, with a view of obtaining expert testimony which shall be entirely unprejudiced as to either side, that all expert testimony should be excluded, other than that which is selected by the court trying the case. No matter how able an expert scientist may be in a particular branch, it is only fair to assume that unless his views coincided with those of the party who retains him, he would not be employed and paid for as expert in any given case." It would be better therefore, for the general cause of justice to have all expert testimony, particularly in patent cases, appointed by the court, better only for the reason that the mental attitude of the jury and the court would be better towards the case itself and the real facts at issue, were it known to them that such expert testimony as they listened to was not paid for by either side.

HOUSEKEEPING BY ELECTRCITY.

Heat, Light and Power, All From The Same Source—No Coal or Gas Bills.

A great electrician and inventor was once known to remark that before many years had passed, electricity would constitute the backbone of industry.

The time is now almost ripe when the truth of this prophesy will become apparent to even the most unlearned in the arts.

A casual glance through the voluminous records of the United States Patent Office discloses the fact that our great inventors have not overlooked the trials and cares of the housekeeper, in their researches but have provided for her every want,

Electric Boiler

Electric Curling Iron

in their perfected electric, heat, light and power producers.

In a modern house equipped with only a very small number of these electric devices, the use of coal and gas is altogether obviated, and the house is heated and lighted and the cooking done by means of the mysterious fluid. In this house you see no apparent heating means and yet there is a diffused warmth evenly distributed throughout all the rooms. This heat is produced by the heavy electric carpets or rugs with which the rooms and halls are provided. Each of these carpets is made of ordinary material but is provided upon its underside with its two loose layers of asbestos. A composition, composed of powdered clay and plumbago is applied between the two layers of asbestos. Suitable contact plates to which wires of an electric circuit are connected, are embedded in the composition at opposite ends of the carpet. The electric current in passing through the plumbago in the composition is resisted by the clay and a gentle heat is thus generated over the entire area of the carpet, heating all parts of the room equally. Each carpet is connected to an independent circuit so that the rooms may be heated to different temperatures as desired and the degree of heat in each may be regulated by a simple rheostadt connection. In such rooms or places as where it is not desirable to use the rugs, electric bracket heaters may be employed. Each of these heaters consists of separated carbon bars that form one continuous zigzag circuit. These bars are encased in a simple ornamental open work bracket adapted to be hung upon the wall. The carbons are connected in circuit by circuit wires.

In the kitchen, electricity reigns supreme.

Hot water is applied for the whole house by a tubular electric boiler. Each water tube of this boiler is wound with wire resistance coils and when the current is passed through the coils by the turning of a switch the water very soon begins to boil because of the heat generated by the resistance to the electric fluid. Along one side of the room is arranged a polished wood bench upon which the electric cooking stove and electric pots and pans

are placed. Above this bench is arranged a smoke trap provided with an electric fan for creating a suction to draw all the smoke from the room and discharge it into the chimney.

The electric stove is heated by a plate of separated resistance bars. This plate can be raised or lowered in the stove to or from the cooking article to give a greater or less amount of heat. The top of the stove is provided with a window and an incandescent light illuminates the interior so that the cooking article is in plain view at all times. The stove is started and stopped by the simple turning of a switch. The degree of heat in the oven is regulated in the same manner as in the electric carpet.

Next to the stove stands an electric broiler. This is composed of two hinged frames provided with hollow spaced bars and a pan beneath the same to catch the drip. Electric resistance wires are passed through the hollow bars and connected to an

electric circuit. When the current is turned through this device the bars become heated and thoroughly broil the meat between them. The electric frying pans, coffee pots, tin pots, gridirons and like cooking utensils cover the remainder of the bench and are each heated separately by electric resistance coils applied on their under sides.

These articles are all detachably connected to the circuit wires that pass along to the rear of the bench so that they may be disconnected and carried about. There is no smut or dirt about any of the devices or any part of the kitchen for the obvious reason that there is no fire or coal to create the same.

In one corner of the kitchen stands the electric dish washer. Rotable shelves are mounted in this washer and are adapted to receive the dishes. Flex-

ible stationary wipers are arranged over the shelves to wash the dishes as they are carried about by the rotating shelves. The shelves are rotated by a small electric motor and hot water is supplied from the electric boiler.

The laundry is a counterpart of the kitchen in neatness and cleanliness. Along one side are arranged the washing tubes provided with rotable beaters or agitators operated by a small electric motor. The tubs are supplied with hot water from the electric boiler in the kitchen. The ironing is done by electrically heated sad irons. These irons are hollow and are provided with electric resistance coils connected to the electric circuit wires by flexible conducting cords so that the process of ironing will not be interfered with. One iron is all that is necessary as it is always hot and its heat can be regulated until the proper degree is secured.

The house is lighted throughout by incandescent electric lamps. The socket of each lamp is provided with a number of small resistances and a duplex switch whereby the light may be turned up or down to give five or six different degrees of illuminating power. All the sweeping throughout the house is done by electrically operated sweepers.

In the bath rooms above are arranged electric baths each consisting of a suitable soft non-conducting base portion in which the body of the bather sinks. Upon this base is laid a flexible sheet made of soft, pliable woven wire. This sheet forms one of the conductors of an electric circuit. The other conductor comprises a similar sheet and is laid loosely over the first. When the bather places himself between the two sheets the current passes through every portion of his body touched by the same and his blood is sent bounding through his veins in a very refreshing manner by the peculiar action of the electric fluid.

In the bedrooms electrically heated mattresses are used on all the beds. These each comprise flexible electric resistance wires embedded in asbestos coverings which are placed in the mattress proper. These mattresses each diffuse a mild gentle heat which can be instantly stopped when so desired by turning the electric current out of the mattresses by a suitable switch. The blankets are also heated in the same manner by electric resistances. All the towels also are heated to a gentle warmth by flexible resistance wires embedded in them and flexibly connected to the electric circuit. The combs and brushes are also electrical and are each provided with their own batteries for generating the current. The bristles are of flexible metal and are connected alternately to the positive and negative elements forming the battery in the back of the brush. The use of this brush invigorates the scalp and prevents falling of the hair and

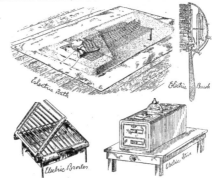

Electric Bath

Electric Broiler

Electric Brush

Electric Stove

like complaints.

If the baby should complain during the night it would only be necessary to put him in the electric cradle which is operated by a small electric motor. The movement of the cradle by the motor also operates a fan so that on hot summer nights baby can be rocked to sleep and fanned at the same time without inconvenience to any one. All the rocking chairs throughout the house are operated by small electric motors arranged under the seats.

W. HARVEY MUZZY.

NEW YORK ELECTRICAL EXHIBIT.

Brief Description of Some of the Features Presented at this Exposition.

At the Centennial in Philadelphia, twenty years ago electrical exhibits were not sufficiently numerous nor of sufficient importance to warrant a separate department and those who attended doubtless remember, that aside from the notable announcement of the telephone and the incipient display of electric lighting systems, the applications of electricity were limited in extent to a few special purposes. The first Exposition in this country at which electricity played a part of consequence was the Southern Exposition at Louisville, Ky., in 1883. A year later an exhibition was held in Philadelphia, under the auspices of the Franklin Institute. The progress in the meantime had been marvelous and in the few brief years since then its industrial growth has been phenomenal until it has become a vast factor in human progress. The Electrical Exhibition at the Grand Palace in New York City,—the past month—afforded a striking presentation of the role that electricity now plays in every day life and though this exhibition was not as extensive as the Electrical display at the World's Columbian Exposition, the character of the exhibits were fully as rich and it will be long remembered as one of the most successful exhibitions of the times. Everything was done to make the exhibition attractive not only to those interested in the subject, but to the general public and as a consequence it soon became popular and thousands thronged the building daily.

The scene presented to the visitor upon entering the Grand Palace was one that must move the admiration of everyone. The building in size and arrangement is solely designed for the purpose of an exhibition and the portion of the building occupied by this Exhibition comprises four immense floor areas two of which range above the main floor to form overhanging balconies about a vast open space. A glance around soon convinced one that the Exhibition was in artistic hands.

The overhanging balconies were appropriately draped with American flags and bunting, the graceful folds of which served as a background for thousands of glowing electric light bulbs that were massed in various patterns about the hall. Extending back under the balconies were booths similarly decorated and in which were displayed every kind of electrical apparatus. During the afternoon and evening the scene was enlivened by music furnished by the Seventy-first Reg't. band, stationed in one of the galleries. During the intermissions the electric pianos could be heard playing the popular airs of the day.

The first thing that attracted attention in the way of an exhibit was the small model of Niagara Falls and the duplicate in miniature of the great power plant. This was situated in the centre of the hall on the main floor and gave one a good idea of the work that has actually been done at Niagara, in the grand scheme for the utilization of the Falls; while the machinery was driven by an electrical current transmitted from the Falls—a distance of 465 miles. To add to the realism of the representation, the roar of the cataract could be heard through telephones arranged about the model. Besides furnishing a vivid portrayal of the engineering features and working of the great plant, it forms one of the most interesting features of the show to the scientific investigator because it indicates the great possibilities of transmitting the current supply gained from natural sources, over wide areas.

Just beyond and extending traversely across the main floor was a model section of the Erie Canal. with miniature canal boats towed by electric motors suspended from trolley wires. These wires were also supplied with current from Niagara Falls. The system here represented in miniature has been adopted and will soon be in operation along the Erie Central Canal and will mark a new era in this form of navigation. The objective point that most attracted the crowd as well as the pickpockets was the dark room at the East end of the main hall, where the public could see a practical illustration of the X-rays phenomena, by the aid of Edison's Fluoroscope. Thousands of people here had a brief chance to view their own anatomy, while passing in front of the fluroroscope, and it was amusing to hear the various expressions of doubt as to the genuiness of what they beheld. Other opportunities to study the X-rays and the method of producing this phenomena were provided in the Practical Electrical Engineer Laboratory while in the historical department there were displayed a fine collection of photo-negatives showing with clearness the bone structure of the different members of the human body. The benefits of this recent discovery were put to a practical test one day upon a orisoner who was taken from the Bellevue Hospital and who was subjected to X-rays for the purpose of determining the location of a bullet in his arm. A picture was taken that clearly disclosed the location of the bullet.

Another great object of scientific interest to the electrical investigator was D. McFarland Moore's booth where that gentleman superintended the manufacture of daylight. Mr. Moore believes that he has found the commercial light of the future, which depends neither upon an incandescent filament in a glass bulb nor upon a glowing crater between two carbon rods. He employs closed vacuum tubes which are exhausted until only about one-millionth of an atmosphere remains. By means of a simple electrical device for disrupting the current—called a vacuum vibrator the few atoms remaining in the tube are rapidly set vibrating and the effect is the purest glow, giving a marvelous light, one that approaches that of sunshine.

The lady visitors were much taken with the domestic applications of electricity which included sewing machines driven by a single cell of battery, lamps lighted by concealed dry batteries and many others too numerous to mention, but the most popular exhibit with them was the Electrical Kitchen where all the cooking was done without kindling a fire, where a steak could be broiled or a pot of coffee made by merely turning a switch. Upon examining the cooking utensils used over the electrical heaters they were found to be free of smut or dirt and by virtue of the refinement it brings to the kitchen it fast met with favor.

HISTORICAL EXHIBIT.

East of the Main Hall were to be found the Historical exhibits which included a collection of early books, papers, models and relics. The exhibit made by the Western Union Company, of Morse's early experimental instruments was most interesting. The Doremus exhibit as it was called included Henry's experimental induction coil; ths famous Tithonometer device by Prof. J. W. Draper, in 1842. Another valuable historical exhibit was that of W. J. Hammer's collection of incandescent electric lamps. It contains the original models of the pioneers in the art, from the earliest beginnings through all the stages of development. This exhibit is an object lesson worthy of study by inventors as it illustrates the trouble and thought that was bestowed in the attainment of a filament for the incandescent lamp. Mr. Hammer's collection also included a notable collection of portraits comprising all of the great celebrities who have been leading spirits in the electric field.

LABORATORY EXHIBIT.

The Laboratory exhibits occupied space in one of the balconies. The apparent purpose of this exhibit was to show as fully as possible the educational facilities afforded by the colleges represented and to give to the investigator an opportunity to acquire all desirable information concerning the equipment in use in illustrating the practical applications. Among the most interesting pieces was an egg shaped globe for the exhibition of the electric arc in vacuo; Hoffman apparatus for the decomposition of water; models showing the operation of the telephone; Erstman galvanometer; ampere table apparatus for development of heat by current and quantive estimation of same.

The Electrical Storage Battery Company of Philadelphia exhibited a horseless carriage that was equipped with 42 cells of storage battery with a capacity sufficient to furnish ample power for twent-five hours while driving the vehicle at an average rate of speed of ten miles an hour. The same company also exhibited a beautiful 21 foot launch propelled by motors driven by their accumulators.

The Edison Company, the General Electric Company, of N. Y. and the Westinghouse Company made perhaps the largest exhibits in the building and each was located in an extremely favorable position on the main floor. Oftentimes it is more interesting and instructive to view the winding and assembling of coils and armatures that go to make up a dynamo or motor than to look at the same in a completed state. The Crocker-Wheeler Electric Company afforded this opportunity to the public and crowds gathered about their space to watch the workmen construct a dynamo. The Metropolitan Telephone and Telegraph Company was installed in a handsome booth. There was to be seen in actual commercial opperation a central office equipped with the latest and most approved Switchboard with all the necessary auxiliary apparatus, and a telephone girl hard at work. An opportunity was also given the public to test the long distance telephone.

It is of course manifest that no adequate account of the many individual industrial exhibits contributed by enterprising manufacturers can be given in an article of this kind. It is proposed, however, that a more detailed account be given in succeeding issues from data secured by a special representative of the INVENTIVE AGE.

The Need of Protection in America.

The development of Japan and China and the effect on manufactures in Europe and America, just now engrosses the attention of mankind generally, and has served to silence every argument heretofore used in the United States in favor of free trade. A correspondent of the *Times of India*, writing of the trades and industries of Japan, gives some interesting particulars from which the following excerpt is taken:

"Upon the cheapness of labor it would be difficult to say anything new. In clocks and watches, for example, they have made great strides. They can sell a clock for 50sen (one shilling) that will keep good time. A Japanese clockmaker's wage is not more than sixpence a day, whereas in England the wage is six shillings a day, and in America much more. Instead of continuing to buy American clocks, as in times past, Japan threatens to send clocks of her own manufacture for sale, at no distant date, to the United States. There are three large watch and clock factories in full operation in Osaka. Japan will shortly monopolize the export of shirtings to Corea, now amounting annually to three millions of dollars. The carpet and rug industry at Sakai, near Osaka, gives employment to 17,500 operatives. A company is being formed to manufacture railway material, locomotives, carraige, and trucks. In Osaka, water-power is being used to produce electromotive energy, which will be applied to industrial purposes, with an output of 15,000 horse-power.

In Fukui district on the west coast of Japan, nine-tenths of the silk piece-goods for which the country has lately become famous are produced by girls who work 12 and 14 hours a day at the rate of half a dollar per piece, or, say, one shilling. There are now nearly 15,000 looms in the Prefecture, with a proportionate number of operatives. A girl can weave from six to eight pieces of silk per month. Japan exports 14 millions gross of matches annually to China, the Straits Settlements, and India."

Ruin for Welsh Tin Platers.

A tale of the ruin impending over the tin plate makers of Wales is told in a report to the State Department by Unitsd States Consul Howells at Cardiff. The American market is felt to be hopelessly lost, and the greatest depression is felt. Of the 491 tin plate mills in the district, no less than 253 are idle, while 91 are being run on a 10 per cent reduction in wages, 40 at 12½ per cent, 91 at 15 per cent, 6 at 20 per cent, 5 at 22½ per cent, and 5 at 25 per cent reduction. There are now twice as many mills as needed, and, while 4,600 men are out of employment in March, the number of idle mills has quadrupled.

If no plate had been produced in America, the productive power of the district would still have exceeded the world's demand. Everything, the report says, now points to continued depression for some time to come, and the day is approaching when only such works as are abreast of the times in turning out plates at the minimum cost can be run.

A queer feature of the situation, pointed out by the Consul, is that the Welsh tin men have overlooked, up to this time, the possibility of cultivating the local market, and have only recently begun to endeavor to induce the house builders to use Welsh tin plate as roofing, instead of the slate which is brought from Pennsylvania.

Motor Vehicles.

It is shown by the report of the Commissioner of Patents, that up to the beginning of 1896 there have been 54 patents granted in this sub-class. A line of invetion in which much interest has been taken by the general public of late is the class of motor vehicles or automobile wagons. This is not strictly a new art, but rather the rival and carrying forward of old ideas, and the adaptation of modern devices, such as electric motors and gasoline engines, to vehicles for the purpose of self-propulsion. The steam engine, while adapted for traction engines or vehicles intended to draw other vehicles with heavy loads like thrashers, sawmills, and freight wagons, is not suitable for light passenger vehicles. The Benz motor was patented in the form adapted for use on a vehicle of this kind July 31, 1888, No. 386,798. The patent of Daimler. January 17, 1888, No. 376,638, shows a motor for this purpose. Duryea, in No. 540,648, June 11, 1895, exhibits a device for this purpose.

A French experimenter, Camille Dareste, has found that the germ in the hen's egg is not destroyed by an electric current that would kill an adult fowl, but that the germ is so modified in most cases that a monstrosity will be hatched from the egg.

Copper Precipitation Plant.

The accompanying illustration gives a fair idea of the largest copper precipitating plant in America. It is located near Butte, Montana. The water as it is pumped from the Anaconda and St. Lawrence mines is carried in a flume several hundred yards to a large number of vats into which all sorts of scrap iron and tin cans are placed. The average time required to precipitate the copper is about one month, and the output averages nearly 100 tons per month of copper. The running expenses are nominal and the entire force employed can be seen in the illustration, consisting as it does of but two men per shift.

Many persons have often been much surprised that any iron tool or utensil left in water in which there is copper in solution, has, as it were, and to all appearances, become copper. This is due to the iron having become coated by chemical action, with a layer of copper, and it was this discovery in the early days of copper mining that led to the process of copper precipitation from the water holding this metal in solution.

Copper in its pure state is very sparingly soluble in water, and in the chemical form in which it exists in the Butte ores, not at all, and it is necessary that it should become a copper sulphate of salt. In this condition it is readily soluble and can be leached out of the matrix and carried up out of the mine in the water. It is not definitely known how this change is brought about in the veins of ore, but it is no doubt due in a great measure to the oxidation: first, of the copper by contact and combination with the inflowing water, carrying more or less oxygen in suspension with carbonic acid gas. All chemical changes and combinations produce more or less heat, and this would tend to accelerate the change in the conditions of the ore. The carbonic acid gas also playing its part by reaction on both the ore and matrix.

The great fire in the St. Lawrence mine in 1889 has, and is now, producing the same chemical change in the ore body, and in the same way that we find taking place in the heaps of burning ore at the Rio Tinto mines, the heat aiding the chemical alteration in the ore mass, and hence its waters are very rich in the copper salt, the assay value of which will probably average one-half of one per cent of the body of water coming out.

We have in this mine water, in solution, the copper salt of sulphate, i. e., sulphuric acid in combination with the metal, and since this acid has a much greater affinity for the metal iron than the first named, immediately the solution is brought into contact with it, the acid at once leaves the copper and combines with the iron, forming an iron salt or sulphate, which is very soluble in water and the copper is precipitated as the pure metal, mixed with some impurities only; the iron salt passing off with the water. The precipitation of the copper held in solution as a salt from the water pumped

from the mines and otherwise, is not a new thing. It has been in operation for many centuries, notably at the great Rio Tinto copper mines in South Spain, and at the long worked mines on the Island of Anglesea, Wales, were immense quantities of old scrap iron are annually consumed. We are indebted to Western Mining World, Butte, Montana, for illustration and data herewith.

Powerful Lifting Magnets.

By means of the electric crane and the electro-magnet, which were introduced into this country and recently exhibited before the American Society of Mechanical Engineers, it is claimed that three men can now do in fifteen minutes the same amount of work which formerly taxed the strength of six men for ninety minutes. It is found invaluable in working with pig iron, heavy castings and immense boiler plates.

It is believed by engineering experts that these lifting magnets will soon replace the present forms of the derrick and traveling crane. Preparations are being made to introduce this device in the great Carnegie works at Pittsburg, and its practical workings are being very carefully watched by at least half a dozen large manufacturers throughout the country.

At first sight, it appears odd that a small coil of wire through which the electric current is carried from the battery at the base of the crane to the magnet runs up along the upright beam and across the upper support to the end of the crane, then over a small wheel to the end of the chain to which is attached the lifting magnet. Duplicate wires are used to prevent any possible accident in case the wire should foul with anything or in any way be broken. The switch-board governing the current is placed at the base.

In lifting a weight of 3,600 pounds a current of from three to four amperes at from 20 to 30 volts is used. The magnets vary in size and weight from comparatively small ones weighing 45 pounds to those having a weight of 350 pounds, but the lifting power of the magnet is not always in proportion to its size.

The body or core of the magnet used in lifting steel shells or circular pieces of metal is shaped like an inverted "U" and closely bound with wire, the

LARGEST COPPER PRECIPITATING PLANT IN THE WORLD.

strength far in excess of any strain which will ever be put upon it. From the base of an upright steel beam a long steel arm projects, first upward at an angle of forty-five degrees; then, with a bend upward and outward, it extends for a distance nearly twice the length of the supporting beam. Two steel rods, reaching from the top of the beam to tie crane, act as supports or holds. One is attached to the crane about a third of the way from the base and just at the bend, while the other is attached to the upper end of the crane.

metal weighing only about forty-five pounds can by that strange force known as magnetic influence, aided by the equally mysterious power of electricity, lift tons of iron with no apparent grip upon the weight to be lifted. It has been proven by experiment, however, that such an electro-magnet can lift seventy-two times its own weight.

In England the electric crane and electro-magnet are in use in a number of places, in particular at the Woolwich arsenal and at the Sandycroft works. Those in use at the Woolwich arsenal were designed by an officer in the British army, and greatly simplify the work of lifting and moving heavy shot and plates of iron and steel. Particularly is the electro-magnet in value of lifting heavy shot, as previous to their use workingmen experienced no end of trouble in getting slings securely around the shot. It was a long and heavy task, and required the labor of many men.

Now the electro-magnet is lowered by the magnectic crane and simply laid on the side of the shell to be raised; the turning of a small lever at the base of the crane switches on the current, and the work is done. Here the new apparatus enables three men to do the work which formerly required nine men.

The construction of the crane and magnet is peculiar, and its inventor claims that there is less chance of a break than when the old-fashioned tackle was used. Indeed, so far as is known, no accidents with the new method have thus far occured.

The electric crane looks very much like an ordinary swinging crane. It is constructed of the best steel, the frame being comparatively light, but of a winding being protected by brass flanges and by a thick covering of brass. The two ends of the wire winding are led to duplicate terminals, where they are joined to the two wires from which they receive the current. Through the center of the magnet run two bars, to which are attached the rings by which the magnet is attached to the hook at the end of the pulley on the lifting chain.

It is said that the monster belt recently turned out by the Chicago Belting Co. is the largest ever made. The dimensions are: Length 150 feet; width, 7 feet; weight, 3,300 pounds, and thickness, seven-eights of an inch. In its construction the selected portions of 450 oak-tanned hides, picked from over 5,000 skins, have been used. From end to end there is not a stitch or rivet, and the figures necessary to enumerate the ingredients that have been made into glue to hold the roughened sides of the three tiers of hides together, and their quantities, would run away into millions. Not the least interesting detail connected with the building of this mammoth strap, is the method by which layers of skins have been arranged so that at every point from end to end there is not a spot where at least two solid thicknesses of leather do not cover the spliced sections of the third layer. It is to be used in the plant of the Louisiana Electrical Light Co., New Orleans.

A fact not generally known is that the cork oak is now being successfully grown in Southern California. The United States now use about $2,000,000 worth of cork annually. In a few years we may expect to raise all our own cork.

VIEWS SHOWING THE CHARACTER OF SOME OF THE NEW COTTON MILLS OF CHINA.

Industrial Development of China.

In speaking of the prospective competition of the East with English manufacturers, the Textile Manufacturer of London, says: "It has been well said that while we are bewailing the fact that cotton mills are being erected in Japan, we preserve a discreet silence as to who is making the machinery with which they are to be equipped; and that while we lament that Japan should realize a profit on the sale of cotton goods, we ignore the fact that the wants of the prosperous artizan expand in proportion as his means increase. But after all, is it not China rather than Japan which will become our greatest competitor in the Far East? In China, labor is cheaper than in Japan. China grows her own cotton, and, to crown all, has a plentiful supply of coal, if it were only permitted to be raised. In general, it may be assumed that the capital for such concerns will be largely provided in this country, and investors may easily find Chinese mills more profitable than those of Oldham. This is, however, about the limit of our compensation, and it cannot be said to entitle us to take a very roseate view of the situation in the immediate future."

The Textile Manufacturer is also running a serial article on "The Cotton Industry of the Far East," in which is shown the progress that has been made during the past few years in that line in Japan and China. From official sources it is found that in 1866, Japan had one cotton mill with 5,456 spindles. On June 25th, 1895, the number of spindles in operation and in course of erection reached the enormous total of 818,741. The illustrations herewith will give our readers some idea of the character of the new plants now being installed in China. These mills have been erected by Messrs. Dobson & Barlow, Lt., Bolton. The recent rapid increase in the number of cotton-spinning spindles—in fact, the inception of the cotton industry itself, is, of course, a result of the late war, which ended by the signing of the treaty of Shimonoseki, whereby foreigners were allowed to open up the country for trading and manufacturing purposes.

The past year has been one of great activity in mill building, especially in the neighborhood of Shanghai.

This mill is known as the Ye Yuen, and has been built in a most charming situation on the Alpha Farm, Shanghai. The mill contains at present 18,200 ring-spinning spindles and all the necessary preparing machinery. This firm have also just completed and delivered to the same mill a further order of over 7280 ring-spinning spindles, with all preparatory machinery, thus bringing up the total number of spindles in the mill to 25,480. There are now over 300,000 spindles at work in the Empire.

An Electrical Aerial Torpedo.

Mr. H. G. Rich, an electrical engineer, of Des Moines, Ia., has invented an instrument of warfare which he calls an "aerial torpedo," designed to be used in the siege of cities, or to scatter large bodies of troops while at rest. The torpedo consists of a small sized gas-filled balloon, capable of sustaining for any length of time from 30 to 40 pounds at an elevation of from 500 to 1000 feet above the earth. Inside of the lower or small end of the balloon is placed a metal cylinder, which contains an electrical device, the purpose of which is to ignite the gas in the balloon at any stated period. Underneath the ballon is suspended a case or basket containing high explosives, similar to dynamite, which explodes with terrific force when striking a hard substance, like the earth or walled embankments. In action, the management of the torpedo is described as very easy and simple, the inventor stating that a corporal's guard can with it accomplish what would require a large force to do by the usual methods now employed in the siege of cities, or the scattering of large bodies of troops. The torpedo complete is small and compact, and a large number can be carried by a few men or a pack animal. The gas to inflate the balloon is carried in light metal cylinders, enough being compressed in one cylinder to inflate a large number of the aerial torpedos.

As illustrating the commercial value of some successful patents, "Industrics and Iron" has this to say in reference to the Dunlop pneumatic tire: "The available working capital of the original company was £22,500. The original company is now to be sold for the enormous sum of £3,000,000 in cash. The share holders have received in dividends and premiums the sum of £658,123, and by the arrangement now concluded they will receive a further sum of £2,887,500, giving a total result in favor of the shareholders of £3,545,623 in return for £260,000."

NEW INVENTIONS.

A Brief Description of Some of the Most Valuable New Ideas.

Elevated Track Cycle.

The invention as is shown in the accompanying cut is a recently patented invention whereby the greatest known speed of travel can be developed. This novel flyer is the product of the genius of William H. Martin, of Mobile, Alabama, who claims that by the application of electricity, 100 miles an hour or more can be overcome and that by man's own power applied the same as in a bicycle, a speed of fifty miles or more an hour can be overcome

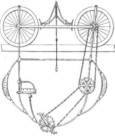

The carriage is operated by foot power, the operator sits on a chair in a light carriage beneath the rail. The lower part of this carriage is supposed to pass above the earth's surface about eight feet avoiding ground collisions and dust and the rails about fifteen feet from the ground. The pedal and sprocket are located in the usual order. Just in front of the seat a sprocket chain engages a second sprocket at the end of a rod extending down from the rear wheel. Another chain reaches up from the second sprocket on the hub of the rear wheel from which extends a stright horizontal bar connecting

the rear wheels. The turning sprocket is balanced upon a U-shaped frame attached at the end to the lower ends of perpendicular rods that are attached to the wheel axles extending downwards. This U-shaped frame also supports the seat. This road and carriage can be built to support and carry at a rapid rate a weight of four or five tons on a single carriage. For regular passenger traffic carried by electricity, the usual carriage will carry say about six passengers; built and tested each carriage to carry a ton, when only an average

weight of less than 1,200 pounds will be required. For pleasure such as running out from small towns or large towns to parks, carriages can be propelled by foot power seating one or more persons. These carriages are propelled with one to three on a level track with much greater ease than on a bicycle on the ground. The road is operated with a double track, all carriages on one side going in one direction, all on other track in opposite direction. This road can be built and equipped for about $2,000 per mile. Will expect this price to be reduced as soon as manufacturing begins in large quantities. The patentee is open for agents in all towns in the United States who will get up stock companies for the purpose of building these roads at different localities. Liberal contracts will be made with the agents for this purpose on application to the inventor, William H. Martin, Mobile, Alabama.

ICE VELOCIPEDE.

An ice velocipede, recently patented, is the invention of Theodore H. Paulson, of Le Grand, Iowa. The machine is built after the pattern of the bicycle, but, without wheels, except one, with teeth on the bearing side of the tire for engaging the ice. This wheel is journaled in the bifurcated end of an arm projecting from the back frame work, and is driven by a chain and sprocket as in the ordinary bicycle. A link at the front end of the bifurcated arm connects with the back frame standard and gives play to the driving wheel, which is governed by a brake-lever situated just before the seat. The standards rest upon skating runners and are guided by the usual handle bar.

A SLUSH DESTROYER.

An apparatus for melting snow has been patented by Mr. Richard Ripley, of Liverpool, England. This useful invention is in the shape of a cart, runs upon wheels and is drawn by horse power. Its principle features are a furnace and cart body or tank resting above the furnace and heating space, which extends along beneath the tank bottom. Both the tank and furnace are detachable; so that the former can be used as a cart upon the running frame.

After a snowfall, the beautiful gift of the skies is shoveled into the receiving tank to be instantly converted into its original element, which runs out through a hose attached at one side of the vehicle.

Considering the inconvenience and "wetness" of snow and slush in cities, where there is so much necessary out door work, it is strange that our progressive Americans have not put in operation a good snow disperser in all the large cities and towns.

ELECTRIC FOOTWEAR.

In the grand march of inventive progress the ladies are taking their position in no mean numbers. Among them is Jennie A. Blair, of New York, who has invented and patented an electric heel and toe protector.

The arrangement comprises a heel and toe connected by an elastic band extending over the instep of the foot. In the respective sockets are pockets containing positive and negative elements the poles of which are brought in operation by a wire running through the elastic bands.

If this apparatus will ease the twinge of gout, cure corns, or call a halt on the swelling bunion, humanity will owe a debt of gratitude to the inventor. Electricity is invading nearly all fields of science and industry. It is applied, as a remedy from head to foot—and some day the physician will prescribe it in pill form to cure our ills.

The inventor claims that this device is adapted for the application of a mild current of electricity to the foot; for prevention and cure of rheumatism, chilblains and other ailments, and also useful as a protection against cold and dampness, and for prevention of wear and tear to hosiery. Being applied only to the heel and toe, it offers no obstruction to the bend of the foot in walking and will not interfere with the fit of the shoe.

A NEW BUNG.

Good inventions that pertain to barrels and liquids, usually have a wide field to harvest. Into this field has entered Gerhard Zeilstra, whose place of residence is Grand Rapids, Mich. The bung has a fastening device, consisting of metal arms spreading upward and outward from the lower end of a central rod, the upper end of which passes through the bung where it engages a cam lever for working the arms. The top and spreading shoulders of the arms rest against the under side of the barrel, holding the bung securely in its place. To remove the bung the cam lever is operated, causing the shoulders of the arms to move inwardly toward the bunghole ; thus the under pressure is relieved and the bung comes out. An elastic packing surrounds the central rod at the hole in which it works, making it

perfectly air-tight. Where fermenting liquids are used in barrels, or kegs, making gases, the value of such a bung is at once apparent.

FRICTION AND THE JOURNAL.

How to overcome friction has been the study of many persons for many years, a research that has not met with, nor will meet with complete success. Yet much has been done to check this wearer-out of matter and retarder of motion. A new journal box is the patented invention of William B. Eichholtz, of New Orleans, La. In this the oil well is removable. Instead of the old cotton waste affair which so often lights the track on account of a "hot box," there is in this new box an oil wick extending down into the reservoir and upwards against and upon opposite sides of the journal. A number of wick-supporting springs extend transversely through the wick and give it the necessary bearing. It will be seen that the oil can be applied constantly and evenly to the journal by this apparatus, which besides the above features, has a dust collar of fibrous material, and other attachments, making a practical invention that recommends itself.

While well adapted for ordinary railroad use, it is particularly intended for street railway motor cars, which have more grit and dirt to contend with. All who have had the experience with the latter will readily see the advantage of having a box which is not only almost dust proof, but also has the oil cellar removable for cleaning and refilling. This is believed to be the only box that can be thoroughly overhauled without the trouble and expense of jacking up the car body high enough to remove the box. Every part of it can be reached in a few minutes by a workman with wrench.

RACK FOR NEWSPAPER FILES.

A convenient rack for holding newspapers is something that has long been needed. The old rack is unhandy, and covers up nearly all of the title of each paper.

The new paper rack invented by Charles W. Mooers, of Towanda, Pa., is a big improvement on the old method. In Mooers' rack the files are

placed horizontally upon hooks in a rack having its top extended from the wall at a greater distance than the bottom and inclined to the left in such a manner that the title of each paper is placed in front and to the left of the one next below the—whole of each title being always exposed to view.

ELECTRIC SIGNAL.

A new electric signal invented by G. E. Painter, of Baltimore, is a unique and practical arrangement for transmitting unwritten knowledge to those afar off. Apart from its main source of power. The machine consists in part of electric connections affording seperate electro-magnetic circuits, movable switch arm, circuit-breaking switch-plate having seperated contact surfaces insulated from each other and a pair of brushes on an arm, which alternate with each other in engaging the surfaces. Other contact plates afford continuous surfaces and intermediate contact or coupling plates, which connect the brushes, and during the movement of the brush arm cause an intermittingly broken current to be supplied to seperate and alternately excite electromagnetic circuits. The arm which governs the mechanism, resembles the break lever of a street car, and the lamp—an upright Edison bulb—is below the governing apparatus.

DYEING.

A new method for dyeing cottons in shades from grayish-blue to deep black, making this material (cotton) in which it is so difficult to get fast colors, hold them as well as wollen, or other fabrics. Moritz Ulrich, of Elberfeld, Germany, is the inventor of this formula, and has assigned his patent to the Farbenfabriken vormals Fr. Bayer & Co., of his town. The cotton to be treated can be put in either hot or cold baths, and the effect is gotten from the reaction of alkali sulfids on dinitrohalogen-napthalenes and subsequently treating the liquid mixture when previously filtered with acids. The product, when dried and pulverized, is a greenish black powder, insoluble in water, alcohol and sulphuric acid, but soluable in hot soda lye and especially in a solution of alkine sulfids in water, being thus capable of making fast colors which resist the action of alkali, acids, soaping and milling.

Fast colors in cottons is something not always a "secure thing," for the fast quality so much sought for, often proves to be that which rubs off fastest.

SPADING MACHINE.

A spading machine worked something on the principle of the lawn mower is the inventive product of Frederick Holzhauer, of Detroit, Mich. This earth digger has upon it roller spades with triangular points, which enter the ground as the roller revolves on a shaft secured to a frame to which is joined the drawing shaft.

Instead of bending over the spade as in the old custom, when earth is to be turned in seedtime, the gardner simply runs this new spader over the ground while the revolving blades do their work—in a quicker and more regular manner than the back-breaking process could possibly do.

The inventor claims the credit for having offered to the world the first improvement in plows that has been made in 800 years. These machines can be made in small, and in large sizes—the small sizes to be used in gardens by hand, and it is claimed the large sizes can be made so that two horses can do more work, than six horses can do with ordinary plow, and will put the land in better condition so as to render the percolation of the rainfall easy and perfect to the depth to which the ground has been stirred. The ordinary plow by its downward draft, presses the bottom of a furrow into a sort of trough, and thus the water is drained off, instead of being held for the coming crop.

A BLOW AT THE MUSK OX.

There is no telling where the uses to which petroleum can be put, will stop. To its adaptability for making dyes, parafine, chewing gum, etc., is now added a still stranger utilization of this product of the under world. This is the manufacture of artificial musk. In accomplishing this kerosine (naptha can be used also) is used in combination with asiphuric acid at low temperature. The solution is nitrated and gradually heated to a certain temperature, after which it is neutralized by an alkali and precipitated with metalic soap.

TO MAKE DIMPLES.

The ladies who consider dimples necessary adjuncts to personal charms, may be interested in knowing that a machine for making those pleasing facial cavities has been invented. Heretofore dimples have been made for those who desired them and were willing to endure the operation required, by gouging from the fair cheek a small portion of flesh and then uniting the skin over the hole thus produced ; thus leaving as good a dimple as one would wish to enhance a smile.

The new dimple-maker comes from the genius of Martin Goetze, of Berlin, Germany, whose dimpler is sure to make an "impression." From its arrangement it cannot do anything else. The dimpling tool—a massage-knob secured to the end of one of two arms which come together at the top and work on a common-pivot—is operated by an old-fashioned brace, such as is used for holding boring-bits. The massage-knob resembles an elongated door-knob, but of course is not so large—one could not imagine a dimple made with a door-knob. By pressing the knob where a dimple is wanted and turning the brace, it is easy to see that an impression will be made, but how long it takes to produce a first-class dimple by this process, is not known. The second arm in combination with the massage knob arm, has at its lower end a massage cylinder, which, by turning the brace rolls over the flesh, and is for doing the kneeding work usually accomplished by hand-pressure. But now the weary can be "braced up" in a hurry.

COPYING PAD.

An improvement in copying pads is that of Thomas M. St. Johns, of New York City. As in the old hectograph a gelatin film is used, but, the pan is dispensed with. The film is mounted on a flexible substance which extends, around the sides and edges of a block. The pad is attached to iron rods by springs, and allows a downward pressure, which will give a quicker result in making copies than the old method, wherein the paper was put in the pan, smoothed out and carefully and slowly pulled off.

BILLARD TABLE.

The billard ball will have a great deal more action imparted to it when the invention of Jacob McIntire of New York, is put in market. The patent was allowed for a billard table cushion, having

a soft flat working face inclined from the upper edge underward. The face or plane is located relatively to the table bed, so that the ball will contact initially at a point intermediate of the top edge of the face and a point on a level with the center of the ball. Slightly in the rear of the cushion face is molded a hardening strip made of some resilient material, which gives stability to the cushion. The patent for this invention has been assigned to the Brunswick-Balke-Collender Co., Chicago, Ill.

TO DISPEL THE MISTS.

A fog dispersing apparatus is the inventive product of Mr. Frank Frey, of Rochester, N. Y. The disperser is a hollow standard provided with inlets for admitting gaseous matter to be forced into a fog for its dispersion.

At the top of the standard, is a semicircular tube, from which projects outwardly and upwardly jet tubes arranged at regular intervals upon its sides.

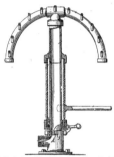

At the bottom of the standard is a large conduit pipe opening into the former which turns in a jacket and is locked stationary, in whatever direction the fog is desired to be charged upon. The appearance of the fog dispeller is not unlike an old-fashioned crossbow.

The fog-dispersing material may consist of any liquid or gaseous body of greater specific gravity than the aqueous vapor of the fog and which will be adapted to mingle with and disperse or precipitate the fog, but the only liquid used by the inventor thus far is water. A fog, says the inventor, is nothing else than small globules of water surrounded by dust, and water washes it down, so to speak.

A FRIEND TO THE LADIES.

The feather boa—that wraps so many fair throats—has found a friend in Theodore W. Austin, who has produced an invention for its manufacture. For this the inventor uses a baseboard having arranged upon it means for receiving the feathers held together in core form. On each side of the quill core the feathers are clamped and afterwards fenly fastened. With a further arrangement of baseboard, and pins in two paraleled series between the apertures, in conjunction with elastic blocks between the two series of pins, the boa is carried on to completion.

INSECT TRAP.

Fred. Smith, of Cincinnati, Ohio, has patented an invention which is intended to insure a sleeper from the attacks of bugs that crawl at night, or in daytime either. In this a shank, inserted into a bed, or furniture leg, is surrounded at the bottom by a cup having two concentric receptacles for holding liquid and pulverulent material, and a fibrous substance, above the pulverulent matter, held in place by an inverted cup.

If any insect passes this trap, it should be welcomed to the feast.

LIFE PRESERVERS.

For those "who go down to the sea in ships," to do battle with the waves, has been invented a new swimming and life-saving device. It is the production of Hans Heckler, of Frankfort-on-the-Main, Germany, and assigned to Charles Heckler, of Missoula, Mont. The life preserver is in shape of a

garment and is provided with a lining made of separate ends of hose. Each of these ends has an inflating tube to which is attached a mouth piece, and nozzle screwed into transverse openings of a block that is enclosed in a sleeve. Openings in the blocks are covered by a valve plate, secured to hollow screw stem, through which air is blown when one wishes to inflate his jacket.

With a preserver of this kind on one could float for a long time on the briny deep while looking for a rescuing vessel. And it seems well adapted for those who wish to learn to swim, as it allows freedom to legs and arms.

WHEN IT RAINS.

The man who desired to make politeness easy and get the thanks of the weary dude by his hat-lifting device, has a follower (in the way of making things easy who) procured a patent in April, for carrying umbrellas without the assistance of one's hands. In this the rain-shedder has its handle held by a vertical wire frame which is attached to a belt surrounding the waist.

Such an invention has its advantages, and there are many who will endorse it. It is conducive to close acquaintanceship, and offers possibilities to

the young man and his "steady" that no other inventor has given to "Romeo and Juliet." All things considered, it seems that the man with the jumping hat should take it off to the umbrella man.

The inventor of this useful article is Mr. Joseph Bolt, of Cranbury, N. J.

BALL CASTER.

George J. S. Collins, of Chicago, Ill., has patented a furniture caster that dispenses with wheel or pivot. The new caster is composed of balls—a large one enclosed within a steel casing open at the bottom from which projects a portion of the ball, and smaller ones at the top of the casing with an antifriction seperator bearing upon the main roller-ball.

The value of ball bearings has long been established, and the latest application (as a caster) of the idea is one of a great many uses to which it might be put. The ball caster is simple and durable, and is especially adapted to heavy furniture, the crushing weight of the smallest one made being 7 tons.

A HOG RINGER.

J. Pleasant Honaker, of Abingdon, Va., has patented a device for ringing hogs. It is a pincher with grooved jaws and handle having a series of threaded holes in which works a screw for holding the jaws in any desired position to accomodate the ring, that is held securely until, with a pressure of the handles the operation is performed. When the hog is caught the pincher is inserted and the work is rapidly and easily accomplished. If the animal is of the razor-back variety the main difficulty will be in catching it.

FOR MUSICIANS.

Every musician knows and deplores the bother and delay connected with turning the leaves of a sheet of music. It requires very quick action, and often when the leaves are refractory—refusing to turn easily—the harmony of the piece being executed is broken, while the performer claws at the sticking page.

This difficulty has been simplified by Messrs. Melvin M. Boothman and Geo. W. Bushong, of Bryan, Ohio, who have invented and patented a machine for turning music leaves. When the performer desires to turn his music he simply operates a button attached at one side of the machine and the turning arms engage the leaf, effectually carrying it to the proper position.

PARTICULAR attention is called to the article on the New York Electrical Exposition appearing in this issue. It is from the AGE'S own special representative, Mr. J. W. Buell, and subsequent articles on the commercial side of the exhibit will prove of interest to exhibitors.

Oldest Implement Now Used Invented by a Woman.

If the mincing knife were to say to the saddler's knife "I am older than you," it would be very hard to decide between them. In fact, they are both the lineal descendants of the same ancestors. But, if the saddler's knife were to say to the mincing knife, "I uphold better than you the ancient trade." that would be true. The woman with the mincing knife might say however, to the saddler, you have not long been in possesion of your knife, for in the olden time your great grandmothers had exclusive use of it. In fact, historically, both mincing knife and saddler's knife were invented longer ago than there is any written record, both of them by woman and both were originally the same implement. It is a favorite doctrine of Mr. Herbert Spencer that in prehistoric times there was the "age of militancy" and later on, including the present time, followed the "age of industrialism." The saddler's knife has a different story to tell and says. "If I were a great philosopher I should say, baring the mythical Amazons, that there has always been a

FIG. 1.—WOMAN'S KNIFE, KOTZEBUE SOUND, ALASKA

sex of militancy and a sex of industrialism. The inventor of me and of a few other useful devices belonged to the sex industrial."

Now the time has come to make good some of the above mentioned assertions, to-wit: that the saddler's knife and the mincing knife are the oldest implements now in use; that woman invented them and that they lie at the very foundation of the industrial history of mankind.

If you will look at figure 1, you will see that it is a spall of flinty or siliceous rock knocked off so as to have a long sharp edge below and thick poll or margin above. To save the hand in using the stone a flexible spruce root is wound about the poll three times, taking advantage of notches and thickness and a very rude warp is formed by working the root cross wise. Nothing could be simpler except a natural spall without handle, but, that would not be an invention. There are specimens in the museum with strips of fur wrapped about the grip or poll. There are rude weapons as old as this implement, but they are no older and they have

FIG. 2.—WOMAN'S KNIFE, BRISTOL BAY, ALASKA.

been superseded by swords, bullets and bayonets. Figure 2, is a later descendent of figure 1, having a retouched blade set in a rough saw cut and held there by natural cement. Figure 3, is a bit of old iron picked up by a savage and set in a rough handle or grip. Much later than these are the Bronze and Iron Age and now the age of steel, in unbroken connection.

How do I know that women invented them? It is a well established law among primitive tribes that the one who uses a tool must make it. Men do not let women even look at them making hunting and fishing apparatus. That would condemn the the things to bad luck forever. On the same ground men will not touch women's devices. Now, reasoning from the known to the unknown, from the fact that this law is in force with every tribe ever heard from, wherein observers have seen women making their own apparatus, the conclusion is reached that the tools found universally in savage women's hands today made by them, were also women's in the past and having always used them, they invented them. Figure 1, represents a tool made and used by an Eskimo woman in Alaska, north of Bering Straits, in the presence of the collector.

Finally the proof must be forthcoming that this simple spall set in a handle of spruce root lies at the foundation of all the industrial activities of the world. In short, it is not a specialized tool at all, but the epitome of industrial edged tools. It is the butcher's knife with which primitive women skinned the animals killed by the men and cut up the meat.

It is the fishwoman's knife with which they scraped off the scales, opened the fish and cut it into slices to dry. It is the builder's, shoemaker's and tailor's knife, with which savage women now and women of the earliest times cut out the hide, or mat, or bark to cover the house withal and the fur, or rawhide for boots and clothing. On a pinch it is

FIG. 3.—KNIFE, QUEEN CHARLOTTE ISLAND, B. C.

the unhairing knife for tawing and curing the peltry and in this function it is handed down to the harness maker. So long as dogs and reindeer were hitched up, the woman's knife was good enough to cut their raw-hide traces and also thongs for a thousand mechanical uses, for nails and screws and bolts, but once the horse yielded its neck to the yoke, its rider borrowed his wife's ancient knife and became a saddler. Here then is a tool, the first used by butchers, fishermen, tanners, shoemakers, tailors, hatters, architects, cabinet makers, vehicle builders for land and water, cooks, gleaners and so on indefinitely, invented and used by early women, tabooed to men, changed in material, but little in function and form even in our day. This was in the personal hand epoch, but, as soon as the physical forces of nature were invoked to move the tool our mothers retired somewhat, and there was not so much a sex of industialism. The discussion of the reasons is not germane to this paper.

O. T. MASON.

A Commercial Tour to South America.

As a practical step towards the establishment of more intimate trade relations between the United States and the most important South American nations, the National Association of Manufacturers proposes to organize a party of representative American business men for the purpose of visiting the Argentine Republic, the Republic of Uruguay and the United States of Brazil in response to invitations which have been extended by the government of those countries, and by which the visitors will be considered as the guests of the nations and will be honored as such. The party is now in process of formation and the arrangements for the trip now being made. The tour will cover the months of July, August and September, which constitute the most comfortable and most pleasant period of the year for such a journey. These months from the southern winter season, when the temperature remains steady around 60 degrees in the Argentine and Urugnay, and when there is no yellow fever in Brazil. The social season is then in full swing in Rio de Janeiro, Buenos Ayres and Montevideo, and the people can be seen at their best.

The object of this trip is to convey to the people of the United States through the members of this party a more thorough and more practical knowledge of the resources of the countries which will be visited, and to indicate the means by which the trade between the nations interested can be enlarged and extended. With these ends in view the party will be made as broadly representative of American commercial, manufacturing and financial interests as may be possible. Each member of the party will be expected to represent some particular branch of business, the interests of which he will consider throughout the trip, and upon returning each member will be expected to prepare a report embodying the results of his observations and study during the tour.

Horseless-Carriage Contest in Providence.

The State Fair Association, of Providence, R. I., has made the announcement that the first attraction in point of importance to be presented at next September's exposition will be an exhibition of horseless carriages, embracing as many of the known makes as it is possible to gather together. The show is to be only one feature of the attraction, which will include a series of races by the horseless carriages. Something like $5,000 in prizes will be distributed.

Wave Motors—The Millennium of Electricity.

Upon arriving at Paris on his business visit to the late great World's Exposition in that city, Edison was of course interviewed by the omnipresent newspaper reporter, and among many other interesting things he was reported as saying—and Edison is always interesting—was, that while whiling away his enforced leisure hours on deck, he was strongly impressed by the immeasurable power of the waves, which tossed the great steamer up and down like a cork. Said he, "the spectacle of so much power going to waste made me wild, and I resolved that at the very first leisure I could find, I would devote to solving the problem of utilizing this illimitable energy, and then would come the Millenium of Electricity."

Several years before these words were uttered, an inventor, and a reader of the INVENTIVE AGE while sojourning in San Francisco, visited the famous "Cliff House" resort, and there saw in operation an apparatus for utilizing the breakers. It was a crude machine, erected in an unfavorable position, and some three or four years since was swept away in a furious storm.

After considerable study and experiment in the same direction, it was finally determined that any apparatus attached to or surmounting any permanent structure, such as piers or wharves, could not be maintained in exposed conditions where the storm waves would almost certainly wreck the apparatus.

Floating vessels, however, properly anchored, would be secure, as disasters seldom occur to lightships stationed on the most dangerous coasts of the world.

An anchored floating wave-motor, then, could only be utilized for the purpose. This being decided upon, what form of apparatus would be most simple and efficient?

As answer to this all-important question, our correspondent devised an attachment to a floating barge, scow, or other vessel, submerged to a depth just exceeding that of the waves, whereby the full force of the oscillations of the floating barge is utilized to operate pumps for elevating water under pressure, or for compressing air, to be employed in operating any kind of machinery, either on the floating vessel, or for generating electric currents, to be conveyed ashore by wire for any of the manifold purposes for which such current is used.

The field of development in this direction is almost illimitable. There is power to be produced from the waves at favorable points along the Atlantic coast compared with which that of Niagara shrinks into insignificance, and the latest experiments show that it is easy and economical to transmit electric power into the very heart of the Continent from the seaboard.

Our correspondent would be glad to correspond with parties who would be interested in assisting to develop this invention, after they had become convinced of its easy practicability. Full information will be furnished to such on application to "Inventor," care of THE INVENTIVE AGE.

New Ideas.

E. W. Dodge, a young musician, of Hudson, Mich., has invented a mandolin attachment for a piano. By the applying of a lever the piano will pass from its natural tone to an almost perfect mandolin effect. The inventor thinks he can make a piano do the work of a full orchestra of guitars and mandolins.

I. E. Burt, of the Burt Art Company, Minneapolis, Minn., is the inventor of a very ingenious and useful portable bicycle stand, something that has been wanting so far. The little apparatus consists of a light bar of nickel plated steel, one end of which is forged into a hook. While being used as a stand, this hook supports the front wheel. When not in use, the stand is attached to the king post by a simple mechanism. This stand can be used on any wheel. The bar is very light, weighing only seven ounces, and is therefore not likely to prove cumbersome to the rider in any respect. The invention of Mr. Burt has another merit, that of being very cheap.

At one of the flour mills at Jamestown, N. D., there is now in use a combined wheat washing and drying machine. The wheat entering this machine passes through water for the first and second washing and then under a number of brushes upon a revolving disc. The grain is then thrown against a perforated cylinder and falls upon a second revolving cylinder. This process is repeated. There are 14 revolving cylinders, with fans between the cylinders to dry off the grain. The wheat passes through the machine to the rollers and is ground into flour. Several attempts have been made to wash wheat or clean it from smut or dirt before grinding it, but there has heretofore in each case been so much dampness in the grain as to spoil the flour.

AROUND THE LOOP

One of the Most Scenic and Interesting Tram-car Trips in the World.

James T. DuBois in an interesting article in his paper, the Hallstead, (Pa.,) *Herald*, says:

"Washington will soon have the most attractive tram-car on earth. For twenty cents, a person will be able to travel the distance of twenty miles and see more sights, both historical and natural, than can be seen for the same price and in the same distance in any part of the habitable world. Beginning at the navy yard, where is located the finest gun factory in this country, one journeys a half a mile to the northwest where the tram strikes the great Pennsylvania avenue, one of the broadest and finest boulevards in the world. This splendid thoroughfare passes the new congressional library building upon which already four millions of dollars have been expended, and which will hold four millions of volumes. A little farther on comes the national capitol, which is recognized by all as being the most notable public building in existence. Then rolling down through the capitol park, the tram passes a fine bronze statue of Garfield, a statue of Chief Justice Marshall, and the well-known Peace Monument. The avenue here turning directly to the west runs in a straight mile to the treasury department. Along this route and easily seen from the street-car window are the botanical gardens, patent and pension offices, the city hall with a statue of Abraham Lincoln, the home where Lincoln died, the general post office, the medical museum, the statue of Benjamin Franklin, the theatres, the Smithsonian, and National Museum, the bureau of engraving and printing, the newspaper offices, fine business houses, elegant markets, the railroad station where Garfield was shot, the Washington monument, the agricultural department, the splendid new city post office, all the great hotels, famous during war time, and treasury department. Here the tram takes a turn, going a short way on New York avenue and thence directly northward along 14th street, passing the equestrian statue of General Thomas and a number of notable residences and foreign legations to U street. Here the fast electric car is taken a long journey of seven miles, passing by the homes of famous men, the Cleveland House, the Episcopal University, and through some of the most beautiful natural scenery in the suburbs of the nation's capital. At a charming suburban place called Chevy Chase, a connection is to be made with the picturesque Glen Echo railroad, which takes one through some very lovely landscapes to what is to become in the future the Bayreuth of America, on the banks of the Potomac, which is called the "Rhine of America." At this point are located the Coliseum the largest amphitheatre in America, which will seat six thousand persons and which in the future is to used for musical festivals, and, together with other handsome buildings for the Chatauquan school. Near this point is the famous Cabin John Bridge, which has the largest single stone arch span in the world. Returning toward Washington on the Great Falls tramway, one passes along the banks of the beautiful Potomac for five miles, passing the Little Falls, the great distributing reservoir, a number of handsome villas, the chain bridge, the acqueduct bridge, the Georgetown University, the home of Mrs. Southworth the novelist, Arlington, where sleep ten thousand of our heroic dead, Fort Meyer, where is stationed the largest body of regular troops at the capital, and finally entering historic old Georgetown, with its quaint streets and old fashioned buildings, full of historic interest. Here the tram car strikes Pennsylvania avenue again, leading by Washington circle, which has the equestrian statue of George Washington. Along this route is seen the old and new Corcoran art galleries, the State, Navy and War building, Gen. Jackson's equestrian statue, the famous statue of Lafayette presented by France to this country, the well known Arlington, and Shoreham hotels, the Blaine mansion where Blaine died, now transformed into a handsome opera house, the Department of Justice, the Columbia university, and the White House, surrounded by its beautiful park. In the future this trip will be called "Around the Loop," and will be one of the great attractions at the national capital, and in no place on earth, for the same amount of money and in the same distance, can one witness such beautiful natural scenery, so many fine public buildings, and so much of historic interest. It will be the thing in the future for all tourists who come to Washington to take a trip "Around the Loop.

The will of Benjamin Franklin was, on March 26, allowed in the Suffolk Probate Court, Boston, as a foreign will, having been probated about a century ago in the Orphans' Court, Philadelphia. The probate of the will in Boston was deemed necessary in view of the legal disposition of the Franklin fund, which amounts to several hundred thousand dollars, and which was created by Franklin's will. The will was drawn up on July 17, 1788.

Diffusions of Roentgen Rays.

Prof. Elihu Thomson, in *Electrical Review*, says:

"I sent you some time ago a short article on the diffusion of Roentgen rays, and at the time I was not aware that any work of a similar nature had been done, though it is true that certain suggestions had been made as to the turbidity of media by, I think Lenard and Professor Roentgen.

I find that Dr. M. I. Pupin has in *Science* of the issue of April 10 given an account of his investigations on the diffusion of Roentgen rays, which accords with my result, in most important particulars. I was not aware that he had done the experimental work which is described in the article, and am now pleased to note that he has gone into the subject very carefully and very thoroughly—in fact, more thoroughly than my time permitted my doing.

These observations on the diffusion of Roentgen rays show that sharp, clear cut shadows of bones or objects imbedded in a considerable depth of flesh can not be expected where the rays have to traverse a considerable depth of flesh after passing the imbedded bone or object, and that blurring or diffusion which wipes out the image may be so great as to make it very indistinct if large masses of semi-transparent material are between the plate or screen and the object the image of which is to be depicted. It is easy, also, for the experimenter to be misled into thinking that the diffused rays back of the bone or object really have come through that object, when in reality the presence of the Roentgen rays back of the solid object, which casts a dense shadow, is due to nothing more than diffusion. I regret to hear that Dr. Pupin's valuable work in this direction has been interrupted by a serious illness."

The X Rays in Court.

The first occasion in which testimony based upon the X rays has been offered in a court of justice and admitted as conclusive evidence happened lately in London. The case in question was in the form of an appeal by the Nottingham Theatre Company, who were recently mulcted in $5,000 damages in a suit brought by an actress who sustained injuries by falling through a dilapidated stairway in the theatre. The plaintiff's case was weak until counsel produced negatives of her left foot taken by Professor Ramsey by means of the new rays. These clearly demonstrated that the bones of the foot had been broken and displaced. With this evidence, the jury brought in a verdict in favor of the plaintiff, and their action was allowed by Justice Hawkins, who is famous as a judge whose decisions have never been reversed. The company have appealed against the verdict, and the Court of Appeals of the Supreme Court will now be called to pass upon its legality. The result is being awaited with considerable interest by lawyers and medical men in all quarters. Should the court sustain the decision of Judge Hawkins, it will amount to a legalizing of the new discovery as competent evidence in law.

To Preserve Oysters.

A stock company, under the name of the Exporters' Locked Shell Oyster Co., has been formed by a number of business men of Philadelphia, with a capitalization of $300,000, and a large preparing-house at Cape Charles, Va., on the Chesapeake bay, is now being equipped with the most improved machinery for the purpose of preparing its goods, prior to shipping them to the Western States of this country, also Europe.

The process employed consists simply in keeping the oysters closed, by means of a non-corrosive rivet securely fastening the shells. This, it is claimed, prevents the loss of the natural liquor, which sustains the life of the oyster. The riveting process requires no chemical or other artificial treatment, the sole purpose being the retention of the natural liquor and flavor continuous, which has hitherto been impossible. It is claimed that oysters so treated have been kept in good condition for sixty days after being taken from the water. Samuel V. Brooman is president of the company.

A foreign exchange says that the autocar race from Paris to Marseilles and back has been postponed till October 1.

The *Electrical Review* has located in its new quarters on the ninth floor of the Times Building, 41 Park Row, New York.

BOOKS AND MAGAZINES.

An interesting feature in "Power" is a page each month devoted to definitions of technical mechanical terms. The dictionary will be published in book form when completed.

Cassier's Magazine of Illustrated Engineering has, in its June number an interesting article on "Compressed Air for Street Railroads," by Whitfield P. Pressinger, with five illustrations.

The Windsor & Kenfield Company, have changed the form of "Brick" and improved its typographical appearance. In this connection it will not be out of place or untruthful to add that in its line "Brick" stands at the head in trade journalism.

It is said that a book is soon to be published in Germany especially devoted to American architecture. The Germans are evidently admirers of the recent agricultural achievements of the Americans and are struck with their peculiar characteristics.

Each succeeding number of "Home Study," an elementary journal for students of mechanics, engineering, electricity, and all technical subjects, confirms the excellence of the project which the Colliery Engineer Co., of Scranton, Pa., has begun. It is a valuable adjunct to the International Correspondence Schools, now so well known throughout the country.

The breezy "saunterings" in Town Topics continue to interest the society folk, not only of Gotham, but of every other community distinguished by the presence of imitators and devotees of New York's 400, while "Tales from Town Topics," a quarterly containing the best sketches, short stories, witticisms and gossip from the weekly issue, is quite a news stand favorite.

D. Van Nostrand Campany announce for early publication, "Elementary Treatise on Roentgen Rays, Theories, Experiments and Applications," by Edward P. Thomson, M. E., author of "The Science of Invention." The work will be illustrated and form an octavo volume, and probably be placed at $1.00. It is a work that ought to meet with much favor on account of the high standing of the author, Mr. Thomson, and his able collaborators, Prof. W. A. Anthony, L. M. Pinolet and Ludwig Gutmann.

One of the most interesting of recent publications is the little work on "Magnetism," by Edwin J. Houston, Ph. D., (Princeton) and A. E. Kennelly, Sc. E. It is intended to meet the demand for reliable information respecting such matters in electricity and magnetism as can be readily understood by those not especially trained in electro-technics. The theories of magnetism are presented and explained in concise and plain language and the phenomena of electro-magnetism, attractive powers of magnates, the Earth's magnetism are fully described. New York: The W. J. Johnston Co.; cloth $1.

The second volume of "Alden's Living Topics Cyclopedia" extends from Boy. to Con., and contains the latest facts concerning the nations, Brazil, British Empire, Bulgaria, Cape Colony, Chile, Chinese Empire, and others, and concerning three States, California, Colorado and Connecticut; also concerning six large cites, Brooklyn, Buffalo, Charleston, Chicago, Cincinnati and Cleveland. The facts are commonly from one year to five years later than can be found in any part of the leading cyclopedias, and commonly a year later than the 1896 almanacs and annuals. Other volumes will follow in rapid succession. It is a useful work.

"Lee's Vest Pocket Pointers for Busy People," contains more useful information in condensed form than any previous publication of the kind. It contains twenty thousand facts of greatest importance. The prominent events of history—arena, population, location and rulers of all nations—States of the Union, population, area, capitals and cities of more than 10,000 inhabitants—all the largest cities in the world, the great battles, chief rivers, lakes, mountains, etc.—postal regulations—rules of order, Constitution of the United States—lexicon of foreign, legal and technical terms—Australian ballot system—patent laws—telegraph cypher, etc., silk cloth, 25c; morocco, gilt, 50c; New York: Laird & Lee, Publishers.

"Wonderland, '96," is the name of one of the handsomest, most interesting and most complete little books ever issued from the passenger department of a railroad. The inspiration comes from a tour of the country traversed by that greatest of transcontinental railways—the Northern Pacific—Mr. Olin D. Wheeler, the author having made several trips through the Yellowstone Park for data and views of that wonderland. The hunting scenes were also taken from life—and experiences—in the great Rocky Mountain regions through which the Northern Pacific runs. An illustrated chapter on Alaska is also included and scenes incident to the greatest wheat regions on earth in North Dakota—where millions of acres of free land still await the land seeker—are interesting features. This little pamphlet is by no means a mere advertising document. It is a work of art and filled with valuable and interesting information especially for tourists who will appreciate the attractions along this great scenic route. General Passenger Agent, C. S. Fee, St. Paul, Minn.,—to whose genius this publication owes its being—will forward one of these pamphlets to any address on receipt of six cents in stamps.

The manufacture of tin plate has become one of the important industies of Indiana. The State Bureau of Statistics, in an official bulletin just published, gives some interesting figures in regard to the industry for the year 1895. According to the statistics presented there were eight tin plate works in operation in the State last year, which turned out a product worth $2,810,566. Raw material used by these works cost $1,556,648 and wages to the amount of $782,676 were paid to 2435 persons employed. The highest wages paid skilled labor was $9.50 a day, while there were 98 men who received from $8 to $9 a day, and 632 men who received between $4 and $7 a day.

Philadelphia merchants propose to send a steamer carrying samples of the manufacturing products of the city to certain foreign ports in order to stimulate trade.

Want to know if your invention can be patented? Write to the INVENTIVE AGE Patent Bureau, Washington, D. C.

The Safety Valve.

[Paper read at a meeting of the Douglass Progressive Association of the Steam Engineers' Union, Washington, D. C., Friday, May 29.]

The earliest approach to a safety valve was used by women in cooking when they covered the pots in which they boiled their vegetables. Though they did not know that the vessel was the generator of an irresistible power and the cover a safety valve, and that the preservation of the contents, and the security of the operator depended on letting the cover alone or not overloading it; hence it no doubt often happened that the confined vapor often threw out the contents with violence.

A safety valve is an appendage to a steam boiler for the purpose of limiting the pressure of steam to that consistent with the normal safety strength of the boiler; it is, in fact the weakest part of a boiler. Everything else resisting steam pressure must be stronger than the safety valve, and its name tells what its duty is.

The safety valve deserves a careful study, for it is not only safety for the steam boiler, but for human life and property. It is supposed that gradual increase of pressure can never take, place if the safety valve is in good working order, though engineers and firemen do not in practice, place their trust in the safety valve alone, but add their watchfulness and attention.

Every safety valve should be so placed that it can be opened at least twice a day to keep it free and in good working order. If at any time the steam gauge should show an excess of pressure and the safety valve does not blow off, one or the other is out of order. In such a case give the gauge the benefit of the doubt, and move the ball of the valve until it blows moderately.

One of the prime causes of boiler explosions is the gradual and insidious increase of steam beyond the endurance of the boiler; but to every boiler there is a limit of pressure within which it is safe. This point should be ascertained by hydraulic test annually, and no excess of pressure beyond this limit, should be allowed at any time.

Owing to the great friction of the parts of a common safety valve, it will not open until the pressure is above what it is set at. It continues to blow off after the pressure of steam has fallen far below the point of opening and wastes large quantities of steam in operation.

So, instead of standing guard over the boiler, a sentry has to be set over it, and should he by accident, ignorance or negligence, not properly attend to his duties, the boiler is without any safe guard whatever. So we see it is of the greatest importance that a safety valve is automatic, certain in its action, prompt in opening and closing at the required points of pressure, and fully reliable to relieve the boiler under all circumstances—in fact to be exactly what its name means.

MORDECAI HARRIS, JR.

Bicycle Patents.

There were 2,621 patents granted in this class up to January 1st last. Among the important steps are the following: The patent to Hood, No. 537,462, April 16, 1895, shows a handle bar which can be adjusted to any height. The patent to Metz, No. 546, 071, September 10, 1895, covers a form of pedal. The patent to Copeland, No. 529,110, November 13, 1894, covers a crank-shaft fastening designed to dispense with the keys which had heretofore been used for connecting the cranks with the crank shaft. The patent to Shire, No. 216,231, June 3, 1879, shows the first hammock saddle, a form which was exclusively used for many years. The patent to Sawyer, No. 222,537, December 9, 1879, shows what is probably the earliest use of a changeable speed gearing in a velocipede. A great variety of these speed gearings are now being invented, with a view to adapting them to the present form of machine. The patent to Pressey, No. 233,640, October 26, 1880, shows the original star machine. The patent to Latta, No. 360,101, March 29, 1887, shows the first drop-frame machine for the use of women. The patent to Smith, No. 403,153, May 14, 1889, shows the first convertible machine, which by the removal of the upper bar of the frame can be adapted to the use of women. The patent to Lawson, No. 345,851, July 20, 1886, contains claims to the modern form of rear-driven safety machine.

The bridge to be built across the Niagara gorge will, when finished, have the largest arch span of any bridge constructed. It will be among the triumphs of modern engineering, of which this age has already an unprecedented record. The Niagara bridge will have a span, between centers of end pins, of 840 feet, and a rise of 150 feet from the level of the pins at the skewbacks to the center of the ribs at the crown of the arch, and this latter point is about 170 feet above low water. The depth of the trusses is 26 feet; they stand 68 feet 7 inches apart, center to center, at the skewbacks, and have a batter of one foot in 8 giving a width of 30 feet between centers of top chords at the crown of the arch.

The Fiske Range Finder.

One of the recent and most attractive features of the Historical and Loan Exhibit of the Electrical Exposition in New York was the set of electric range finding apparatus designed by Lieut. Bradley A. Fiske, U. S. N., and to be placed on the big new battleship "Iowa." It is intended to measure with exactitude the range or distance of the enemy. Two telescopes, at the ends of the ship, are fitted with contacts, which move along arcs of resistance wire, as the telescopes are directed at any object. The wires are connected together with a Wheatstone bridge, the galvanometer of which is placed in a secure place below the water line of the ship. The act of directing the telescopes towards any object disturbs the "balance of the bridge" and makes the galvanometer needle deflect by an amount proportional to the convergence of the telescopes, and inversely proportional to the distance. The scale of the galvanometer is divided into yards, so that the needle automatically points to the graduation representing the distance of the object. The operation of the device is perfect, and the simple ingenuity of the scheme pleases every electrician.

THAT most enthusiastic representative of Southern business interests and industrial development, the *Manufacturer's Record*, of Baltimore, drops into a meditative mood and in a very able article on the future possibilities of the South, says: "The South now comes upon the field of human activity, the heir of all that science and art have accomplished without the costly experience through which other sections have had to pass in the development of their natural resources. The hundreds of millions of dollars, spent by other sections in learning how to build railroads, how to make iron and steel, how to manufacture cotton, how to extract gold from refractory ores, all at the lowest cost, have brought these industries to a point of perfection scarcely dreamed of a few years ago. The South inherits all of these advantages, and can do in a decade what it has taken other sections a quarter of a century to accomplish. It is not unlikely that the South will become the greatest manufacturing country on the face of the earth. What that means can be faintly comprehended when it is remembered that England's wealth and power are based on her manufactures; that the United States now has $6,000,000,000 capital invested in factories, whose annual product is valued at over $10,000,000,000, or three times the total annual value of all agricultural products of the country, and that the 5,000,000 hands now employed in the industrial interests of this country receive $2,300,000,000 a year in wages. Of this vast industry, the foundation of our marvelous national progress, the basis on which New York, Philadelphia, Boston, Chicago and the other great cities of this country rest, the South has now less than one-tenth. The middle-aged man of today may live to see the South's industrial interests exceed in capital, in production and in wealth-creation the entire business now represented by these stupendous figures for the whole country."

Prof. Elihu Thomson, on May 3, fell from a tandem bicycle, which he was riding with a friend, and broke one of the small bones in his right ankle. The accident occured near his home at Swampscott, Mass.

Beware of the sharks who offer to sell patents. Who are the sharks? If you are a subscriber of the INVENTINE AGE, $1.00 a year, it may save you hundreds of dollars.

The output of the Minneapolis flour mills during the year ending April 30, was by far the largest on record, exceeding 12,000,000 barrels.

As an aid to scientific research $2,500 has been granted to the St. Petersburg Medical Academy for carrying out X-ray experiments.

It is said that the coal fields of China are exceeded in extent only by those of America.

Let Us Restrict Immigration.

Congress has at last awakened to the necessity of restricting immigration to the United States, especially from Southern Europe, which, if not checked by more rigid laws, promises to flood the labor markets and great cities of this country with a most degraded and undesirable people. The time has come when we must consider with more seriousness than heretofore the effect of continued "opening of our doors to the oppressed of all nations." We cannot afford to lower the standard of American wages and American living, and that is what unrestricted immigration of that class whose wants and necessities are few means. The United States consul reports from personal observation conditions which exist in Athens. A blacksmith earns from 8 to 66 cents per day; a bricklayer, 33 to 55; carpenters, 44 to 66; coopers, 11 to 33 painters 33 to 55; printers, 44 to 55; tailors, 5 to 33, street car conductors, $6.60 to $11 per month, etc.

These wages are paid in depreciated currency, which makes them really lower than they appear on their face. Imported goods are a luxury seldom or ever enjoyed by the working people. Meat is seldom eaten; many only once or twice a year know the taste of flesh. A usual meal is bread 1.1 cent; olives 1.1 cent; cheese, 1.1 cent: wine, 0.5 cents—a total of 3.8 cents.

Yet, with these low wages the working people celebrate thirty-three holidays each year; besides each trade has its especial holiday, and no one will work on the saint's day after whom he was named. A four-room laborer's cottage rents from $3.20 to $6.50 per month, and begging is comparatively unknown.

Death of a Famous Inventor.

Charles Goodyear, oldest son of the late Charles Goodyear, inventor of the process for making vulcanized rubber, died in New Haven, Conn., on the 22d ult. Mr. Goodyear was president of the Goodyear Company and an inventor of many important devices, among them the lock stitch, for sewing soles on shoes, in use at the present time.

The Treasury surplus in Great Britain at the close of the fiscal year is without a precedent. In a financial sense, the country is at present impregnable, and this is in vivid contrast to the rest of the leading nations of the world. In spite of a long period of depression and the manifest decay of certain industries, the sources of revenue have been unimpaired, and the deposits in savings banks round out in splendid totals. This would indicate a better state of things than might be inferred from dismal announcements heard more on this side the Atlantic than on the other. Whether temporary or otherwise it means an increased purchasing power on the part of the best buyer in the world of American products.—Age of Steel.

It is stated that the first trolley railroad in Persia will be built from Teheran to the summer resorts, about ten miles to the north of the city, where everybody lives during the hot season. The summer on the Persian plateau is very hot and dry, and it is only in the neighborhood of the mountains that Europeans can stand the great heat. A concession for ninety years has been granted to a German contractor, who will start the building of the road at once.

When passing across the waters in a steam ship, we wake after a night's repose, and find ourselves conducted on our voyage a hundred miles, we exult in the triumphs of art, which has moved the ponderous vessel so swiftly, and yet so gently as not to disturb our slumbers. But with motion vastly more quieter and uniform, we have in the same time, been carried with the earth over a distance of more than half a million of miles.

An oak tree of the average size, with seven hundred thousand leaves, lifts from the earth into the air about one hundred and twenty-three tons of water during the five months it displays its foliage. This evaporated water sooner or later falls as rain, and by the action of gravity begins to flow downward. Thus the great rivers are fed.

It is now claimed that sea water can be converted into a pleasant, wholesome, and palatable drink by citric acid, which precipitates chloride of sodium.

Edward Flemming, of Mauguireville, Md., has invented a corn-harvesting machine which cuts off corn and stands it up in shocks.

A Cleveland undertaker is introducing horseless carriages in place of teams in conducting funerals.

Aftermath.

It is now reported that the Rothchilds have secured a half interest in the great Anaconda copper interests in Montana, paying therefor $17,500,000.

COL. F. E. Hain, general manager of the Manhattan Railway Company, of New York city, was killed by being run over by a freight car at Clifton Springs, N. Y., on the 7th ult.

S. B. Husselman, late of Wooster, Ohio, Pittsburg, Philadelphia and Washington and under indictment for swindling Col. Thos. R. Reeves, on a patent wire fence, in addition to other crimes is found to have from three to five living wives.

THE annual product of the Baltimore tin factories amounts to over $10,000,000. Some idea of the magnitude of the business may be had when the fact is taken into consideration that over 2,800 hands are employed—notwithstanding the fact that the great bulk of the work is done by machinery.

WE are in receipt of a neat little folder from the Watertown Engine Co., illustrating some of their new conceptions in the way of steam engines. This firm has recently opened an office at 39 Courtland street, New York City, with Mr. L. Copleston, formerly with Messrs. E. P. Hampson & Co.—a very capable representative—in charge.

THE press dispatches from the Pacific coast show that the battleship Oregon, built by the Union Iron Works, of San Francisco, for the U. S. Government is a wonder—being the fastest battleship of her size afloat. She made an average of 18.49 knots an hour and by her work earns a premium of $175,000 for her builders. She carries 160 pounds of steam and the revolutions were from 128 to 130 per minute.

FOLLOWING the recommendations of the INVENTIVE AGE, the American Bar Association, the American Association of Inventors and manufacturers and the urgent request of the Commissioner of Patents, Congress recently passed an act reducing the price of uncertified printed copies of specifications and drawings of patents. The INVENTIVE AGE will after July 1st charge but 10 cents for extra copies.

THE Baltimore & Ohio Railway Company has received the last of the great electric locomotives that will hereafter handle all the freight and passenger traffic in and out of Baltimore through the tunnel which passes under the city—which by the way is the largest soft earth tunnel in the world. Each of these locomotives weighs 96 tons and are of something like 1,200 horse power. A speed of 80 miles an hour has been easily reached by these locomotives.

The annual meeting for election of officers of the American Institute of Electrical Engineers was held at the Electrical Exposition Building on Tuesday evening, May 29, at 8 o'clock. The following officers were elected: President, Louis Duncan ; Vice-presidents, Chas. F. Steinmetz, Harris J. Ryan, Wilbur M. Stine ; Managers, John W. Lieb, Jr., F. A. Pickernell, William L. Puffer, L. B. Stillwell ; Secretary, Ralph W. Pope ; Treasurer, George A. Hamilton.

THE contract for printing the Patent Office Gazette for the coming year, has been awarded to the Norris Peters Co whose bid was a little below that of Andrew B. Graham. This firm had the contract for this printing for many years prior to the present administration and in the experience of the work gave universal satisfaction. There is, however much merit in the suggestions of the Commissioners that the contract then ought to be for a more extended period than one year and some modern means of producing the Gazette ought to be adopted. No firm can afford to put in the special machinery required for one year's contract, as was the case with Graham. If he could have had the contract for three or four years on the terms of his one year's contract he could have made suitable profits. The Norris Peters Company have the advantage of having a complete outfit on hand.

STANDARD WORKS.

"The Electrical World." An illustrated weekly review of current progress in electricity and its practical applications. Annual subscription.....................................$2.00

"Electric Railway Gazette." An illustrated weekly record of electric railway practice and development. Annual subscription.........$3.00

"Johnston's Electrical and Street Railway Directory." Published annually.........$5.00

Dictionary of Electrical Words, Terms and Phrases. By Edwin J. Houston, Ph. D....$5.00

The Electric Motor and its Applications. By T. C. Martin and Jos. Wetzler...........$3.00

The Electric Railway in Theory and Practice. The first systematic treatise on the electric railway. By O. T. Crosby and Dr. Louis Bell....................................$2.50

Alternating Currents. By Frederick Bedell, Ph.D., and Albert C. Crehore, Ph.D......$2.50

Gerard's Electricity. Translated under the direction of Dr. Louis Duncan.............$2.50

Theory and Calculation of Alternating-Current Phenomena. By Charles Proteus Steinmetz..................................$2.50

Practical Calculation of Dynamo Electric Machines. By A. E. Wiener.............$2.50

Central Station Bookkeeping. By H. A. Foster......................................$1.50

Continuous Current Dynamos and Motors. By Frank P. Cox, B.S................$2.00

¤Electricity at the Paris Exposition of 1889. By Carl Hering..........................$2.00

The Elements of Static Electricity. By Philip Atkinson, Ph.D................$1.50

Electricity and Magnetism. By Edwin J. Houston, Ph.D.....................$1.00

Electric Measurements and other Advanced Primers of Electricity. By Edwin J. Houston, Ph.D............................$1.00

The Electrical Transmission of Intelligence and other Advanced Primers of Electricity. By Edwin J. Houston, Ph.D...........$1.00

Electricity One Hundred Years Ago and Today. By Edwin J. Houston, Ph.D.....$1.00

Alternating Electric Currents. By Edwin J. Houston, Ph.D. and A. E. Kennelly, D.Sc. $1.00

Electric Heating. By E. J. Houston Ph.D and A. E. Kennelly, D.Sc..............$1.00

Electro-magnetism. By E. J. Houston, Ph.D and A. E. Kennelly, D.Sc...........$1.00

Electricity in Electro-Therapeutics. By E. J. Houston, Ph.D., and A. E. Kennelly, D.Sc.....................................$1.00

Electric Arc Lighting. By E. J. Houston. Ph.D. and A. E. Kennelly, D. Sc.......$1.00

Electric Incandescent Lighting. By E. J. Houston, Ph.D. & A. E. Kennelly, D.Sc....$1.00

Electric Motors. By E. J. Houston, Ph.D. and A. E. Kennelly, D.Sc..............$1.00

Electric Street Railways. By E. J. Houston, Ph.D., and A. E. Kennelly, D.Sc.......$1.00

Electric Telegraphy. By E. J. Houston, Ph.D and A. E. Kennelly, D.Sc.............$1.00

Alternating Currents of Electricity. By Gisbert Kapp..............................$1.00

Recent Progress in Electric Railways. By Carl Hering..........................$1.00

Original Papers on Dynamo Machinery and Allied Subjects. By John Hopkinson, F.R.S.....................................$1.00

Davis' Standard Tables for Electric Wiremen. Revised by W. D. Weaver.......$1.00

Universal Wiring Computer. By Carl Hering....................................$1.00

Experiments with Alternating Currents of High Potential and High Frequency. By Nikola Tesla..............................$1.00

Lectures on the Electro-Magnet. By Prof. Silvanus P. Thompson.................$3.00

Dynamo and Motor Buildings for Amateurs. By Lieutenant C. D. Parkhurst......$1.00

Reference Book of Tables and Formulas for Electric Street Railway Engineers. By E. A. Merrill...............................$1.00

Electric Railway Motors. By Nelson W. Perry..................................$1.00

Practical Information for Telephonists, By T. D. Lockwood........................$1.00

Wheeler's Chart of Wire Gauges.........$1.00

Practical Treatise on Lightning Conductors. By R. W. Spang.........................75

Tables of Equivalents of Units of Measurement. By Carl Hering..................50

Copies of any of the above books or of any other electrical book published will be sent by mail, postage prepaid, to any address in the world on receipt of price.

THE INVENTIVE AGE,
Washington, D. C.

THE INVENTIVE AGE wants live agents in every part of the country. Readers of this notice will confer a great favor by sending us names of persons who would likely take an interest in the matter. Our cyclopedia premium offer is a great winner. The book alone is worth double the amount. $1. charged for the two.

ANY person sending us the names and addresses of five inventors who have not yet applied for patents will receive the INVENTIVE AGE one year free.

The Inventive Age
AND INDUSTRIAL REVIEW
A JOURNAL OF MANUFACTURING INDUSTRY AND SCIENTIFIC PROGRESS

| Seventh Year. No. 7. | WASHINGTON, D. C., JULY, 1896. | Single Copies 10 Cents. $1 Per Year. |

THE AIR MOTOR.

A Practical Possibility—To be Used in Washington.

The passage of a bill by the last Congress compelling the Belt Line street railway to adopt some system other than horse power has caused considerable comment from press and public. There seems to be some doubt as to the ability of the Belt Line company to put on motor power within the time specified, as it is well known that no other known system but storage could be used within the required three months limit. There has been but little faith put in the air motor idea, and much has been said against it. But for all this we are assured by competent authority that this motor is not only a practical possibility, but that it will be in operation on the Belt Line within three months. The motors are now being built, and when one is completed it will be brought to Washington whose citizens will have the pleasure of riding by air—without a flying machine. The air motor is not a new idea, for it has been used in France, at Nantes, in Brussels, and in this country, at New Orleans. At first the air motors in Paris were compelled to return to a central station for storage, requiring time and expense for this purpose; but an improvement was made whereby the motor gets its air supply from an under ground conduit, having convenient plugs located along the route for supplying the air cylinders.

The Washington Belt Line will use the storage system taking its supply immediately from the stations, of which there will be three located at different points. The motors will make a round trip, except possibly on the East Capitol division, with one charge, being reloaded at each return to the storage station. There will be no change in the roadbed or rails other than a little grading and straightening up.

Ease and good speed is expected from the new power, and from a standpoint of safety, there is nothing against it. In cheapness in operating, it compares favorably with the over head trolley; there are no poles or wires connected with it to obstruct and disfigure the streets, nor electricity with its dangerous and uncertain current.

The air motor which has been used at Rome, N. Y. resembles in many respects—if not altogether, that which will be employed by the Belt Line Company.

It is of the same style and size of the ordinary electric car, and of the same seating capacity. The motor mechanism located under the car, and taking up no paying space, consists of two simple link-motion, reciprocating engines having cylinders six inches in diameter and fourteen inch stroke, with valves cutting off at 1-10 to 1-6, and applying the power by connecting parallel rods direct to crank-pins of the drive wheels, which are four in number, twenty-six inches in diameter, running on a wheel base of seven and a half feet. Upon the wheel truck rests the entire weight of the car and mechanism, evenly distributed upon elliptic springs, enabling the car to pass smoothly over bad tracks and crossings. The mechanical features of the motor are almost identical with that of the steam locomotive. The storage reservoirs, located beneath the car floor and the seats, have a capacity of thirty-five

COMPRESSED AIR MOTOR CAR.

cubic feet, enough to run the car twelve to fifteen miles continuously; but double this amount of air could be carried, with which a run of twenty miles could be made. The reservoirs in use are seamless steel flasks eight inches in diameter and of varying lengths, in which air is compressed to 2000 pounds to the square inch. Between these flasks and the motors is placed a small tank containing six cubic feet of super-heated water, at a temperature of about 325° F., which is supplied from a stationary steam boiler, working at a pressure of 80 pounds to the square inch. The tank is jacketed with non-conducting material, to prevent external radiation.

In operation the compressed air, after passing through a reducing valve and being lowered to 150 pounds per square inch, circulates freely through the hot water, and a mixture of heated air and vaporized water passes to the motors. The controlling apparatus of a car of this kind is very simple, taking up about the same platform space as does the ordinary horse car break handle, and requiring no especial skill for its operation. The motor is noiseless, odorless and free from any offensive feature; there is neither electricity, steam or smoke from it, and as it responds promptly and easily to the starting mechanism, there is no jerking or disagreeable motion.

Since Washington has been the experiment ground for so many new ideas in street railways, it is strange that the gas motor has not claimed the attention of the industrious "promoters." This system has been successfully operated in Switzerland and Germany, where cars are run by gas at a cost that compares favorably with the horse lines. They (the gas cars) make a maximum speed of eleven miles an hour, and by taking their gas at the street mains, do away with the necessity of going to a supply station when the quantity carried is exhausted.

The cost of running a gas motor is said to be less than the overhead trolley—about $1.05 to $1.09 per 1000 feet is the charge. But this is not as cheap as compressed air, and there is not the necessity (in the air motor) for constant motion, as in the gas motor. The speed of the air cars is practically unlimited, and by carrying reservoirs on trailer cars, distance will hardly be a matter for consideration.

When the extensive plans of the Belt Line Company are carried out, they will have a splendid system, consisting of many miles of street railway, ramifying the capital city, extending into the suburbs, making connection with the great Washington and Baltimore electric road, and giving to the traveling public a rapid air-transit, which, if things turn out as planned, will be all that can be desired by those who ride in cars.

Mr. Adolph Pessl, of Vienna, has invented an improved apparatus for making round ropes composed of co-axial layers of textile threads twisted in alternately opposite directions. The rope is made by means of a series of discs or arms arranged one behind the other, to have a common axis and carrying the thread reels or bobbins, the arrangement being that the first disc makes the core and the following disc makes the several concentric layers of threads, whilst a winch slowly draws the finished rope from the apparatus and coils it up.

The Inventive Age

Established 1889.

INVENTIVE AGE PUBLISHING CO.,

8th and H Sts., Washington, D. C.

ALEX. S. CAPEHART. MARSHALL H. JEWELL.

The INVENTIVE AGE is sent, postage prepaid, to any address in the United States, Canada or Mexico for $1 a year; to any other country, postage prepaid, $1.50.

Correspondence with inventors, mechanics, manufacturers, scientists and others is invited. The columns of this journal are open for the discussion of such subjects as are of general interest to its readers.

Technical matter is particularly desired. We want practical information from practical men.

Nothing will be published in the editorial columns for pay.

The INVENTIVE AGE is thoroughly independent.

Advertising rates made known on application. Special facilities for furnishing cuts of any patented article together with descriptive article. Business specials 25 cents a line each insertion, 7 words to the line. No advertisement less than 50 cents.

Address all communications to THE INVENTIVE AGE, Washington, D. C.

Entered at the Postoffice in Washington as second-class matter.

WASHINGTON, D. C., JULY, 1896.

Among the papers in the District of Columbia devoted to inventions and patents, none has credit for so large a regular issue as is accorded to the "Inventive Age," published monthly at Washington.—*Printers Ink.*

A SPECIAL commission has arrived in this country, authorized by the Japanese Government to study the workings of electric power and the telephone systems in the United States. The commission is composed of S. Mine, R. Natayama and Y. Wadachi.

AMONG the army of individuals, to whom Congress turned a deaf ear last session, was an inventor named James Selden Cowden of Virginia, who modestly asked for $15,000 to assist him in demonstrating the feasibility of his airship. He cites the $30,000 given to the developement of the Morse system of telegraphy as a precedent.

THE Tennessee centennial ceremonies at Nashville, June 1st., were of an imposing nature despite the discouraging climatic conditions. Work on the exposition buildings is progressing rapidly and a year hence will witness a really wonderful exhibition, inter-state and international in its character and interesting in all its features.

ON the 21st ult, the Lake Street Elevated Railway Co., Chicago began running their cars by electricity. This road is about 6½ miles long and of double track. The current is carried upon a third rail just out side the guard timber. This rail is supported by pillar insulators set every six feet, and is protected by two planks set on edge. This provision is made to prevent accident from carelessness.

THE next meeting of the American Street Railway Association will be held at St. Louis, October, 20, in the building used by the national republican convention. The larger portion of the building will be devoted to displays and space therefor must be applied for to J. C. Pennington, 2020 State street, Chicago, before August 15. It is expected that this convention will eclipse all previous meetings of the kind.

THE Atlanta Telephone Company is the name of an organization that has jumped into prominence because of its successful opposition to the old Bell Company in Atlanta. It has just let the contract for furnishing 10,000 telephone instruments, the largest telephone order ever given. The old company now has but 1,400 subscribers at $84.00 a year. The new company has 2,800 subscribers under contract for five years at a yearly rental of only $36.00 a year and expects to have 7,000 subscribers before the year is out. And Atlanta is not the only place where war is being waged on exhorbtant telephone rates. The Carnegie Telephone Co. is another organization having in view a reduction of rates under the state charter limited to $2.00 a month for business purposes and $1.50 for residences.

IN the struggle of inventive geniuses, on the one side to produce an armor plate that cannot be broken or punctured and on the other hand to increase the penetrative power of projectiles, the former seem just now to be ahead. This was proven in a recent test at Indian Head, conducted by government officials. Two shells, one an eight-inch Carpenter, and the other a twelve-inch Wheeler-Stirling, were broken up on reforged plates. The shells did well according to the old standards, but were clearly over-matched by the plates. These particular shells were treated by an improved process, and the requirements had been increased at the makers instance, but the tests demonstrated that the requirements now are too severe, as was supposed by the ordinance officers.

THE remarkable growth of cotton manufacturing in Japan, which has attracted so much attention, is discussed in an elaborate letter written by Mr. Robert P. Porter, superintendent of the last census, who has just returned from an investigating trip to that country. Mr. Porter's letter, written at Tokio under date of May 13, gives in detail the progress of this industry. The manager of the Kanegafuchi Spinning Co., of Tokio, in an interview with Mr. Porter, stated that by the end of this year Japan would have 1,500,000 spindles. His company is now building a branch mill at Shanghai for 40,000 spindles, and also another mill in Japan. "Japan," he said, "expects, after supplying the home demand, to control the market for the 400,000,000 of people in China.

MAX COHN of Milwaukee, Wis., writes to correct a statement in the last issue of the AGE to the effect that in the perfection of his flying machine he received information from Prof. Langley's efforts in that direction. He claims priority of discovery of some of the scientific principles involved in the successful air ship and mechanical flight. The inventor says:

Prof. Langley arrived at about the same theory, regarding mechanical flight, but his mechanical demonstration of this theory, in my opinion is positively wrong. This may seem a very bold statement to make, but it is a case where history repeats, itself and you need not look far and long for living proofs. Edison or Berliner and others discovered and invented wonders, while Professors' sought in *vain* to solve the same problems by adhering to old worn out "University" laws of nature. etc.

THERE has been considerable interest of late in the possibilities of compressed air as a motive power, says Age of Steel. Its efficiency in many forms of service has long been demonstrated, but in late years is has assumed a new importance. Some of this is due to the more rapid progress made in scientific investigation, and not a little to the pressure of industrial necessities. Mechanical devices have largely displaced hand power. The old rate of production is practically no where in the race of trade. It has multiplied beyond precedent and is not likely to call a halt till its abuse calls for a bridle. Every kind of motive power has been promptly challenged as to economy and efficiency.

Science has been sleepless, and ingenuity never tires in exploring and utilizing all the known forces of nature. These are harnessed and directed by the skill of man. Steam, gas, air, electricity, water, and even the magnet are all utilized in the movement of machinery. What is yet possible to be utilized from the sun and the sea is alread yon the program of progress and experiment, and the end is not yet. What may be accomplished the wisest of men would fail to forecast, but that much as yet is practically below the line of vision, few if any would have the hardihood to dispute. As each application of power comes into favor or prominence it has its enthusiasts and its critics and not until it is followed by another and its special virtues are demonstrated by comparison is the practical value of each or any established. Compressed air will rehearse the same old story. It will have its crude and premature claims, and also its suprises. In the hurry and hair-raising rush of modern methods it will suffer from faulty and immature devices, but with it as with other sources of power time and experience will sift the chaff from the wheat and let merit decide all disputed points.

Photography in Colors.

It is at last announced in an authoritative manner that photography in colors is an assured fact. The patent has been issued to James D. McDonough, a Chicago man, who has been at work upon the problem for many years. The discovery was not made public until 500 color photographs had been taken. Among these were portraits in which the flesh tints were reproduced with exactness, but perhaps the most remarkable of all the inventor's achievements was a photograph of a Japanese panel, which reproduced seventeen shades of color with absolute fidelity.

The new discovery will rank as one of the most remarkable and far reaching in its effects of any that the scientists in the photographic field have made for many years. Daguerre had hardly astonished the world with the daguerreotype before these scientific operators were at work in all parts of the world trying to photograph colors. Some of them abandoned it in despair, others thought many times they had found the secret, but all conceeded that the solution of the problem was attended with difficulties of the gravest description. At last, however, the mystery has been disclosed and the patient labor of years has been rewarded.

The value of the discovery will be enormous. Objects in life or nature photographed without color are at best only half seen. It is simply form without expression. The color is the striking part of the picture. It makes the difference between a great painting and a black and white reproduction. The future value of photography now can hardly be overestimated. The art has made great progress since the days of Daguerre and its uses have been largely increased. No village is so small that it does not have its photographer, and in cities they may be found in every block. It has become a necessity. It covers almost every action of man and animal and every phase of nature. It has some imperfections in perspective growing out of the manner in which the lines of light strike the object glass and the tendency to exaggerate what is nearest the camera, but even with these drawbacks photography has become indispensable.

The new uses of photography in colors will be almost limitless. It will reproduce the portraits of men, women, and children, the animals of the farmer, the varying phases of the landscape, the architecture of cities, the fleets upon the ocean, the billows of the sea, the procession of the clouds, the majesty of the mountains, the verdure of the trees, the beauty of the flowers, the lightings and the stars, all in their natural colors. It must inevitably be substituted for much of the work of the painters, even if it does not make their art practically useless, and the best of it all is that the process is cheap, though the syndicate which controls it, with the customary greed of human nature, will probably attempt to overcharge to the limit of extortion.

NOTES.

The Fastest Vessel in the World.—According to "Invention" England has got the fastest vessel in the world. The Desperate, a torpedo-boat destroyer made by Thorneycroft, ran a measured mile at the rate of 35¼ statute miles per hour. That boat can get away from anything on earth or catch anything that floats if it starts after it.

A Vegetable Meat.—In Japan, that land of gentle manners and other queer things, they have invented vegetable meat. The substance is called in the vernacular, "torfu." It consists mainly of protein matter of the soya bean and is claimed to be as easily digestible as meat. Torfu is as white as snow and is sold in tablets; it tastes like fresh malt. What with mineral wool, wood silk, and vegetable meat, and other articles of food and wear made by science, Nature may as well go out of business at once.

The Ferris wheel, so conspicuous at the World's Columbian Exposition, has been re-erected in the northern part of Chicago, not far from Lake Michigan. The company operating it is composed of substantially the same stockholders as during the World's Fair. Robert W. Hunt is president,

SCIENCE AND INVENTION.

The Inventor of the Electric Ocean Telegraph Cable.

The Scientific World may not know that the ocean telegraph cable was invented by William Gordon of Providence, Rhode Island, in 1848. Mr. Gordon conceived the idea of a cable formed of three or more spirally wound, rope like, strands about a central electrical conducting wire core, and the whole insulated by Gutta Percha. Only two other persons, George B. Simpson, and James Reynolds came into the field about this date. Simpson's plan was to directly coat a central core, or conducting wire, with gutta percha. The Simpson and Gordon inventions were patented, the former May 21, 1867, No. 65,019, and the latter April 4, 1876, No. 175,693. The Reynold's application was abandoned, a patent never issued thereon.

The torturous experiences of Mr. Gordon before the United States Patent Office, after he had filed his application May 13, 1848, is enough to deter inventors of limited means from entering upon the work of securing a patent for a meritorious invention. When Gordon's application was first examined by a distinguished electrical examiner of the United States Patent Office, he was told that his invention was not patentable, because in the art of making *clothes lines and picture cords* a central strengthening wire of steel or copper had been wound with spiral strands. With this set back, the inventor for a time yielded the contest. Had he been a patent lawyer or employed one the examiner would have told that the clothes line, or picture hanging cord lacked the essential feature of his invention, viz: insulation with gutta percha; and also, that the ocean cable device was a new invention to the world, by reason of the discovery that messages could be transmitted under water, from country to country by electricity; furthermore that it was new, by reason of combining three essentials: 1st., insulation of the electric core wire, 2nd Relievance of the core wire in a great degree from the strain to which it would be subjected when suspended or laid under the water of the ocean; and 3rd Rendered still flexible, to prevent breakage under tension from various causes. Not until after a special act of Congress dated July 8, 1870 which relieved all pending rejected applications from the charge of abandoment, was the effect of the first erroneous rejection of Gordon's application removed; and, not until after the expiration of six years was he successful in getting his patent. Any one familiar with the hardships of this inventor will have no need of reading Dicken's "Bleak House", or that portion which tells of the lives and hopes of those who are heirs to property, and kept out of the same by chancery proceedings.

It may here be of service to the inquirer as to the origin of cable inventions, to mention that it is stated that Dr. O'Shaughnessy made experiments in Europe in 1839; that he wound cotton thread around a central wire core and applied other means than gutta percha by which to secure insulation, without destroying flexibility. It, however, will be seen that this plan did not come up to the level of the Gordon invention of winding strong, rope like strands around a central wire for giving essential strength to the wire. This crude invention of Dr. O'Shaughnessy was not patented until 1858.

Cooke and Wheatstone took out a patent in 1837. This patent was a very different thing from Gordon's ocean cable. It was made at an early date in telegraph history, at which time it was not known that a naked wire suspended in the air for a distance of hundreds of miles by means of poles would answer the purpose sought. It was proposed at that time, that the wires should be placed in trenches in the ground, after enveloping them in some insulating substance, and enclosing them in metallic tubes. Cooke and Wheatstone proposed to construct a continous series of wooden rails, supported on posts and extending through the entire route. On the upper side of each of these rails was to be formed a deep continous groove, in which was to be laid the telegraph wire after having surrounded it with some non conducting substance. It was then to be covered over, within the groove, in such a way as to be protected from injury and to secure its insulation. One mode of protecting the wire preparatory to its being laid in this groove was to wrap it with coarse thread which was then to be covered with varnish. It was further stated that it might, in addition be covered with a spiral fillet of hemp, and then covered with pitch. This fillet of hemp was intended as a "woolding". It was not proposed that is should be laid up in the manner of the rope cable of Gordon so as to have it possess longitudinal strength, for there was no need of such strength in their construction. This invention viewed in its entirely, lacked flexibility.

Professor Morse in 1842-3 made the invention of the electric telegraph, and he provided insulation and flexibility, but the provision of strength by spirally winding strong, rope like, strands, whereby a ocean telegraph cable was produced, was not in his mind.

One Gilpin took out a patent in England in 1854 and Balestrine in 1856 but Gordon's invention extended back to 1848.

In acting upon Gordon's application, on Jan. 21, 1876, the Primary Examiner of the United States Patent Office said, "The examiner is willing to add as an expression of his opinion, that at the date of the original filing (1848, by Gordon) there was nothing on record in the office, and no publication, which should have prevented applicant (Gordon) from obtaining the patent then asked for, excepting the then pending application of George B. Simpson, since patented No. 65,019, May 21, 1867, for the use of gutta percha as an insulating medium, and James Reynolds."

Now by refering to the Patent Office record files of the respective parties, it is found that Reynold's application was filed in June 1848, and Simpson's June 1849.

It is a remarkable fact that on an examination of Gordon's drawings, specification and model, as originally filed it is found that they had in them all the essential elements of the successful submarine cable of the present day.

In Gordon's original specification filed in Patent Office May 1848, the use of Gutta Percha is set forth as the substance by which the insulation of the electric conducting wire was to be effected. That substance was never before used or suggested for this purpose. In the Journal of the Society of Arts, a work of recognized authority in matters of this nature, (See number for April 23. 1858) it is stated that telegraphy must have proved impracticable but for the discovery of gutta percha." It has been said that of all insulating substances employed for submarine cables, gutta percha proves itself the most lasting, as insects or the terreda do not appear to feed upon or destroy it.

Improvements of various kinds have been made upon the Gordon invention but they are all in furtherance of his general idea. The fibre of iron, in flexible form, has been substituted for that of hemp, and improved modes of securing the more perfect insulation of the core within, have been devised, but all having in view the strengthing and insulating the central core—leaving the cable sufficiently flexible to answer the end in view. In a word, the submarine cable of to-day differs in no essential principle from that devised by William Gordon, and for which he sought a patent in 1848, and which he only succeeded in securing in 1876—a period of 28 years having elapsed between his application and the date of the patent; and that too, when it can be truly said of his first development at the Patent Office, that the very model and drawings filed by him in 1848, with out the least essential change, would, probably, have effectually served the purpose of an efficient telegraph cable now. That this is not an extravagant assertion we have only to refer to the address of Sir James Anderson before the London Statistical Society found in the popular Science Monthly for May 1873, page 44, where among other things he says:

"The very light cable invented by Mr. Varley (No. 21) admits of being laid by having the strain taken off the core by two hempen strands—the core itself being the third strand of the cable. As a light cable to be manufactured in a great hurry, and laid to meet some emergency, it has a good deal of merit, but for a deep sea cable, I am of the opinion that it would be found incomplete."

Now this Varley cable which possessed a great deal of merit in 1873 is substantially the same as that for which William Gordon had sought a patent twenty three years previous, and was the first practical contrivance of the kind that had entered into the mind of mortal man. It was the offspring of genius, the vital spark in which had long been arrested and left smouldering, but not extinguished, through the errors and mistakes of the Patent Office.

The Varley cable differed from Gordon's in having but two hempen strands instead of three as Gordon represented. In the Varley cable the wire or core formed the third strand, while in the Gordon the wire or core was in the centre and was wound about by three strands, where it could be more effectually insulated and protected. The Gordon cable was manifestily far superior to the more modern invention of Varley. It was in fact the progenitor of the whole succeeding race of submarine cables.

The Patent Office granted the underlying or subordinating patent to Gordon in 1876, but alas—it was to late, for the great telegraph companies invention, of the United States who had long unjustly appropriated his invention, defied his claims and he nor his posterity ever received any reward for this great benefaction to the world.

Would it not be well for the Administrators of the Patent Office to halt in making rules outside of law, which deprives inventors of the right, in such cases where six or nine months delay may have occurred on account of the erroneous decisions of the Examiner upon their application, or from other causes. The very fact that Congress enacted a special remedial law July 8, 1870. which enabled parties whose cases had elasped by reason of neglect to follow them up and have them legally patented shows that the legislators of our country were opposed to forfeitures of men's rights, and especially the products of their brains.

R. W. Fenwick.
Counselor in Patent Causes.

Inventors and Inventions.

There is a doubt existing amongst many as to the true meaning of the word "inventor." To orignate a new article of popular or technical interest is sufficient to cause the expression to be used. Few of so-called inventive power have more than made additions to the large stock of inventions of a similar if not identical nature. A patent that is granted is not always proof that an invention has been made. An army of men make it their business to simply improve what already exists.

Few create entirely or make a radical departure from the beaten track. The discoverer of a new principle rarely makes an application of it himself. The ability to discover a new law requires a perfect equipoise of the reasoning abilities; it may not suffice at all for the construction of a machine making use of the discovery. Again it is a familiar fact that the original inventor of a new machine makes but few changes in its crudity, but the host that follow perfect it in detail until we have as an example, side by side, the first rough design of Watt and the elaborate triple expansion engine of a modern marine equipment.

The unfinished idea does not lose in value because it is not perfected. Like a rough casting it needs finishing and polishing. The general structure does not pass through other than detailed processes insignificant individually in comparison with the whole, but of equal importance in the concrete. An inventor may enter into either fields as his native qualities appear. It is therefore necessary to divide inventors, and therefore their inventions, into two classes; the first greater than the second. The first headed by such men as Watt; the second greater class, those that added their mite to the pile. Inventors may therefore invent the entire device; that is originate the base of a great superstructure, or he may simply add or improve some necessary adjunct. The steam-engine represented the original idea; its further application to rail and water required comparatively few changes in other than its external form. But the use of the governor, the thousand and one details, each of which was and are the subject of special study, give a perfect illustration of the variations and striking differences between "inventors and inventions."—*Electrical Age.*

REVIEWS.

Justice Walter Clark, LL. D. of the Supreme Bench of North Carolina, contributed an instructive and delightful paper on Mexico, to the June Arena, the interest of which is enhanced by several excellent illustrations, including a recent portrait of the President of the Mexican Republic.

"Photographic Amusements," is the title of a publication just issued by the Scovill & Adams Co., of New York, publishers of the Photographic Times. This is No. 56, in the Photographic series issued by this firm. It is edited by Walter E. Woodbury and is a compilation of novel effects to be obtained with the camera. It is an intensely interesting little book for photographers. New York: The Scovill and Adams Co., Price $1.00

McClure's Magazine for July will have an illustrated paper by Cleveland Moffett, showing the exact status, at the present moment, of the horseless carriage, and indicating the immense revolution that impends in travel and traffic, now that the horseless carriage has practically passed the experimental stage.

Many very attractive and beautifully illustrated articles are given in Frank Leslie's Popular Monthly for July, and also several excellent short stories. The leading feature is a description of General Robert E. Lee's part in the battle of Fredericksburg and Chancellorsville, written by Colonel John J. Garnett, of the Confederate States Artillery, and forming the sixth paper in the magazine's great "Lee Series." The article is profusely illustrated with portraits and battle scenes.

Prof. Hinton, of Princeton, has invented a mechanical device for which no better name can be found than the "dummy pitcher." It is intended to take the place of a human pitcher in base ball practice. The device is not yet perfected, it being Prof. Hinton's idea to have it light enough to be raised to the shoulder and aimed like a rifle.

Why He Patents His Inventions.

As Thomas A. Edison watched the pumping of the air from a glass tube in his laboratory a day or two ago a man said to him: "You patent every little thing you discover, don't you, Mr. Edison?" "I do," said Mr. Edison. "And do you know why I do it?" "I suppose you do it so you will reap the benefit of your discovery," was the reply, "I thought you'd say that," said Mr. Edison, "and I don't suppose you will belive me when I tell you it isn't so, nevertheless. I discover a great many things that I would be very glad to give to the public for nothing, but I don't dare. I patent these things to save myself from defending law suits. There are a lot of sharks in this world who are continually on the lookout for new things, and when one of them hears of something new he hustles to the Patent Office to see if it is patented. If it isn't he claims it as an orginial discovery and files his claim. Then he will turn right around and, like as not, begin a suit with the man who invented the things for making or using it. The inventor will say: 'But I discovered this thing first; I am the inventor.' He is referred to the Patent Office, where he finds the official claim of original invention. The fact that the papers are filed long after he made his discovery does not help him, for all the other man does is to hire a fellow to swear that he made the discovery a month or two prior to the date the inventor claims. It sounds ridiculous, probably, but it is a fact, that there are often races between the inventors and the sharks to reach the Patent Office, the sharks having had early information about the inventor's discovery. There are many such races, and thousands of dollars depend on each one. What I say is literally true.

The Proposed 16-Inch Gun for Coast Defense.

With the comparative failure, some years ago, of the 16¼ inch 110 ton guns, and the 17 inch 105 ton guns, respectively mounted in the English and Italian fleets, the manufacture of these monster weapons ceased altogether, and it was predicted at the time that no more of them would be built. The tendency of late years has been to reduce the weight of the main battery, the heaviest guns of the United States battle ships being 62 tons weight; of the English, 46 tons; of the German, 43 tons; and of the French, 44 tons. The reasons which led to the adoption of the lighter guns were the great difficulty of manufacturing guns of over 100 tons weight that would stand the actual test of firing; their destructive racking effect upon the ships in which they were mounted; the large amount of weight which had to be allotted to their mounts and protection; and the slowness of their discharge. It was found, moreover, that by increasing the length of the guns, reducing the caliber, and using smokeless powder, a much greater speed of firing and an equal amount of penetration could be obtained for the same total weight of guns and mounts.

But, while the argument in favor of lighter and more handy guns is a powerful one, as applied to battle ships, it is not so strong as applied to land fortifications. The mounting and protection of a 110 ton gun in an earthwork redoubt would cost far less than it would to place the same gun with similar protection upon a battle ship. The unsteady platform afforded by a ship's deck, combined with the slowness of firing, makes the probability of scoring a hit very remote; but such a gun mounted in a fort and trained across a channel, such as the entrance to New York Bay or San Francisco Harbor, where the ranges are short and accurately known, would have every chance to get home a shot normal to the water line belt armor of a passing ship. One such penetration of the vitals by an 1,800 pound shot would do more to wreck the ship than a continued battering by lighter shot and shell. It is mainly for this latter reason that General Flagler, of the United States army, advocates the building of a certain number of 16 inch guns for coast defense.

The destructive force of a shot may be expended in penetrating the armor or in racking and crushing in the sides of a ship. While it is true that modern 45 or 50 ton guns have a high power of penetration, they fall far below the 110 ton guns in the crushing force of the blow delivered. Thus the 12 inch 45 ton United States gun has a muzzle energy of 26,000 foot tons, whereas the 16 inch gun would probably develop not less than 60,000 foot tons. The racking effect of such a blow, squarely delivered on the belt armor of a passing ship, would be terribly destructive, even if it should fail to penetrate.

The recent improvements in the material and manufacture of guns make it possible to turn out 110 ton guns that would be free from the defects of the early English and Italian guns. The drooping which occurred at the muzzle of these guns after firing a limited number of rounds was due to the short length of the outer hoops, which robbed the gun of its necessary transverse strength. By employing longer hoops and disposing them to better advantage, as is dope in the United States guns, there is no question but what a 110 ton gun could be turned out which would be thoroughly reliable.

The energy of the blow from an 1,813 pound Holtzer projectile, fired fnom one of the 110 ton guns built for the battleship Sanspareil was 54,320 foot tons, and the shot bored a 16¼ inch hole through 20 inches of compound steel and iron plate, 8 inch of iron, 20 feet of oak, 5 feet of granite, 11 feet of concrete, and finally buried itself in a 6 foot wall of brick masonry.

The shell of a 110 gun contains a bursting charge of 187½ pounds of powder, and as it would hurl in all directions nearly a ton of flying fragments, its destructive effect in a boiler or engine room, or in a crowded battery, would be incalculable. Some idea of its effect may be gathered from the havoc wrought at the battle of the Yalu, when a 12 inch 725 pound shell struck the barbette of the Japanese admiral's ship, putting the 66 ton gun out of action, killing 30 and wounding 40 of the crew, besides wrecking all the internal fittings on that deck.

The moral effect of two or three 110 ton guns, mounted at Sandy Hook, New York, or at the Golden Gate, San Francisco, upon an attacking fleet would be well worth the cost of their manufacture; and should a ship attempt to cross their line of fire, it would be at the risk of almost certain disablement.—*Scientific American.*

According to a patent taken out by Messrs. Burrows and Radclyffe, of London, ramie, rhea, China grass, or other fibers are spun wet, so that the yarn may be stronger, smoother, and brighter than by the dry process. This is obtained, it is said, by passing the roving through a trough of water, which may be pure, or may contain some ingredient to dye the rove, or to aid its brilliandy, and afterwards through the feed rollers acting in conjunction with carrier rollers having V-shaped grooves in them to prevent by wedging any untwisting of the rovings. The drawing rollers operate as heretofore to draw away the fibre in the wet state, which is then twisted into yarn.

Index Classification.

GENERATION.

METHOD OF GENERATING THE FIELD.

1. Magneto-electric machines in which there are permanent magnets.
2. Dynamo-electric machines, which rely on the residual magnetism of the cores to start the current.
  (a) The Series Dynamo in which the coil of the armatures and electro magnets are joined up in series.
  (b) The Shunt Dynamo in which a portion of of the current is shunted for use.
  (c) The separately excited Dynamo, in which the electro-magnets are fed by an auxilliary machine.
  (d) Combination methods:
      Series and separate.
      Series and shunt.
      Series and long shunt.
      Series and separate coils, etc.

GENERAL TYPES.

Bi-polar.   Multipolar.   Uniform Field.

FIELD MAGNETS.

Fixed.
Rotating.
Cores.
Pole pieces.
Winding.

COMMUTATORS.

Brushes and holders.
Armature rubbed by brushes.
Arrangement of connections.
Construction of—
  Multiple.
  Enclosing in case.
  Materials.
  Insulation applied in stripa.
  (See Ins Proper) plastic.
  Reciprocating.
  Lubricating.

CONTACTS.

Maintained by—
  Bayonet joint.
  Cam or wedge.
  Screw.
  Spring (See Switch.)

ARMATURES.

Ring,—anchor and flat.
Drum.
Disc.
Pole.
Induction.
  Adapting for Alternating.   (See Motors).

ARMATURES,

DEVELOPMENT OF PARTS.

Bobbins (See Armatures Winding
Coils on poles of magnets.
Conductors for
Cores
  Cooling laminating, ventilating.
  Safety devices for over heating.
  Binding wires.
Driving Spoke and Spider.
Insulation—(See ins. proper.)
Mounting.
Multiple.
Unconnected with external circuit.
Winding.
Bearings.
Testing.
Utilizing, Secondary impulses
Protecting Cases
Vibrating strips.

DESIGNS.

FORM OF ARRANGEMENT.

FRAME WORK.

BED PLATES. (See CASTINGS.)

Dyanmos—Driven by—
  Compound Air and Gas engines.
  Axles of Vehicles.
  On Trains.
  Hand (See hand machines.)
  Steam Engines.
  Steam Engines direct.
  Rotary Engines.
  Tide Mills.
  Water Power.
  Water Wheels.
  Wind Mills and Wind Motors.
  Springs.
  Locomotive and Tramway Engines.
  Gearing for
  Motors stated to be applicable but not specially modified (See Motors.)
  Reversing
  Vibrations of Vehicles.
  Fly wheel on shaft.
  Regulating speed.
  Regulating by—

REGULATION.

For regulating in External Circuits.
See Electric Circuits Regulating by—
  Gearing, varying speed.
  Means for operating. Regulatiug devices.
  Moving Brushes (See Brushes.)
  Shifting Commutators.
  Operating throttle or cut off valves.
  Special Winding.
  Special Pole Pieces.
  Varying Current in field Coils
      in external circuit.
      in magnetic circuit.
  Electric Motors.
  Subsidiary Armatures.
  Screening Armatures.
  Subsidiary dyanmos.
  Induction appliances.

NOTICE.

Particular attention is called to the Index Classification as above presented, which is the first installment of a close and scientific sifting of the subject-matter of the United States Patents relating to the subject of Electricity. The prescribed 'Office' classes and unclasses are employed as a ground work, and are extended by outling the contents of each in appropriate order, indicating at proper points, such allied Cross-references as would necessarily need investigation to reveal obscure references.

The object in presenting to the readers of THE INVENTIVE AGE this detailed classification or tabulation of subjects of invention is to enable them when ordering copies of United States Patents relating to the State of Art on a given subject to restrict their order to such patents as pertain to, or have a bearing upon the subject of invention which they seek to collect.

We are prepared to furnish copies of United States Patents according to the above tabulation, at the regular rates plus a reasonable charge for the service.

NEW INVENTIONS.

A Brief Description of Some of the Most Valuable New Ideas.

FIREMAN'S HAT.

A hat for those who follow the dangerous occupation of firemen, is the invention of James R. Hopkins, of Somerville, Mass. The hat has an outer crown composed of aluminum, a cape, of the same material, to which the rim is secured. The rim or ring is bent throughout its length in the projecting part to form a groove, combined with a flexible inner crown or head rest and a resilient hoop adapted to be sprung into the groove over the skirt of the inner crown for holding the latter in place.

There are several advantages in a fireman's hat of this kind. It is fire-proof, light, and is capable of resistirg the impact of heavy blows, which firemen often receive while performing their perilous duties.

FOR KILLING THE THISTLE.

Wilbur F. Kintner, whose home is in Leadville, Colo., has had a patent for killing the emblem of the Land o'Cakes—the thistle. But in this instance, no doubt the inventor had his mind turned toward the thistle that has become such a blooming trouble to the American farmer—that of the Russian variety, which in some localities seems to hint at the possibilities of spontaneous generation.

For destroying this unprofitable weed the inventor uses a preparation of muriatic, strong lye and salt brine. There is no way for the thistle to escape destruction when this compound is applied to it; in fact there are few, if any, weeds that would not quit living, in short order if treated to a sample of the composition. The latter undoubtedly has its strong qualities—but from the way it defies the farmer, the thistle is by no means a weak foe.

AUGER FOR POST-HOLES.

An auger for boaring post-holes is the invention of Ambrose H. Kite, of Stillwater, Okla. This boring apparatus consists of a cylinder core-drum, and attached cutters at the end of operating rods together with a centering rod carrying guide plates.

The time and labor employed in digging holes with the spade that causes so much exertion and back bending, is very much reduced when one can plant his improved earth lifter where he desires to place a post, and by turning a horizontal bar cause the dirt to wind out rapidly and easily, leaving a hole just the required size. Some day the man with a patent grave-digger will come along—to "dig your grave while you wait"—but it will hardly be used on the stage when poor Yorick's head-piece plays its part.

STATION INDICATOR.

Alvin W. Chorman, of Washington, Pa., has patented an indicator whereby a traveler on railroad trains can tell quickly where a train is going and how far it must travel before arriving at certain points. For this an apparatus is used consisting of a box frame containing two apron rolls journaled therein, in connection with driving shafts, gearing and slots in the box, at which the printed information appears when the shafts are turned.

In going anywhere, if there is much distance to be covered, one usually desires to know how far his or her destination is from the starting point. The indicator gives this information. besides showing the names of the various stations on the route before they are reached—if the railroad man does not forget to operate the machine.

MINING MACHINE.

Since mining in its various features presents many difficulties and hardships, the invention of a good practical machine that will lessen toil and bring quicker results should be welcome knowledge to those who break into mother earth for her hidden wealth.

A mining machine has been patented by Edward S. McKinlay, of Denver, Colo., and if living in a state where mines are plentiful has anything to do with the inventive product in that line—and it should—Mr. McKinlay's idea ought to be of value. This machine is a front-thrust or breast, undercutting portable arragement, with a bed having side bars situated above the ground, bars extending downward from them, crossbars resting upon the ground and longitudinal bars at or near the center connected to the cross-bars. A carriage having a cutting apparatus across its front end is fitted to the bed, across which it moves back and forth, propelled by chains geared to motor or steam engine.

WATER MOTOR.

New methods for furnishing power are always interesting. The utilization of water power dates far back, but still in the march of progress new ideas continue to develop in connection with this ancient force.

Allen R. Lewis, of Shelton, Wash., has had a patent issued to him for a runway water motor possessing some features which seem to suggest its power to "mote" successfully. This motor consists principally of a runway having a flared chute, or mouth, a gate located in the throat of the chute, a fender arranged in front of the mouth, water wheel in the runway and cylinders situated upon opposite sides of the runway. The cylinders are connected with openings by conductors, and are operated by plungers working within them; the plungers by connection with the wheel, receiving their impetus when the latter is acted upon by the water. A reservoir has valve communication with the cylinder and a service pipe with the reservoir.

NEW VALVE FOR PNEUMATIC TIRES.

This valuable device comprises a shell or casing made with an internally-arranged valve chamber; a neck forming a valve seat at one end of the chamber and throat leading from the valve extending centrally through the neck of the valve seat. In the chamber is a valve cup containing elastic packing arranged with its face below the rim of the cup. The chamber contains also a spring which seats the valve, and a valve stem extending from the bottom of the cup through the elastic packing, works within and engages the throat which leads from the valve chamber.

This improvement on the old air compressor is the patented product of Henry W. Adams, Jr., of Elgin, Ill., who has assigned it to Fred W. Morgan and Rufus Wright of Chicago, Ill.

PASSENGER HOLDING STRAP.

Since there is "no-seat no fare" law governing the street car systems of this country, as there is in that well-ordered foreign city, Paris, Jermain P. Barnes, of Rock Stream, N. Y., has sought to mitigate in some degree the discomfort, and sometimes danger, of swinging by the strap which serves to keep the unseated car passenger out of the lap of the more fortunate.

The inventor of this strap device seeks to keep the traveller as nearly in the middle of the car as possible, while he is compelled to stand. To accomplish this, a winding case contains a spring to the end of which is attached the holding strap, is secured to the suspending pole of a street car, so that the side of the winding case from which the strap depends, projects toward the middle of the car. This allows the passenger to stand without bending over the seated, as if he (it is nearly always "he") were deeply interested in the person below him. The spring in the device is a good idea, for it eases the strain on the arm when a sudden jerk from lurching in turning curves or starting puts bone and muscle to the tension test. The apparatus is simple and seems to "fill the bill."

"SELF HEALING" TIRE.

George H. Chinnock, of Brooklyn, N. Y., has patented a pneumatic tire, which, when wounded by any of the many sharp foes of the wheel, has the property of closing up the orifice without the services of wheel doctor or plaster. The tube is of vulcanized rubber provided with thickened portions, non-elastic portion between the thickened portions, and a band of "self-healing" material upon the non-elastic material. Beside these, there are provided ribs having outer covering in which they fit. The ribs are held in position by the inner air pressure. It would seem that a tire of this nature has in it qualities that recommend themselves, for anything that makes it impossible for air to escape, by closing up immediately when punctured, thus doing away with any attention from the rider, is a valuable adjunct to the many bicycle improvements. This patent was assigned to the Self-Healing Pneumatic Fire Company of New York.

VEGETABLE CUTTER AND GRATER.

The small inventions that concern the busy housewife—those that assist in lightening labor and do their work well and economically—are the things that often bring reward to the persons who conceive them.

For a useful kitchen utensil, Adam J. Hofman, of New York City, has patented a vegetable grater and cutter, which should be welcome to those who employ the little adjuncts that help lift the trials of the kitchen.

The utensil is of an inverted-funnel shape, with a perforated portion ranging up and down one side to be used as a greater. On another side is a trasverse slot, in which is a cutter. Surmounting the conical part is a cylindrical neck to be employed as a handle. In using, the vegetables are rubbed downward (for cutting) and the slicing is accomplished easily and neatly, by a curved blade which gives the required thickness. In grating the deposit is kept within the cone and can be gotton at quickly and conveniently.

OIL-CARRYING PROJECTILE.

"Pouring oil on troubled waters" has its application in a literal sense in the invention of S. O'Day, of Muskegon, Mich., who has patented a projectile to be filled with oil and cast into the sea when the angry waters are desired to be stilled. The wave-quieter is bottle-shaped and has two aperatures, a large one at the base-end for letting in water and a smaller at the neck for automatically letting out oil. The projectile is made thicker on one side for weighing it in the water and thus preserving the correct balance. The position taken by the shell when thrown into the water, is horizontal. It floats about "trimmed" like a ballasted ship, distributing its smoothing influence over the frothing waves, which find it impossible to break when they become well greased.

One method of quieting storm-tossen waves has been to use a distributing bag suspended from ship. This will answer when the vessel is driving before the storm, but when she is facing a head sea, the projecting at some distance ahead of an oil filled shell would seem to be the best plan; for the on-coming waves would then receive their "quietus" before getting where damage could be done.

CHALK-LINE REEL.

Those who have watched a carpenter chalk his marking line are aware that that manner of applying the marking substance is rather a slow and untidy process. Besides this, the line is usually lying about "loose" (unwound) and often tangled, causing time and patience to get it straightenen out.

A remedy for this has been patented by Gustaf E. Jahnsan, of New York City. The arrangement is a circular casing whithin which is mounted a shaft carrying a reel; upon the reel is wound the marking cord with its end projecting through an opening, and in one side of the casing is an opening for inserting a piece of chalk. The reel is simple and effective. When something is to be lined, the cord is drawn out, chalked and ready to be snapped upon the wood to be used.

BLACKING BRUSH.

Joseph C. Wikers, of New York City, has patented an improved blacking brush wereby one can apply the material so necessary for respectable-looking foot-wear, without putting the soiling stuff on ones hands and clothes. This brush is of the fountain kind, and is composed of a reservoir having a handle with an annular flange and head with bristles secured in and by the flange. A spindle extends through the reservoir and is provided with a valve for closing a discharge-opening in the head. Within the reservoir are radial stirring-blades projecting from the spindle.

AUDIBLE SIGNAL.

The utilization of sound for signals is an old idea (as are the majority of ideas, for that matter) but any new arrangement for helping "those who go down to the sea in ships" to sail amidst fogs and in darkness, or anything that will trausmit information on land or sea, is of interest and value.

A unique signaling apparatus has been patented by John F. Barker, of Springfield, Mass. This sound-maker is worked by means of material, which is exploded in a chamber and acts upon a sounding device through the medium of a cylinder connected with the exploding-chamber and sounder. Automatically-seperating electrodes are introduced into the exploding-chamber, and there is provided a source of electricity for operating. The cylinder carries a piston-head, which normally closes an opening leading to the sounder and which is driven backward by the explosion past the opening allowing the force produced to escape into the noise-maker.

DOOR CHECK.

A pneumatic door check is the recently patented invention of Rufus Wright, of Chicago, Ill. In this there is an elastic air-cushion, abutment and a dog hinged to the latter so as to swing in one direction independent of the abutment. The abutment is forced against the air-cushion by the dog when the latter opposes the closing of the door, and the dog is swung independently of the abutment when it opposes the opening of the door.

There have been a number of door checks patented, but there is always room for a good thing. The pneumatic apparatus is playing a big part in the march of invention, and there is much room for the application of the air idea to little things, many of which find their uses in our everyday affairs.

CLEANING SHIP'S BOTTOMS.

When we consider the cost, time and labor required in attending to a ship's bottom when it is in need of cleaning—putting the vessel in a dry dock, etc., it is rather remarkable that some progressive American long ere this has not invented an apparatus for this purpose, something that would obviate the necessity for taking the ship from the water.

The inventive genius of Patrick Reilley, of Philadelphia, has tackled this problem, and the Patent Office has given him a patent. This new ship-bottom cleaner is adapted to conform to the bottom of the vessel in cross section, and is composed in part of a flexible frame and an endless train of cleaning brushes mounted in the frame. These brushes travel around shafts with sprocket-wheel attachments, and when operated—from above—are brought in contact with the foul hull, sweeping it free from seaweed, barnacles, etc., with a swift rotary motion.

It is easy to see the value of a good ship cleaner; for besides the expense and trouble of dry-docking a foul craft, there is the necessity of making port for this work. An apparatus that can be used at sea for cleaning bottoms would be of immense value to cruisers and naval vessels generally.

TO TEACH BABIES TO WALK.

A new convertible baby walker, a patent for which has recently been issued, is something that commends itself to mothers who wish to amuse the little toddlers and help them to learn the uses of locomotion.

The baby walker is the joint product of Geo. W. Weiland and Miles Thomas, of Priceburg, Pa., and is a combination of rocking chair and walker. When baby desires to rock he sits on a folding seat with his body projecting through the hole at the chair top. For walking the rest is folded away and the rockers which are joined in the middle by hinges, are lowered until their ends, upon which are rollers, come in contact with the floor.

The arrangement is convenient and better than the simple go-cart as a walker and safer than the ordinary chair for rocking.

COATING PICTURE MOULDINGS.

A new idea in treating picture mouldings has been brought out by Augustus Spar, who is a resident of Chicago, Ill. This invention relates to putting plastic material upon moulding by means of machinery and in the same process produce embossed patterns. In a hopper, converging at its lower end, a tapering spiral is worked by machinery. The spiral forces the material from the hopper down through a nozzle to the moulding, over which a stamping wheel revolves. This wheel conforms to the shape of the moulding and when in motion stamps the design it carries, upon the plastic substance.

A UNIQUE PASSENGER CHECK.

A railroad ticket that will be of mnch service to passenger, conductor and anditor, is a patented design, by William J. Perdue, of Fort Smith, Arkansas. At one end of the ticket is the passenger's check, and at the opposite end, connected by four slit strips or slots, is the railroad auditor's check. The seperate strips are arranged in ruled columns for entering sums of money, and the passenger's check contains spaces for the name of the railroad, number of the train, date of trip and direction in which the train is moving. The auditor's check contains also the trip date, number of train, direction and the color of the passenger, and the car in which he is located.

FOR GETTING SILVER.

The low price of silver has not prevented the busy mind of the inventor from thinking out new ways for its extraction from the ore. The latest idea in this respect is to procure the virgin metal by means of a solution of hot acidulated brine, by which the crushed chloridized ore is leached—the brine being allowed to percolate through the crushed material. A second leaching of the ore, with sulphate of copper added to the solution, is followed by a third, that searches out the precious particles left by the two first processes.

A good, economical metal extractor should reccommend itself to the owner of mines—and the process mentioned here—formulated and patented by Geo. A. Schroter, of Denver, Col.—seems to possess these qualities.

A MARBLE SHOOTER.

The inventor who employs his genius in producing new things for boys, not only deserves, and gets their well wishes, but, when he brings forth a good boy-amuser, the latter feels that he just must have one, and so the man of ideas reaps his reward. William J. Sleffel of Columbus, Ohio, is the inventor of a spring pistol for shooting marbles—those old friends of the boys. The pistol consists of a cast iron stock with ring for index finger, a ball guide of two parallel wires having their inner ends secured to the upper portion of the stock. To these are added a ball holder, ejecting spring and trigger. It is only necessary to aim at the ring and pull the trigger. But, we are not saying anything of the awful crack the marble would give to an unlucky knuckle.

BICYCLE TRAINER.

Those who wish a good thing on which to practice in bicycle training, will find in the patented device of Chas. W. Fox, of St. Louis, Mo., an ingenious apparatus for that purpose.

In this the bicycle rests on small grooved sprocket wheels and a standard. The grooved wheels are situated near the ends and in the longitudinal apertures of an oblong rectangular frame, and are connected by a sprocket chain. The pedal driving gear is supported upon a standard having an open upper end for receiving an axle bearing. It is only necessary for a rider—who desires exercise—to mount, start his machine and pedal away. He can work until tired out, without getting an inch from the "starting" point.

An arrangement of this kind should recommend itself to beginners on the wheel, for the standard in the apparatus would assist in balancing, while one was getting accustomed to the saddle, leg-motion act, etc.

A CONVENIENT TRUNK.

Those who travel much know the inconvenience of packing and unpacking—the labor of lifting heavy trunk trays, etc. There is an improvement in this respect—patented by Paul Girand, of Dallas, Texas—which causes the trays by the operation of levers to rise from the trunk and assume an extended position, being held together by the attached levers. When the lowermost tray rises to the top of the trunk it is automatically locked there. By this simple mechanism the contents of a trunk are brought within easy reach, and the stooping over which brings so much discomfort to the back is avoided.

COLLAPSIBLE VEHICLE TOP.

A vehicle with a top that will fold up snugly and easily and that can be put out of sight, has many advantages over the old carriage and buggy top that so often persists in remaining in a half-folded position despite the efforts of the person who desires to ride without a top covering. Pierre Pellefigues, of San Francisco, Cal., has sought to do away with the objectionable features of the usual folding vehicle top by an invention that folds the top back into a horizontal slotted box situated at the back of the conveyance. Instead of working on a pivot at the side of the vehicle, as in the ordinary carriage, this apparatus has telescoping tubes in which the rods of the frame work for its closing before being lowered into the receptacle.

GLASS PICTURE FRAMES.

As time goes on inventive genius is turning its attention more and more to productions in glass. The transparent material lends itself to a numerous variety of things, and is only waiting for the lucky chemist to divest it of its brittle quality to become the greatest factor in constructive adaptibility.

A glass picture frame is the invention of Jacob A. Booher, of Pittsburg, Pa., which while simple in construction, offers possibilities in decorative glass for pleasing effects.

This frame has transparent double panels between which the picture is inserted by means of a slot. By this method the picture is preserved from moisture and dust, and can be set off to advantage by having the frame made in various patterns and colors.

ELASTIC TIRE.

Many and various have been the efforts to improve on the elastic wheel tire. The pneumatic tire is considered almost perfection by the riders of the silent steed; but, the readiness of air to escape through the smallest puncture, leaving the tube looking like a chrushed snake, is something that all are aware of and something that needs a remedy.

The patented idea of Joseph D. Prescott, of Boston, Mass., while not dealing with inflated tires, recommends itself as a good method of obtaining the advantages offered in the elasticity of metal for wheel use.

In this idea a metalic elastic tire is used over a wheel rim having a concave channel in its rim, in which are a series of interposed elastic strips with their ends curved towards each other and fastened to the under surface of the outer band, that forms the bearing surface of the wheel. This forms a desirable combination of rim and spring that is effective and comparatively economical.

BICYCLE REST.

Le Roy B. Thomson and Walter Burke, of Portland, Oregon, have had a patent issued them for a bicycle rest. This is an upright rod with pivoted legs that when not in use can be closed up. When the legs are to be opened, a block, with spring jaws, sliding on a horizontal tube—that which connects the bicycle standards—grasps the pivoted legs and their bottom ends come open and make a support for the wheel. The legs are situated in front of the rider and just behind the forward wheel.

This arrangement makes it possible for a cycler to stop his wheel without getting off and keeps it in an upright position when left where there is no place for it to lean against.

A DISTANCE AND RANGE FINDER.

Among the interesting patents granted for April is one, to Mr. Ora A. Bell, of New York, for a device for determining ranges and distances. This consists of standards arranged at predetermined points, on a fixed line. At the top of each standard is mounted a revolving glass, and in connection with theses glasses are cylinders containing fluids, which, under pressure by means of a piston, act upon a recording guage, giving the observer an automatic record of his work.

NEW SUSPENDERS.

It is seldom that fair woman invades the masculine dominion with an invention concerning man's wearing apparel. But, Ruth D. Harvey, of Boston, Mass., patented a pair of suspenders for the benefit of the male population. There is no hint anywhere in the invention that bloomers are in any way concerned with these suspenders. They are for the trousers of the *genus homo*, and will do the work required.

The invention consists of shoulder straps terminating in tabs having wire loops, with rollers passing through the tabs. Bifurcated lipped clasps, to which are attached two elastic cords, swing on the loops, the independent cords terminating in similar clasps below which also secure the buttonhole loops. At the back the straps cross and form in diamond shape, coming to a point above the loop.

A MAGNETIC TORPEDO.

A torpedo that is intended to be attracted to and held against the iron or steel hull of an enemy's ship, is the inventive product of Louis F. Johnson, Walter J. Slacke and Howard Lacy, all of Easton, Pa. The torpedo shell contains in its forward portion an electro-magnet with curved poles, conforming to the converging lines of the shell.

There is good reason for believing that a shell thus equipped, would find its way to a metal-clad vessel more readily than would the ordinary torpedo; and it would be more difficult it seems, for a ship to get out of the way of a magnetized destroyer than to escape from one governed only by the rudder.

Like a shark that turns in various ways to meet and match the efforts of fleeing prey, the electric engine of destruction would follow its doomed victim.

A NEW MANDOLIN.

This is one invented by Mr. Neil Merril, of Oskosh, Wis., who has produced a maker of dulcet sounds, from wood and metal. The sides and back of the mandolin are of metal, to be pressed, spun, or cast. The sounding board, or face, and neck are of wood, the former having a reinforcing attaching strip secured to the under side at the edge, arranged so as to be removed at will from the body of the instrument.

Although the lovers of wooden instruments—the tones of which are supposed to be improved with age, and to be imbued with sound qualities unknown to any musical production in metal—may decry a metalic inovation in the dominion of strings, there are possibilities in this direction which have been more than foreshadowed long in the past.

The Japanese understood the sound-producing qualities of metal, centuries ago. It is only necessary for one to strike softly, with the hand, one of their small temple bells, to hear swelling forth sweet, resonant sounds that linger wonderfully long, dying away like the sound of music drifting to us from afar off.

GAS AS YOU WANT IT.

There is considerable doubt in the minds of most people who use gas, as to the reliability of the average gas meter. They seem to think that it partakes of the disposition which controlled one Ananias. To remedy this seeming short-coming, William N. Milsted has invented and assigned to the Prepayment Appliance Co., of Burlington,

N. J., a slot machine into which one simply drops a coin and receives gas to the amount of deposit.

The mechanism is run by clock-work operated with valves in connection with gas pipe. The weight of the coin starts the machinery and the gas is cut off by the stopping of the clock arrangement when the quantity paid for is obtained.

There is no fraud in this. The gas company gets what is due it, and the consumer knows that money must go into the slot, or he must go in the dark.

A BICYCLE TOW.

Those who object to riding the bicycle a-tandem and yet wish to avail themselves of the speed developed by an obliging bicycler, will find in the apparatus invented by Jean Francle Armand, a native of Toronto, Canada, something that may meet their wants.

To the bicycle Mr. Armand has added an attachment for towing. It is a forked reach extending on both sides of the rear wheel of the bicycle and fastening to the axle by means of tubular brackets. The outer end of the forked reach is a disc shaped and fits into a slot of the same shape, working therein on a pivot. From the disc extends a forked bracket which is secured to a small two-wheeled vehicle, to be drawn by the rider as he speeds along. The disc pivot allows the vehicle to follow the turnings of the machine with ease.

This is an idea that should meet with the approval of the fair sex.

A PATENT REEFER.

The old-fashioned method of reefing the sails of boats and ships—especially during a storm—is one that is often accompanied with great danger; besides it is slow and very laborious. To making reefing less hazardous and arduous, Benjamin G. Cahoon, of Marshfield, Mass., has invented a reefer that will no doubt ease the burdens of poor "Jack." Heretofore the sail has been made smaller by tying it down by the reef points, but now it will be pulled down to a device which is attached to the boom, and which will close upon and engage the lower portion of the sail. The device will hold the bolt-rope at the foot of the sail, also the lacing cord and will have attached to it hooks for securing the reef-points.

MORTISE AND TENON.

A good thing in fastenings for tenon joints, is that for which a patent was granted Lucius F. Arnold, of Phenix, R. I. The tenon has its end split along two intersecting planes and is provided with an annular flange or shoulder. The mortise has a contracted entrance and a squared shoulder opening behind the entrance. When the tenon is pushed into the mortise, its end is compressed until it reaches the larger shoulder space when it assumes its first shape, becoming immovably fixed. It will be easily seen that a structure put together after this manner could not come to pieces without a great breaking up.

NEW VEHICLE WHEEL.

A new wheel hub, recently patented, is something which promises improvement in vehicle progress. It is of the ball-bearing variety and consists of a tube, working within a cover, ball-races, annular dust guards that impinge against the ball-bearing axle cones, and caps on ends of spoke barrel. The spokes are of wire extending through radial openings. The invention is the combined product of Messrs. Bolte, Thomas and Donlevy, of Milwaukee, Wis.

AN AUTOMATIC UMBRELLA HOIST.

Those who have had experience with refractory umbrellas and spent precious, not to say anxious moments when the rain was attesting its wetness, while trying to get up sail—will appreciate the invention of Frederick Miller, of Scranton, Pa., who has produced a self-opening article of this kind.

This umbrella has two sets of radial braces—upper and lower—for operating the ribs by means of an actuating coiled spring surrounding the stick between an upper and lower runner which holds the inner ends of the series of braces. A sleeve near the upper middle part of the stick covers the actuating spring, having attached a sliding finger grip. When you desire to hoist your patent umbrella it is only necessary to turn the combination loose, and the spring does the rest.

THE FLOWER GROWERS.

A useful little article is that not long since patented by G. W. Parmley, of Shamokin, Pa., for the use of those who grow potted plants. It is very simple, consisting of a wire ring for encircling the top of the flower pot, and entwined wires with cross sections upon which plants and vines can be supported. The wire around the pot forms a brace, and the whole is convenient and graceful. Its utility and apparent cheapness should recommend it to flower growers.

A NEW PEDAL.

A new bicycle pedal is the invention of Samuel O. Jones, of Stillwater, Minn. It is a combination with a bicycle pedal of foot plates centrally pivoted to each other, and provided with inwardly-turned prongs for securing them to the rider's shoe. The plates are held to opposite sides of the pedal by spring catches.

This should meet with favor from lovers of the silent steed, for it affords a secure footing and should facilitate travel.

Mr. Jones has also patented an apparatus for measuring liquids. This consists of a tightly-enclosed case from which extends outlet ports, and a measuring device arranged longitudinally, adjusted to register quantities. As there are no measuring vessels required with this apparatus, all running over and spilling on the floor is prevented. It is only necessary to determine the measure of liquid wanted, opperate a piston, and the amount predetermined, coming into one port allows the discharge of a similar quantity at an other.

This is a useful invention for store keepers, as it is convenient, economical, and tends to cleanliness.

A CANDLE LAMP.

Considering the danger of setting things on fire, depositing grease and the small light obtained in using the ordinary candle, it is rather strange that some convenient substitute for the tallow, sperm and paraffin article has not been invented long before this. But now we have in the invention of Sarah P. Bancroft of Boston, Mass., an oil burning candle-lamp which seems to be a good thing.

This lamp has a tubular oil fount with an open upper end on which is a cap having its upper end contracted to form a conical tapering shape. On this is a rotating burner threaded to fit into the cap. A convenient arrangement for filling adds to the value of this lamp as does its attractive chimney and shade fixtures.

SAW.

The saw invented and patented by Peter M. Dahl and Robert Poindexter, and assigned to E. C. Atkins of Indianapolis, Ind., will no doubt meet with the approval of mill owners. In this invention the saw-plate has tooth shanks and recesses with overhanging edges of something more than half a circle. Bifurcated tooth points fit into the recesses, their consequent rings being extended into points adapted to be bent inwardly behind the tooth shanks for holding the points in place.

There has been no more valuable improvement in saw mill machinery than the adjustible tooth; and a clean cutting substantial saw—one that can easily be repaired when its "dental" arrangment is out of order—is a boon to those who convert logs into building material.

A NEW CARVING FORK

Every one is not compelled to carve—a great many have nothing to exercise this peculiar genius upon. But those who preside at the table or in the kitchen, know the value of good carving knives and forks. A newly patented invention concerning the latter utensil is the combined idea of George J. Fanner and John H. Kirkwood of Cleveland, Ohio, who have assigned their patent to the Fanner Manufacturing Co., of the same city.

This fork is made of coiled wire-tines, handle shank and guard. It is durable and of course light. The handle is arranged for a firm hold, as the openings between the wound wire effectually prevent slipping. It does'nt make any difference as to the age of the goose or steak when this fork is used, for it seems to be a fast sticker.

VELOCIPEDE BRAKE.

Thomas B. Jeffery of Chicago, Ill., is the inventor of a recently patented brake for bicycles. The brake is situated in the fork which strides the front wheel obtaining its bearings in the former and bears upon the outer periphery of the wheel. Rigid supporting arms extend downward on opposite sides of the wheel, and clips for fulcrum supports are detachably secured upon the fork arms. The supporting arms are provided with projections to be engaged by the foot when the brakes are to be put on.

The value of a good bicycle brake is evident. For the lack of a good apparatus of this nature many a rapid rider has found his silent horse running away with him. In large cities where the streets are crowded the means for stopping the "machine" quickly to avoid danger to the rider and those with whom he sometimes come in sudden contract, is a thing of absolute necessity; and the more perfect a brake can be gotten up the better it will be for rider and pedestrian.

DRESS-SHIELD FASTNER.

Inventions that concern ladies' dress are very apt to claim attention, for the fair sex beleive in keeping up with the time in every thing that makes to their comfort or enhances their charms. Every women knows the boon that was conferred upon her when the dress-shield was brought before the public, and all are interested in anything which relates to that very necessary adjunct of female attire.

A neat and useful clasp for fastening dress shields is the invention of Pauline W. Neffen of New York, N. Y. This clasp is of the safety pin variety consisting of a long, flat blank pointed at one end and having an integral ear at one side of the opposite end. The blank is doubled near the middle, and the ear is bent over to guard the pointed end. The clasp is small, its two parts being in close proximity.

MAIL POUCH.

John E. Quinn of Toledo, Ohio, is the inventor of an improved mail pouch, which has in the idea that which should recommend it to those who are at the head of Uncle Sam's post office affairs. The letter and parcel carrying business is something that comes very close to the heart of a people and every precaution, should be taken for safety in sending over land and water the vast amounts of mail matter that belong to this and other countries.

The rough usage to which mail pouches are subjected, causes them to wear out comparatively quick, and if there is a lack of careful inspection there is a possibility of injured mail matter, especially so in the train catch-pouch where there is considerable strain on the pouch both in taking on and throwing off the mail car.

The invention mentioned here is for strengthening the bottom of the pouch. The lower end of the bag is turned up at the end of the bottom piece, making three thickness of material; outside of these is a band, rising from the bottom upward, which is riveted to the other pieces of leather, thus making a bag in which could be carried cannon balls, if necessary.

BLOTTER.

Inventors who use their genius on things that immediately concern those who write much, have a large and appreciative class to be interested in their productions. A blotter may be considered an insignificant affair by many; something to be obtained cheap, and often gotten without cost. But a great point about blotters is their "disposition" to be absent when most wanted, usually hidden under the biggest pile of papers at hand; and when found

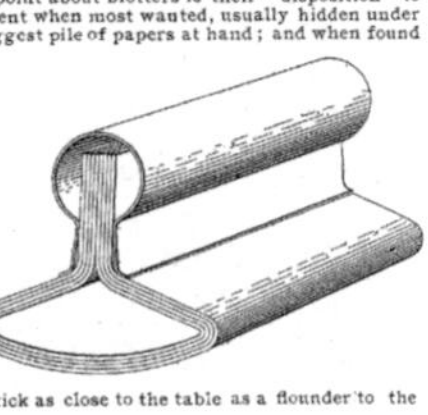

they stick as close to the table as a flounder to the mud.

Alfred B. Gawler of Washington, D. C. has recently patented a blotter that cannot well loose itself on desk or table, and has the advantage of being easily grasped when wanted.

In this the ends of the blotting paper are brought together, and inserted into spring arms having supplementary spring catches. The spring top forms a convenient hand-hold, while retaining the blotting paper, which is kept in a semi-oval widened position at the bottom, receiving from the manner in which it is bent, an elastic quality.

SOFT-TREAD HORSE SHOE.

In listening to the pounding of horse's feet, especially of those employed to haul street cars and hammer their wretched way over miles of cobble-stones every day, one cannot help thinking of the injury to the poor brutes that results from such continual beating of their feet against an unyielding substance. The life of an animal thus used is not very long. And in view of this fact it seems that the owners of these horses would adopt some other method of shoeing their equine property, or else put down a different roadbed.

As a remedy for the shock given to the feet of horses from violent contact with hard substances, James Freyne of Philadelphia, Pa., has invented a horseshoe composed of rubber. The shoe has lateraly projecting plates of metal formed with nail holes or slots and apertures that are filled with the rubber of the shoe, which holds the plates firmly in their places.

A horse thus equipped could do better and more traveling and finds itself at the end of a day's work in much better condition than the jaded nags now present when they are allowed a little respite from their killing toil.

DUST PAN.

The energetic housewife who has a decided antipathy to dirt and dust, and who knows the back-breaking result of much stooping while in performance of her daily avocation, cleaning up house—will look with favor upon the invention of George H. Gerow of Port Huron, Mich. This new friend of the household is a dust pan.

Heretofore one had to bend over with the little short handled pan while laboriously raking in the dust (which pans out nothing but toil.) But in the new dust pan a long, upright handle does away with stooping and allows the dust to be swept in while the sweeper stands in the position usually assumed while sweeping the floor. The pan rests on legs with its forward end just touching the floor so that a broom can be drawn to and upon it without difficulty.

FLOOR ROCKING CHAIR.

The American people find more pleasure in rocking chairs than do any other people in the world. This is not because they are the laziest people; not by any means, but they like comfort as well as the rest of mankind. Where there is so much rocking there must be a consequent wear on carpets and floors, and an occasional overturning when the rocker gets up to much motion.

With an eye to wear, tare and upsetting possibilities, William I. Bunker of La Grange, Ill., has invented and patented a floor-rocker that has a number of advantages over the ordinary rocking chair. With this new chair are independent wear-plates adapted to rest upon the floor, while their upper surfaces are arranged to receive the rockers. Springs provided with end brackets connect the rockers and plates near their longitudinal center to balance the plates equally on the rockers in all positions of the chair.

The danger of upsetting in a chair of this kind is reduced to a minimum as is the wear on the floor or its covering.

ARTIFICIAL ASPHALT.

The importation of asphalt into this country has been going on for many years and is still a profitable business notwithstanding the fact that there are in the United States great quantities of natural gilsenite that could be mined and used to as good purpose as the Trinidad article. It makes good street paving, but is rather expensive, too much so it seems for small cities and towns.

There is room for a good substitute for asphalt—something cheaper but which approximates it in utility and durability. John A. Just no doubt had this idea in mind when he invented his recently patented formula for making a composition resembling asphalt. In carrying out his idea this inventor uses rosin and sulphur until the former has been thoroughly changed by the chemical action of the sulphur. To this is then added a heavy hydrocarbon and sulphur, after which the compound is heated.

The patent for this formula has been assigned to the Asphaltina Company of America, Syracuse, N. Y.

INKSTAND.

When an inkstand is left open (and many of them are), the ink gets muddy, gets full of dust, and dries up if left open long, owing to the action of the air upon the fluid. Besides this certain kinds of ink bottles and inkstands have a way of painting everything around them black, blue or red, accord- to the color of their contents, when overturned.

And arrangment that will prevent all these disagreeable things should be of value to the writing public. Such a desirable apparatus has been invented by Carl H. Schwiets of Denver, Colorado to whom a patent was recently issued. Mr. Schwiets' inkstand contains an automatic closing device consisting of a V shaped guiding frame extending downward into the inkstand and holding sliding-plates which meet at their lower ends normally closing the inkwell aperature. Being pressed by the pen when it is inserted, the plates slide back, but resume their positions when it is withdrawn.

WIRE FENCE.

There have been many inventions in wire fencing, and many among them that were not of any great value. But among the good things in this line, is that for which a patent has recently been granted to Edwin A. and Hugh F. Barker, of Athens, Ky.

This fence has longitudinal wires, posts, and pickets constructed of flat metal provided at its edges at intenvals with tongues formed by straight longitudinal cuts bent completely around the fence wires, forming eyes. The tongues are placed at different elevations and bend the fence wires, whereby they are adapted to take up the slack between the fence posts and lock the stay, creating a tension on the fence wires, while preventing their vertical or longitudinal movement, but permitting the necessary contraction and expansion.

ARM-REST FOR TELEGRAPH OPERATORS.

To those who are not familiar with the work of a telegrapher it may seem an easy matter to operate the key since there is so little apparent effort made by the person who turns the electric fluid loose. But if one will sit at an instrument all day at work, it will be found that the strain on the arm and fingers of the operator makes this work something more than child's play.

Earl L. Brownson and Henry W. Goodro of Hinesburg, Vt. have patented an arm-rest for making the work of the telegrapher less arduous. The rest is a frame or shoe with a flat base having triangular sides mounted on the base, converging forwardly, and inclining outwardly. An inclined cushion for supporting the arm is secured to the sides of the frame and suspended so as to form a curved channel that diminishes toward the rear end. The cushion is adjusted to support the wrist of the operator above and clear of the base of the frame and on a line with the telegraph key, allowing a free use of hand and wrist.

FRUIT GATHERER.

Henry Edgarton of Shirley, Mass., has been granted a patent for a fruit gatherer. In this a flexible chute, with its upper end surrounded with upright V shaped wire fingers, is attached to a long pole, at the upper end of which is a pair of cutters operated from below by a wire attachment, much as the tree-trimmer is worked. When fruit is to be gathered from trees, the gatherer is pushed up to the branch containing the former; the wire fingers engage the branch, with the fruit in the mouth of the chute, the cutters clip the stems, and the fruit falls down through the chute to a receptacle at its lower end.

The advantage of this apparatus will be seen at once; for by its use all climbing or beating trees and promiscous scattering of the fruit is made unnecessary, and time labor and expense are saved. Such an invention should be welcome to farmers and fruit growers everywhere, as something useful, convenient and profitable.

PIPE WRENCH.

A good and simple pipe-wrench is the recently patented invention of Erasmus B. Frizelle of Sterling, Kansas. This wrench is a lever having its engaging end bent rearwardly. To the curved end is bolted another curved section. with its outer and under part serrated to prevent slipping when pressure is brought on the pipe. The engaging portion of the lever is V shaped, and by having a working bolted part a powerful grasp can be obtained upon the pipe-joint.

There is value in an effective tool of this nature, and no one knows it so well as the man who has to deal with refractory pipe-joints where rust has made them almost immovable.

BLACKBOARD RULER.

The inability to make a straight line is something that belongs to the majority of people—most people can't tell a straight line when they see it. The eyes of children should be trained in this respect, especially if they are to follow any mechanical occupation.

For making straight lines on blackboards, Malana A. Harris of Akron, Ohio has invented an apparatus which does the work effectually. For this a perpendicular shaft is mounted on a base supplied with two sets of rollers—one set on the floor, for moving, and the other set on the side of the base for guiding. The upright shaft contains clips for holding pencils horizontally, which make their marks as the apparatus is rolled along the face of the blackboard. When this liner comes into school use, the blackboard will not look so much like an illustrating place for geometric designs as it usually does now. The liner will no doubt please the children, who are always pleased with anything on wheels.

AUTOMATIC BICYCLE PUMP.

James K. Tomilson of Terre Haute, Ind., has invented a bicycle pump that will not require the usual manipulation of the piston by hand power when a tire needs inflating. In Mr. Tomilson's idea, the wheel automatically fills its tire by having within the latter a pump, the piston of which is pushed upward by pressure from the surface portion of the tire, and downward by a spring coiled about the piston. On the sides of the pump cylinder are a sreries of air-jet tubes, which distribute the air, that is taken in from the inlet valves situated at the upper outer portion of the tire.

The value of an arrangement of this kind is at once apparent, for it not only does away with the necessity of attaching and working the hand pump, but it guarantees a supply of air at all times, allowing the rider to go long distances without being subject to the often disagreeable consequences of forgetting or losing his pump. And besides this advantage there is that of overcoming the "flattening effects" of a small puncture, if the automatic pump is made powerful enough.

ELEVATING AND FOLDING BED.

In houses where economy in space must be practiced, and there are a great many of such, the folding bed plays a considerable part in practical utility. It yields up its position in the day time, when the occupant desires more room, and at night, when floor-space is not so much needed, and "appearances" not so necessary, it comes down like an absent friend whose return is welcome. One of these folding household friends is the recent invention of Warren Cole of Knoxvlle, Tenn. The bed has a supporting frame arranged to be secured against a wall, or other support, in this is a bed frame held up with cords or chains by means of counterbalanced weights which pass over pulleys in the upper corners of the frame. A bed is hinged in the bed frame and can be vertically folded or extended for use, and the bed-frame can be raised to the top of the supporting frame or lowered to its bottom. The balance-weights employed in operating the bed-work in housed channels.

The bed when elevated is six feet from the floor thus giving all the floor space for other purposes, it is perfectly balanced and a child of 10 or 12 years can handle it easily.

CAR CUSPIDOR.

Joseph F. Miller, of Tonawanda, New York, has patented a cuspidor to be used in cars. This spittoon is mounted on an arm, which in turn is mounted on an upright secured at its lower end to a horizontal bar near the car floor. The whole apparatus is to be situated under the car seat, and is arranged so that the cuspidor can be swung out from the seat, in front of the user when wanted, and back immediately under the seat when not in use. It has a cover, and is altogether an improvement on the old article which is usually an offense to sight and smell.

Since it seemes to be impossible to prohibit the habit of many men, not a few of them tobacco chewers of spitting on car floors, when there are no receptacles for this purpose at hand and sometimes when there are, this new cuspidor should find a welcome acceptance from the traveling public, and its presence should be announced by a printed invitation to careless expectorators, telling them to "look under the seat for the cuspidor."

SPARK ARRESTER.

The millions of dollars worth of property destroyed every year by fire, and loss of life by the same means, should make anything that will help prevent these calamities more than welcome to public attention. It is well known that sparks from locomotives and mills are responsible for many fires, and is surprising that a good spark arrester has not long ere this been in use. A patent for one has been granted Joseph T. Bright, of Midway, Ky., who should meet with reward for his good work.

This spark-arrester is a cylinder, in which is a number of corrugated plates and screens placed alternately so that the corrugations will run across the cylinder from side to side, knuckling into each other with space between them for smoke to pass through. The cylinder is to be fitted inside the smoke-stack of an engine, and when so adjusted will no doubt prove an effectual obstacle to dangerous sparks. The invention is simple and easily worked, qualities commendable in any apparatus.

REVOLVING PLOW.

The evolution of the plow has been a slow process. For centuries the horse and patient ox have drawn by direct application of strength, the plow share through the resisting soil. But now Ferguson Marshall of Philadelphia, Pa., has thought out and patented a plow that, instead of turning up earth with the share, digs it up somewhat after a spading style while rolling over the ground.

This revolving plow has a series of shafts carrying

cutter-blades provided with a shear for moulding over the soil. The blades are spaced on the shafts so that one row cuts in the space of ground left by the preceding row alternately. The blades can be set to any desired length so that the soil can be dug deep or shallow at the will of the plowman. More ground can be turned up in a shorter time by a plow of this kind than can be plowed in the old fashioned manner; and if steam or electricity is used, it is easy to see what rapid progress can be made with such an improved earth-turner.

NAILING-HATCHET.

A magazine hatchet is the lately patented invention of Magnus Ortenbald of Sandstone, Minn. The hatchet contains a nail-holding chamber in its body just in front of the handle, and a door for admitting the nails. The chamber is so arranged that by a spring its contents are pushed downward at the will of the workman, allowing one nail at a time to slide into a channel and out at the head or driving part of the hatchet, where it is held, point downward until driven partly into the lumber.

When using small nails for lathing, box making, etc, the labor saving and rapid work, capable of being accomplished with a tool of this nature amounts to a great deal. The workman does not have to stop to pick up the nails; he simply presses a spring and strikes.

COMMUNION CUP.

There has been a great deal of talk concerning communion cups and contagion in the usual method of using them, in which every communicant drinks from the same cup. As an improvement on this, Addison Ballard and James M. Harker of Chicago, Ill., have patented a communion apparatus that gives to each partaker a separate cup. These cups are covered and have a vent through which extends a tube that projects into the bottom of the receptacle. As there must be a number of the cups for each church, they are to be carried on a tray which is included in the invention.

UNIQUE FIRE ESCAPES.

What Some of our Inventors Think on the Subject Ideas not all Practical.

If one is interested in the subject he will, upon visiting the United States Patent Office, be attracted by the size and importance of that class or division of the office devoted exclusively to fire escapes.

Here can be found every imaginable form and type of escape from the complicated machine weighing many thousands of pounds to the small portable escape adapted to be carried about in the pocket. But all this wild chaos of inventions does not, by any means, represent practical working machines founded on the laws of nature and mechanics.

One very sanguine inventor thinks that a parachute made of waxed cloth and kept extended by a light bamboo frame, would when attached to a persons head, permit him to jump from a great elevation as a high burning building and land in safety. He further provides for the comfort of the escaping victim, by providing him with shoes having enormous elastic soles that absorb all shock upon his striking the earth.

Still another of the genius inventor, species crank, is opinioned that a long constructed elastic tube or chute attached at its upper end to the window sash and floor and extending to the ground upon the outside, is the ideal escape. The escaping person in this case jumps into the tube feet foremost and as the tube is of less diameter than his body he decends very slowly stretching the elastic tube as he passes through it. Suitable perforations are formed in the tube to permit the escaping victim to breathe while en route to the pavement below. The tube is also provided with suitable self closing doors at each floor whereby it can receive passengers at these different points along its length. Another of the same species has provided an escape tube which is contracted about every ten feet of its length by suitable elastic belts. The escaping victim in this case is treated to a succession of falls of about five feet each; the elastic contracted portions of the chute breaking each fall and starting a new one.

A tube provided with a rigid bottom arranged in stair step fashion has also been patented. This tube is lowered from the window of the burning building and its lower end anchored at some dis-

tance away so as to give the tube an inclination. The escaping person enters the tube at its mouth feet foremost and slides down the stairstep bottom just as a boy would slide down ordinary steps. Tubes with ladders therin of every imaginable form have been registered in the office. The tubes in these cases are also to prevent nervous persons from falling away from the ladders and to prevent them becoming dizzy by shutting off their view of the earth.

Long canvas sheets have also been mounted on reels below the windows so that when lowered and their ends anchored at a distance from the burning building, the escaping persons could slide down the same.

In the bed escape the bottom of the same is made up of a series of ladders that can be removed, hooked together and suspended from the window sill by suitable hooks and thus furnish a means of escape. A great many other articles of furniture that can be converted into fire escapes have also been patented, among them the chair escape. This consists of an ordinary chair having a rope wound upon a reel and concealed in its bottom. The rope passes up through the back of the chair so as to always keep it in an upright position and is provided with a hook for attachment to the window sill. A strap secures the escaping person in the chair and by turning a crank attached to the rope reel he is gradually lowered to the ground.

The friction pipe escape is probably the simpleat and one of the most practical portable escapes patented.

This comprises a stiff metal friction tube coiled once or twice upon itself and having a foot stirup pendent from it. A fireproof rope is passed through the tube and attached to the sill of the window. The escaping person by placing one foot in the atirup and grasping the tube with one hand and the rope with the other can graduate his decent by moving the rope to one side or the other and thus increase the friction at the flaring mouth of the tube. The ordinary pulley tackle is also a very simple escape but depends upon the strength of the escaping person to operate it. The pulley is mounted on a suitable bracket and has a lowering

The Trolly Escape.

The Hand Spool Escape.

Adjusting The Mortar Escape.

The Parachute Escape.

The Friction Reel Escape.

Lowering Himself.

The Friction Pipe Escape.

The Rubber Chute Escape.

rope passed over it. A belt is attached to one side of this rope and is adapted to fit beneath the arms of a person; he grasping the other side of the rope to lower himself.

The friction reel escape consists of a rope reel provided with a belt to pass under the escaping persons arms. A friction band brake surrounds the periphery of the reel and is provided with an operating handle by means of which the friction retarding the revolution of the reel can be graduated. The hand spool escape is very similar to the last mentioned escape. The wire reel in this instance is mounted on a shaft having extended ends. These ends are provided with loose non heat conducting friction sleeves. Suitable leather hand pieces are applied over these sleeves and support a foot stirrup below. The escaping person places his foot in the stirrup and grasps the leather hand pieces with both hands. He now graduates his decent by increasing or decreasing the friction between the rotatable shaft and the non heat conducting sleeves by his hands.

The trolley escape is a very unique and rapid means of decending from a burning building with little or no danger. A wire is first lowered from the window and its lower end anchored at some distance away or in a neighboring building. The escaping person places a belt about him to which is attached a divided trolley wheel. A hand lever which he grasps is adapted to move the two parts of the trolley wheel toward each other so that the friction on the wire can be increased and thus graduate the decent.

All these devices are such as the entrapped victims have under their own control but there are many others to be used where the victims are helpless.

The spring supported platform upon which the victim jumps and the pneumatic mattress from the same purpose have both been patented as have also many other forms of life nets to catch the terrified victims as they jump or fall from the roofs and upper stories.

The motar escape is rather more interesting than practical. It consists of a small motar mounted upon the fire wagon and adapted to throw a grappling hook either over the building or through one of the windows. The grapple is of very peculiar construction and is adapted to automatically clutch anything it comes in contact with when thrown by the motar. A suitable line is attached to the grapple so that a flexible rope ladder may be raised and adjusted by the victims after the motar has been discharged to throw the grapple to their aid.

Innumerable shooting and swivel ladders, lazy tong platforms, shooting towers, elevated endless carriers, endless chains carrying escape cages, externally adjustable elevators have also been covered by a multitude of patents.

A rather startling invention because of its departure from the ordinary line of thought consists of a telescoping tower normally concealed beneath the pavement in front of a high building. Upon a fire occuring in the building an electrical fuse connection automatically starts the machinery of this tower and a trap door in the pavement opens and the different sections of the tower rise one above the other until the uppermost has reached the top of the building. At each section is projected a bridge drops therefrom and connects with one of the windows on that floor. When the tower has been fully extended an endless carrier within the tower and provided with passenger baskets is automatically set in motion to carry the escaping victims safely to the pavement below.

W. Harvey Muzzy.

PECULIARITIES OF ASBESTOS

A Connecting Link Between the Vegetable and the Mineral Kingdom.

To the question. "What is asbestos?" it is not altogether easy to find an answer. Geologists classify it among the horn-blendes. In itself asbestos is a physical paradox, a mineralogial vegetable, both fibrous and crystalline, elastic yet brittle, a floating stone, but as capable of being carded, spun and woven as flax, cotton or silk. It is apparently a connecting link between the vegetable and the mineral kingdom, possessing some of the characteristics of both. In appearance it is light, buoyant and feathery as thistledown; yet, in its crude state, it is dense and heavy as the solid rock in which it is found. Apparently as perishable as grass, it is yet older than any order of animal or vegetable life on earth. The dissolving influences of time seem to have no effect upon it. The action of unnumbered centuries, by which the hardest rocks known to geologists are worn away, has left no perceptible imprint on the asbestos found embedded in them.

While much of this bulk is of the roughest and most gritty materials known, it is usually as smooth to the touch as soap or oil. Seemingly as combustible as tow, the fiercest heat cannot consume it, and no known combination of acids will destructively affect the appearance and strength of its fiber, even after days of exposure to its action. It is, in fact, practicatly indestructible. Its incombustible nature renders it a complete protection from flames, beyond but this most valuable quality its industrial value is greatly augmented by its non-conduction of heat and electricity, as well as by its important property of practical insolubility in acids.

Asbestos has been found in all quarters of the globe. It comes from Italy, China, Japan, Australia, Spain, Portugal, Hungary, Germany, Russia, The Cape, Central Africa, Canada, Newfoundland, Texas and other parts of this country. Scarcely a week passes without the opinion of experts being asked on some new discovery of this mineral substance.

The asbestos generally found in the United States, especially in Virginia, the Carolinas and Texas, also in Staten island, New Jersey and Pennsylvania, is one of the woody form, in appearance like fossilized wood. The veins range in length from a few inches to several feet. The fiber can be split off like soft wood, the appearance being woolly, and when separated it has no strength or cohesion. It cannot be spun or even pulped. At one time it was thought it might be profitably used as a filler in paper making, but virtually it is of no commercial value. This and kindred classes of asbestos have often been instrumental in the creation of visions of wealth that were never to be realized. A sensation was recently caused by the announcement that a whole mountain of asbestos had been discovered in Oregon. The fiber was reputed to vary from one-half an inch to two inches in length and to be of excellent quality, The discovery was made through the herding of sheep on a mountain side. The value of the find, however, was discounted in the eyes of experts by the supplementary statement that where the flock had trampled the rock the asbestos threads showed up like bunches of wool.

Notwithstanding this wide distribution of asbestos, the only varieties which at present appear to demand serious consideration from a commercial point of view are the Russian, the South African, the Italian and the Canadian.

Before the developement of the Canadian fields the Italian asbestos was supreme in the market. For nearly twenty years Italy has been looked to for the best grades of the fiber.

The method of mining is entirely different from that followed in Italy. It is, in fact, quarrying more than mining, as the face of the rock is stripped and the cut is carried down until it reaches the asbestos-carrying serpentine, which is then removed and sent to the top of the quarry.

On arriving at the factory the crude asbestos is placed under a huge roller, which instantly reduces it to a clinging mat-like fibrous mass. This is rapidly passed through a succession of sifters and separators, which tear, strip and clean the fiber until it is ready for the different departments for which it is to be graded. It is then taken up by blowers and shot into canvas bins. Nothing more beautiful than the material at this stage can be imagined. What a few moments before was dark, sheen rock has been transmuted into a white, shining mass of delicate, quivering down.

The process of manufacture is intensely interesting, more especially from the fact that as the industry is constantly entering upon novel phases, new methods of treatment and special machinery have to be devised. One of its special uses is for wall plaster. This is a new application which will have a distinct effect in modifying the practice of indoor plastering. Instead of the ordinary tedious and elaborate preparation of studs and strips, and the use of inferior and dust-creatng mortar, with its after-scoring, which is necessary to give cohesion to the final coat of plaster of Paris, a single coating of the asbestos is laid on. It has a glossy surface that will not crack, as, while firm, it is perfectly flexible. It can be put on the raw brick, and a room of which the walls have been built in the morning can before night have a smoothly finished interior surface, shining like glass and hard as a rock. A kindred application of asbestos is now coming into vogue in the shape of uninflammable decorations for walls and ceilings. These are used a great deal for the saloons of steamships. They are embossed in very beautiful designs, and can be treated with gold, varnish, lacquers or any other substance, for the enhancement of their ornamental effect.

The applications of asbestos are now so infinite that it is impossible to enumerate them here; but a few of the more important of them may be mentioned.

Firemen clad in asbestos clothing and masks, as are those of London and Paris, can walk through the hottest flame with comparative impunity. Asbestos fireproof curtains have reduced the morality of theater fires in a very appreciable degree. In torpedoes, the difficulty of dealing with the charges of wet gun cotton is overcome by inclosing them in asbestos, the employment of which has also, in a great measure, brought the dynamite shell to its present efficiency. Asbestos is made into a cloth available for aeronautical purposes. A balloon made of this uninflammable material escapes one of the most terrible dangers to which an ordinarily constructed balloon is liable. Probably one of the first applications of asbestos in this country was to roofing. To buildings covered with this material the shower of sparks from a neighboring conflagration involves no danger. One of the largest branches of asbestos manufacture is that of sectional cylinders for pipe coverings, for retaining the heat of steam and other pipes, felt protective coverings for boilers, frostproof protections for gas or water pipes, and cement felting, which can be laid on with a trowel, for the covering of steam pipes, boilers or stills.

An interesting innovation in this class of manufacture is asbesto-sponge. It is not generally known that sponge has great powers of fire resistance. The discovery was made accidentally not long ago, and the result was that a consignment of scraps of sponge picked up on the southern coasts was ordered for experimental purposes. The sponge was finely comminuted and mixed intimately with asbestos fiber. The combination was found so successful for any covering which had to be fireproof as well as heat-proof that the material has become standard. Being full of air cells, it necessarily makes an excellent non-conductor. Another very extensive department in asbestos manufacture is that of packings. Of these there are an infinite number of forms.

To the electric engineer asbestos is absolutely indispensable. Many parts of electrical devices and machinery and wires through which the electric current passes become heated, and were it not for the electrical insulating and heat-resisting qualities which asbestos possesses, the apparatus would be completely destroyed, particularly in the case known to electrician as "short-circuiting." For such purposes it has been found advisable to combine asbestos with rubber and other gums, and this combination is now used universally for not only electrical, but also steam and mechanical purposes.

A considerable part of an asbestos factory is devoted to weaving, the asbestos being first drawn into thread for that purpose. Here again is an apparently endless diversity. There is the fireplace curtain blower, which, with an automatic spring roller attachment, takes the place in the frame of the fireplace of the less sightly sheet-iron blower; and filtering cloths for many purposes, from straining molten metal to clarifying saccharine juices in beet-root sugar refineries. A cloth is made for straining and filtering acids and alkalies in chemical laboratories. This is specially useful when the liquid to be treated is of a caustic or strongly acid nature. The filter can be thrown in the fire and after the residual matter has been consumed the web is as good as new. For filtering purposes generally asbestos has a unique adaptability, and in tropical countries it is held in greatful estimation as a cooler and purifier of water.

The newest departure in the asbestos field is the construction of electro-thermic apparatus. The heating effect of the electric current is utilized by imbedding the wire in an asbestos sheet or pad. The pad is used by physicians and nurses for maintaining artificial heat in local applications, and is said to be already largely used in hospitals. Another application of the same principle is to car heaters. A sheet of asbestos, with the imbeded wires, is clamped between two thin steel plates, and the portable heater thus provided, or a series if need be, is connected to the car circuit quickly and easily. It gives an even and healthy heat, and can be so regulated as not to overheat the car.—*Washington Star.*

A New Swiss Bicycle.

The American consul at Geneva transmits the drawing of a bicycle which has been invented in Geneva, and which is to be exhibited at the Swiss National Exposition. It is claimed for this machine that the position which the rider occupies upon it is not only infinitely easier, but that by means of the support for the back, his forces are far more effectively utilized and with considerably less fatigue. His position, as shown by the drawing, is held to be the normal position of a man in a sitting position, and the bicycle is therefore called "La Bicyclette Normale." The inventor says in his prospectus:

"The principle of the machine is the utilization of the considerable amount of force, very little known, which is afforded by a point of support. Without this point of support, the only force a man

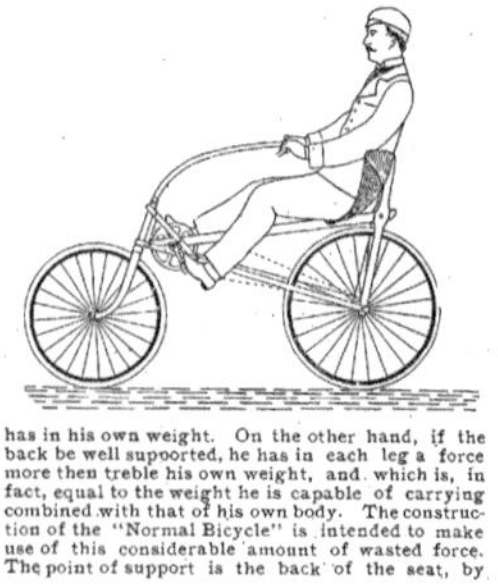

has in his own weight. On the other hand, if the back be well supoorted, he has in each leg a force more then treble his own weight, and which is, in fact, equal to the weight he is capable of carrying combined with that of his own body. The construction of the "Normal Bicycle" is intended to make use of this considerable amount of wasted force. The point of support is the back of the seat, by means of which the cyclist's body is thrown back and his legs lifted up, owing to the position of the pedals. The body is thus placed in a "normal" posture—hence the name of the machine—he is upright or leaning slightly backwards. The "Normal Bicycle" presents the advantages of greater safety, perfect comfort, healthy position, a greater power over the machine, greater speed, both up-hill and on level ground, and less fatigue."

It is also claimed for this bicycle that, being much lower than the ordinary so-called "safety" bicycle, it is much easier to mount.

A Curious Glacier.

The fall of a glacier in the Bernese Overland last autumn from an altitude of 10,823 feet above sea level is thus described by *Engineering:* The whole mass, estimated to be half as large again as the largest of the pyramids of Egypt, leaped down 4,600 feet to the bottom of the valley, then up 1,300 on the other side, and back into the valley just far enough not to destroy the watercourse through it. It appears to have jumped the watercourse, moving as a solid mass. It took only about twenty seconds in its first downward plunge, ten in its leap upward and ten in falling back, so that at the end of forty seconds the mass had changed its place from near the top of the mountain to the farther side of the valley, where it buried nearly one square mile of rich pasture to the depth of six feet. A similar ice avalanche is recorded as having occurred at the same spot on the same day of the year in 1782.

Niagara's New Bridge.

The new metal arch bridge at Niagara Falls will be noteworthy in two respects. The new bridge is to be built over the old suspension structure without interruption to the traffic on the latter. The span from end pier to end pier will be 840 feet, making it the largest arch span in the world. The principal existing all-metal arches, on the authority of the Railroad Gazette, are:

	Span.	Rise.
Louis I., Oporto, Portugal	566	146
Garabit, France	542	170
Pia Maria, Portugal	525	121
Eads' St. Louis Bridge	520	47
Washington Bridge, New York	510	91.7
Paderno, Italy	492	123
Rochester Driving Park	428	67

The suspension bridge now in use which has been familiar to all visitors to the great natural wonder for forty years will be kept in place until the new arch is ready, as it would be impossible to construct false works over the Niagara gorge to sustain a structure of this class.

The span will have a rise of 150 feet from the level ol the piers at the skewbacks to the centre of the ribs at the crain of the arch, which point is 170 feet above low water. The depth of the trusses is 26 feet, and they will be 68.7 inches apart.

The bridge will carry one floor, 46 feet wide, divided longitudinally into three parts. On the middle portion, which is 22 feet 9 inches wide, will be two trolley tracks. Each side of these tracks will be a roadway for carriages 8 feet wide, and outside of these, raised 6 inches from the level of the roadway, will be footpaths.

The construction of this remarkable span is from plans of L. L. Buck, engineer of the new East River bridge, between New York and Brooklyn, and the author of the plans by which the railroad suspension bridge at Niagara was replaced by an arched bridge.

The approaching or flanking spans will be 190 feet long on the American side, and 210 feet on the Canadian side. The total metal in the new structure will be about 4,000,000 pounds. Every confidence is expressed in Mr. Buck's ability to carry out his plans. The replacing of the railroad bridge by another without an hours interruption of business was one of the engineering feats of the decade.

Those who have not seen the great structure at Niagara which is intended to replace will hardly realize the stupendous character of the undertaking. Imagine the task of replacing the simplest sort of bridge without interrupting traffic, and then add about 1,000 per cent to the difficulty. This will give something of an idea of what confronts engineers and builders.

In an undertaking of this nature the slightest error might be productive of infinite disaster. Every measurement must be accurate to a hair's breadth. Every portion of the great arch must perform its particular share of the great combination that will be one of the marvels of the world. All that is done must be accomplished quickly, for in affairs of this nature time is indeed money. Every man who can be utilized will join the army of construction. Perhaps no work of recent years has required, or will yet need, more skilled labor. In fact, in bridge building it is becoming unsafe to utilize labor of any other class. The bridge when complete will in truth be a work of genius in point of construction, as well as point of conception.

The work of preparing the material for the great structure has been in progress for some time, as little can be accomplished in an enterprise of this nature until the preliminaries are complete. When the effort of placing the different parts of the bridge in position is begun, Niagara will be one of the busiest of busy places.

Profit-Sharing.

From extensive correspondence with all the firms and corporations known to have tried profit-sharing in the United States, Paul Monroe has made an important resume of results up to date. The reports, given chiefly in the language of the concerns themselves are summarized as follows: Of the 50 firms which have adopted the system, 12 continue it, five have abandoned it indefinitely and 33 have abandoned it permanently. Those which continue the plan have an experience extending on an average through seven years. The second class average but one year, and, recognizing the insufficiency of such a trial, have not decided it a failure. The third class vary in length of trial from a maximum of eight years to a minimum of six months; the majority having tried it for a period of from two to three years.—*Iron Trade Review.*

Child's Inventions.

That many children have great ingenuity of mind in fashioning toys of various kinds is well-known. That they have very frequently turned this quality to good use in the invention and construction of some of our most mechanical appliances is attested by the following instances:

The children of a Dutch spectacle maker happened to be playing one day with some of their father's glasses in front of the shop door. Placing two of the glasses together they peeped through them, and were exceedingly astonished to see the weather-cock of the neighboring steeple brought within a short distance of their eyes. They were naturally puzzled, and called their father to see the strange sight.

When the spectacle maker looked through the glasses he was no less surprised than the children had been. He went indoors and thought the matter over, and then the idea occurred to him that he might construct a curious new toy which would give people a good deal of amusement. He did so, and Galileo, hearing of this instrument that was said to make distant things appear close at hand, saw at once what a valuable help it would be to the heavens. He set himself to work upon it, and soon produced the telescope.

A poor Swiss, named Argand, invented a lamp with a wick fitted into a hollow cylinder, up which a current was allowed to pass, thus giving a supply of oxygen to the interior as well as the exterior of the circular flame. At first Argand used the lamp without the glass chimney, the invention of which important adjunct would doubtless have been delayed for some time had it not been for the thoughtless juvenile experiments of his little brother.

One day when Argand was busy in his workroom, and sitting before the burning lamp, the boy was amusing himself, by placing a bottomless oil flask over different articles. Suddenly he placed it upon the flame of the lamp, which instantly shot up the long, circular neck of the flask with increased brilliancy. Argand did not happen to be the man to allow such a suggestive occurrence to escape him.

The idea of the lamp chimney almost immediately came into his head, and in a short time his invention was perfected.

One of the early difficulties with the steam engine was that of condensing the steam in the cylinder. Savery dashed cold water on the outside, but Newcomen afterward invented a method of directing a stream of cold water into the inside of the cylinder at every rise of the piston. This was accomplished by two stop cocks, which were turned by hand, and the whole action of the machine depended on the attention of the person who watched these two cocks.

Humphrey Potter, a boy employed to tend one of Newcomen's engines, belonging to Mr. Beighton, found the constant watching so troublesome that he set himself to contrive a way by which the cocks might be turned at the right time, and yet allow him an opportunity of playing with the boys in the street. Observing the particular moment at which the valve required to be opened for the admission of the steam was that at which the pump-rod of the beam was raised to its highest, and that the moment at which the other cock required to be opened was when the piston rod end was at its highest, he saw that, by attaching strings to the stop-cocks, and connecting them with various parts of the beams, the rising and falling of the two ends would turn the two cocks as necessary. This rude gear of a skulking boy was discovered and practically, adopted, rods being substituted for the strings.

Making Camphor.

One of the principal products of the territory which has come under Japanese administration as a result of the war with China is camphor. In the Scottish Geographical Magazine, Mr. John Dodd, writing on Formosa, tells how this product is cultivated. Small shawles are scattered over the hills where the camphor trees grow, and in all directions the clearing of the woods is going on at a rapid rate. On the hillsides are built distilleries, consisting of oblong-shaped structures, principally of mud bricks, and about 10 or 12 feet long, 6 feet broad and 4 high. On each side are five or 10 fire-holes, about a foot apart and the same distance above the ground. On each fire-hole is placed an earthen pot full of water, and above it a cyclindrical tube, about a foot in diameter and two feet high, passes up through the structure and appears above it.

The tube is capped by a large inverted jar, with a packing of damp hemp between the jar and the cylinder to prevent the escape of steam. The cylinder is filled with chips of wood about the size of the little finger, which rests on a perforated lid covering the jar of water, so that when the steam rises it passes up to the inverted jar, or condenser, absorbing certain resinous matters from the wood on its way.

While distillation is going on an essential oil is produced, and is found mixed with the water on the inside of the jar. Whem the jar is removed the beady drops solidify, crystallization commences, and camphor in a crude form, looking like newly-formed snow, is detached by the hand, placed in baskets lined with plantain leaves, and hurried off to the nearest border town for sale.

Growth and Adaption of Cork.

Cork is very generally used in England, but most probably few of us know from where much of it comes. There are two places in the Barcelona Consular district from which cork is very largely exported. These are San Feliu de Guixols and Palamos. In the latter place the preparation of the bark is undertaken. One of our British Consuls during a late visit saw the cork in its native state as stripped from the trees, and he noticed that it was steamed or boiled, then pressed into sheets, and cut into stoppers; the shavings which escaped were collected, and after being pressed into bales were shipped off to different parts of the country for use in the manufacture of linoleum. The preparation of this cork forms the staple industry of Palamos; the greater part of the machinery used in this industry comes from France, though there is one large firm owned by Germans whose engines and driving machines, as also the electric lighting plant, are of German manufacture. It is extremely difficult to obtain correct information as to the amount of cork yearly exported from Spain, but the Consul gives the following statement as to its production, saying however, that the statement is open to correction. In the province of Gerona, about 198,000 acres are devoted to the cultivation of the cork tree. These produce about 20,000 tons, valued at about £1,610,000. Cork ranks third in importance in the list of exports from Spain.

Lord Kelvin's Jubilee.

The affection and esteem in which Lord Kelvin is held by scientific people in all parts of the world is fittingly shown in the splendid celebration in his honor at Glasgow, June 15th in which the city, the university and delegates from scientific bodies took part. Not the least ardent of Lord Kelvin's admires are Americans, and it was thoughtful of the Commercial Cable Company to place a number of his American friends in direct communication with Lord Kelvin during the cermonies in Glasgow. The honor and glory of the celebration were undoubtedly very gratifying to Lord Kelvin, but his greatest monument consists in the thousands of students whose minds have been trained in science under his able guidance.—*Electrical Review.*

Photographing in Colors.

A new development of photography in colors was explained by M. Lippman, the distinguished French investigator. He has now succeeded in reproducing perfectly all the colors of Nature on a sensitive plate. Light, as every one knows, rushes through the camera, as through all space, at the rate of about 186,000 miles per second. With this velocity it leaves traces of its energy in the photographic picture in light and shade, but it is colorless, because the forms of the individual waves or vibrations are not depicted.

To secure this result M. Lippman places behind the thin, transparent gelatine film a mirror of mercury. This stops the rays of light and reflects them, thus rendering the vibrations practically stationary, as the result showed. They then leave on the film the impress of each separate prismatic color and shade. The flim remains transparent and its hues are like those of a soap bubble and other substances in themselves colorless. In other words, the photographic plate must be held at a certain angle in order to see the chromatic effects.

M. Lippman shows by reflected electric light a number of pictures produced in this manner, several being the simple colors of the spectrum, and others photographs of natural objects and scenes, including portraits.—*Popular Science News.*

Economy of Electrical Cooking.

Mr. R. E. Crompton, a well-known electrical engineer, has given some trustworthy data as to the economy of electrical heating and cooking. The convenience and cleanliness of electricity in the home and kitchen are generally recognized, but it is not so well-known that an electrical oven, that is to say, an oven heated by the electric current, is "practically twice as economical as any other oven, whether heated by gas or solid fuel." This arises chiefly from the great waste of heat in gas and coal fires and ovens. According to Mr. Crompton, the gas current carries away some 80 per cent of the total heat generated, whereas only 10 per cent of the electric heat is wasted. Again, probably not more than 2 per cent of the heat given out by coal in kitchen fires is utilized in cooking food, whether in private houses, clubs or hotels, showing an enormous and expensive waste.

Property in an Idea.

The editor of the London Truth is decidedly opposed to the patenting of inventions. He describes a patent as "attempt to create a peculiar and artificial kind of property in an idea," and declares that "the attempt can never be altogether successful." While it must be admitted that patents are a fruitful source of litigation, and that no patent system that has ever been devised has been free from objections, it does not by any means follow that the policy which underlies all such systems is a mistake. If perfection were the test, all human creations would have to be set down as failures. There is no plan of government, no code of laws, no educational system, no method of taxation, no any thing else, in fact, that human wisdom, aided by human experience, has devised that is "altogether successful." All property rights are, to some extent, "artificial," being created by law, and always subject to limitations. A man cannot do as he pleases with his own house, his own lands, or his own domestic animals. He cannot even buy and sell goods in most communities without paying for the right to do so; and there are many kinds of goods that he cannot give away without liability to prosecution. It will be seen, therefore, that property in patents is not exceptional in being likely to entail trouble on its owner.

Truth says, but not very truly, that "there is a general notion that if you did not protect inventions by means of patents, inventors would cease to invent, and material progress would come to a stand still." But the editor submits that history disproves this. He asserts that men with great mechanical gifts do not exercise them solely with a view to commercial profit, any more than astronomers search the heavens for new worlds with an eye to registering patents and floating companies on the results of their discoveries.

No intelligent person contends that there would be no inventions or that material progress would be forever stopped if all patent laws were repealed. But it is the opinion of nearly all who have given thoughtful attention to the subject that invention would be checked and progress retarded by such change. We have, in the United States, the best patent system in the world, and no American doubts that it has been one of the chief contributors to the unexampled development and growth of the country, materially, morally, and intellectually. Hope of reward is the greatest incentive to effort. The patent laws encourage men and women to devote time and money to inventions. Our laws aim to do justice to both the inventor and the people, and our patent system, now more than a hundred years old, is intrenched in the confidence of the country by reason of the influence that it has exerted.

Primitive Machinery.

Among the objects sent to the United States National Museum from the Vale of Kashmir by Dr. W. L. Abbott, of Philadelphia, is a very primitive form of spinning wheel. Evidently the maker has had instruction from Chinese sources but owing to his poor resources has been driven to utilize his folk ingenuity to eke out his device. Practically the machine is the common spinning wheel consisting of a large wheel with string for band working over the shaft of the spindle itself. The posts and handle of the main wheel resemble chair legs rudely turned. The spokes of the wheel are broad pieces of roughly hewed, thin boards resembling shingles. These are notched on the edges near the ends. There are two sets of these spokes about six inches apart and they do not stand opposite to each other but are alternate. A coarse twine of yak hair is wound continously backward and forward between the notches on the spokes so as to form a quasirim. The alternating of the spokes are to allow for this continous weaving backward and forward, the right hand notch at the end of one spoke coming opposite the left hand notch on the edge of the opposite spoke and vice versa. This form of band wheel or driving wheel is very common in China.

The spindle is a long needle of iron. The posts supporting it are also on the chair leg style but the interesting part is the bearings which are made of coarse barley straw, woven, as in the footing of a Chinese sandal; doubled and drawn through large holes in the upright posts. The spindle is simply passed through these straw bearings which are saturated with oil. A wooden stop prevents the spindle from being drawn through the straw bearing. The band string is prevented form deviating by a long notch or cut in a post standing between the spindle bearings. A straw is fitted over the outer or working end of the spindle and the yarn is first twisted and then woven on this straw. When the straw, acting as a spool, is filled up it is drawn off.

The United States National Museum possesses a part of a spindle from New York, gift of Mr. S. F. Hawley, of Broadalbin, Fulton County. It is labeled A. Minor's improvement. Patent warrented; accelerating wheelhead made by William W. Hopkins, Chesterfield factory, New Hampshire, with the advice that "Particular care must be taken to keep the apparatus dry. When used it must be kept well oiled and when new bands are required they must be made smooth and even. Spindle warrented of cast-steel." This apparatus is not unlike many others in the United States National Museum collection excepting that the bearings of the spindle itself are made of corn husks neatly doubled around the shaft, drawn through an auger hole in the horizontal piece and held in by a dowel pin driven down from above. Both of the corn husk bearings are saturated with oil.

There are also in the Museum old spindles with leather bearings, but it would be interesting to know why in Kashmir and in New Hampshire the husk and straw should be used in place of metal to form the bearings of a revolving spindle. Perpaps some of the readers of "THE INVENTIVE AGE" can tell me something about the history of spindle bearings made of vegetable fiber, what advantage they have over metalic bearings, and what has been their limitations in time and geographic distribution? O. T. MASON.

The first stage of the great Watkin Tower, at Wembly Park, near London, has been completed and opened. It is 160 feet high, and consumed in construction 2700 tons of steel. Work is now beginning on the second stage, which will carry the structure up 300 feet higher. Eventually the tower will be raised to a hight of 1100 feet above the ground. Its total cost is estimated at about $1,000,000.

Mr. Edison has of late been in receipt of bushels of letters relative to the use of the fluoroscope, and many of them are of an amusing or pathetic nature.

In one case, his correspondent wrote from Pottstown, Pa., as follows: "Will you please send me one pound of X-rays, and bill as soon as possible." In another instance a man with weak sight asked whether he could send his spectacles to Mr. Edison and have them so "fixed up" that they would enable the wearer to see better by means of utilizing the X-rays.

Electric Eyes for the Blind.

In a recent meeting of teachers and other experts on educational matters, in Russia, at Moscow, Dr. Noishewski demonstrated an instrument invented by him several years ago, but considerably improved recently, which he calls the "electrophthalm," or "electric eye." This very senstive instrument is intended to enable the blind, by the sense of feeling, to observe objects which we see with our eyes. Noishewski uses in the construction of his instrument two metalloids, selenium and telliurium, both of which will change their quality as conductors of electricity with different conditions of light.

The instrument constructed by Noishewski denotes these changes, and, while it does not enable the blind to see, they will feel the various effects of changing light by means of this apparatus. The "Wjedomosti reports, in a recent number, that Dr. Noishewski has succeeded in having a totally blind man find the windows in a room, and after a little practice, to distinguish approaching men from approaching animals. The inventor is still at work upon improvements to the "electrophthalm," and hopes to perfect it to such an extent that the blind will be able to tell almost with certainty when approaching opaque or transparent objects, which would be a great stride toward bettering the condition of the blind.

A Piece of Iron 2,000 Years Old.

S. T. Wellman, the well known metallurgist, of the Wellman Seaver Engineering Co., Cleveland, has a portion of a round bar of iron—and a few like pieces are held in the United States—that antedates the Christian era by two or three centuries. The iron, which had been originally hammered into plates, and was deeply rusted from age, was found a few years ago by Dr. Karl Humann, in the ruins of the Temple of Artemis Leucophryne at Magnesia, Asia Minor. Dr. Humann sent it to Hallbauer in Germany and the latter made from a portion of it a memorial tablet. This was presented to Bismarck in April, 1894. It bore this inscription, in German: "For you, Prince Bismarck, the iron chancellor, Hermogenes forged this iron at Magnesia, 200 B. C. Humann found it in the Temple of Artemis after 2,000 years and sent it to Hallbauer, who gave it the form in which it shall bear witness that your deeds shall outlive millenia."

At the time of the presentation to Bismarck *Stahl und Eisen* gave a photographic reproduction of the plate and an accouut of the discovery of the iron. The Temple of Artemis, one of the most magnificent of ancient monuments, was rebuilt about 300 B. C., though by some the date is put at 200 B. C.

The metal is described as approximating steel in its composition, though closely akin to malleable iron. It was made at a low temperature and great care was necessary in the forging. It was found rather difficult to roll the pieces that were preserved as relics, these having a diameter of about ½ inch. One analysis showed carbon 0.20 per cent; phosphorus, 0.016 per cent: iron, 92.71 per cent. Another gave carbon 0.23 per cent; phosphorus, 0.0223 per cent; sulphur, a trace, with no distinguishable amount of manganese of silicon. An analysis in the laboratory of Prof. Ledebur showed 1.01 per cent of slag, 0.025 per cent of phosphorus and 0.061 per cent of carbon.

The Rothchild's have secured the controlling interest in the great Anaconda copper mines at Butte, Montana, on a basis of $45,000,000 for the whole.

Col. C. R. Lesher, of Lansing, has obtained a patent on an invention which harrows and rolls the ground, and drills and covers the seed, all in one operation.

A Swiss scientist has been testing the presence of bacteria in the mountain air, and finds that not a single microbe exists above an altitude of 2,000 feet.

A shipment of 500 tons of light steel rails was recently made to Japan from the Milwaukee works of the Illinois Steel company. The rails were sent by lake to Buffalo, there transferred to cars for New York, and then loaded on a steamer for Japan.

Franklin who drew the lightening, was a professional politician, not a professional scientist. Morse, who invented the telegraph, was merely an amateur. Watts, who invented the steam engine, was not a professional machinist. Nearly all the great discoveries in photography have been made by amateurs. And so in nearly all lines it is the man who loves the work, not the man who lives by it, through whom progress comes.

NEW PATENTS FOR SALE.

Advertisements inserted in this column for 20 cents a line (about 7 words) each insertion. Every new subscriber sending $1.00 to THE INVENTIVE AGE will be entitled to the AGE one year and to five lines three times FREE. Additional lines or insertions at regular rates.

"HARNESSING THE WAVES."—This will usher in the "Millennium of Electricity," as Edison declares. See the article in this month's AGE. The field is broad and "white with the harvest." Who will help me reap it? Address, for particulars, "WAVE POWER," care of THE INVENTIVE AGE.

FOR SALE.—My Patent, No. 554,065—Sand-Paper Holder—both for the United States and Canada. A very useful device. For further information address, A. Hornig, Sandusky, Ohio. 6-8

WANTED.—Some one to join as co-partner with me to bring out Link and Pin Car Coupler No. 485,333. Rev. J. W. Postbn, Holly Springs, Miss. 5-7

FOR SALE.—Good patent. Easy and cheap to manufacture. Thoroughly tested. Many in use. Recommeded by all using it. Ready seller. Price reasonable. Good investment. Pat. Jan. 9th, 1894. For references and price, terms etc., address P. O. Box, No. 1072, Tacoma, Wash. 5-7

FOR SALE.—Patent No. 501,800 issued 18th of July 1893; flash light apparatus; being compact light simple, and durable; always ready, can be used with all kinds of Cameras. Will sell the entire patent, or manufacture on royalty. A. T. Mallick, Jamestown, N. Dak. 3-5

FOR SALE.—Or manufacture on royalty United States patent 518,165, Canada patent 46,029; also several unpatented inventions; one-third to party that will furnish money to have them patented. Enclose stamp. E. H. Thalaker, Petersburg, W. Va. 3-5

WANTED.—A Partner to take an undivided one-half interest in the best 25 cent household article on earth. Is patented in both U. S. and Canada, have sold several State and County rights, also over 5,000 of the goods. I have the machinery and have reserved the right to manufacture all goods and 4 cents will cover the cost of each article. I will give the right man a snap. Address, J. E. Hyames, Gobbleville, Van Buren Co., Mich. 3-4

FOR SALE.—Patent, 492,875, Folding Hoe, entirely new principle. Will sell State or County rights cheap. Any live man can make money right in his own state or town handling the hoe. Inexpensive to manufacture, and sells on sight. For full information, address Dr. F. A. Rice, 62 Main street, Lockport, N. Y. 2-7

FOR SALE.—Valuable Patent No. 532,324 for improvement in metal carriers or baskets for bottles and other articles. Address Patentee, H. M. Kolb, 732 Sansom street, Philadelphia, Pa. 2-4

FOR SALE.—Canadian Patent, the Highwine and Whiskey Distillers Continuous Beer Still. Manufacturers and capitalists who wish to purchase valuable patents can make very large profit. Operated in U. S. Distilleries. Address, Hermann Binz, Frankfort, Ky. 2-4

FOR SALE.—Or on royalty, Patent No. 554,643; a folding suit or garment hanger; simple and cheap; big profits in it. Address, Chas. Behrend, Jr. Oconomowoc, Wis. 4-6.

FOR SALE.—Patent No. 517,919 A Chimney Regulator. Best ever invented. Is sure to be a blessing to every home. Will take some land in exchange. Address Jurgen Hurichs, Danforth, Ill. 7-9

FOR SALE.—A half interest in my "Hair Grower," I have made a discovery whereby I can grow a full head of hair on the baldest head; cure any case of dandruff or scalp disease to dead certainty. I want a partner with money to put this new discovery upon a large scale. A fortune for some one. Sample bottle sent upon receipt of $1.50. Address, Geo. W. Schoenhut, Eldora, Iowa. Box 98.

FOR SALE.—A complete set of Patent Office Reports and Official Gazette of the U. S. Patent Office from 1790 to 1894. H. B. Dickinson, Ravenna, O., Adm'r of Estate of Bradford Howland. tf.

BUSINESS SPECIALS.

Advertisements under this heading 20 cents a line each insertion—seven words to the line. Parties desiring to purchase valuable patents or wanting to manufacture patented articles will find this a valuable advertising medium.

WANTED.—Publisher to publish and introduce Walsh's Perpetual Calendar and Almanac on royalty; or I will sell the copyright. A complete calender, absolutely perpetual. No complicated rules or computations. James A. Walsh, Helena Mont. 2-4

WANTED.—Reliable manufacturer to manufacture on royalty, or will sell at a moderate figure. Caliper Adjustment. Described elsewhere. Address, Arthur Munch, 653 East 5th, st., St. Paul, Minn. 4-7

Information Wanted.

A subscriber of THE INVENTIVE AGE wants information that possibly that some reader of this item may be able to give. He says:

The silver mines of Cerro de Pasco, Peru, South America, have the greatest attraction for me, although the only reliable information I have of them dates back to the exploration of Lieutenants Herndon and Gibbons, 1850 to 1852. I desire to confer with parties in the United States who have the information as to how close Meigs' trans-Audiene railroad runs to Cerro de Pasco and how near those extensive coal fields are to these mines—and if Meigs' or others tunneled them to make thorough drainage. My plan is to gasify the coal and pipe it to the mines. It costs $12.50 per ton far coal in 100 pound loads on Lamas.

Better Jump off the Dock.

Referring to an item in last months issue descriptive of a method of finding the length of a belt, Mr. W. Lee Cheney, of Meriden, Conn., writes: "F. Casey of Harrisonville, Ohio, better go and jump off the dock. If the shafts are near together and one pully much larger then the other, his method will not be right.

More than all that, it is necessary to measure the distance between the shafts anyhow, and while F. Casey of Harrisonville, Ohio, is measuring this, why don't he carry his tape line around the place the belt is to go and get the length without any figuring."

Expert Train Running.

For the month of May the record of train movement on the B. and O. R. R. eclipsed the record breaking record for April, when the passenger trains arrived at their destinations as per schedule ninety-five times out of a possible hundred. The B. & O. Fast Freight Trains between New York, Philadelphia and Baltimore on the east, and Cincinnati, St. Louis add Chicago on the west, are being moved with an equal degree of percision.

THE MOUNTAIN CHAUTAUQUA.

Mountain Lake Park, Md., on the Main Line of the Picturesque B. and O.

The most superb and sensible summer resort in America. $300,000 expended in improvements; 200 beautiful cottages; hotel and cottage board at from $5.00 to $12.00 per week—cheaper than staying at home. The mountain air and the mountain views simply indescribable. Session August 5th to the 25th. Three superb entertainments daily. The best music and the best lecturers which money can procure. Dr. T. De Witt Talmage, Gen. John B. Gordon and Bishop J. H. Vincent already secured, with 100 others. Dr. W. L. Davidson, the great Chautauqua manager, in charge.

SUMMER SCHOOLS.—20 departments of important school work in charge of leading instructors from the prominent universities. A wonderful chance for teachers and students desiring to make up studies. Tuition insignificant. Wishes of students gratified. Low rates on railroads. For full detailed information and illustrated programme, address A. R. Sperry, Mountain Lake Park, Md.

Reduced Rates to Chicago.

Account of the Democratic National Convention, Chicago, Ill., the B, & O. R. R. will sell excursion tickets from all ticket stations on its line east of the Ohio river, for all trains July 3, 4, 5 and 6, good for return passage until July 12 inclusive, at one single fare for the round trip.

Tickets will, also, be sold by all connecting lines.

The B. & O. maintains a double daily service of fast vestibuled express trains, with Pullman Sleeping and Dinning Cars attached, running through to Chicago solid without change or transfer.

Synonyms, Antonyms, and Prepositions.

A new volume on "Synonyms, Antonyms, and Prepositions" will shortly be issued by the Funk & Wagnalls Publication House. This has been prepared with great care by the Rev. James C. Fernald, editor of the department of Synonyms in the Funk & Wagnalls Standard Dictionary. The editor has carefully discriminated the chief synonyms of the English language, some 6,000 or 7,000 in number, by the same method that has won so much approval in the Standard Dictionary. Taking one word in each group as the basis of comparison, Mr. Fernald defines this clearly and then he proceeds to show how the other words agree with or differ from it, thus the whole group is held to one fixed point. The treatment is in popular and readable style. The book also contains a large number of Antonyms as well as prepositions, and its closing pages are devoted to Questions and Examples of service to both teacher and student.

The type, brevier, is pleasing to the eye and the key-words at the top of each page enhance the value of the book for the purposes of ready-reference.

ANY person sending us the names and addresses of five inventors who have not yet applied for patents will receive the INVENTIVE AGE one year free.

IDEAS DEVELOPED. Absolute secrecy. Send for particulars. Advice and suggestions given free when asked for. Correspondence and sample orders solicited. 25 years in business.

GARDAM & SON,

96 John Street. New York.

Aftermath.

THE American Association for the Advancement of Science will hold its 45th meeting in Buffalo during the week begining August 22.

BARON Roentgen has had the honorary citizenship of his native town, Lennep, Rhenish Prussia, conferred on him, in honor of his discovery of the X ray.

THE directors of the American Bell Telephone Company on June 10 declared a regular quarterly dividend of three per cent and one and one-half per cent extra.

THE stockholders of the Westinghouse Electric aud Manufacturing Company, at a special meeting held on June 5, formally voted to increase the capital stock of the company from $10,000,000 to $15,000,000.

It is estimated that the telephone system of St. Louis suffered during the recent tornado to the extent of $80,000. Over 3,500 miles of wire were knocked out, 30 aerial cables were broken, and the company lost about 400 poles.

The Pilots' Association of New York have completed plans for a steam pilot boat, the first in the world. She will be equipped with all conveniences, including electric lighting, and will have a speed of 14 knots an hour.

THE eighth annual convention of the American Boiler Manufacturer's Association convened in Cleveland Tuesday June 2nd, and continued two days. H. S. Robinson of Boston is president of the association and Col. E. D. Meier Secretary.

It is announced that the Niagara Falls Power Company has ordered from the Westinghouse Electric and Manufacturing Company, of Pittsburg, seven dynamos to be used exclusively for the Buffalo transmission. These machines will have a capacity of 5,000 horse-power each. Contracts have also been given for a power house and wheel pit for the dynamos. The officers are reported as saying that before summer closes Niagara power will be furnished to Buffalo.

The contract for the entire development of 20,000 horse-power on the Richelieu River, the outlet of Lake Champlain, has been let for $550,000, the electric machinery not being included. This power is to be carried to Montreal, the distance being about 12 miles. This is the second electric water-power development furnishing current to Montreal, the first being for 12,000 horse-power at the Lachine Rapids, five miles above the city in the St. Lawrence River. The investment in both powers will be about $3,500,000, all subscribed for by Montreal capitalists.

A curious method of producing platinum is practised by the inhabitants of the villages on the Tura river in the Russian government district of Tomsk in Siberia. They call this method "plowing the water." A raft is constructed and an inclining gutter of boards fastened to it, which at its lower end is provided with an iron plow. While floating down the river they scrape or plough its bottom. The sand scraped out falls into the gutter and passes into a tub filled with pine boughs upon which platinum is deposited. The sand of the Tura river and its tributaries is so rich in platinum and its primitive production so profitable that the peasants are abandoning agriculture and devoting themselves to "plowing the water."

Recent Australian Patents to Americans.

The following list of applications has been specially p epa ed for by Mr. George G. Turri, Certified Patent Agent, Melbourne.

W. D. Asten, Attorney of the American Tobacco Co. of Victoria Limited. (Assignee of H. Bilgram, Philadelphia, for improvements in machines for making conical cigarettes.)

W. D. Asten, Attorney of The American Tobacco Co. of Victoria Limited. Communicated by J. A. Bonsack, Philadelphia, for improved method of and machine for making cigarettes tubes.

W. D. Asten, Attorney of The American Tobacco Co. of Victoria, Limited, improvements in cigarette machines,

E. Waters, (Communicated by F. R. Keys, and C. A. Baker, Binghampton, for improvements in cigar making machines.)

L. P. Jacobs, and M. M. Levinson, (Assignees of J. R. Williams, East Orange,) for improvements in machines for cutting out cigar wrappers or binders.

E. Waters, (Communicated by The Hall Signal Co. of New York.) Assigness of H. E. Booth of Stoneham, for Railway Signals.

B. Baron, New York, for improvements in cigarettes and in method of and apparatus for manufacturing the same.

W. D. Asten, attorney of The American Tobacco Co. Limited of Victoria, (Communicated by K. H. Carper, Salem), for improvements in filler forming mechanism for cigarette machines.

W. D. Asten, attorney of The American Tobacco Co. Limited of Victoria, (Communicated by D. B. Strouse, Salem,) for improvements in cigarette machines.

W. D. Asten, attorney of The American Tobacco Co. Limited of Victoria, (Communicated by D. B. Strouse, Salem,) for improvements in machines for securing the seams of paper tubes especially designed and adapted for cigarette wrappers.

G. G. Turri, (Communicated by W. C. Sherman, of Orlando,) for non-fillable bottle.

P. J. Schlict, summit for improvements in relating to art of and apparatus for producing combustion.

A. C. Shuttleworth, Philadelphia, for Electro magnetic induction system of propulsion.

M. L. Warson and E. F. Pickett, Buffalo, for improvements in wheel tires.

A. S. Weaver, Hamilton and W. J. Goold, Toronto, for improvements in bicycles.

TO INVENTORS.

We make a specialty of manufacturing goods for inventors. Make models, dies, do experimental work, make special tools, machinery, etc. INVENTOR'S HANDBOOK FREE.

Cincinnati Specialty Manufacturing Co.,

Third Street, Cincinnati, Ohio.

For Mchanics and Engineers.

Every mechanic and engineer should have a copy of "Shop Kinks and Machine Shop Chat," by Robt. Grimshaw. It is the most complete work of the kind ever published. If any "kink" is discovered in the machine shop—it can be straightened out—after consulting this valuable work. Sent to any address for $2.50.

A Thousand Opportunities.

The Racine Fire Engine Co., of Racine, Wis., are the manufacturers of every kind of fire-fighting device, engines, hose, carts, fire suits, etc. Probably 1,000 readers of this item live in towns where additional supplies are wanted for the protection of the property of the tax-payers. As will be seen, by their announcement in another column, this gives an opportunity for some one to make $25 where conditions of that nature exist.

Saturday and Sunday Trips to The Country.

Commencing Saturday, May 30th, and continuing until further notice, the B. & O. R. R. Co. will sell excursion tickets, at rate of one fare for the round trip, for regular trains of Saturday and Sunday, to points on the Metropolitan Branch and Main Line between Washington, Harper's Ferry and Charlestown, and to points on the Washington branch between Washington and Laurel. 3

A Boston dispatch says: S. H. Roper, of Roxbury, a mechanical engineer, while pacing Tom Butler, the professional bicyclist, with a steam bicycle, which Roper invented, was killed at the Charles River Park, on the 1st inst. Roper's machine became unmanageable, and he was thrown, striking upon his head. He died a few moments later. Roper, who was seventy years of age, had been at work on his motor for twenty-five years.

A Great Novelty.

ELECTRIC LIGHT

FOR THE NECKTIE.

$1.50.

Complete with Powerful Pocket Battery and all accessories, postpaid.
The Lamp an Efficient Beauty.
Battery 3 1-2 by 3-4 by 1 3-4 inches.
This elegant outfit is practical, brilliant and satisfactory, and every part is guaranteed.
Electricians who want a piece of jewelry that is emblematical, or anybody who appreciates a real novelty, can, by ordering now, secure a complete outfit for $1.50. Order at once of
THE INVENTIVE AGE,
Washington, D. C.
Given FREE for four new subscribers.

HOW TO MAKE A DYNAMO.

By ALFRED CROFTS.

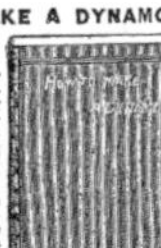

A practical work for Amateurs and Electricians, containing numerous illustrations and detailed instructions for constructing Dynamos of all sizes, to produce the Electric Light, containing 96 pages of genuine information which will enable anyone to construct a Dynamo either for pleasure or profit.

Large 12 mo,
Cloth, 75 Cts.

GREAT PREMIUM OFFER! = =

Valuable
Sensible
Economical

Zell's Condensed Cyclopedia
Always Sold at $6.50 Complete.

Complete in One Volume. A Compendium of Universal Information, Embracing Agriculture, Anatomy, Architecture, Archeology, Astronomy, Banking, Biblical Science, Biography, Botany, Chemistry, Commerce, Conchology, Cosmography Ethics, The Fine Arts, Geography, Geology, Grammar, Heraldry, History, Hydraulics, Hygiene, Jurisprudence, Legislation, Literature, Logic, Mathematics Mechanical Arts, Metallurgy, Metaphysics, Mineralogy, Military Science, Mining, Medicine, Natural History, Philosophy, Navigation and Nautical Affairs, Physics, Physiology, Political Economy, Rhetoric, Theology, Zoology, &c.

Brought down to the Year 1890. Substantially Bound in Cloth with the Correct Pronunciation of every Term and Proper name, by L. Colange, LL.D. Handy Book for Quick Reference. Invaluable to the Professional Man, Student, Farmer, Working-Man, Merchant, Mechanic. Indispensable to all.

In this work all subjects are arranged and treated just as words simply are treated in a dictionary--alphabetically, and all capable of subdivision are treated under separate heads; so that instead of one long and wearisome article on a subject being given, it is divided up under various proper heads. Therefore, if you desire any particular point of a subject only, you can turn to it at once, without a long, vexatious, or profitless search. It is a great triumph of literary effort, embracing, in a small compass and for little money, more (we are confident) than was ever before given to the public, and will prove itself an exhaustless source of information, and one of the most appreciated works in every family.

It cannot in any sense--as is the case with most books--be regarded as a luxury, but will be in daily use wherever it is, by old and young, as a means of instruction and increase of useful knowledge.

Thousands of leading men offer the highest testimonials. It is not a cheap work it is standard.

Never before, even in this age of cheap books, was such a splendid offer made, viz:

$2 For Cyclopedia, Postage Paid. For Inventive Age, One Year. **$2**

An Extra Copy of Cyclopedia to any person sending us Six New Subscribers. The Same Offer is made to old Subscribers renewing for another term.

Address THE INVENTIVE AGE, Washington, D. C.

Inventive Age
AND INDUSTRIAL REVIEW
A JOURNAL OF MANUFACTURING INDUSTRY AND SCIENTIFIC PROGRESS

WASHINGTON, D. C., AUGUST, 1896.

Single Copies 10 Cents.
$1 Per Year.

ntennial.

xposition, which will
7, and continue six
nd counties of Ten-
money secured from
opriations, and to the
itizens of Nashville,
nty of Davidson, in
have added $325,000.
ngs is going steadily

State Department of the United States, has created material interest in the enterprise in many foreign lands.

Application for space and for concessions have been received from many parts of the world, and the committees having this part of the work in charge are at work classifying and adjusting them. At the Exposition Park the walks have been completed except for the final dressing, and as each building is completed the beautifying of that part

lakes are brimming with water, and new features are being begun every day. No detail in the preparation of the Tennessee Centennial Exposition is being slighted, and there is sufficient time between now and May, 1, 1897, for the completion of every idea undertaken and the smoothing of every uncomely place. The Tennessee Centennial Exposition is the expression of patriotic citizens, and those citizens are determined that this enterprise shall reflect the greatness and the thrift and glory

The Inventive Age

Established 1889.

INVENTIVE AGE PUBLISHING CO.,

8th and H Sts., Washington, D. C.

ALEX. S. CAPEHART. MARSHALL H. JEWELL.

The INVENTIVE AGE is sent, postage prepaid, to any address in the United States, Canada or Mexico for $1 a year; to any other country, postage prepaid, $1.50.

Correspondence with inventors, mechanics, manufacturers, scientists and others is invited. The columns of this journal are open for the discussion of such subjects as are of general interest to its readers.

Technical matter is particularly desired. We want practical information from practical men.

Nothing will be published in the editorial columns for pay.

The INVENTIVE AGE is thoroughly independent.

Advertising rates made known on application. Special facilities for furnishing cuts of any patented article together with descriptive article. Business specials 15 cents a line each insertion, 7 words to the line. No advertisement less than 25 cents.

Address all communications to THE INVENTIVE AGE, Washington, D. C.

Entered at the Postoffice in Washington as second-class matter.

WASHINGTON, D. C., AUGUST, 1896.

Among the papers in the District of Columbia devoted to inventions and patents, none has credit for so large a regular issue as is accorded to the "Inventive Age," published monthly at Washington.—*Printers Ink.*

A BOOM in oil wells is now reported from Tennessee, the Standard Oil Company being heavily interested.

PARTICULAR attention is called to the article on another page headed "Secrets of the Human Life." It is an abstract from an interesting paper read by Prof. C. Cole, editor of "Storms and Signs," Kingston, Pa., before a meeting of the Pennsylvania Editorial Association.

"SOMETHING About X-Rays" is the title of a little brochure by Edward Trevert. It is from the press of the Bubier Publishing Company, Lynn, Mass. It is a compliation from the leading electrical magazines of the day and is profusely illustrated. It can be had through the INVENTIVE AGE for twenty-five cents.

THE report of the Illinois railroad commission for 1895 shows that of the 63,485,413 passengers carried over the 10,500 miles of line in that state during the year only 12 lost their lives by accident. In other words, only one out of every 5,290,451 passengers was killed. The report also shows that only one out of every 409,583 passengers carried was injured.

THE recent special issue of the Scientific American, signalizing the fiftieth anniversary of its publication was one of the most interesting reviews of industrial progress ever issued. Its special articles on the triumphs of inventive genius during the last half of the century are intensely interesting and of great value to inventors and advanced thinkers.

THE fact that we have in circulation in the United States about 50,000,000 silver dollars and 378,000,000 more coined that the government would be glad to put in circulation, lends much force to the declaration of Senator Proctor, that the people of the country—especially the laboring people—are more interested in starting the mills to work than in starting the mints.

AN electric barb is one of the recent inventions which a Halifax whaler proposes to experiment with on his next expedition. It is proposed to place a dynamo on the ship to generate a current of 10,000 volts. This current will be carried over several thousand feet of wire to the harpoon and after the harpoon is properly thrown—it's expected the electric current will do the rest. It's a fascinating proposition to say the least.

IN order to call particular attention to the great progress of Southern commercial interest, a special supplement to the regular issue of the Manufacturers' Record devoted mainly to New Orleans and the growth of the commerce of that city was issued on the 17th ult. It is truly stated that many thousands of general investors, capitalists and manufacturers throughout the North and West and in Great Britain will have a clearer insight into the advantages of the South, and a better knowledge of what it is accomplishing after reading this issue than they have ever had before.

THE greatest nuisances in railway travel, with the possible exception of the "newsy," are the smoke and cinders from the locomotive. And now a woman inventor, Mrs. Mary J. Wyatt, of Raleigh, N. C., claims to have invented a successful conveyor that will do away with these inconveniences. The records of the Patent Office during the past few years show that the inventive genius of women is no less acute than that of the sterner sex.

CARBONIC acid gas as a substitute for ice in the preservation of fruit in freight cars has been given a thorough trial and has been found to be impracticable. A car containing fruit, after being on the way from San Jose, Cal., to Chicago, Ill., during a period of seven days, was opened and the fruit was in bad condition. The temperature of the car en route was kept at 72 to 74 degrees.

UNDER the authority of a resolution passed by Congress, Director Preston, of the United States mint, will shortly experiment with pure nickel and aluminum as substitutes for the present nickel pieces and one and two-cent bronze pieces. There may be advantages in the change but they are not apparent at this time.

THE columns of the INVENTIVE AGE are open to communications from inventors, manufacturers, mechanics, engineers and others, and to the discussion of all proper subjects. Questions will be answered and information given on any subject of general interest.

A FLORIDA boy, Robert G. Bidwell, of Orlando, has invented a device which he claims meets all requirements of a non-fillable bottle—but there are so many failures in this line the public will be rather skeptical.

Traps for Inventors.

In this 19th century the profession of patent solicitors is degenerating from the professional to the commercial. Inventors and patentees have their attention arrested by flaming announcements. One class of these agents offer medals and large lottery prizes amounting to thousands of dollars to inventors who will place their applications for patents in their hands. However, before a medal or prize is awarded, these inventors are compelled in order to become competitors, to pay into the hands of these agents from $50 to $75. These competing inventors are lead to believe that a scientific and mechanical corps of experts, in the employ of these agents, make crucial examinations of their inventions, comparing them with all others that are competing for the prize, and in due time they receive a communication, from their agents, accompanied by a medal—certifying that the medal has been awarded by a corps of experts, on the ground that the invention is the best of all others presented to them for a patent. At some subsequent period it is announced that the money prize has been awarded.

It would seem that intelligent men would not fall into such traps in this enlightened age, but alas! they like innocent lambs are fleeced, or dealt with in the same manner as are rural citizens who fall in the hands of "green goods" merchants.

For many years the story of the GOLD (gilded) medal awarded by a French scientific society to United State patentees, has been well known, and yet new victims are constantly being made. When the announcement is received from Paris that the GOLD (gilded) medal has been awarded to a United States patentee for his invention, because on examination by its Savans it had been found to be the best of the kind patented, the medal is accompanied with a demand for a *considerable sum of money to pay the expense of its transmission to this country.* This sum of money is the secret of all the interest that this French association manifest in regard to United States patentees. This bold attempt to get money, ought to be understood by intelligent patentees, when they read the word "gilded" in small letters and enclosed in brackets following the word "GOLD."

Recently an inventor applied to one of the medal awarding United States patent agents and received a medal, *but no patent*, and after he had expended about a hundred and seventy-five dollars as fees, to this agent, and to the Patent Office, made a visit to Washington, D. C., and called on the chief of police in respect to his patent business, and requested him to refer him to an honest, reliable and capable patent counsel or solicitor. Being given the name of a reputable house in Washington, he visited the same, and on entering the door he said, "I am referred by the chief of police to you as the kind of patent solicitor I am seeking. I do not want a medal awarded me, for my medal has cost me $175, and no patent has been granted me. I want an honest, reliable attorney who, when he takes my case, and I pay him my money, I can go home and feel satisfied that all will be done squarely, and I shall get a patent for my invention from the United States Patent Office, instead of a mere medal from my agent." The experience of this inventor ought to be a warning to others, and the course that he pursued should be followed by them.

Some years ago an adverstisement appeared in the papers as follows: "Wanted—An invention for sawing stone to a taper form; $5000 reward offered for the best invention of the kind for this purpose."

In response to this announcement, made no doubt by some designing, hungry patent agent, in conspiracy with an outside accomplice, for the purpose of increasing his income, several hundred inventors sent models of stone sawing machines to the Patent Office for patents. Nearly every one of these models represented two saws set to form an acute angle, and as the saws descended, cut the stone to a taper form. One agent filed so many applications in the United States Patent Office, all like one another, that the principal exminer of the patent office in charge of this class finally became disgusted with such proceedings on the part of this agent, and wrote a letter to each of the later applicants substantially in these words: "Your application for a patent on a machine for sawing stone to a taper form has been examined and rejected on application of A. B.; C. D. and E. F. filed through the same agency that has your case in charge." This was a sockdologer to the agent, and an eye-opener to his clients.

At the termination of the period set for awarding the $5,000 prize offered for the best stone sawing machine, these expectant inventors carried their models of stone sawing machine to a place designated in Vermont, and alas! on exposing them to the supposed generous citizen who had advertised for the invention, were told that none of the plans were as good as one which he had made himself, and therefore the prize, would not be forthcoming. Sad hearted, and disappointed, they returned home with an experience which ought to last a lifetime.

By this trap inventors were lead to expend thousands of dollars for models, traveling expenses, and agency and government fees, with no profit to themselves, simply benefiting an unscrupulous patent agent and his accomplices.

Inventors ought to look carefully before they bite at such bait.

Another trap set for patentees is the one that the "INVENTIVE AGE" has for many months been warning patentees against. This trap is the patent right selling agent who sends to every patentee a letter which says: "Your patent has been examined by our scientific board or corps of mechanical experts, and it has been pronounced to be worth $25,000 or $50,000 or $100,000. We would like to have the agency for selling your patent." Furthermore offers are made to take out foreign patents on already issued United States patents for one half the usual fees, etc. It is only necessary to say that patents in many foreign countries, for United States patented inventions, which have been published in the United States Patent Office Gazette fully enough to be understood by practical mechanics, are invalid.

The cloth of the old Egyptians was so good that, though it has been used for thousands of years as wrappings of the mummies, the Arabs of to-day can wear it. It is all of linen, the ancient Egyptians considering wool unclean.

Sir William Turner has compiled a table which shows that a whale of fifty tons weight exerts 145 horse-power in swimming twelve miles an hour.

NEW INVENTIONS.

A Brief Description of Some of the Most Valuable New Ideas.

ARM-REST FOR TELEGRAPH OPERATORS.

To those who are not familiar with the work of a telegrapher it may seem an easy matter to operate the key since there is so little apparent effort made by the person who turns the electric fluid loose. But if one will sit at an instrument all day at work, it will be found that the strain on the arm and fingers of the operator makes this work something more than child's play.

Earl L. Brownson and Henry W. Goodro of Hinesburg, Vt. have patented an arm-rest for making the work of the telegrapher less arduous. The rest is a frame or shoe with a flat base having triangu-

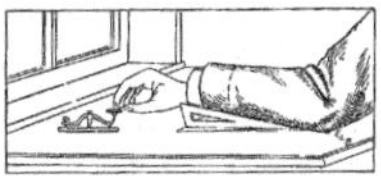

lar sides mounted on the base, converging forwardly, and inclining outwardly. An inclined cushion for supporting the arm is secured to the sides of the frame and suspended so as to form a curved channel that diminishes toward the rear end. The cushion is adjusted to support the wrist of the operator above and clear of the base of the frame and on a line with the telegraph key, allowing a free use of hand and wrist, but preventing movement of the muscles above the wrist that are so frequently affected with paralysis, thus preventing that disease, and effecting a cure in those who have it.

BOAT PROPELLER.

There have been many inventions for getting speed out of the ordinary row boat, other than those in which the back-straining process is employed. But the most recent, and one that has merit, is that for which a patent has been given to Charles D. Augur of Turin, N. Y.

In this the boat is propelled by side wheels consisting of paddle blades attached to a shaft that have their bearings in journal boxes on the gunwales. At the inner end of the shafts are wheels with hand-holds for turning, arranged so that a person can sit either before or behind them while engaged in circular motion. One man can work both wheels or two men can operate them at the same time.

A boat equipped after this fashion has its advantages, for it does away with oars, which, when long and heavy, require much labor in handling, take up room in the boat, and in un-experienced hands often endanger the lives of a boat's occupants.

MATCH BOX.

William C. Bower, of Birmingham, Ala., has patented a match receptacle, which effectually prevents the injury to walls and wood-work that is so often done by those who scratch their matches on the most convenient substance. And not only does it do this, but it keeps the matches in position so that they are not apt to become ignited, as sometimes happens when they are left in promiscuous heaps or packed in close receptacles.

This new match-box sits on a pedestal having beveled edges to which the box is hinged. Across the top of the box are a series of bars with corrugated edges, between which the watch is inserted from underneath, with its fire end downwards. When a light is desired it is only necessary to pull the match out, as its head coming in contact with the corrugated bars, will ignite from the friction consequent upon the process.

TOOTH BRUSH.

The spreading disposition of the tooth brush and its disagreeable willingness to part with its bristles when in use, is known to all. A tooth brush attachment useful in this and other respects has been patented by Llewellyn W. Jones of Allegheny, Pa., This brush has a rubber band surrounding the bristles, and is detachably fastened to the brush-head by loops passing under it. The rubber band, which has openings at its bottom edge, rises to near the tops of the bristles, thus holding them in position and preventing too much spreading, while affording a convenient means for retaining paste or powder, which in the ordinary brush is very rapidly scattered, loosing much of its intended effect.

The Surrounding rubber being flexible allows the bristles to move freely in any direction when in use and in a six months trial has absolutely prevented the

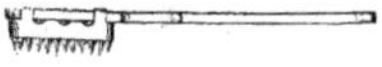

bristles from being worn or broken off close to the head of the brush thus increasing the life of the brush several hundred per cent, the opening in the band where it fits the head of the brush allows all liquid to drain out and a free circulation of air through the bristle to thoroughly dry them thus preventing the brush from becoming foul or the bristle from cutting out.

STRING CUTTER.

The cutting of leather strings may seem a simple thing to the casual observer, something that would hardly require anything but the simplest kind of machinery. Yet when the great quantity of strings used, their evenness, pliability and strength, is considered, it will be seen that a good string cutter plays an important part in the productive field. A useful apparatus of this kind recently patented by William Kootz of Milwaukee, Wis., is one which has a series of cutting knives with guiding fingers at their outer ends for engaging and guiding slitted strips of leather to the knives in straight continuation. The guiding fingers are attached to a comb, extending from one of its sides, and on the opposite side are arranged projecting pins. The comb is adopted to be applied to the outer ends of the cutting knives.

CARPET STRETCHER.

The putting down of carpets is a work which usually tries the patience and not infrequently records the results of such work in sore hands and aching joints. There is always a great deal of pulling and stretching required to get the floor covering in proper shape, and when this is done there is many a slip 'twixt the tack and strip before the latter is securely fastened in place. To make this disagreeable labor easier, Warren T. Reaser of Altoona, Pa., has invented and patented a carpet stretcher which is convenient and simple. It consists of a band strap, on the under side of which is secured a pair of spaced hook-plates containing fowardly curving hooks. When in use the hand is inserted into the strap and forced forward after the hooks have engaged the carpet. The apparatus has no saw-like teeth to tear the carpet, as is the case with some stretchers; it is light and easily applied, and should be a welcome addition to the things that make the work of the housewife lighter.

FOR CHECKING HORSES.

Joseph A. Mullen of New York, N. Y. is the inventor of a device for checking runaway horses, which has in it that which should recommend it to those who enjoy the pleasure and profit of wheeled vehicles.

This horse check is arranged so that by the operation of a lever in connection with suitable gearing located under the seat, a toothed drum working with a sleeve, is moved to the inner side of the wheel hub, which contains corresponding teeth, into which the toothed drum is locked. This effectually prevents further motion of the wheel and brings the whole weight resistence of the vehicle on the horse.

The frequency of runaways and their often disasterous consequences, should have long before this brought forth a good apparatus for their prevention. In crowded cities the danger from uncontrollable horses is always very great, and as many of the latter are driven by people of little muscular strength, a check that will quickly and securely lock wheels is not only of value, but is a necessity.

MECHANICAL MOVEMENTS.

A useful invention is that for which a patent has been recently granted William Livingstone of Flushing, N. Y., whose genius has produced an engine attachment for overcoming the inertia or machinery when it has reached the dead-point. This consists in part, of an auxiliary crank in combination with the main crank. The differential crank is operated independently of the main crank, so that when the engine approaches the dead-point it is kept in action by its auxiliary.

The value of this is at once apparent, for not only is time and power saved by continuous movement of the engine but in ships the unbroken action of the machinery will be assured, which is of extreme importance when there is danger of collision, running aground, etc.

BICYCLE CANOPY.

When the invention of George C. Omerod, of Asbury Park, N. J., is brought into public use the bicycle will present a different appearance from that which it now has. This addition to the wheel, is a canopy, and is supported by telescoping uprights attached to the handle-bar. At the top of the uprights is the canopy frame extending from the former back and front and bent so that the ends of the frame rods meet at the upper ends of the standards. The canopy is arranged to fold and can be adjusted at any angle to suit the rider.

With a covering of this kind the wheelman can travel without fear of rain or sun. It should commend itself especially to lady lovers of the wheel, who care so much about their complexions, for with a canopy over them they can defy old Sol, whose rays have no respect for any color but tan.

MARINE AND LAND VELOCIPEDE.

There have been a great many things patented wherein were included several uses to which the invention could be put. But it remained for Charles L. Rhone of Brooklyn, N. Y., to invent a velocipede that can be run on land or water. This dual machine is made with pararlelled buoyant tubes, having transverse braces with a saddle mounted on one of them and a paddle wheel mounted in another. A sprocket is attached to the paddle wheel and is geared to a sproketed treadle shaft by a chain. A rudder and steering devise is added for water use, so that the marine cycle can plow the raging main, or a main that doesn't rage, without fear of loosing its course. For running on land the velocipede has a wheel at "bow and stern."

An apparatus of this kind should be popular wherever there is good riding ground by river or lake, for when in such places one could launch on the cool water when tired of the shore, and by carrying fishing tackle along, the pleasure of that sport could be added to the outing.

FOR HOLDING RUBBER HEELS.

Albert N. Barrett of Los Angeles, Cal., has patented a device for securing rubber heels to boots or shoes. The invention consists of a rubber heel having a horizontal slit, and a retaining plate for the heel, which surrounds the latter and holds it in place by means of a vertical flange having teeth projecting inwardly for securing the heel. The plate only, is attached to the sole; the heel is removable. An arrangement of this kind is of use for perserving the shoe heel as well as for preventing the jar to the foot when walking, which follows when hard-heeled shoes are worn.

SHOE FASTENING.

The difficulty of keeping shoe strings tied is an annoyance which afflicts almost every one who uses them; and the proneness of shoe buttons to suddenly depart is well known to those who prefer that kind of footwear. There have been a number of inventions gotten up for shoe fastening, among them that recently patented by William H. Benford of Lamar, Mo., which does away with both buttons and strings. In this fastening an elastic band and hooks are used. The band is interlaced through eyes at one side of the shoe instep to form loops, and on the other side are the hooks, which engage the loops when the shoe top is closed This makes a very convenient and secure fastening, and by employing an elastic band the leather over the instep is left less rigid upon the foot than it would be if cord or leather strings were used.

TROUSERS CREASER.

It is the earnest desire of the men who dress well, to keep their pantaloons creased and to keep the "pants" from the least suspicion of bagging at the knees. To do this one must have quite a number of pairs of trousers, or remain pretty much in a perpendicular position, unless he employs a good creaser.

An apparatus for creasing trousers has been patented by Edward J. Boyd of New York, N. Y., whose invention will no doubt be of service to the lovers of neat leg wear. In this there is a center board having mortises through it, and outer boards with tongues that fit into the mortises. The pantaloons to be creased are put between these boards arranged so that the front edge of the former can be pressed by the tongue board. When this is done, clips are applied for pressure.

SAFETY POCKET.

Despite the fact that so much has been spoken and written by the joker concerning the difficulty of finding a woman's pocket, one of the fair sex has invented and patented a safety pocket. The pocket can be used for frocks, bloomers or pantaloons, will prevent things from falling from it, and will make it less easy for the pick pocket to ply his trade. The invention is by Charlotte M. Johnson of Charleston, W. Va., and consists of a pocket made of a single sheath of flexible material surrounded by a constricting, unattached elastic band, which is held in place by flexible longitudinal keepers.

When the hand is inserted into this pocket, the band expands, but the pocket immediately resumes its hour glass shape when the hand is withdrawn.

PEN-WIPER—AND ORANGE PULP REMOVER.

As the wheel of progress moves on, the field of invention is being entered more and more by women. While usually not turning their attention to the more complicated arrangements of art and science, the fair sex produce many ideas of practical use. Two of this sort are a pen-wiper and an orange pulp remover, the inventions of Elizabeth G. Bouton of Pittsfield, Mass. The ordinary pen-wiper is a simple thing, easily made, but is soon soaked with ink, torn and unfit for use. The one here mentioned is composed of elastic material, having jaws for clasping and taking the ink from the pen. The jaws are worked by pressure applied to the body of the pen-wiper, and contain absorbent material.

The pulp remover is a simple, but very convenient apparatus for quickly and nicely removing the edible parts of an orange.

COW-TAIL HOLDER.

A cow's tail is not usually considered a very important part of that animal; nor is it something to which inventive genius has largely turned its attention. Yet there is much "business" in a cow's tail—especially in "fly-time"—that makes it an object to be looked after when the bovine is to be milked. The patented invention of Lorenzo D. Corser of Bridgeton, Me., is intended to put an effective check on the switching propensity of the cow, by securing her tail in a circular metal band attached to a device by which the whole arrangement is secured to the animal.

A NEW PUZZLE.

Junius H. Flint of Salem, Mass., has patented a new puzzle. This consists of a rectangular board, provided with flanges, and pockets arranged at the corners and near the middle of the board. The corner pockets are made to face each other while those at the center face lengthwise. The openings in the pockets are of different sizes made for receiving balls of corresponding dimensions.

Many puzzles have been fruitful sources of income to their inventors, for there is something very taking with the public in anything in this line that is difficult to perform or that possesses in any degree an element of mechanical mystery.

THILL-COUPLING.

There have been a number of thill-couplings invented, but there is room for a good device of this kind. The thill-coupling recently patented by William A. Gowen, of Portland, Me., possesses that which should make it acceptable to the users of vehicles, and it should find a ready market In this invention there is a clip having a forwardly extending flange, a keeper for the clips, coupling links, and a vertical coupling bar, which passes through the flange, coupling links and keeper and is provided with a tightening nut. The coupling links have on their adjacent faces comical projections in combination with a thill iron with comical rockets in their sides for receiving the projections.

NEW RAILROAD SPIKE.

The danger accompanying the use of poor railroad spikes is evident. To the employment of poor material has no doubt been justly attributed the cause of many accidents. If a spike has a poor holding capacity there is liability to draw from the tie and thus cause the rail to become loose, allowing it to spread, which, when this is the case, means a wrecked train.

A railroad spike, recently patented by William S. McClay of Uniontown, Pa., has some improvements which should recommend it to the managers of railways. This spike has a head convexed on top and formed with an offsetting lip and lateral ribs. Its sides are tapering, with the penetrating point in the rear line passing centrally and longitudinally through the spike. The front side of the point is of a longer bevel than the rear side. The latter has a longitudinal depression curving inwardly between its extremities, forming an anchoring point, in proximate relation to the spike point, and when diven into the tie said spike cannot work loose. The shape of the spike is such that in driving it will draw towards the foot of the rail, and take firm hold of same thereby avoiding the necessity of driving the head of the spike over until it takes hold of the rail, and leaving a hole back of the spike to take in water causing the spike to corrode and the tie to rot. This spike fills the hole made by drivings and does not leave any hole to absorb water. Said spike has an extra strong head and tee shaped lip, and extra strong lateral ribs along the head for drawing purposes.

ELECTRIC CIGAR LIGHTER.

Considering the numerous applications of electricity for power, heating, lighting, etc., it is a wonder that some one has not before this used it in connection with cigar lighting. This idea has now been carried out by Augustus C. Gruhlke of Waterloo, Ind., and William F. Kessler of Auburn, in the same state. The cigar lighter is a tube containing an oil receptacle gas chamber, an insulated plate, contact piece, a coiled spring and conducting wires. The insulated plate is secured to the tube, with which one of the wires is connected, and the oil receptacle is electrically connected with the other wire. When a light is wanted, the pivoted contact is pressed, and the mysterious current performs its duty. This patent was assigned to Mr. Kessler, one of its inventors.

CLOD-CRUSHER.

The value of an effective apparatus for crushing clods is well known to farmers who have to deal with stiff soil that persists in turning up in lumps, which must be broken before the harrow can smooth them for seed planting.

Andrew J. Harlow of Sparksville, Ind, has patented a clod crusher, which from its admirable construction will no doubt be well received by those who look to the sod for their prosperity. This machine consists principally of a beam provided with fixed teeth arranged at intervals; a drum rotating on the beam, and arms—having bent outer portions to form shovels, and straight inner portion for rotating between the teeth mounted on the drum.

The patent for this invention has been assigned to William A. Holland, of Fort Ritner, Ind.

COMBINATION GUN-ROD.

Chauncey M. Powers and William E. Surface, of Decatur, Ill., have invented and patented a gun-cleaning rod that should be a welcome addition to the outfit of both sportsman and soldier. The rod has attached to its handle-end a combination of screw driver and oil can arranged in compact and convenient form. There is always the necessity for keeping a gun well oiled, especially in wet weather, and as a hunter does not often wish to carry a separate oil can with him, the little oil-holding attachment is a thing to be desired. The screw-driver, for taking the gun apart when it must be cleaned, could not be dispensed with.

MUSTACHE GUARD.

For men whose upper lips are covered with a hirsute adornment, a drinking cup has been invented, and recently patented. This cup has a trefoil-shaped bridge over its top, and a circular oval shield with notched edges depending from the bridge into the cup. In the bridge is a round hole for receiving the end of the drinker's nose when the cup is tilted. With a cup of this kind a man can drink without carrying away on his mustache the flavor and perfume of the liquid that passed under it.

The patent for this invention was issued to Abram Vandersall of Columbus, Ohio.

BALE TIE.

Benjamin Adams of Charleston, S. C., has had a patent issued to him for a bale tie, something, which with its other uses, is of especial value in cotton-baling. The tie is of metal having a frame or buckle to which one of its ends is attached by passing it through the buckle and folding it back. The free end of the tie is also passed through the buckle when it is to be secured, and a sliding band slips over the folded end of the tie for clamping it and securing the buckle. The sliding band carries a hook adapted to engage perforations in the tire and can be adjusted to any degree of tightness.

CIGAR MOISTENER.

Those who love to burn the fragrant weed (tobacco) and to be soothed by the influence of good cigars, will find in the invention recently patented by Henry T. Sidway, of Chicago, Ill., a useful adjunct to a smoker's outfit. It is a cigar moistener, and consists mainly of a shell made in form of a cigar, having a perforated chamber for holding absorbent material, and another for holding liquids. The chambers are divided by a partition in which is an aperture containing a wick that takes moisture from the liquid chamber to the absorbent material.

Cigars kept in receptacle of this kind should retain their flavor much longer than if allowed to become dry. And this idea holds good in regard to chewing tobacco, for an apparatus of the same nature could be used for plug and fine cut, keeping them in the moist condition so much desired by those who chew.

ACCELERATING CARTRIDGE.

In these days of wars and rumors of wars, inventive genius is turning in no small degree towards the production of effective killing apparatus; for the materilization of an idea which is useful to governments for war purposes, is apt to bring the inventor much profit.

Among the patents for destructive devices recently issued is that of Harris P. Hurst of Summit, Miss. This invention is a cartridge, having secured to its base a tube for holding a primary charge an accelerating charge out side the tube. The tube contains an independent body attachable to the projectile between the latter and the primary charge, seperating the same from the main charge until the projectile is in motion from the force of the explosive in the tube.

The advantage of this arrangement is apparent, for by it quick and complete combustion is obtained and greater force is developed.

FAN ATTACHMENT FOR ROCKING CHAIRS.

Thomas Cunniff of Brookline, Mass., has been issued a patent for a fan attachment for rocking chairs, which will give those who love to rock the pleasure of keeping cool while so doing. A chair equipped with this apparatus has at its sides standards joined at the top by a cross piece, on which revolvers a number of fans, directly over the person rocking. The power is supplied of course, by the motion of the chair, and is transmitted to the fan shaft by running gear attached to the rockers and operating in connection with pulley, springs, pawls and ratchers.

This invention is a very convenient utilization of waste-power, and no doubt will find its way into public favor. The rocking chair is used for its comfortable qualities, and the more comfort one can get from it, the better it is liked.

THIMBLE.

To the many thousands who use the sewing needle, the patented invention of Uriah A. Knauss, of Bethlehem, Pa., will come as a welcome friend. It is only a thimble attachment, but one that is intended as a help to fingers that too often suffer from tugging at the needle. In this the thimble is provided with a needle grasping clamp secured to its outer wall, and comprises a resilient thumb-piece movable to and from the surface of the thimble, and a portion fastened to the latter, between which the needle is grasped. This makes it unnecessay for the fingers to come in contact with the needle, which, after it has been pushed partly through the material by the thimble, is caught with the clasping device and easily drawn forward.

SKIRT PROTECTOR.

Deborah Owen, of of VanWert, Iowa, has had a patent granted her for a skirt protector. In this there is an overskirt, comprising back, front and side breadths, with the side breadths united, with the side breadths gathered and hemmed at their bottoms and provided with elastic cords in the hems. The front and back portions of the skirt are brought together and fastened at the bottom, and the side breadths are secured at and around the ankle.

A skirt-protector of this kind should be well received by women who go much on the street in inclement weather, for it insures protection against both water and mud, and is conducive to health.

HEATING BY ELECTRICITY.

The problem of heating houses conveniently and economically, is one that has been engaging the attention of the public for a long time. The furnace (hot air), steam and hot water, have their drawbacks, and the stove is expensive, troublesome and obsolete. It seems that the best heating apparatus of the future will be of the electric kind, for it represents the minimum of work, the maximum of convenience, and the only point to be met concerning it, is that of cheapness. A move in the right direction is the patented invention of Charles H. Minchew of Taunton, Mass., who has produced an electric heater. In this there is a rectangular metalic box provided in the ends with a series of openings and filled with a non-conducting compound. There are a series of flanged insular buttons removably fitted in the openings in the ends of the box; supporting hooks, having their shanks removably and adjustably fitted in the inner ends of the buttons, and a heating wire, looped at the ends of the box over a pair of the supporting hooks, and provided with a series of parallel, longitudinal heating coils embedded in the non-conducting substance.

BOTTLE COOLER.

An apparatus for cooling bottles is a recently patented invention of Louis P. Bachand of Southbridge, Mass. In this the bottles are placed in a case or box having a removable ice tray situated in the bottom of the box. There is also a tray for the reception of bottles, mounted upon cleats secured to the front and back of the case above the ice chamber, and containing compartments for each bottle. The tray is perforated so that the cold air can circulate freely around the bottles. An arrangement of this kind should be welcome to householders particularly, for it does away with the necessity for putting bottles in the family refrigerator with the various vegetables, meats, etc., and is portable, economical and convenient.

MEMORANDUM BOOK.

Alexander J. Gallager of Monroe, Pa,, and Gustav Langrien of Westville, N. J., have invented and patented a memorandum book, which recommends itself to the many who find those useful records things of daily necessity. This book has a cover with extended flaps at its outer end, one of the flaps being flexible and the other conparatively stiff. The flaps are fastened on their under sides, and by extending over the ends and corners of the book prevent it from becoming dog-eared, while preserving the writing from being rubbed, which would be the case in a loose cover.

MACHINE FOR MINING.

A mining machine is a new invention for which a patent has been issued to James E. Lee, David A. Lee, aud Thomas E. Lee, of Centerville, Iowa. This machine runs on a track, and is operated by means of motor. With gear wheels and other attachments, there is connected a cutter comprising a core and a continuous spirally-coiled band wound around the core, arranged flat against it, and provided at its front edge with outwardly-projecting teeth, formed integral with the band and arranged at intervals at each coil. These teeth lie beyond the outer face of the band and have front cutting edges and rear edges convexly curved.

It is evident that a mining machine supplied with a convenient motor power and means for easy conveyance, is of vast superiority over the pick and muscle usually employed. In the former it is only necessary to put the cutters in operation against the material to be broken into and the work is carried on with dispatch.

PENHOLDER FINGER REST.

For those who do a great deal of writing, the finger rest patented by John T. Ahrens of Wilmington, Del., will be of much service. This invention consists of a skeleton rubber sleeve having a flange with angular facets at its lower end, and thin flat ribs extending along, and projecting from the sleeve.

The finger rest not only affords relief to the writer's fingers, helping to prevent cramps that so often accompany prolonged use of the pen, but it also prevents the ink from smearing the penstaff and consequently the writer's hand. It should be a welcome addition to household and office.

STRAW HARPOON.

One would hardly look for a harpoon on a farm, the usual place for those instruments being the homes of men who catch whales and other large inhabitants of the sea. But John A. McGreevey, of Burrows, Ind., has patented a harpoon to be used on a farm, for pulling straw. This straw puller resembles very much the harpoon used in killing whales. It has a head with concave edges and diverging going barbs, and when plunged into a pile of straw will engage it more effectually than could be done with the old fashioned wood or iron hook or the straw-knife. This invention should meet with the approval of all farmers, for it facilitates work, and that is of main importance to those who till the soil.

DRYING CLOSET.

A closet for drying clothes, etc., is the patented invention of John McGlone, of St. Louis, Mo. This dryer has a central hot-air chamber and heating apparatus; drying chambers, cold air inlet, escape flue, and fan for circulating the air. There is no doubt of the usefulness of an apparatus of this nature. The quick and effective drying of clothes is a matter of much importance to the housewife, who often must wait for fair weather or else utilize the kitchen for drying purposes. The latter method is usually very inconvenient and anything that will obviate this necessity should be a boom.

A NEW CHURN.

The churn is something which enters largely into the economy of the farm household. It not only provides the butter for the farmer's bread, but adds no inconsiderable portion to the income of those who till the soil.

William Sparling, of Little Rock, Ark., is the patentee of a churn, in which the advantageous power of the lever is employed. The lever is fulcrumbed in an upright, in connection with a dasher rod, making a convenient and easily operated arrangement, whereby much force can be exerted with little labor.

MEASURING THE THICKNESS OF LEATHER.

Marshall Tidd, of Woburn, Mass., has been granted a patent for a machine for measuring the thickness of leather. This consists in part of a frame of horizontal parallel bars set one above the other; a shaft containing a coiled spring; a numbered indicating plate and an indicating pointing-arm. When a slab of leather is put under the upright shaft, the latter bears upward against the pointing arm, which follows the scale up or down as the inequalities in the leather come under the shaft. While good for measuring leather this apparatus can be used for showing irregularities in the surfaces of other materials.

LOCKING DEVICE FOR RAIL JOINTS.

There are many devices for fastening the joints of rails, but there is always room for a good improvement. There is so much dependent on the security of the rail—which upholds such ponderous weight—that anything that helps in this direction should be doubly welcome. Le Anderson, of Paris, Texas, has had a patent given him for a locking device for rail joints, which consists mainly of a locking finish plate with key hole bolt slots, the narrow portion of the slots having inclined sides. Double headed bolts are used, one of the heads being tapered, whereby the bolt is subjected to a double wedging action. A perforated face-plate receives the tapered heads, which are secured by rivets that are passed into the fish plates.

DRAWING SLATE.

A drawing slate is the newly patented invention of Henry A. Mishler, of Mechanicsburg, Pa. In this there is a transparent writing surface and a frame having longitudinal slots, in which are slide-carrying rollers, in combination with rollers or rods journaled in bearings secured to the frame. A flexible belt, working beneath the transparent surface carries views in such a manner that they can be observed and reproduced by the draughtsman.

This slate will be of service for practical work, and will prove a source of pleasure and instruction in home and school.

MAGNET FOR HOLDING IRON AND STEEL.

The practical use of the magnet is daily becoming more and more apparent. It has already been applied to a number of things, among them the lifting of heavy pieces of steel and iron in foundries. The latest idea in this connection is embraced in the patent allowed Oakley S. Walker, of Worcester, Mass. The invention is a magnetic chuck for holding iron or steel while it is being machined or otherwise manipulated. A magnetizing coil is used, an inner core forming one pole of the magnet, and an outer shell forming the other pole. There is an air-gap between the two poles, and the meeting faces or edges of the poles are formed in zigzag points. An adjustable magnetic strip, the edges of which form a raised surface from the chuck face, extends the action of the magnetic lines at an angle to the face of the chuck.

An apparatus that will hold on securely to a piece of steel or iron simply by being brought in contact with it, is a big improvement over tongs and other grasping tools, and it will not be long before its value is fully recognized. The bright day of electricity has been with us for sometime, we will now begin to get well acquainted with its sister, the magnet.

LAMP-FLUE HOLDER.

Chas. P. Johnson, Eureka, Ks., has a patent for an apparatus for holding lamp chimneys. For this wire is used, making a ring, which encircles the upper portion of the flue and extending downward on two sides to the burner, which it engages. The ends of the wire forming the ring are bent to form eyes which embrace slidingly the body portion of the ring, making it adjustable to any size flue. Besides this the side wires are coiled near their middle to form springs that bear against the large part of the chimney.

It is impossible for a lamp chimney supplied with this invention, to fall off, and considering the liability to this accident, the new holder should have a wide and useful field.

PNEUMATIC TIRE.

John F. Seiberling, of Akron, Ohio, was recently granted a patent for an improvement in bicycle tires, which consists mainly of the usual rubber tube, and a continuous tube being a segment of a circle in cross section. This fits inside the first tube near its bottom, and has a metallic plate on its inner flat side and a series of webs extending from its curved face upward to the metallic plate. The latter contains a rubber diaphragm which is joined to the walls of the tube on both sides of the plate. With this tire the danger from puncture is reduced to a minimum, for although the tire may be cut, yet the air above the plate will still be held securely.

PORTABLE SAWING MACHINE.

A machine for sawing rail road ties in the track, is the lately patented invention of Jacob B. Whitfield, of New Brunswick, N. J. In this there is a base piece, from which rises three uprights, two of them carrying a transverse shaft, on which is mounted a band wheel. From this wheel extends a link band to a shaft that carries a circular saw. The latter is at the end of two swinging parallel supports and is operated by a crank leaver in connection with the band and pulleys. For security the base piece is attachable to the rail by means of a suitable clamp.

It seems that though this saw was invented for cutting off stationery rail road ties, it could be used to advantage for various other useful purposes.

The frame projects beyond the end of the base plate so that the saw can drop down in between the rails when necessary. Heretofore this work has been done by the ordinary slow and inconvenient process of the cross-cut saw, and the invention of Mr. Whitfield overcomes all the objections of the old fashioned methods. The sawing of switch timbers at angles is practically an impossibility by means of the cross-cut saw, but the new appliance will cut off the end of the timber or the tie at any angle as easily as straight through. One man with the machine can do the same work that it now requires two men to perform, and in a third less time, without taking into consideration the saving of time in not having to remove the ballast around the ties.

A flanged pulley will be furnished with each machine so that it may be attached to a hoisting machine by belt and run by steam if desired to saw heavy bridge timber. This is one of the most useful inventions of the year.

NEW PRINTING PLATE.

The inventions in type setting and printing apparatus heretofore have always been made with a view to the use of type metal. But now Alfred Patek, of New York, N. Y., has come forward with a patent for printing directly from prepared fibrous material.

In the process for this a matrix is first formed with the characters depressed in it. Then sheets of paper are united with paste to form a pad, which is treated with a non absorbent compound and forced into the matrix. After the pad has taken its shape from the matrix, it is dried, hardened, and is ready for the press.

This is quite a departure from the present method of reproducing type forms, and if practicable will bring its inventor much of this world's goods.

PORTABLE FURNACE.

William T. Underwood and Leroy C. Underwood, of Itasca, Tex., have invented and patented a furnace or camp stove that can be carried from place to place with ease, and is useful for cooking and heating. The stove is a hollow cylindrical body provided with a door opening and a series of inwardly projecting studs for grate-rests arranged midway between top and bottom. The grate, to be used when food is to be cooked, is removable, so that when the stove is required for heating alone, large fuel can be used.

Those who have camped out, hunting, etc., know the comfort of a good fire and the pleasures of a good cooking arrangement.

HAT PIN.

Since the hat pin is prone to work itself loose and leave a lady's head gear, which has very little grasp in the crown, if any, free to wander in the wind, it is strange that some one has not before this provided a means for its security. To meet this necessity, Malley H. Coltharp, of Tallulah, La., has invented and patented a hat pin that will no doubt gain the thanks and dollars of the fair sex generally.

In this invention the pin is bent at one end to form a supplemental spring which extends obliquely to the shaft and is arranged to engage the hat in such a manner that it (the pin) cannot pull out.

A NEW SCRUBBER.

The inventor who devotes his attention to productions that concern the cleaning work of a household, has a large field from which to reap his harvest. And anything that will clean, and clean well, carries its own recommendation.

In floor scrubbing the mop and brush have had and still have extensive use. But there is no reason why an improvement in this line should not meet with a big success. The scrubber recently patented by Joseph S. Dunham of Camden, N. J., will make a bid for public favor, which it should receive, for it is practical, easy to manage, and is adapted to "get down to business" in a thorough manner. This brush has a channeled head of elastic material with transverse braces in the wall of the head and a longitudinal bar on the crown of the former. There is little to get out of order in it, and being of elastic material—in its scrubbing portion—it should require less water and clean quicker than a hair brush.

CAR-JACK.

There have been a number of lifting devices patented, which do their work more or less well, but the recent invention in this line, the work of Emery E. Taylor, of Minneapolis, Minn., possesses some advantages over those of his predecessors. Mr. Emery's patent is for a car-jack with a device connected thereto consisting of two intermeshing serrated members, one pivotally connected to the jack and the other co-operating with it in varying positions of adjustment. When in operation a spring holds this in position and the car wheel is held down while the journal-box is lifted. There is so much lifting in connection with railroad cars, that it is easy to see the commercial value of a good jack.

A NEW GATE.

Gates are more liable to get out of order than any other portion of the fence, due of course to the frequent swinging in and out. We seldom find a large gate especially the kind used on a farm, that swings easily and without considerable effort on the part of opener, such gates are usually either sagged or otherwise out of shape.

Elisha G. Holder, of Marquez, Tex., has patented a gate that has in it much that should recommend it for convenience and wear. It is of the rotary pattern, with latch posts at the ends and spring latches on the ends of the gate for engaging the posts. Levers are fulcrumed on the top of the gate, connected with each other; connecting rods join the levers with the spring latches, and vibratory braces rise from the sides of the gate and are secured to the levers.

MAGNETIC PEN-HOLDER.

It is strange that the retaining power of the magnet has not been brought more into general use than it has. There are many things to which this, however, could be applied for doing away with mechanism and simplifying construction.

A useful invention in this line has recently been patented by Herman Moyer of Fort Worth, Texas, and consists of a base-plate for receiving ink wells and magnetic rack for pen-holders. The rack is a metal band, having a series of grooves for the pen-holders, and extends from a support curving downward and outward toward the ink wells. At the ends of the magnet are non-conductors, which prevent the loss of power. When a pen-holder is put on this rack, it sticks there without the aid of clamps or springs, and as the magnet can be utilized for pen points and other small metallic articles, its usefulness is apparent.

OVER-HEAD BICYCLE.

Another adaption of the bicycle has been brought forth. It is an aerial wheel for over-head tracks, and was patented by Hiram B. Nickerson, of Stoughton, Mass. The wheels of this bicycle are grooved for embracing the top of the track, and are connected by a rigid tie at their axles. From the axle of the larger wheel a seat support extends downward, curving rearwardly, and forwardly at its lower end, where is located the propelling sprocket wheel. From this an operating band extends upward to the axle of the large wheel, which is put in motion by the operator, who sits upon a saddle attached to the lower portion of the seat support.

An arrangement of this nature should prove a great source of pleasure and a splendid means for exercise. There is no doubt that much speed can be gotten from it, and this being so there is no reason why it should not have a commercial value since rapid transit is something always desired and much sought for. It is possible to have a passenger carrying attachment to this aerial wheel, and it could be used in transmitting mails in places where there are no railroad facilities.

The patent for this invention has been assigned to the Aerial Bicycle Co. of Maine.

EGG TESTER.

Heretofore the average human has found out the condition of eggs only after they were broken and ready to be eaten. But now an invention has been brought forward, whereby the bad egg can be shown up in an entirely satisfactory manner without having its covering broken. A patent for this has been granted to John L. Ritter, of Shenandoah, Va., and was assigned to James G. Whitlock and Sallie McAdams of Richmond, Va.

The egg tester is a cabinent containing a tray for holding the eggs, which are so placed that the light from a lamp and reflector can shine upon them from beneath the tray. The intensified light makes the eggshells almost transparent, so that any change in the egg can be marked with accuracy. As there are millions of eggs used in this country alone, an apparatus for testing their fitness for use will no doubt find a ready and wide market.

BICYCLE STAND.

Harvey N. Timms, of DesMoines, Iowa, has been granted a patent for a movable stand for bicycles. In this is a U-shaped horizontal cross piece with wheels on its ends, an upright fixed to its central portion and a clamp at the top designed to be secured to the rear braces of a bicycle frame between the crank-bracket and rear wheel. Braces are attached to the end portions of the cross-piece to engage the axle of the rear wheel.

Of course the advantages of this apparatus are that the wheel can be moved at will to any desirable place; and when it is left at one's dwelling, the ease with which it can be pushed about out of the way without falling adds much to the value of the invention. To the unitiated the bicycle is a cranky affair, to be handled gingerly, especially by the weaker sex.

CAN-OPENER.

They who have labored with a can opener, forcing the cutter through the obstinate tin and often receiving wounds from its ragged edges, will appreciate the invention of Lyman W. Merriam of Fitchburg, Mass. to whom a patent has been recently issued.

This patent was for a can opener consisting of a wire embeded in the top of the can and adjusted so that by withdrawing the wire the central portion of the can top will be opened. This is a very convenient arrangement, one that will save labor and render a quick and satisfactory service.

TROLLEY FINDER.

Although the trolley system has been so much abused on account of the infrequent disastrous effects of its charged wires, it seems to have come to remain with us as a means of street car power. One difficulty in operating the trolley system has been the liability of the trolley wheel to slip from the wire, causing frequent delays. This trouble however finds a remedy in the recently patented invention of Harry H. Blanchard of Augusta, Me., whose trolley finder seems to fill the bill. The apparatus consists principally of two flaring arms situated on either side of the trolley wheel and extending outward and upward so as to make it impossible for the current wire to be missed when the "finders" are raised. When the arms are to be raised an operating cord is pulled and the former extend their shafts encompassing the wire and causing it to glide at once to the trolley wheel. When the arms are not in use, a spring keeps them in a downward position.

IRONING BOARD.

Clara J. Cramer, of Griswold, Iowa, has been granted a patent for an ironing board for sleeves or garments of a like nature. The board is concavo-convex in longitudinal section, having straight end portions, pins in the end, and uprights for supporting the ends of the board.

A sleeve is connected with each support to receive one end of the board, and contains notches in its inner edge and yielding pressure devices for forcing it forward on the board.

Ironing sleeves, as well as many other things that cannot be ironed flat, requires much care, and, for perfect work, there should be special apparatus employed for some garments. For the work for which it is designed the ironer mentioned here seems all that could be desired.

Worthless Patents.

Among those who have been attracted by the prize offering of John Wedderburn & Co., patent attorneys, is Mr. David Coswell, of North Fork, Pa. He received the monthly prize for "the best and most practical invention," etc. Of course he thought he had something possessing great merit and he forthwith requested the American Machinest to publish an engraving and description, which the publishers did for the purpose of showing how utterly worthless the patent was, and concludes its criticism as follows:

"We point out these facts only because it seems to be a case easily understood and typical of a great number of patents that are taken out, that are utterly worthless from the mechanical and the legal as well as from the commercial point of view. The worthlessness of such patents is by no means always the fault of the attorney, however. Many people demand patents upon their inventions, and must have them. In many cases it is impossible to get good patents, and then worthless ones are obtained to supply the demand. It is, however, equally true that worthless patents are in many cases obtained where it would be just as easy to get good ones, if the attorney were competent and honest. Some attorneys think it their duty to advise against application for patents which they know will be worthless; others do not believe this is any part of a patent attorney's duty, and that his sole duty is to go ahead and get the best patent possible. One of the former class would have advised against this application."

Ninety-five Miles an Hour.

In these columns has been described an invention of Wm. J. Holman of Minneapolis, Minn., calculated to increase the speed of a railway locomotive. In the recent test on the tracks of the New Jersey railroad a speed of 94 1-2 miles was attained in a run of 12 miles. The engine drew a tender and two ordinary passengers coaches. The inventor expects to be able, under the most favorable track conditions, to reach a speed of 120 miles an hour. The trucks on this locomotive are interchangeable and made up of five smooth or friction geared wheels placed under each driver. Three of them rest on the track, while the other two bear upon the former, and each driver in turn rests upon the two.

The arrangement, the inventor thinks, affords natural pockets for all wheels above those on the track, and makes it an impossibility for any of them to become displaced during any kind of running. The track wheels are held in position by the side bars, which hinge around the center middle wheel. The hinging renders the trucks flexible and provides for the engine a sort of endless track of moving wheels, which are always smooth. It is said that the vibrating motion commonly imparted to the locomotive in high running is done away with by this device.

Each truck wheel has a large and small diameter, and works on the cog principle, but friction takes the place of the cogs. The larger rim rests upon the rail and the small rim extends outwardly over the tires.

The death of Sir John Pender, one of the men who made ocean telegraphy an accomplished fact, occurred on the 7th inst. at Manchester, Eng. He was 80 years old.

According to the San Francisco *Journal of Electricity*, a wave-motor installed in a bay at Capitola is running in good order and is developing as high as 180 horse-power.

A press report is authority for the statement that a household ice machine has been invented that can be had for $15 and with which every family can make ice at home at a cost of two dollars for the season.

The capital invested in the beer-brewing business in this country amounts to $250,000,000, according to statements made at the National convention of brewers held recently; and sales of beer amounted to $36,000,000 last year against $8,500,000 in the year 1876.

On June 15 a decision was rendered in the New Jersey Court of Errors and Appeals in regard to the right of corporations to set poles in front of private properties without the consent of the owners. The decision answers the question in the affirmative.

Progress of Invention During the Past 50 Years.

PRIZE ESSAY BY "BETA" (EDWARD W. BYRN, A.M.)

[Receiving the $250 Prize and Published in the 50th anniversary Number of Scientific American.]

If the life of man be threescore years and ten, fifty years will about mark the span of ripe manhood's busy labor, and the sage of today, turning back the pages of memory, may, as the times pass in review, enjoy the rare privilege of of personal observation of, direct contact and positive knowledge concerning the events of this prolific period. To him what a vista it must present; what a convergence of the perspective; for the past fifty years represents an epoch of invention and progress unique in the history of the world. It is something more than a merely normal growth or natural development. It has been a gigantic tidal wave of human ingenuity and resource, so stupendous in its magnitude, so complex in its diversity, so profound in its thought, so fruitful in its wealth, so beneficent in its results, that the mind is strained and embarrassed in its effort to expand to a full appreciation of it. Indeed, the period seems a grand climax of discovery, rather than an increment of growth. It has been a splendid, brilliant campaign of brains and energy, rising to the highest achievement amid the most fertile resources, and conducted by the strongest and best equipment of modern thought and modern strength.

The great works of the ancients are in the main mere monuments of the patient manual labor of myriads of workers, and can only rank with the buildings of the diatom and coral insect. Not so with modern achievement. This last half century has been peculiarly an age of ideas and conservation of energy, materialized in practical embodiment as labor-saving inventions, often the product of a single mind, and partaking of the sacred quality of creation.

The old word of creation is, that God breathed into the clay the breath of life. In the new world of invention mind has breathed into matter, and a new and expanding creation unfolds itself. The speculative philosophy of the past is but a too empty consolation for short-lived, busy man, and, seeing with the eye of science the possibilities of matter, he has touched it with the divine breath of thought and made a new world.

It is so easy to lose sight of the wonderful when once familiar with it, that we usually fail to give the full measure of positive appreciation to the great things of this great age. They burst upon our vision at first like flashing meteors; we marvel at them for a little while, and then we accept them as facts, which soon become so commonplace and so fused into the common life as to be only noticed by their omission.

Perhaps, then, it will serve a better purpose to contrast the present conditions with those existing fifty years ago. Reverse the engine of progress, and let us run fifty years into the past, and practically we have taken from us the telegraph, the sewing machine, the bicycle the reaper and vulcanized rubber goods. We are no telephone, no cable nor electric railways, no electric light, no photo-engraving, no photo-lithographing nor snapshot camera, no gas engine, no web perfecting printing press, no practical woodworking machinery nor great furniture stores, no passenger elevator, no asphalt pavement, no steam fire engine, no triple expansion steam engine, no Giffard injector, no celluloid, no barbed wire fence, no time lock for safes, no self-binding harvester, no oil nor gas wells, no ice machines nor cold storage. We lose the phonograph and graphophone, air engines, stem winding watches, cash registers and cash carriers, the great suspension bridges, iron frame buildings, monitors and heavy ironclads, revolvers, torpedoes, magazine guns and Gatling guns, linotype machines, all practical typewriters, all pasteurizing, knowledge of microbes or disease germs, and sanitary plumbing, water gas, soda water fountains, air brakes, coal tar dyes and medicines, nitro-glycerine, dynamite and guncotton, dynamo electric machines, aluminum ware, electric locomotives, Bessemer steel, with its wonderful developments, ocean cables, etc. The negative conditions of that period extend into such an appalling void that we stop short, shrinking from the thought of what it would mean to modern civilization to eliminate from its life these potent factors of its existence.

As the issue of patents in this country is based upon novelty, it will aid us in the effort to appreciate this great movement to note the increase of United States patents in the past fifty years. Beginning in 1846, and dividing the time into periods of five years, the increase is shown most graphically in the scaled diagram No. 1.

If the growth of United States patents and the progress of the last half century can be taken as fairly correlated, what an insignificant thing is the little attenuated triangle back of 1846 compared with the swelling curves of the later period! It is probably safe to say that fully nine-tenths of all the material riches and physical comforts of to day have grown into existence in the past fifty years.

It is interesting to observe how closely the grant of patents and the prosperity of the country are related. Referring to scaled diagram No. 2, the zigzag line marks the increase or decrease in the patents issued from year to year. We note the depression of the civil war, followed by the rapid reaction and growth of reconstruction. Again, the depression caused by the financial panic of 1873, and again in 1876, the unsettled and dangerous conditions of politics incident to the contested presidential election. This was followed by another wave of prosperity, indented with depressions in the presidential election years, while the stringency of the times from 1890 to 1894 shows a marked influence in the corresponding depression in the line, all of which indicates a most sympathetic relation.

Passing now to the chronological development of the period, Morse had just harnessed the most elusive steed of all Nature's forces, and put it in the permanent service of man; when nitro-glycerine, discovered by Sobrero, in 1846, for the first time lent its terrible emphasis, and seemed to bring an awakening of the dormant genius of man.

Within the first decade (1846–1856) came the sew-

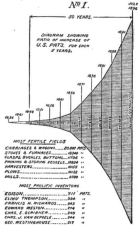

Nᵒ 1.

50 YEARS.

JULY 1896

DIAGRAM SHOWING RATIO OF INCREASE OF U.S. PATS. FOR EACH 5 YEARS.

MOST FERTILE FIELDS

CARRIAGES & WAGONS......20,000 PATS.
STOVES & FURNACES.........15,340 "
CLASPS, BUCKLES, BUTTONS...11724 "
PACKING & STORING VESSELS..10654 "
HARVESTERS................10125 "
PLOWS....................10132 "
MILLS....................9720 "

MOST PROLIFIC INVENTORS

EDISON..................711 PATS.
ELIHU THOMPSON..........334 "
FRANCIS H. RICHARDS.....343 "
EDWARD WESTON...........294 "
CHAS. E. SCRIBNER.......248 "
CHAS. J. VAN DEPOELE....244 "
GEO. WESTINGHOUSE.......217 "

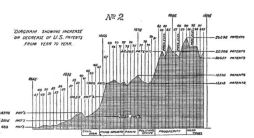

Nᵒ 2.

DIAGRAM SHOWING INCREASE OR DECREASE OF U.S. PATENTS FROM YEAR TO YEAR.

ing machine, Bain's chemical telegraph, the Suez Canal, the House printing telegraph, the McCormick reaper, the discovery of the planet Neptune, the Corliss engine, the collodion and dry plate processes in photography, the Ruhmkorff coil, the Bass time lock for safes, the electric fire alarm of Channing & Farmer, Gintle's duplex telegraph, the sleeping car of Woodruff, Wilson's four-motioned feed for the sewing machine, Ericsson's hot air engine, the Niagara suspension bridge, and the building of the Great Eastern.

The next decade (1856–1866) brought with it the Atlantic cable, the discovery of the aniline dyes by Perkin, the making of paper pulp from wood, the discovery of coal oil in the United States, the invention of the circular knitting machine, the Giffard injector, for supplying feed water to steam boilers; the discovery of caesium, rubidium, indium and thallium; the McKay shoe sewing machine, Ericsson's ironclad monitor, Nobel's explosive gelatine, the Whitehead torpedo, and the first embodiment of the fundamental principles of the dynamo electric generator by Hjorth, of Denmark.

The next decade (1866–1876) marks the beginning of the most remarkable period of activity and development in the history of the world. The perfection of the dynamo, and its twin brother the electric motor, by Wilde, Siemens, Wheatstone, Varley, Farmer, Gramme, Brush, Weston, Edison, Thomson, and others, soon brought the great development of the electric light and electric railways. Then appeared the Bessemer process of making steel; dynamite; the St. Louis bridge; the Westinghouse air brake; and the middlings purifying and roller processes in milling. That great chemist and probably greatest public benefactor, Louis Pasteur, added his work to this period; the Gatling gun appeared; great developments were made in ice machines and cold storage equipments; machines for making barbed wire fences; compressed air rock drills and the Mont Cenis tunnel; pressed glassware; Stearns duplex telegraph, and Edison's quadruplex; the cable car system of Hallidie, and the Janney car coupler; the self-binding reaper and harvester; the tempering of steel wire and springs by electricity; the Lowe process for making water gas; cash carriers for stores; and machines for making tin cans.

With the next decade (1876–1886) there arose a star of the first magnitude in the constellation of inventions. The railway and telegraph had already made all people near neighbors, but it remained for the Bell telephone to establish the close kinship of one great talkative family, in constant intercourse, the tiny wire, sentient and responsive to the familiar voice, transmitting the message with tone and accent unchanged by the thousands of miles of distance between. Then come in order the hydraulic dredges, and Mississippi jetties of Eads; the Jablochkoff electric candle; photography by electric light; the cigarette machine; the Otto gas engine; the great improvement and development of the type writer; the casting of chilled car wheels; the Birkenhead and Rabbeth spinning spindles; and enameled sheet iron ware for the kitchen. Next the phonograph of Edison appears, litterally speaking for itself, and reproducing human speech and all sounds with startling fidelity. Who can tell what stores of interesting and instructive knowledge would be in our possession if the phonograph had appeared in the ages of the past, and its records had been preserved.

The voices of our dead ancestors, of Demosthenes and Cicero, and even of Christ himself speaking as he spake unto the multitude, would be an enduring reality and a precious legacy. In this decade we also find the first electric railway operated in Berlin; the development of the storage battery; welding metals by electricity; passenger elevators; the construction of the Brooklyn bridge; the synthetic production of many useful medicines, dyes, and antiseptics, from the coaltar products; and the Cowels process for manufacturing aluminum.

In the last decade (1886–1896) inventions in such great numbers and yet of such importance have appeared that selection seems impossible without doing injustice to the others. The graphophone; the Pullman and Wagner railway cars and vestibuled trains; the Harvey process of annealing armor plates; artificial silk from pyroxyline; automobile or horseless carriages; the Zalinski dynamite gun; the Mergenthaler linotype machine, moulding and setting its own type, a whole line at a time, and doing the work of four compositors; the Welsbach gas burner; the Krag-Jorgensen rifle; Prof. Langley's aerodrome; the manufacture of acetylene gas from calcium carbide; the discovery of argon; the application of the cathode rays in photography by Roentgen; Edison's fluoroscope for seeing with the cathode rays; Tesla's discoveries in electricity, and the kinetoscope, are some of the modern inventions which still interest and engage the attention of the world, while the great development in photography, and of the Web perfecting printing press, the typewriter, the modern bicycle, and the cash register is beyond enumeration or adequate comment.

Looking at this campaign of progress from an anthropological and geographical standpoint, it is interesting to note who are its agents and what its scene of action. It will be found that almost entirely the field lies in a little belt of the civilized world between the 30th and 50th parallels of latitude of the western hemisphere and between the 40th and 60th parallels of the western part of the eastern hemisphere, and the work of a relatively small number of the Caucasion race under the benign influences of a Christian civilization.

Remembering, furthermore, that most of this great development is of American authorship, does it not appear plain that all this marvelous growth has some correlation that teaches an important lesson? Why should this mighty wave of civilization set in at such a recent period, and more notably in our own land, when there have been so many nations far in advance of us in point of age? The answer is to be found in the beneficent institutions of our comparatively new and free country, whose laws have been made to justly regard the inventor as a public benefactor, and the wisdom of which policy is demonstrated by the growth of this period, amply proving that invention and civilization stand correlated—invention the cause and civilization the effect.

This retrospect, necessarily cursory and superficial, brings to view sufficient of the great inventions as mile stones on the great roadway of progress to inspire us with emotions of wonder and admiration at the resourceful and dominant spirit of man. Delving into the secret recesses of the earth, he has tapped the hidden supplies of Nature's fuel, has invaded her treasure house of gold and silver, robbed Mother Earth of her hoarded stores, and possessed himself of her family record, finding on the pages of geology sixty millions of years existence. Peering into the invisible little world, the infinite secrets of microcosm have yielded their fruitful and potent knowledge of bacteria and cell growth. With telescope and spectroscope he has climbed into limitless space above, and defined the size, distance and constitution of a star millions of miles away. The lightning is made his swift messenger, and thought flashes in submarine depths around the world, the voice travels faster than the wind, dead matter is made to speak, the invisible has been revealed, the powers of Niagara are harnessed to do his will, and all of Nature's forces have been made his constant servants in attendance. We witness a new heaven and a new earth, contemplation of which becomes oppressive with the magnitude and grandeur of the spectacle, and involuntarily we find ourselves asking the question, "Is it all done? Is the work finished? Is the field of invention exhausted?" It does seem that it is quite impossible to again equal the great inventions of this wonderfully prolific epoch; but as these great inventions, which now seem commonplace to us, would have seemed quite impossible to our ancestors, we may indulge the hope of future possibilities beyond any present conception, but onward and upward in the great evolution of human destiny.

Rejoicing in our strength and capabilities, the new light of man's power and destiny breaks more clearly over us, and content with the infinite quality of mind and matter, the teachings of philosophy, and the facts of evolution, we rest in the assurance of positive knowledge that all that has been done in the past is merely preliminary, that human ingenuity knows no limit, and so long as man himself remains hedged about with the limitations of mortality and the conditions of growth, so long will his strivings and attainments be infinite.

BETA.

Machinery Compared with Muscular Power.

Speaking of prime movers before the Association for the Advancement of Science, at London several years ago, Sir Frederick Bramell drew an interesting picture of the puny thing that muscular power, whether animal or human, really was when compared with the vast efforts exerted nowadays by machinery. Contrasting a galley, for example—a vessel propelled by oars—with a modern Atlantic liner, and assuming that prime movers were nonexistent and that this vessel was to be propelled after galley fashion, he proceeded thus: Take the the length of the vessel as 600 feet, and assume that place could be found for as many as 400 oars on each side, each oar is worked by three men, or 2,400 men, and allow that six men under these conditions could develop work equal to one horse-power. We should then have 400 horse-power. Double the number of men and we should have 800 horse-power, with 4,800 men at work, and at least the same number in reserve if the journey is to be carried on continuously. Contrast the puny result thus obtained with 19,500 horse-power given forth by a large prime mover of the present day, such a power requiring, on the above mode of calculation, 117,000 men at work and 117,000 in reserve, and these to be carried in a vessel less than 600 feet in length. Even if it were possible to carry this number of men in such a vessel, by no conceivable means could their power be utilized so as to impart to it a speed of 20 knots an hour.

This illustrates how a prime mover may not only be a mere substitute for muscular work, but may afford the means of attaining an end that could not by any possibility be attained by muscular exertion, no matter what money was expended or what galley-slave suffering was inflicted. Take again the case of a railway locomotive, in which we have from 400 to 600 horse-power developed in an implement which, even including its tender its tender, does not occupy an area of more than 50 square yards and that can draw us at 60 miles an hour. Here again the prime mover succeeds in doing that which no expenditure of money or of life could enable us to obtained from muscular effort.—*Cassier's Magazine.*

The Third Rail Electric System.

There are three principal methods of conducting electricity along a railway route so that the current may be taken off to operate motors on the cars. One is to string a copper wire overhead, and make contact therewith by means of a metallic wheel at the end of a pole, rising from the car roof. Another is to bury the wire in a conduit, and thrust some sort of a hanger or shoe down into the latter from the car. The third plan is to lay a third line of iron or steel rails, from which the current can be taken off by a sliding device, very much like that used in a conduit. The extra rail is cheap and very convenient, but until within a few weeks it has been employed only upon elevated railways.

It is customary, nowadays, to have the return current go back to the powerhouse through the rails on which the car rides. If, therefore, some one were to lay an iron bar across from the supply rail to one of these other rails, the current would leap across the bridge and sneak home by the easiest route, instead of going up through the motor and doing some work. As the electricians would say, the bar would "short circuit" the current. If the connection between the supply and return rails be made by a horse or man, instead of an inanimate metal bar he would receive a severe shock, especially if the parts coming in contact with the rails were wet. For this reason, it has been deemed unwise to employ the third rail down at street level in cities. Up on an elvated structure there would be little chance of an accident. But an experiment is now being made with the third rail on a surface road; a road orignally operated by steam. And the chief object of the test is to see whether the plan is practicable under such circumstances. If so, our great electric long-distance railways may, in the near future, dispense with the overhead wire and trolley altogether.

Last summer, on a branch of the Old Colony road southeast of Boston, known as the "Nantasket Beach Line," passenger cars were run for several miles by eleetricity, and the familar trolley system was employed. At the present time an extension of this same branch (Bostonward) is being operated by the third rail. When the cars get to the point where the old overhead wire is; the latter is used instead. As will be readily seen, a railway track running out through sparsely settled regions and fenced in on either side like any ordinary steam line is not very likely to be the scene of an accident from a "short circuit." Men and animals have no business on the tracks, except at regular cossings. On the Nantasket Beach line, this third rail is laid between the other two, instead of outside, as in Chicago. It is supported upon a narrow wooden stringer, and is shaped something like the letter A. The overhang on either side provides a dry undersurface, even in a rainstorm. In this way leakage of electricity is guarded against. The result of the Nantasket experiment this summer will be closely watched by railroad men.

It is said the first article made of aluminum was a baby-rattle, intended for the infant Prince Imperial of France in 1856.

Major McKinley heard the last day's proceedings of the Republican convention over the long distance telephone at his home in Canton, Ohio.

The Bank of England has 1,600 officials on its rolls, and 1,000 clerks. If a clerk is late three minutes he receives a warning; the fourth time he is discharged at once.

The whaleback John Ericsson has been launched at the yards of the American Steel Co., at Superior, It is 204 feet long between perpendiculars, has a molded breadth of 48 feet and a depth of 27 feet at the center.

For a fee of from two to eight cents a message, one may talk from even the smallest of Swiss towns over a long-distance telephone system to any part of the country. The instruments are kept in perfect repair and the service is said to be excellent.

There are nearly 90,000 barmaids in England. More than 1,000 in London are daughters of gentlemen; 400 have fathers, brothers or uncles in the church; 200 are daughters of army officers, 200 daughters of physicians and surgeons, 100 daughters of navy officers.

The Modern Sky Scraper.

The INVENTIVE AGE has in former issues illustrated and described the manner of building the modern so-called fire-proof tall office buildings that now adorn our large cities. Through the courtesy of "Brick" we herewith present an illustration of the completed Fisher building, Chicago—a twenty story building which for rapidity as well as thoroughness of construction has no equal. Fifteen stories of this building were completed in fifteen days, the work being continued nights by the use of electric lights. In speaking of tall, steel-frame office buildings in general and the Fisher building in particular, Brick says:

"In regard to fire-resisting properties, no very conclusive tests have been made, principally because these high office buildings are very carefully looked after, the tenants are a superior class of merchants and companies, manufacturing operations are not permitted, nor the storage of combustible materials.

The two important points, stability and resistance to fire, command the most attention and interest in connection with high office buildings. Stability has not yet been subjected to the severest test.

It seems, however, reasonable to believe that the steel skeleton, or bolted frame construction, would withstand a cyclone better than any other construction. The St. Louis disaster has given a great number of instances of walls falling outwards and not blown in. This supports the theroy that the chief mischief was caused by the vacuum produced by the sweep of the wind. In such case the steel-frame building, with its steel-frame floors, each one bolted to the network of upright columns, would be the strongest construction that exists.

The cost of the buiding was about $675,000 and the value of the 7,000 square feet of ground on which it stands $700,000. It is estimated that the rentals will net 6½ per cent on the total valuation as above. The earlier buildings of this kind cost, complete, 40 cents per cubic foot.

Everything has been done inside that could make the building attractive to tenants. An important novelty is a third water supply. Commonly there are two sets of water-pipes one for hot the other for cold water, but here we find, over each lavatory basin, three taps, from one we draw hot water, from the other ordinary cold water and from the third, filtered, ice-cold drinking water.

Another novelty is that a system of pneumatic clocks has been installed. An air-pump in the basement, which is controlled by a master clock, sends an impulse every half minute to every part of the building. The arrangement is very simple, the air-pump and air reservoir, are in the basement and pipes are carried to each floor, these were put in at the same time as all the other hundreds of pipes and wires, which are distributed through the building like nerves through a human body. The master clock has on its seconds-hand arbor a small cam. This, every thirty seconds, lifts a lever, which opens a valve allowing the air to pass through the system of pipes, arrived at the clock it lifts a small diaphragm which is connected to a pawl and ratchet that drive the hands of the clock, so that every ½alf minute the hand moves. The pneumatic timepieces are the property of a company, which charges a rent of 50 cents per month per timepiece. The equipment is heated by steam and the total length of steam piping is about 3 1-2 miles.

Can You Invent?

A self-locking hat pin.

A machine to put a mourning border on stationery.

A pocket match that can be relighted, and that will not blow out in the wind.

A frame to slip over or clasp on an ordinary razor to make it a "safety" razor.

A method of making kerosene oil oderless without injuring its lighting qualities.

A fan attachment for baby carriage operated by sproket and chain from wheel hub.

A method of turning cooking or baking tins in an oven without opening the oven door.

A can opener that works with a crank, one turn of the crank revolving a cutter around the top of the can.

A magazine toy pistol in which the strip of percusive fulminate revolves on a wheel, so that the act of cocking the pistol throws a cap under the hammer.

A device to fasten to a table or chair in a restaurant to hold a customer's hat and cane or umbrella while dining.

A speed recorder or governor to attach to bicycles that will indicate continously the degree of speed at which they are running, or that will ring a bell when going faster than the "regulation" speed.

THE FISHER BUILDING, CHICAGO.

Chicago's High Tower.

The Chicago scheme for the errection of a tall steel tower has not been abandoned, although it has dropped out of sight for some time. Anonncement is now made that the City Tower Company have purchased a site on the west side of the city, consisting of a block 350 by 600 feet, bounded by Throop, Loomis, Harrison and Congress streets. The Metropolitan Elevated Railroad runs so close to it that a station can be built at the entrance to the tower park. The tower is to be built on the plan designed by D. R. Proctor, who is also the promoter of the company. It is to reach 1150 feet from the ground, this being nearly 200 feet higher than the Eiffel tower. The equipment contemplated will comprise 35 elevators and the tower will have seven landings, the lowest to be 235 feet from the ground.

It has been supposed that nothing could take the place of ivory in the maufacture of the sensitively elastic balls used in playing billards. But the scarcity of ivory has set inventive wits at work, and now in Sweeden hollow balls of cast steel are found to be a satisfactory substitute.

A Word of Caution to Inventors.

For years the INVENTIVE AGE has waged war on the numerous swindling patent selling agencies and as a result scores of these frauds, being unable to withstand the X rays of investigation, have quietly folded their tents and gone out of business. There are still scores of them left and the AGE is glad to see that reputable patent attorneys take up the matter with their clients and warn them of the pitfalls. This subject is cleverly handled in the handsomely engraved thirty-fifth anniversary guide book recently issued by the well known firm of Mason, Fenwick & Lawrence, of Washington, D. C., under thecaption of "A Word of Caution," as follows :

"We are not in the patent selling business, though we frequently are employed to, and do attend to drawing agreements for sales and licenses, etc., for our clients, but it has become customary for patent selling agencies, often of irresponsible character, to flood the country with inquiries in regard to the price of the lately issued patents and requests for the right to sell them on more or less reasonable terms. These inquiries suggest that the parties inquiring have observed the probable great value of the patent referred to, but neglect to state that the same inquiries and statements by them are made by them simultaneously in regard to every patent issued from the Patent Office. Many of these agents are totally unreliable, and in justice to our clients we advise them to be very cautious in signing any contract with such partics or paying them money.

If you are an inventor and wish to apply for a patent, we would advise that you exhibit the invention to one of your own townsmen, and agree to give him an interest in the invention and patent upon his furnishing the necessary money for securing a patent.

If you are a patentee, get up an attractive illustrated circular clearly describing and exhibiting the invention, and send the same to manufacturers in the United States engaged in manufacturing your class of invention, offering the invention for sale outright or on royalty—don't refuse a good offer when you get one. Hundreds of inventors have missed opportunities to dispose of their inventions, simply because they had exaggerated imaginary ideas of the real value of their invention. Another good plan for a patentee would be to organize a local company among reliable men, and get the article or machine on the market in this way."

The $1 Edison Motor.

This cut shown herewith represents one of the most satisfactory and perfect of all small dynamos, capable of running toys, fans, light models and for

innumerable uses. To the youth scientifically inclined, it offers great satisfaction and benefit.

It is a marvel in price, and if not made in large quantities would cost from $5 to $10. We can furnish the motor for $1 and will send one free to any person forwarding us three new subscribers,

Minute Workmanship.

In the twentieth year of Queen Elizabeth, says an English contemporary, a blacksmith named Mark Scaliot made a lock consisting of eleven pieces of iron, steel and brass, all of which, together with the key to it, weighed but one grain of gold, consisting of forty three links, and having fastened this to the before mentioned lock and key, he put the chain around the neck of a flea, which drew them all with ease. All these together, chain and flea, weighed only one grain and a half.

Oswaldus Northingerus, who was more famous even than Scaliot for his minute contrivances, is said to have made 1600 dishes of turned ivory, all perfect and complete in every part, yet so small, thin and slender, that all of them were included at once in a cup turned out of a pepercorn of the common size. Johannes Shad, of Mitelbeach, carried this wonderful work with him to Rome, and showed it to Pope Paul V, who saw and counted them all by the help of a pair of spectacles. They were so little as to be almost invisible to the eye.

Johannes Ferrarius, a Jesuit, had in his possession cannons of wood, with their carriages, wheels and other military furniture, all of which were also contained in a peppercorn of the ordinary size.

An artist named Claudius Gallus made for Hippolytus d' Este, Cardinal of Ferara, representations of sundry birds setting on the tops of trees, which by hydraulic art, and secret conveyance of water through the trunks and branches of the trees, were made to sing and clap their wings, but at the sudden appearance of an owl out of a bush of the same artifice, they immediately became all mute and silent.

Dr. Oliver gives an account of a cherry stone on which were carved 124 heads, so distinctly that the naked eye could distinguish those belonging to popes and kings by their miters and crowns. A Nuremburg top maker inclosed in a cherrystone, which was exhibited at the French Crystal Palace, a plan of Sebastopol, a railway station, and the "Messiah" of Klopstock.

Myrmecide wrought out of ivory a chariot with four wheels and as many horses in so little room that a small fly might cover them all with her wings.

Peter Bayle, a clerk of chancery in the time of Queen Elizabeth, once wrote the Lord's Prayer, the creed, the commandments, two prayers, and his own name and office in addition to the year, month, and day of the Queen's reign, in characters so small as to be enclosed "in the head of a ring" which ring was afterward accepted by the Queen, and was worn on the august finger.

We have Pliny's statement that in his time there existed a copy of Homer's "Iliad" small enough to go into a nutshell, and a German, Professor Schreiber, produced only a few years back, by the stereographic process, a copy of the German translation, extending to 600 pages, of both "Iliad" and the "Odyssey," so small that a nutshell held the whole comfortably.

Water Raising Wheel.

At the last meeting of the Agricultural Society of France, Mr. H. de Coursac exhibited a new type of water raising wheel, used by himself for irrigation, adapted only for flowing waters. The wheel is of very simple construction, and consists of a number of flat paddles, to the outer edge of which cylindrical troughs are attached. The cylinders lie horizontal; the open end is bent; from the closed end a fine syphon tube starts, which is bent back in a plane at right angles to the cylinder. When the paddle with its cylinder dips under the water, the air escapes through the syphon. Emerging from the river, the water is held by the syphon arrangement until the open end is just above the distributing conduit. The wheel which the inventor has in use at his Chateau de la Planche has a diameter of 10 ft; each of the 12 paddles is provided with a trough which holds 13 litres, about 11 quarts. 500 cubic metres, more than 600 cubic yards, of water can be raised in 24 hours. The lift would be a little less than the diameter of the wheel.

Bullets Swerved by Electrical Currents.

The Journal de Geneve is authority for the statement that at some recent trials of the Swiss Federal Rifle meeting in practice shooting, curious deviations were noted in the results of the shots, the cause being finally attributed to the electrical conductors paralleling the range. In order to corroborate this supposition the authorities established parallel with the range and at a distance of about 120 feet, four steel cables carrying heavey currents. It was found that for a range of 260 yards the bullet was laterally deviated about 70 feet. Beyond this range the deviation became still more apparent, and when using artillery and a range of 3000 yards the deviation amounted to 14 degrees. It would be difficult to prophesy the final bearing which these experiments may produce in the art of future warfare.

Unpatented Inventions.

If you look back on the history of human progress you will find that none of the great epoch-making inventions has ever been patented, says London Truth. The man who lit the first fire—whether Prometheus or the party from whom he stole the idea—did not get a patent for it. Neither did the man who made the first wheel—in every sense one of the most revolutionary inventions in the history of man. The same thing may be said of the invention of soap, candles, gunpowder, umbrellas, and the mariner's compass; or, to come down to our own day, of the steam engine and the electric telegraph. Patents are mostly concerned with small mechanical details and improvements—it may be in candles or umbrellas, or it may be in the application of steam and electricity—and by means of these patents enormous profits have been secured to second-rate inventors; but the great ideas and discoveries which underlies these details have been given to the world gratis. There is a general notion that if you did not protect inventions by means of patents inventors would cease to invent and material progress would come to a standstill. But history does not bear this out in the least. Men with mechanical gifts do not exercise them solely with a view to commercial profit, any more than astronomers search the heavens for new worlds with an eye to registering patents and floating companies on the results of their discoveries.

Compressed Air Motors.

In the last issue of the INVENTIVE AGE appeared an article on compressed air as applied to the Belt Line street railway in Washington. The following from Age of Steel will be of interest in this connection:

The Worcester "Gazette" tells of the progress of the work on the new system of street cars propelled by compressed air which is going on in that city. It is known as the high pressure system. The cars run in Europe have a pressure of a little over 400 pounds, while on these it is proposed to have a high pressure of 2000 pounds to the square inch on the cylinders. This would reduce the size of the cylinders and make it possible to entirely conceal them under the seats. In European lines there are huge tubes carried under the cars, which give an ungainly appearance, add to the weight and increase the chance of accident. Thus far compressed air cars have been charged sufficiently for runs of about 8 miles, while it is proposed to give the storage in these a capacity to run 20 miles without refilling. The charging will only take a minute or two, and it is proposed to have compressed air conduits along the track, with hydrants at intervals, so the cylinders can be charged by stopping a moment on the route. The regulator and brake are to be separated as in trolley cars, and will be of such a simple pattern, that an ordinary motorman can manipulate them. Instead of using the compressed air for the brakes, as in previous systems, it is proposed to reverse the power when necessary to stop. The motor operated by compressed air follows the steam engine more than an electric motor, but the impulse is given by springs which are squeezed by the air and free themselves. The promoters are satisfied that the element of danger has been eliminated by constructing tough cylinders.

A Chicago paper states that Frank A. Hecht, of Chicago, has completed arrangements for the funds necessary to erect a $2,000,000 hotel in Washington. Eastern and Western capital is interested, and the structure is to be located on the site of the present Willard Hotel. It will be twelve stories high, fire-proof, granite and marble, 800 rooms; have theatre, promenade and roof garden attached to accommodate 1,500 people; will have a rotunda 65x185 feet, costing $200,000; work to be commenced in 1897.

The Illinois appellate court for the fourth district has handed down an opinion declaring that where a secret society expels a member, the courts may inquire into the expulsion and see whether it is just.

Secretary Morton states that 30,000 acres of timber is consumed every day in the United States, in factories, railroads, fences, farms, and buildings.

There are 160,000 bicycles in use in France, each of which pays a tax of $1.93. A great many American machines are imported.

To the Pole by Balloon.

There are features of novel interest in the latest effort to reach the north pole. The explorer who is now claiming attention is Prof. Andree, and, although he proposes to make the trip by balloon, it cannot be said that his plans are any more impracticable than those of many of his predecessors. When last heard from Andree was on Danish island, off the northwest coast of Spitzbergen, and he may already be started on his aerial journey. If not he is expected to be under way between the present date and Aug. 24.

His starting point, Danish island, is located very near the line of the 80th parallel and therefore is about 690 miles from the pole. This, of course, is a considerable distance south of the latitude attained by the more successful explorers, but it is a reasonable probability that if Andree makes any headway at all he will at least advance into the region of the highest known exploration. His direction, if winds favor, will be in a straight line due north, and in a balloon 690 miles is not a very protracted journey. He will be amply supplied with provisions, carrying in addition ice sledges and a small boat in case his mammoth balloon collapses and he has to resort to other means of travel.

The adventure seems quite as daring and facinating as are most of the polar explorations. Apparently, Prof. Andree will have the usual perilous lot of the pole hunters if his balloon comes to grief, but if he succeeds his success should be won with considerably less than the ordinary hardships. With favoring winds and no mishaps he might float across the pole and on into British North America, experiencing only a slight part of the perils and rigors met by those who go in boats and sledges to the artic zone.

The Utility of Inventions.

It is, no doubt, true that when a new invention is introduced which revolutionizes some particular art or branch of business, it at first decreases the number of persons employed in that particular line; but that is only temporary, for in a short time the result is a cheapening of the product, a greatly increased demand for it, because of this cheapening, and then necessarily an increased demand for laborers in that line, and almost universally at increased wages. The statistics show this to be true beyond the possibility of a question. The records of the labor bureau of the United States show this to be true beyond the possibility of a question. The records of the labor bureau of the United States show that from 1860 to 1880, the most prolific peroid of inventions, and the most intensified in all directions of their introduction, the population increased 59.51 per cent while in the same peroid the number of persons employed in all occupations manufacturing, agriculture, domestic service and everything—increased 109.87 per cent; and in the decade from 1870 to 1880 the population increased 30.08 per cent while the number of persons employed increased 30 per cent. As shown by the investigation of a committee of the United States Senate, wages have increased 61 per cent in the United States since 1860. And, as we all know, during that same period the cost to the people of nearly all manufactured articles has been decreased in as great if not a greater ratio.—*Canadian Journal of Fabrics.*

New England's Extensive Fishing Industry.

Few people realize, or even pause to think, concerning the vast importance of the New England fishing interests, and the vast amount of capital invested in them. It will be safe to call the capital invested in the fish business, $5,500,000, with 400 vessels engaged in the different branches of the business, the latter valued at $2,000,000. The wharves, buildings, etc., are valued at $1,250,000, and there is a working capital of some $2,250,000. The annual sales of salt fish amount to $3,590,000; of fresh fish, $1,000,000; glue, guano and isinglass, $500,000; fish oil and guano, $250,000; boxes and packages, $160,000. The production of boneless fish annually amounts to about $1,500,000. In the manufacture of glue 2,825,000 pounds of fish skins are used, and 7,062,500 pounds of waste are used for fertilizer. To carry on this business there are imported 871,432 bushels of salt.—*Bangor Industrial Journal.*

What is said to be the largest plate of glass ever produced in this country was turned out recently at the works of the Pittsburgh Plate Glass Company, Charleroi, Pa. It measured 148x244 inches.

The longest distance a letter can be carried within the limits of the United States is from Key West, Fla., to Ounalaska, 6,271 miles.

To Revive Played Out Oil Wells.

The result of a useful invention, which is just about to be placed on the market, will, it is claimed, be to create a new era in the history of oil and to start fresh life into the regions which have flourished in the past, but are now regarded as played out. It will turn the desert regions of the oil sections into bustling towns and will furnish an enormous supply of petroleum that would be sufficient of itself to do the work of the greater part of the world. The dead wells, it is claimed, will become as fountains and the springs of the deep will give up their wealth in an inexhaustible quantity.

This invention is the work of a Washington gentleman, Tapley W. Young—and consists of an electric heater, which can be lowered into the well.

The electricity is generated in large quantities and so powerful is the heat that the refuse matter, which clogs the pores of the stone, will be melted and run out, thus allowing the fresh upward flow of oil.

To understand the principles upon which this invention is founded it will be necessary to explain that the theories as to the exhaustion of so many wells is that the oil, in passing upward through the stone, has clogged the porous stone with paraffine in such quantities that the further flow is stopped and the well ceases to produce. Some think that it is because the supply in the earth has given out. The generally accepted idea is that the oil is still in abundance and only ceases to flow when its exit is stopped.

The stone through which oil passes is of a very porous nature, and as the liquid is in a crude state, the thick matter becomes as dregs, settling in the rock and at the edges of the bottom of the well. It has been common to use torpedoes to shatter the stone at the bottom of the well, thus breaking up the clogged matter, but this is an expensive process.

By the Young method the machine, which is about three feet long and resembles an iron cartridge, is placed at the bottom of the well, and the electricity turned on so powerfully that it receives just enough volts to produce an enormous heat without melting the metal. The current goes down with the wires, and by the peculiar construction of the carbon-packed chambers the intense heat is radiated about in the rock in all conditions. Thus the paraffine and other refuse is softened and melted up so that it runs, and when the well is started a fresh flow takes place just as strong as it did when the well was first sunk.

The invention made by Mr. Young is in the hands of the Standard Oil Company, and will be thoroughly treated. As that company owns so much oil land, the machine will be a source of saving of great wealth to them, as well as to the many hundreds of people all over the oil districts.—*Paint, Oil and Drug Review.*

Modern Facilities of Transportation.

In no more emphatic manner are modern facilities of transportation emphasized than in the safety with which perishable food is conveyed from a great distance. In this particular Australia, South America, and the United States are no further removed from Europe than a single province formerly was from the capital of the country of which it formed a part. Algeria is now supplying Paris markets with camel meat. An extensive plant has been created in that French colony for the killing and refrigerating of those animals, and daily shipments are made to Paris. The meat of the camel is described as not unlike beef, with the tenderness of veal. The hump is the choicest portion. Eggs that formerly were gathered near the localities where they were sold now come from distant points. Four million daily are received in London from foreign countries. Most of them come from Russia. They command in England twice the price they bring in the home market. The export of eggs from Russia, that in 1885 amounted to 235,000,000, increased in 1895 to 1,250,000,000. These are official figures. In addition, great quantities of dressed fowl are annually exported from Russia to all European countries.

Dr. S. A. Haguman, of Cincinnati, has applied for a patent on an invention which promises to create a sensation among musicians. It is a mechanism to be applied to the piano for the purpose of correcting the false tones of that instrument, which have hitherto been regarded as a necessary evil. The invention makes a piano as true as the violin, and it accomplishes the result by a sort of mechanical fingering similar to the human fingering on the violin. Competent musicians say the problem has been successfully solved. The inventor is a half brother of the late Professor David Swing, of Chicago.

Seeing by Wire.

Will it ever be possible to enable the persons who speak with each other by telephone to see one another at the same time, as "in a glass darkly," perhaps, but still "face to face?" Will it ever be feasible for a man in London to see opera in La Scala, or the Falls of Niagara, or the Feast of Lanterns, in Canton, without stirring from home? It is a captivating idea, and, although we cannot pronounce with certainty, there is a good deal to be said in favor of the possibility of its realization.

To begin with, it is known that light is merely a form of energy, or, as the late Prof. Tyndall would call it, a "mode of motion." It is, in fact, a wavelike motion in the exquisite medium that we call the luminiferous ether, which is understood to permeate all bodies. The waves resemble those set up in water when a stone is dropped into it; that is to say, they are transverse rays, the particles of water rising and falling alternately across the line on which the waves travel. In this respect light differs from sound, in which the particles of air conveying the sound vibrate to and fro along the course of propagation of the sound. Now it has been found of late years that waves similar to those of light in all but size can be set up in the luminiferous ether by oscillatory discharges of electricity, and there is growing evidence to show that some well-known effects of electricity are the result of wave motions in the ether of the same kind as those of light.

If, therefore, we could find a means of transforming the waves of light into corresponding electric waves and transmit these to a distance by wire, or even without wires, then retransform them, back again into light, the problem would be solved. The progress of electrical research appears to tend in that direction.

A Marvelous Churn.

Mr. W. C. Lott, of the Haim Bros. Manufacturing Company, of St. Louis, was in Memphis, recently and gave some interesting tests with his aerating, separating and fusifying churn. Butter was made from sweet cream in 2¼ minutes in one case and in less than 2 minutes in the other, and that, too, when the cream had been carried in the hot sun and was at a temperature of nearly 90 degrees when the test was made. Mr. Lott is willing to wager $5,000 that he can make butter in less than half a minute under favorable circumstances. The churns are made all sizes, and will be sold strictly to the trade. It is a separator, aerator and churn combined, and will produce butter from sweet, sour or clabbered milk in from one-half a minute to ten minutes time.

Ice in Three Minutes.

A machine that weighs less than thirty pounds can be purchased with which solid crystal ice can be made in three minutes, says the Kansas City Journal. It is expected to prove a boon for mountain camping or yacht cruising parties. The machine is constructed on the well known principle that water will freeze when rapidly evaporated by means of a vacuum pump and a powerful absorbent, and consists, therefore, mainly of a powerful vacuum pump, a vessel called the absorber, which contains sulphuric-acid of the common or commercial description, and a vessel to contain whatever is required to be frozen or cooled. The only working cost of the plant is for the supply of common sulphuric acid, which can be bought of any chemist, and will last a considerable time.

The South seems to be disposed to take the lead in the use of horseless vehicles. A company, consisting of Judge John B. Jones, Mr. C. B. Meyers, Col. J. N. Smithee, Mr. T. L. Cox, Col. P. Raleigh and Mr. George R. Brown, all of Little Rock, Ark., has been organized to put in a line of horseless carriages to run in opposition to the street cars. They have asked the city for a franchise, and offer for it 5 per cent of the net receipts. As stated elsewhere, a Montgomery company is negotiating for horseless vehicles to be used as delivery wagons.

The width of the Suez Canal is 825 feet.

The world's railroads added reach 407,566 miles.

An average of 100 new words are annually used by the people in the United States.

The human system can endure heat of 212°, the boiling point of water, because the skin is a bad conductor, and because the perspiration cools the body. Men have withstood without injury a heat of 300° for several minutes.

Electric Vehicles in Chicago.

Electrically propelled carriages are being sold and rented by the American Electric Vehicle Company, of 447 Wabash avenue, Chicago. They are turning out handsomely finished vehicles, furnished with storage batteries and motors and especially devised braking and steering gears. Henry Potwin is president, F. S. Culver vice-president, C. E. Corrigan general manager and treasurer, and C. E. Woods secretary. They have completed for use an electric mail phaeton which has an equipment of two 2 horse-power motors at a normal working rate which can be momentarily run up to 6 or 7 horse-power. The motors are connected, one on either hind wheel, independently, and are run in series or parallel, giving variations of speed. The

special feature of these motors is that they are dust and water proof, and are especially designed for this class of work. The battery equipment consists of thirty-two 180 ampere-hour cells, made by the Syracuse Storage Battery Company, which are grouped in series or parallel by a controller, and, working in conjunction with the motors, make possible five different speeds.

The vehicle is supplied with two brakes—one a band brake applied to the sheave on the shaft of the sproket wheel, the other an emergency brake which operates directly on the tires. The arrangement of the mechanism in the latter brake is such that the act of putting it on not only cuts the current off from the motors, but automatically sets the controller back to the starting point. The maximum speed of the vehicle is 15 miles an hour. The batteries weigh 1000 pounds and will give a continous 6 or 7 horse-power service.—*Electrical Engineer.*

Difficulties of Chinese Telegraphy.

According to the "Statesman's Year Book," all the principal cities of China are now connected with one another and with Pekin, the capital by telegraph. Recent visitors to China say, however, that telegraphing there is a laborious and an expensive process.

The Chinese have no alphabet, and their literary characters number many thousands, so it is simply impossible to invent sufficient signal to cover the written language. This difficulty was obviated by inventing a telegraphic signal for each of the cardinal numbers, and so numbers or figures might be telegraphed to any extent. Then a code dictionary was prepared, in which each number from 1 up to several thousands stood for a particular Chinese letter or ideograph. It is, in fact, a cipher system. The sender of the message need not bother himself about its meaning.

SPECIAL OFFER TO PHOTOGRAPHERS.

THE SECRET OF HUMAN LIFE.

Abstracts From Prof. Coles' Famous Lecture at Atlantic City, Before the Pennsylvania State Editor's Association.

Mr. President, Gentlemen of the Press, Brother Editors of the Great Keystone State, and Ladies :

I am now going to tell you something that you never heard before, something that you will never forget. I shall give you the secret of life, or what makes the arm move. I shall give you some of the results of years of study through the Electric Eye, an instrument of my own construction and invention, an instrument that magnifies 250,000 times. Let the following go down on record as the greatest discovery of the nineteenth century, and revealed to the Pennsylvania State Editors' Association, this second day of July, 1886, assembled at Atlantic City. The human body is but a structure which has been reared by myriads of living animalcula. These animalcula have a language of their own which is understood by their Spirit Master, God. These minute creatures are as thoroughly organized and equipped, as finely divided into muscles, nerves, stomachs and bones as the largest animal that ever trod the surface of the earth. Each part of our organism has independent life. The animalcula that construct and repair the lungs take no part in the construction of the liver or other parts of our organism, each part being built of different material by different trades of animalcula. The stomach is the storage house in which all the material for the building up and repairing of the body is kept. One mouthful of beef contains thousands of beef animalcula, and, when driven into the stomach for slaughter, are to the eyes and stomachs of the human animalcula as great as ten thousand fatted beeves would be to the human eye.

Edison, the great scientist, once said: "There is a world of unknown power about us of which at present we are entirely ignorant; yet I am sure that these forces exist as I am that I am alive. The grandest truths, too, are the simplest. Let me ask you a question." Here he fluttered aloft his brown right hand and exclaimed. "Did you see that? Well, what made that finger move?" Then, as if answering the question himself, he said, "The finger is moved by the energy of the beefsteak. Now, if we could only know what that is! It is infinitely finer, infinitely more comprehensive, infinitely more potent, than any other form of energy of which I am cognizant; and yet—well it is beyond us all." We shall soon see.

As previously stated, there are different trades of animalcula and they build the structure according to their knowledge of building. The same breed of animalcula, you must remember, that build up your father's and mother's body are now building your body, and will help build your children's bodies; therefore, the diseases extending down to the third and fourth generation. With God all things are possible. Yes, the Electric Eye proves that, it is possible for a complete muscular and nervous organism to be packed away into a space only the one-thousandth of a pin's point and our own bodies are covered with a colony not one of which is half that measure. They run with a speed that would a hundred times distance any race horse of a comparative size. The mystery of their organism is greatest, however, in the minute brain. How do they know when danger comes? Why do they run, or leap, or fly, so quickly when they are pursued by an enemy? What sort of thought behind the motion? Is it possible for them to comprehend their relation to other things as I do? Do they know how to find their natural food? Yes, they have a language of their own; they are living, breathing, thinking, acting life germs.

After years of careful study and thorough investigation, by the aid of the Electric Eye, I now believe that there is not a thing growing on this earth to day, in the animal, vegetable or mineral kingdom but what is being built by live animalcula, after their own kind, and that these living animalcula, no matter how infinitesimally small, have a language that God understands as well as He understands our language.

Therefore he can send the disease microbes on their mission of destruction just as easy to day as he could send the plagues on Pharoah and his hosts in the days of Moses.

Yes, all human, animal, vegetable and mineral life is merely the aggregate animation of myriads of smaller lives, which in turn may represent the sums of inconceivably finer gradations in the scale of sentient existence, until the mind loses itself in the maze of its own complications.

In Job, the 25th chapter and 6th verse, we read that "man is a worm," and "the son of man which is a worm." This mundane sphere on which we live is nothing but a large worm, and all mankind, animal kind, vegetable kind, and mineral kind, feed upon it just as the bacteria and parasites feed upon our bodies. The two great oceans—Atlantic and Pacific—are its lungs that heave and breathe, causintg he tides to ebb and flow, sending the blood (water) through its veins, from the lowest valley to the highest mountain peák. The Gulf streams are but the warm breath that flows from her lungs. We all know and believe that the ocean contains thousands of different species of animal life that feed upon the fruits of old mother ocean; and we have reasons to believe that the bottom of the ocean contains as much fruit, of its kind, as does the earth's surface, and, in each cubic inch of its fruit or in one single ounce of ocean water there are more living creatures than there are human beings on the face of the globe!

The surface of our bodies is covered with scales, like those of a fish; they are like great, overlapping defenses—armor plating—more effectually than any warship. A small grass seed will cover a thousand or more of these scales and under these powerful protectors are pierced myriads of little pores which serve as canals. Nobody has any conception of the number of them, and nobody depending upon the rough, haphazard disclosures of the ordinary microscope can get any clew to an adequate idea of the number of canals and passageways that exist in the human body; neither does labyrinthian begin to describe the intricacies of the net work formed by them. But if it were possible, there isn't one of us who, though we were spared ten thousand years of active existence, could complete a journey of the aggregate distance traveled by these various channels, even though he travel all his life in a trolley car going at the rate of sixty miles an hour. Yet, through these pores the sweat exudes like water through a sieve. How minute, then, must be its particles! and yet each drop of this stagnant water contains a world of animated beings, swimming with as much liberty as whales in the sea. Under the epidermis are millions of indefatigable insects, who dig into the hide, so to speak, like the earth worm digs into the soil in one's garden. These parasites crawl over our surfaces, burrow beneath our skin, and riot and propagate their kind in every corner of our frame, producing such disturbances as to require the use of medicine to destroy them. And, every dose of medicine given is a blind experiment. Yes, and it will continue to be a blind experiment until we learn just what to feed the living animalcula and how, and when, and where. In Job, XIX chapter, 20th verse, we read, "And though after the skin worms destroy this body," etc. Under the Electric Eye one of these curious fellows becomes an eight-legged monster several inches long, having brownish sides, red eyes, a head that ends in a large, scissor like bill, something like the bill of a gar pike fish, and an appetite that would put to blush an Atlantic City school girl during vacation. And a strange but very interesting fact in connection with the creature's make up is that each of these eight three-jointed legs ends in a five pronged hoof with web attachments not unlike the rudimentary hand that the evolutionists speak of. Just think of it; forty prongs constantly clawing one's flesh, and two saw like blades of a capacious mouth cutting one's person up into canabalistic delicacies for his own wormship to feed on! That's pleasant, isn't it? And your whole body is literally full of them. The minutest speck of saliva from a diseased mouth, or of that white substance which accumulates on uncleaned teeth, seems like a great mess of coroup-tion in which thousands of loathsome worms are crawling and twisting. The subject is barbarously repulsive, I know, but in the interest of health one must allude to it.

When the Creator placed Eve beside Adam, He did not place her on a pedestal above him or on a footstool beneath him, but on a level with him. He no more intended that she should be a sickly, nervous, narrow waisted, trouble faced weakling, than that Adam should be. Both of them were pronounced "very good," and we have fallen from our high estate through ignorance. We inherit debility and disease from our parents because they did not furnish the animalcula proper material to build with. Health is a science just as much as astronomy and algebra are, and its principles can be mastered by study and practice.

An electric dynamo does not generate electricity, but by its rapid motion or vibration through a magnetic field creates a vacuum and thereby abstracts from the air a strange fluid that we call electricity. The lungs, the human dynamos, consists of millions of vibratory cells, which also abstract from the air a peculiar fluid that I call magneticity, which flows through the arteries and veins just the same as water flows through the veins of old mother earth, and, by this fluid, the blood corpuscles are carried to all parts of the body. The Electric Eye discovers ten different kinds of corpuscles. In a cubic inch of magneticity there are thousands and thousands of corpuscles, and within each tiny corpuscle—which is neither more nor less than a ship of conveyance that conveys the living animalcula to all parts of the body—are hundreds, yea, thousands of smaller corpuscles, different in form and color, filled with living animalcula, epuipped with all necessary materials for the building up of the body. Now we have it—Life! Wonderful!

Now we understand! Go tell it to Edison—that the beef animalcula gives the human animalcula (blood) strength. Tell him that the human animalcula are endowed with instinct, if not reason, and that the combined forces of count less millions of human animalcula, at the command of their (spirit) commander, seize the muscles and up goes the arm; out goes the finger; indeed, what ever command is given by the spirit, is immediately obeyed.

We have the testimony of their voice and understanding in Genesis: "And the Lord said unto Cain, where is Abel thy brother? And he said I know not; am I my brother's keeper? And the Lord said, "What hast thou done? The voice of thy brother's blood crieth unto me from the ground." Cain could not tell the Lord what he had done, but the living animalcula that had been spilled from Abel's body told him what had happened. Now I understand how the spirit and the living animalcula in our bodies are on speaking terms. Now I understand why those innumerable, silvery, thread nerve fibers are spun out with such an infinite degree of nicety all over and through our bodies. Now I understand why every part of the skin has its own separate connections with the center of the nervous system and unite there just the same as telephones and telegraph wires at a terminus. Now I understand why there are two distinct sets of fibers in the nerve—the sensory fibers, which conduct toward the brain, and the motor fibers, which conduct to the muscles. They are but a double line of telegraph, telephone, and seeascope wires, one for inquiries, the other for responses. These fine nerves all go to the brain, the home of the spirit (or spirits) who direct the movements of the body and become wholly accountable for all its work and deeds. God imprisons the spirit within the brain at birth, once for all, then closes the case, and gives the key to the Angel of Death.

Running on Time.

As illustrating the degree of efficiency to which the present management of the B. & O. R. R. has brought its motive power equipment and *esprit de corps* of the operation staff, we call attention to the fact that during the months of April, May and June the passenger trains and fast freight trains have almost invariably arrived at their respective destinations on schedule time. The very few exceptions to the general rule were due to causes inseparable from railway operation, and against which no forethought can wholly guard. It may be safely said that during the period named no road in America, comparable in magnitude to the B. & O. can surpass its record for punctuality in train movement.

BOOKS AND MAGAZINES.

The second volume of "Alden's Living Topics Cyclopedia" extends from Boy. to Con., and contains the latest facts concerning the nations, Brazil, British Empire, Bulgaria, Cape Colony, Chile, Chinese Empire, and others, and concerning three States, California, Colorado and Connecticut; also concerning six large cites, Brooklyn, Buffalo, Charleston, Chicago, Cincinnati and Cleveland. The facts are commonly from one year to five years later than can be found in any part of the leading cyclopedias, and commonly a year later than the 1896 almanacs and annuals. Other volumes will follow in rapid succession. It is a useful work.

"Wonderland, '96," is the name of one of the handsomest, most interesting and most complete little books ever issued from the passenger department of a railroad. The inspiration comes from a tour of the country traversed by that greatest of transcontinental railways—the Northern Pacific—Mr. Olin D. Wheeler, the author having made several trips through the Yellowstone Park for data and views of that wonderland. The hunting scenes were also taken from life—and experiences—in the great Rocky Mountain regions through which the Northern Pacific runs. An illustrated chapter on Alaska is also included and scenes incident to the greatest wheat regions on earth in North Dakota—where millions of acres of free land still await the land seeker—are interesting features. This little pamphlet is by no means a mere advertising document. It is a work of art and filled with valuable and interesting information especially for tourists who will appreciate the attractions along this great scenic route. General Passenger Agent, C. S. Fee, St. Paul, Minn.,—to whose genius this publication owes its being—will forward one of these pamphlets to any address on receipt of six cents in stamps.

NEW PATENTS FOR SALE.

Advertisements inserted in this column for 20 cents a line (about 7 words) each insertion. Every new subscriber sending $1.00 to THE INVENTIVE AGE will be entitled to the AGE one year and to five lines three times FREE. Additional lines or insertions at regular rates.

"HARNESSING THE WAVES."—This will usher in the "Millennium of Electricity," as Edison declares. See the article in this month's AGE. The field is broad and "white with the harvest." Who will help me reap it? Address, for particulars, "WAVE POWER," care of THE INVENTIVE AGE.

FOR SALE.—My Patent, No. 554,065—Sand-Paper Holder—both for the United States and Canada. A very useful device. For further information address, A. Hornig, Sandusky, Ohio. 6-8

WANTED.—Some one to join as co-partner with me to bring out Link and Pin Car Coupler No. 485,333. Rev. J. W. Poston, Holly Springs, Miss. 5-7

FOR SALE.—Good patent. Easy and cheap to manufacture. Thoroughly tested. Many in use. Recommeded by all using it. Ready seller. Price reasonable. Good investment. Pat. Jan. 9th, 1894. For references and price, terms etc., address P. O. Box, No. 1072, Tacoma, Wash. 5-7

FOR SALE.—Patent No. 501,800 issued 18th of July 1893; flash light apparatus; being compact light simple, and durable; always ready, can be used with all kinds of Cameras. Will sell the entire patent, or manufacture on royalty. A. T. Mallick, Jamestown, N. Dak. 3-5

FOR SALE.—Or manufacture on royalty United States patent 518,165, Canada patent 46,029; also several unpatented inventions; one-third to party that will furnish money to have them patented. Enclose stamp. E. H. Thalaker, Petersburg, W. Va. 3-5

WANTED.—A Partner to take an undivided one-half interest in the best 25 cent household article on earth. Is patented in both U. S. and Canada, have sold several State and County rights, also over 5,000 of the goods. I have the machinery and have reserved the right to manufacture all goods and 4 cents will cover the cost of each article. I will give the right man a snap. Address, J. E. Hyames, Gobbleville, Van Buren Co., Mich. 3-4

FOR SALE.—Patent, 492,875, Folding Hoe, entirely new principle. Will sell State or County rights cheap. Any live man can make money right in his own state or town handling the hoe. Inexpensive to manufacture, and sells on sight. For full information, address Dr. F. A. Rice, 62 Main street, Lockport, N. Y. 2-7

FOR SALE.—Valuable Patent No. 532,324 for improvement in metal carriers or baskets for bottles and other articles. Address Patentee, H. M. Kolb, 732 Sansom street, Philadelphia, Pa. 2-4

FOR SALE.—Canadian Patent, the Highwine and Whiskey Distillers Continuous Beer Still. Manufacturers and capitalists who wish to purchase valuable patents can make very large profit. Operated in U. S. Distilleries. Address, Hermann Binz, Frankfort, Ky. 2-4

FOR SALE.—Or on royalty, Patent No. 554,643; a folding suit or garment hanger; simple and cheap: big profits in it. Address, Chas. Behrend, Jr. Oconomowoc, Wis. 4-6.

FOR SALE.—Patent No. 517,929 A Chimney Regulator. Best ever invented. Is sure to be a blessing to every home. Will take some land in exchange. Address Jurgen Jurichs, Danforth, Ill. 7-9

FOR SALE.—A half interest in my "Hair Grower," I have made a discovery whereby I can grow a full head of hair on the baldest head; cure any case of dandruff or scalp disease to dead certainty. I want a partner with money to put this new discovery upon a large scale. A fortune for some one. Sample bottle sent upon receipt of $1.50. Address, Geo. W. Schoenhut, Eldora, Iowa. Box 98.

FOR SALE.—A complete set of Patent Office Reports and Official Gazette of the U. S. Patent Office from 1790 to 1894. H. B. Dickinson, Ravenna, O., Adm'r of Estate of Bradford Howland. tf.

BUSINESS SPECIALS.

Advertisements under this heading 20 cents a line each insertion—seven words to the line. Parties desiring to purchase valuable patents or wanting to manufacture patented articles will find this a valuable advertising medium.

WANTED.—Publisher to publish and introduce Walsh's Perpetual Calendar and Almanac on royalty; or I will sell the copyright. A complete calender, absolutely perpetual. No complicated rules or computations. James A. Walsh, Helena Mont. 2-4

WANTED.—Reliable manufacturer to manufacture on royalty, or will sell at a moderate figure. Caliper Adjustment. Described elsewhere. Address, Arthur Munch, 653 East 5th, st., St. Paul, Minn. 4-7

Aftermath.

The city of Cleveland, Ohio, celebrated its centenary last month.

The city of Mexico advertising for proposals to light the city with electricity 600 arcs and 150 incandescents.

The longest ride for 5 cents on a street railway can be had in Brooklyn—18 miles—with Chicago second, 15 miles.

The Hallet and Davis Co., of Chicago, music dealers, have made an assignment. causing failure also of the main concern in Boston.

The banana is said to be the most prolific of all food products, being 44 times more productive than potatoes and 131 times more than wheat.

It is said that 1,913,000 out of a total of 2,500,000 cotton spindles in the South have agreed to shut down 38½ per cent of the time between July and October.

Colonel J. M. Wilson, commissioner of public buildings at Washington, states that 1,244,227 persons have ascended the Washington Monument since 1888 without an accident.

Bids for building two fast cruisers for the Japanese navy have been opened in Washington, but not made public. It is understood that Cramps will build one and the Union Iron Works, San Francisco, the other.

Official statistics of the Bureau of Immigration show that 343,267 immigrants arrived in the United States during the fiscal year ending June 30, 1896, of whom 212,466 were males and 130,801 females. There were debarred from entry 3037, of whom 2010 were paupers and 776 contract laborers.

Frank Mason, U. S. consul at Frankfort, in a report to the state department strongly urges a concerted movement on the part of our shoe manufacturers to control the German market. He points out how this can be done and says that the way has already been opened by the great success that has attended the importation of American shoes in Germany.

The New York Herald's correspondent in Montevideo, Uruguay, telegraphs that a Spanish resident, an electrician, declares that he has discovered means whereby he can maneuver balloons in any direction in the air, and the inventor has been asked by the Spanish Government to visit Madrid; that if his invention is found to be practicable, it may be applied to use in Cuba.

The Pillsbury-Washburn Milling company of Minneapolis has begun an action for infringement upon patent against the American Wired Hoop company of West Superior and has directed that papers be served on the Duluth Imperial Mill company. It is claimed that a process for printing on boards which is owned by the plaintiff has been used by the defendant. The defendent company manufactures the barrels which the Inperial company uses. The suit is begun in the United States court and an injunction restraining the company from using the printing device is asked.

ACCORDING to the Phœnix (Ariz.) Gazette there are probably 1,000 camels now roaming over the deserts of the southwestern portion of the United States. This will be a surprise to the majority of people who have never connected the camel with this country except with circuses and menageries. It seems that during the civil war the government purchased 150 of these animals for the use of the troops then operating in the Arizona region. The character of the sand and gravel cut their feet and rendered the animals useless as beasts of burden. They were turned loose to graze and roam as they pleased. It seems they have become acclimated and are now increasing—a hunter recently counting over 500 in a herd.

THE MOUNTAIN CHAUTAUQUA.

Mountain Lake Park, Md., on the Main Line of the Picturesque B. and O.

The most superb and sensible summer resort in America. $300,000 expended in improvements; 200 beautiful cottages; hotel and cottage board at from $5.00 to $12.00 per week—cheaper than staying at home. The mountain air and the mountain views simply indescribable. Session August 5th to the 25th. Three superb entertainments daily. The best music and the best lecturers which money can procure. Dr. T. De Witt Talmage, Gen. John B. Gordon and Bishop J. H. Vincent already secured, with 100 others. Dr. W. L. Davidson, the great Chautauqua manager, in charge.

SUMMER SCHOOLS.—20 departments of important school work in charge of leading instructors from the prominent universities. A wonderful chance for teachers and students desiring to make up studies. Tuition insignificant. Wishes of students gratified, Low rates on railroads. For full detailed information and illustrated programme, address A. R. Sperry, Mountain Lake Park, Md.

Engraving and Electrotyping.

The city of Washington now possesses, in the firm of W. C. Newton & Co., one of the most complete and extensive printers' supply houses in the country and under the immediate charge of one of the most efficient managers in the east there has recently been added an electrotyping and engraving department, which is now prepared to turn out all classes of work in that line. A speciality will be made of fine engravings from Patent Office drawings and photographs of machinery. Patent attorneys, model makers, promoters and inventors will be specially interested in this matter because they are guaranteed the most reasonable rates.

STANDARD WORKS.

"The Electrical World." An illustrated weekly review of current progress in electricity and its practical applications. Annual subscription.......... $3.00

"Electric Railway Gazette." An illustrated weekly record of electric railway practice and development. Annual subscription........ $3.00

"Johnston's Electrical and Street Railway Directory." Published annually.......... $5.00

Dictionary of Electrical Words, Terms and Phrases. By Edwin J. Houston, Ph. D.... $5.00

The Electric Motor and its Applications. By T. C. Martin and Jos. Wetzler.............. $3.00

The Electric Railway in Theory and Practice. The first systematic treatise on the electric railway. By O. T. Crosby and Dr. Louis Bell.......... $2.50

Alternating Currents. By Frederick Bedell, Ph.D., and Albert C. Crehóre, Ph.D........ $2.50

Gerard's Electricity. Translated under the direction of Dr. Louis Duncan............ $2.50

Theory and Calculation of Alternating-Current Phenomena. By Charles Proteus Steinmetz.......... $2.50

Practical Calculation of Dynamo Electric Machines. By A. E. Wiener.............. $2.50

Central Station Bookkeeping. By H. A. Foster.......... $2-50

Continuous Current Dynamos and Motors. By Frank P. Cox, B.S.................... $2.00

Electricity at the Paris Exposition of 1889. By Carl Hering.......................... $2.00

The Elements of Static Electricity, By Philip Atkinson, Ph.D.................... $1.50

Electricity and Magnetism. By Edwin J. Houston, Ph.D.......................... $1.00

Electric Measurements and other Advanced Primers of Electricity. By Edwin J. Houston, Ph.D.......................... $1.00

The Electrical Transmission of Intelligence and other Advanced Primers of Electricity. By Edwin J. Houston, Ph.D................ $1.00

Electricity One Hundred Years Ago and Today. By Edwin J. Houston, Ph.D........ $1.00

Alternating Electric Currents. By Edwin J. Houston, Ph.D. and A. E. Kennelly, D.Sc. $1.00

Electric Heating. By E. J. Houston Ph.D and A. E. Kennelly, D.Sc.................. $1.00

Electromagnetism. By E. J. Houston, Ph.D and A. E. Kennelly, D.Sc.................. $1.00

Electricity in Electro-Therapeutics. By E. J. Houston, Ph.D., and A. E. Kennelly, D.Sc.......................... $1.00

Electric Arc Lighting. By E. J. Houston, Ph. D., and A. E. Kennelly, D. Sc. $1.00

Electric Incandescent Lighting. By E. J. Houston, Ph.D. A. E. Kennelly, D.Sc...... $1.00

Electric Motors. By E. J. Houston, Ph.D., and A. E. Kennelly, D.Sc.................. $1.00

Electric Street Railways. By E. J. Houston, Ph.D., and A. E. Kennelly, D.Sc.......... $1.00

Tables of Equivalents of Units of Measurement, By Carl Hering...................... .50

Copies of any of the above books or of any other electrical book published will be sent by mail, postage prepaid, to any address in the world on receipt of price.

THE INVENTIVE AGE,
Washington, D. C.

RECENT investigations by Margot have established the fact that an alloy, composed of 95 parts of tin and five of zinc, becomes firmly adherent to glass, and is unalterable and exhibits an attractive metallic lustre, says the *Engineering and Mining Journal*. An alloy, consisting of 90 parts tin and 10 of aluminum, melts at 390 centigrade, becomes strongly soldered to glass, and is possessed of a very stable brilliancy. With these two alloys it is possible, it is claimed, to solder glass as easily as it is to solder two pieces of metal, and this operation may be done by soldering the pieces of glass, when heated in a furnace, by rubbing their surface with a rod of the solder, the alloy as it flows being evenly distributed with a tampoon of paper or a strip of aluminum, or an ordinary soldering iron can be used for melting the solder.

Expert Train Running.

For the month of May the record of train movement on the B. and O. R. R. eclipsed the record breaking record for April, when the passenger trains arrived at their destinations as per schedule ninety-five times out of a possible hundred. The B. & O. Fast Freight Trains between New York, Philadelphia and Baltimore on the east, and Cincinnati, St. Louis add Chicago on the west, are being moved with an equal degree of percision.

ANY person sending us the names and addresses of five inventors who have not yet applied for patents will receive the INVENTIVE AGE one year free.

IDEAS DEVELOPED. Absolute secrecy. Send for particulars. Advice and suggestions given free when asked for. Correspondence and sample orders solicited. 25 years in business. Models, patterns and all kinds of castings and machine work.

GARDAM & SON,
96 John Street. New York.

TO INVENTORS.

We make a specialty of manufacturing goods for inventors. Make models, dies, do experimental work, make special tools, machinery, etc. INVENTOR'S HANDBOOK FREE.

Cincinnati Specialty Manufacturing Co.,
Third Street, Cincinnati, Ohio.

G. A. R. National Encampment, St. Paul.

The B. & O. R. R. will sell tickets from all points on its lines east of the Ohio River to St. Paul at one single fare for the round trip, good for all trains, August 29th, 30th and 31st, valid for return passage until Sept. 16th, with the privilege of an additional extension until September 30th by depositing ticket with Joint Agent.

The Rate from	Philadelphia,	will be	-	$25 00
" " "	Baltimore	"	-	25 00
" " "	Washington	"	-	25 00
" " "	Lexington	"	-	24 75
" " "	Cumberland	"	-	23 00
" " "	Grafton	"	-	2[illegible] 00

And correspondingly low rates from other stations. Tickets will also be placed on sale at the offices of all connecting lines.

The B. & O. maintains a double daily service of through solid vestibule trains between the East and Chicago, with Pullman sleeping and dining cars attached.

G. A. R. Veterans will remember that all B. & O. trains run via Washington, Harper's Ferry and the Potomac Valley.

For Mchanics and Engineers.

Every mechanic and engineer should have a copy of "Shop Kinks and Machine Shop Chat," by Robt. Grimshaw. It is the most complete work of the kind ever published. If any "kink" is discovered in the machine shop—it can be straightened out—after consulting this valuable work. Sent to any address for $2.50.

Saturday and Sunday Trips to The Country.

Commencing Saturday, May 30th, and continuing until further notice, the B. & O. R. R. Co. will sell excursion tickets, at rate of one fare for the round trip, for regular trains of Saturday and Sunday, to points on the Metropolitan Branch and Main Line between Washington, Harper's Ferry and Charlestown, and to points on the Washington branch between Washington and Laurel. 3

A Great Novelty.

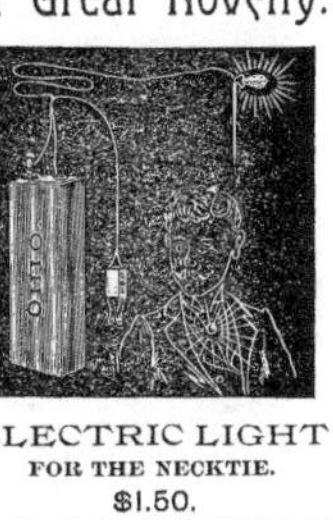

ELECTRIC LIGHT
FOR THE NECKTIE.
$1.50.

Complete with Powerful Pocket Battery and all accessories, postpaid.
The Lamp an Efficient Beauty.
Battery 3 1-2 by 3-4 by 1 3-4 inches.
This elegant outfit is practical, brilliant and satisfactory, and every part is guaranteed.
Electricians who want a piece of jewelry that is emblematical, or anybody who appreciates a real novelty, can, by ordering now, secure a complete outfit for $1.50. Order at once of
THE INVENTIVE AGE,
Washington, D. C.
Given FREE for four new subscribers.

HOW TO MAKE A DYNAMO.
By ALFRED CROFTS.

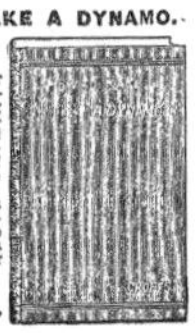

A practical work for Amateurs and Electricians, containing numerous illustrations and detailed instructions for constructing Dynamos of all sizes, to produce the Electric Light, containing 96 pages of genuine information which will enable anyone to construct a Dynamo either for pleasure or profit.

**Large 12 mo,
Cloth, 75 Cts.**

GREAT PREMIUM OFFER = =

Valuable
Sensible
Economical

Zell's Condensed Cyclopedia

Always Sold at $6.50 Complete.

Complete in One Volume. A Compendium of Universal Information, Embracing Agriculture, Anatomy, Architecture, Archeology, Astronomy, Banking, Biblical Science, Biography, Botany, Chemistry, Commerce, Conchology, Cosmography Ethics, The Fine Arts, Geography, Geology, Grammar, Heraldry, History, Hydraulics, Hygiene, Jurisprudence, Legislation, Literature, Logic, Mathematics Mechanical Arts, Metallurgy, Metaphysics, Mineralogy, Military Science, Mining, Medicine, Natural History, Philosophy, Navigation and Nautical Affairs, Physics, Physiology, Political Economy, Rhetoric, Theology, Zoology, &c.

Brought down to the Year 1890. Substantially Bound in Cloth with the Correct Pronunciation of every Term and Proper name, by L. Colange, LL.D. Handy Book for Quick Reference. Invaluable to the Professional Man, Student, Farmer, Working-Man, Merchant, Mechanic. Indispensable to all.

In this work all subjects are arranged and treated just as words simply are treated in a dictionary--alphabetically, and all capable of subdivision are treated under separate heads; so that instead of one long and wearisome article on a subject being given, it is divided up under various proper heads. Therefore, if you desire any particular point of a subject only, you can turn to it at once, without a long, vexatious, or profitless search. It is a great triumph of literary effort, embracing, in a small compass and for little money, more (we are confident) than was ever before given to the public, and will prove itself an exhaustless source of information, and one of the most appreciated works in every family.

It cannot in any sense--as is the case with most books--be regarded as a luxury, but will be in daily use wherever it is, by old and young, as a means of instruction and increase of useful knowledge.

Thousands of leading men offer the highest testimonials. It is not a cheap work it is standard.

Never before, even in this age of cheap books, was such a splendid offer made, viz:

$2 For Cyclopedia, Postage Paid. **$2**
For Inventive Age, One Year.

An Extra Copy of Cyclopedia to any person sending us Six New Subscribers. The Same Offer is made to old Subscribers renewing for another term.

Address THE INVENTIVE AGE, Washington, D. C.

The Inventive Age
·AND INDUSTRIAL REVIEW·
A JOURNAL OF MANUFACTURING INDUSTRY *AND SCIENTIFIC PROGRESS*

Seventh Year. No. 9. WASHINGTON, D. C., SEPTEMBER, 1896. Single Copies 10 Cents. $1 Per Year.

Largest River Steamboat in the World.

No more striking evidence of the inventive, mechanical and aesthetic genius of this progressive age can be found than in the modern American steamboats, built, not only for trans-Atlantic service but for river and inland lake traffic. The St. Paul and St Louis present the highest type of seagoing craft while the Northwest and her sister ship are some novel features in the architecture of this vessel.

In the grand saloon there are two galleries, fore and aft, and in completeness of finish and application of modern luxuries this vessel towers over all previous efforts in that direction. The People's Line, first to introduce the single gallery style of architecture in steamboat building—the New World, back in 1855—is now the pioneer in the double gallery and an upper tier of staterooms. The Adirondack, therefore, has five decks, known as the main, saloon, gallery, upper gallery and dome decks. If the floor in the hull was included there would be six distinct decks.

The new vessel was modeled and designed by Mr. John Englis, vice-president of the company, who, owing to his association for many years with this and other navigation lines, is familiar with the requirements of the traveling public. The entire construction and equipment of the vessel was supervised by Mr. Charles M. Englis, of the firm of John Englis & Sons, of Brooklyn, N. Y.

The boat is propelled by a vertical beam engine of the standard American pattern of 4,000 h. p.

From "Seaboard," Smith & Stanton, Pub's.] HUDSON RIVER STEAMBOAT ADIRONDACK, OF THE PEOPLE'S LINE.

This vessel is 412 ft. in length, of 3644 gross tons, provides sleeping accommodations for 1000 passengers, the largest river steamboat ever built, and runs at a speed of 20 miles an hour. The steamer on the right is the towboat Norwich, of 255 gross tons, built in New York sixty years ago, and the oldest steamboat in existence—in continuous service since 1836. As a contrast to the mighty Adirondack she forms an interesting object lesson, showing the size and type of half a century ago with those of the present day.

on the Great Lakes, the Priscilla of the Fall River line and the new Adirondack of the People's Hudson River line combine all the attributes of floating palaces. The latter steamer plies between New York and Albany and is the largest steamboat ever built for river navigation. The vessel is 412 feet long; 50 feet beam and 90 feet wide over guards. She draws 8 feet of water, and is of 3644 gross tons. There built by the W. & A. Fletcher Company, at their North River Iron Works, at Hoboken, N. J.

The seating capacity of the dining room is 250 persons and the grand saloon is "a dream in white, green and gold," decorations. On the second, third and fourth decks are located 350 state rooms, in which there are two berths, including 24 parlor

(*Continued on page 135.*)

Inventive Age

Established 1889.

INVENTIVE AGE PUBLISHING CO.,

8th and H Sts., Washington, D. C.

ALEX. S. CAPEHART.   MARSHALL H. JEWELL.

The INVENTIVE AGE is sent, postage prepaid, to any address in the United States, Canada or Mexico for $1 a year; to any other country, postage prepaid, $1.50.
Correspondence with inventors, mechanics, manufacturers, scientists and others is invited. The columns of this journal are open for the discussion of such subjects as are of general interest to its readers.
Technical matter is particularly desired. We want practical information from practical men.
Nothing will be published in the editorial columns for pay.
The INVENTIVE AGE is thoroughly independent.
Advertising rates made known on application. Special facilities for furnishing cuts of any patented article together with descriptive article. Business specials 15 cents a line each insertion, 7 words to the line. No advertisement less than 25 cents.
Address all communications to THE INVENTIVE AGE, Washington, D. C.

Entered at the Postoffice in Washington as second-class matter.

WASHINGTON, D. C., SEPTEMBER, 1896.

Among the papers in the District of Columbia devoted to inventions and patents, none has credit for so large a regular issue as is accorded to the "inventive Age," published monthly at Washington.—*Printers Ink.*

A PERFECT bicycle brake has not yet made its appearance. There are many of them, but the successful inventor has not yet put in an appearance.

IN Great Britain the trunk telephone lines are being absorbed by the government and they will soon be operated by the government the same as the telegraph system.

THE death of John J. Hogan, inventor of the famous Hogan boiler, is announced. He died of Bright's disease, at Middletown, N. Y., where he had recently established a boiler works.

SNOW in August is a scene that few men have ever witnessed, yet on Monday, August 17th, snow fell in Philadelphia, Pa., and in the Catskill mountains, New York state.

ACCORDING to Street Railway Journal it has been demonstrated by the Detroit street railway, that a first-class street railway system cannot be maintained under ordinances restricting the rate of tariff to 3 cents.

THE Providence, twenty-five mile horseless carriage race, to occur at Narragansett Park from the 7th to the 11th of this month is attracting much attention and promises to be the most interesting event of the kind ever held.

THE Electrical Review feels that something is wrong—has heard of no new inventions in the street car fender line for three weeks. It would have been fully as well for the public if four-fifths of those invented already had never been heard of.

AND now comes the report that Prof. Andree has abandoned his balloon trip to the north pole—or rather postponed it till next year on account of unfavorable conditions and the advanced season. In the meantime the earth will continue to revolve on its axis.

PUGET SOUND coal mines are about to have an object lesson in Japanese competition by the placing of coal from Japanese mines on the market in San Francisco—the market heretofore occupied by the operators of mines in Washington and British Columbia.

THE comparative safety of compressed air motors is just now being discussed and in New York and Washington, where experiments are being made, the fact that passengers are expected to ride in the vicinity of reservoirs stored with air at the enormous pressure of 3,000 pounds per square inch,

leads many to believe that electric motors are the safest.

The new one-dollar silver certificate is a handsome piece of engraving. The border represents a laurel chain wreath, in the links of which appears the names of illustrious Americans. The artist did not overlook the inventors. There are the names of Franklin and Morse.

A SYNDICATE of New York and Philadelphia capitalists headed by Mr. J. Canby of Philadelphia, has been formed to construct and maintain an electric trolley railway between New York and Philadelphia, a distance of about 100 miles. Frank A. McGowan and J. Henry Darrah of Trenton, N. J. are chief promoters.

THE contract to transmit electric power from Niagara Falls to Buffalo for the purpose of running the street railway cars has been let to the White, Crosby Company of New York. The pole line will be of sufficient capacity to transmit 40,000 horse power to Buffalo, although only a portion of that amount is now needed.

IOWA passed an anti-cigarette law, prohibiting absolutely the manufacture or sale of cigarettes in the state or their importation into the state. Judge Sanborn, of the United States circuit court, following the decision of the Supreme Court in the famous prohibition case—interfering with interstate commerce—declared the law unconstitutional.

THE new American liner, St. Paul, has established a new record from Southampton to New York, making the distance between the Needles and the red light-ship off Sandy Hook in six days and thirty-one minutes—arriving August 14. The record was formerly held by her twin companion, the St. Louis. The average speed of the St. Paul was 21.08 knots per hour.

As an indication of how rapidly the tin plate industry is growing in this country it is only necessary to remember that for the first seven months of this year England's exports to this country amounted to only 73,552 tons against 124,839 tons for the same period last year. And the English authorities admit that there is no reason to expect a recovery of former conditions.

THE filteration of water by electricity is just now engaging the attention of scientists. Experiments show that bacteria and microbes are effectually destroyed by electrical currents and the water completely clarified but a cheaper process than that under which the experiments have thus far been conducted must be found before electrical filteration can become a commercial success.

THE death of Josephus F. Halloway, removes one of the most prominent engineers of the country. In 1885 he was president of the American Society of Mechanical Engineers, and at the time of his death was vice-president of the American Institute of Mining Engineers. He was born in Uniontown, Ohio, in 1825. During his life he was associated with many important engineering enterprises, and was a frequent contributor to the technical press.

CONSERVATIVE Paris has stubbornly refused to allow rapid transit companies to occupy her streets either with an overhead electric trolley or underground conduit system, but the demand for improved transportation facilities has been so urgent of late it has now been decided to test the closed conduit system of M. Vuillenmier. It is doubtful if this system will prove anything like as efficient as the new underground electric systems of Washington.

THE Russian loan of $200,000,000 recently issued by the Paris Rothchilds was covered twenty-five times over. Russia is about to adopt the gold standard and her credit is good for any amount she requires. A considerable number—it is hoped not one-half of the people of the United States want to go to a silver basis, in which event we will have

forfeited the confidence of all leading civilized nations and taken position among the degenerating powers of the earth.

IT is not particularly cheering to the iron trade in the United States, says Iron Trade Review, to contemplate at this time an unwonted activity in iron and steel among our European competitors. It is a fact, nevertheless, that the United Kingdom, Germany, France and Belgium have been outdoing their records of 1895, while on this side the turmoil over finance and the uncertain status of all business have quenched enterprise and reduced industrial operations within narrow limits.

THE new light and power plant at Fresno, Cal., receives its supply by wire from the San Joaquin river in the mountains thirty-five miles distant, which makes it the largest commercial electric transmission line in the world. There is a fall of 1400 feet—in a distance of 4100—in the pipe line and the receiving reservoir, 30 inches by 57 feet, is built to withstand a pressure of 800 pounds to the square inch. This is said to be the highest head of water ever used for power transmission purposes.

AT this time, when both the leading parties and a healthy public sentiment everywhere is favoring the policy of discriminating duties, as applied to American shipping interests, and as a means of building up the American merchant marine, it is surprising that a marine journal should be found faltering in its advocacy of a truly American policy. Yet this is what the New York Marine Journal is doing. Its contemporary, Seaboard, is, however, steadfast in its advocacy of that policy, the triumph of which means the restoration of American shipping to the foreign carrying trade.

THE recent meeting of the American Association for the Advancement of Science, at Buffalo, was well attended and the papers presented of unusual interest. The officers elected for the coming year were : President, Prof. Wolcott Gibbs, Newport, R. I.; vice-presidents, who form the chairmen of the various sections, mathematics and astronomy, W. W. Beman, Ann Arbor, Mich.; physics, Carl Barus, Providence, R. I.; chemistry, W. P. Mason, Troy, N. Y.; mechanical science and engineering, John Galbraith, Toronto, Canada ; social and economic science, Richard T. Colburn, Elizabeth, N. J.; permanent secretary, F. W. Putnam, Cambridge, Mass.; general secretary, Asaph Hall, Jr., Ann Arbor, Mich.; secretary of council, D. S. Kellicott, Columbus, Ohio; treasurer, R. S. Woodward, New York. The next place of meeting was left in the hands of the council for final action. Secretaries of various sections were elected as follows : Mathematics and astronomy, J. McMahon; physics, F. Bedell; chemistry, P. C. Freer ; mechanical science and engineering, J. J. Flather; social and economic science, A. Blue.

Attachment for Saws.

Richard H. Gowan, of Coralcana, Tex., has recently patented a guage attachment for saws, wherby the cutting depth of the saw is regulated. The guage is in shape of a bar divided by a narrow slot which allows it to fit over the saw, downwards. The guage is held to the saw by clamping devices, and has at its foot portions horizontal pieces, with

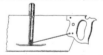

upwardly curved ends, for resting on the material sawed and checking the further downward progress of the saw.

This is a very useful arrangement for the carpenter, for with it he can work without the use of guage marks. He simply saws away until the foot of the guage strikes the board or timber thereby enabling him to do his work more accurately than he could without the guage.

Worthless Patents.

In reviewing some prevalent Patent Office criticisms and causes therefor, the Electrical World says: "The enormous number of patents which are from week to week being granted fully demonstrates another weakness in the system. Ideas are embodied in the specifications which, if not worth the paper they are printed on, constitute merely minor details which enable the applicant to practice a routine system of blackmail on legitimate concerns. It is not infrequent that the latter, rather than be subjected to a threatened suit, will pay a royalty or other fee to the blackmailer." The Electrical World overlooks one of the greatest evils or rather abuses of our patent system—the activity of quack patent attorneys, who, through various attention-catching devices and representations are monopolizing the patent work, and increasing the number of patents regardless of value or utility. Scores of adventurers have invaded the field of the conservative, conscientious patent solicitors, and a noble profession is being debased by Cheap John, penny-catching fakirs. There are now a large number of patents granted that are patents in name only; they possess no value whatever when put to the test—in law or in the realm of mechanical skill. It is the rule that one never gets more than he pays for and the inventor, who is influenced by prize offerings and reduced fees will find that he is no exception to the rule. Inventors cannot be too cautious either before or after applying for a patent. If he has an invention that is worth patenting at all it is worth a good patent; if his invention has been anticipated by others and no claims can be allowed on essential features he as entitled to this information from his attorney before proceeding with his case. Unless a client is well posted as to the state of the art we invariably recommend that a preliminary examination be made to ascertain if the invention is patentable. In three cases out of five inventions submitted are not patentable—that is, only a worthless patent can be secured.

It must be remembered that the Patent Office does not assume responsibility for the acts of attorneys, and the selection of competent, and honest counsel should, therefore, be the first consideration of applicants for patents.

Warning to Inventors.

In one of his reports to Congress, Commissioner S. S. Fisher gave to inventors a most important warning against unskilled, unreliable, and dishonest patent attorneys: He said:

"The tendency of many agents to be more solicitous about the number than the quality of patents is aggravated by those who solicit patents on contingent fees, or who without special training and qualification adopt this business as incident to a claim agency, and press for patents as they do for back pay and pensions.

"Such men are often more desirous of obtaining a patent of any kind and by any means than they are of obtaining one which will be of any value to their clients. Inventors are often poor, uneducated, and lacking in legal knowledge. They desire a cheap solicitor and do not know how to choose a good one. They are pleased with the parchment and the seal and are not themselves able to judge of the scope and value of the grant.

"Honest and skillful solicitors, with a thorough knowledge of the practice of the office and of patent law, and who are able and willing to advise their clients as to the exact value of the patents which they can obtain for them, may be of much service to inventors. There are many such. But those who care for nothing but to give them something called a patent, that they may secure their fee, have in many instances proved a curse. To get rid of their client and of trouble they have sometimes been content to take less than he was entitled to and in many cases they have, with much self-laudation, presented him with the shadow when the substance was beyond his reach. Between such men and the office strife is constant."

Those Rascally Patent Agents.

THE INVENTIVE AGE, of Washington, D. C., being at "the seat of war," and thrice armed like one whose cause is just, is waging a relentless conflict with the swindling patent agencies and in behalf of the honest inventor and owners of patents. The rascally patent agent and broker, is a festering nuisance, and should be driven out of the capital of the nation. To that purpose the INVENTIVE AGE is committed, and should succeed.—*Stone.*

Why Patents are Worthless.

Whether or not college professors should take out patents seems to have awakened some discussion in the columns of the Electrical Review. Just why there should be any question about the propriety of a college professor taking advantage of the patent laws, which were designed to stimulate inventors and benefit all, is not quite clear and public opinion will doubtless decide in favor of no discrimination as to the benificaries of the patent system.

Mr. Edward C. Weaver, of Washington, D. C., is one of those who look at the matter in this light and he also makes some significant observations regarding worthless patents—and reasons therefor—which will interest inventors and other readers of the INVENTIVE AGE. We quote a portion of Mr. Weaver's communication as follows:

The patent laws were designed to stimulate invention. Let them benefit every man. Only those who from personal motives so desire, should be permitted to evade their wise provisions.

And from an ethical standpoint, the philanthropic desire to benefit the public by renouncing the monopoly of a letters patent is erroneous. It fails more often than it succeeds, and in many cases the first inventor, by his failure either to publish his invention over the land or to patent it, only lives to see some one with a less amount of philanthropy, calmly reaping the benefits of a patent.

While on the face this seems a hardship to the public, what help is there for it? The Patent Office can not deny a patent, unless they have positive evidence of prior invention. A former patent, or a definite publication, accessible to the examiners, is a decisive anticipation.

The scientific man has almost without exception been responsible for inventions of a generic nature, practical and mechanical, for applied improvements. The two work hand in hand, and are mutually dependent for the production of great industrial advances.

Professor Stine's saying that "The reason that so large a percentage of patents are either worthless or faulty, is that the average inventor is so greatly lacking in scientific and mathematical knowledge," may perhaps not frown at a slight amendment from one who breathes "patent" air.

One reason why the majority of granted patents are worthless is, that attorneys encourage inventors to take out patents on anything and everything, trusting to luck to secure an infinitesimal *claim*, and hence a fee. To a large extent this is caused by the leaning of the average inventor to cheapness, to "no patent, no pay," thus driving the good old fashioned, reliable attorney from the field. There is no such thing as a *cheap attorney* in patents. Neither do reliable attorneys run *prize* concerns nor grab-bag competitions. No man would care to have a quack lawyer carry out the commands of his will for the benefit of children he had loved in life. Why should he deliver the child of his intellect into the hands of a patent quack?

Again, there are numberless inventions, each but a minute atom in some pronounced step of progress. Indeed, history has often shown that the tortoise of minute and apparently valueless invention has won the race as against the hare that would at once leap at results. Thus, years and years were spent by the most learned men in the attempt to perfect the telephone. They all ran too fast and stepped far over the mark. Again 10 years of persistent invention on the sewing machine produces the magnificent mechanism of today.

A New Can Opener.

Those who have labored with a can opener, forcing the cutter through the obstinate tin and after receiving wounds from its ragged edges, will appreciate the invention of Lyman W. Merriam, of Fitchburg, Mass., to whom patents have been issued. One patent is for a can opener consisting of a wire embedded in one end of the can and adjusted so that by withdrawing the wire the central portion of the can end will be opened. This is a very convenient arrangement and one that will save labor and render a quick and satisfactory service.

The other patent consists in having a wire embedded between a narrow strip of tin and the edge of one end of the can head. It is only necessary to take hold of the end of the wire with the fingers and strip the wire out. This patent can is intended more for paint, meats, etc. The inventor will give any further particulars desired.

INVENTORS who receive propositions from patent brokers and schemers can assist us in our work of exposing humbugs if they will forward such documents and correspondence to the INVENTIVE AGE.

How Inventors are Swindled.

The gullibility of the average inventor is probably no greater than that of the average citizen who has never exhibited a particular penchant for invention, but certain it is that no class of people are beset by a larger horde of fakirs and humbugs than they. Indeed, so large is the army of "Patent Brokers," "Development Companies," "Patent Agents," and the like it would seem that the chief patrons, the majority of the subscribers of the Official Gazette of the United States Patent Office, were from among this class. The issuance of the Gazette each Tuesday, giving the name and address of each person to whom a patent has been granted, is followed on Wednesday by the mailing of as fine an array and as complete an assortment of "fake" and misleading literature as it is possible for the most vivid imagination to conjecture. So bold and fraudulent have some of these schemers become that it is necessary, in order to evade the vigilance of the United States authorities, to frequently change, not only their post office address, but the name under which they carry on their business as well.

THE INVENTIVE AGE has taken occasion to refer to this matter in previous issues and the accumulation of damaging evidence against many of the so-called "Patent Agents," and "Patent Brokers," justifies us in continuing and repeating the warning to inventors.

There ought to be and must be a reform in the manner of doing business on the part of patent agents claiming to be legitimate. The inventors of America have long enough been the prey of unscrupulous and designing patent sharks, whom an indulgent government allows the use of the mails for fraudulent purposes. It is a fact, that will not be denied by any intelligent investigator of their business methods, that four-fifths of the so-called patent agencies and patent brokers are humbugs. Only the co-operation of reputable attorneys of the country and their clients, the inventors for whom they have obtained patents, is necessary to drive scores of these vampires out of their nefarious business.

The First Step Towards a Patent.

The following advice from that excellent magazine "Stone:"

It is very seldom that the inventor in working over the creation of his inventive thought and perfecting his idea or machine gives any thought to his position and duties both to himself and to the public. As a rule he is inspired by one or two motives, either to accomplish great good for humanity or to line his own pocketbook. And it is equally true that he may be influenced by both of these motives at one time, they preserving a just equilibrium.

Following natural laws, probably the inventor's first duty is to himself. At the very moment his invention has reached such an embryonic stage that he has a thorough conception of it he should inform himself thoroughly of the state of the art before him. It is a fact that the industrial world is fifty years behind the world of the inventors. Thousands of devices absolutely unknown to artisans and to manufacturers are old and well known to the examining corps of our Patent Office. It is absolutely necessary that an inventor should avail himself of this knowledge by seeking the advice of a competent Washington patent attorney and informing himself of the state of the art. This in many cases will save many years of labor in perfecting and money spent in experimenting.

The inventor's duty to the public is to so perfect his invention and reduce it to its simplest mechanical principles that there is nothing left for succeeding inventors to improve and to extend the monoply of the patent right. It is eminently fair that the first inventor should have this monopoly for seventeen years; it is also just that the public should have the untrammeled use at the expiration of that period.

Therefore, it is the inventor's duty to so clothe his own monopoly that none others shall follow. This duty also enjoins upon him the selection of the ablest and most competent patent lawyers at his convenience, for they alone can aid him in the thorough performance of his obligation.

INVENTORS who have "bought their experience" with alleged patent selling agencies, or who have received quantities of "shark" literature, glittering offers and inducements, can assist in "smoking out" the frauds by sending all information in their possession to the INVENTIVE AGE which is the only legitimate, independent magazine published in the interest of inventors—the price of which, $1 for a whole year, has saved many times that amount to hundreds of inexperienced inventors.

THE INVENTIVE AGE publishes each month a list of attorneys who for any reason have been disbarred from practice before the U. S. Patent Office. There are swindling attorneys as well as patent sellers and inventors are warned against them by the INVENTIVE AGE.

The Evolution of the Cycle.

During the year of 1804 the public were greatly interested and amused by the patenting of a machine or vehicle capable of carrying several persons and adapted to be propelled by the riders themselves. This velocipede, for such it was, consisted of a strong iron and wood frame provided with seats for the riders and mounted upon four wooden wheels; the front wheels being movable for guiding purposes. One of the rear wheels was provided with an annular toothed ring engaged by a train of power gearing mounted on the frame and adapted to be actuated by the riders through the medium of a crank attached to the shafts of one of the gears. Could the good citizens of that period have looked forward to 1896 and beheld the mammoth industries that have grown out of the pursuit of this invention, as represented in the featherweight "scorcher" of the present, they would certainly have given the odd looking vehicle more serious attention.

The introduction of the so called "dandy horse" (1) a number of years later, did not tend to increase the popularity of the cycle with the public at large for the very obvious reason that its usefulness was very much in question. This "dandy horse" was in effect a bicycle having a frame carrying a wheel at either end and provided with a seat for the rider. The front wheel was movable so that the machine might be guided. The rider's seat was so located that his feet would rest upon the ground and as he walked forward the machine was carried with him and supported a portion of his weight. Picture if you can a scorcher of the present day breaking a record on this cycling monstrosity.

Next in 1853 we get from the inventive genius, a tri-cycle provided with two large rear wheels arranged side by side with a seat between them, and a small front wheel; the latter journaled to the frame on one side only by means of a swivel pin whereby it might be turned to guide the machine. This front wheel was rotated to pull the machine forward by a rod connected to a crank pin on said wheel and provided with a handle to be grasped and operated by the rider. During the year 1866 Pierre Lallement, a Frenchman, introduced and patented in the United States, a safety bicycle provided with two wheels, one behind the other; the front wheel being arranged for guiding. Foot pedals were mounted on the spokes of the front wheel on opposite sides and were to be operated by the rider's feet. The rider of this machine was not supported in an upright position but met with the same difficulty as the rider of the present day in balancing himself by the turning of the front wheel.

Up to the year 1869 the interest in cycling seemed to be confined to the so called "cranks" and the inventors of the several machines, but during this year the public interest was awakened at last and eyes opened to the fact that the cycle afforded a useful and convenient means of locomotion and was something more than a mere plaything. A great many patents were taken out during this year but all these efforts were as yet very crude. The oscillating seat whereby the weight of the rider was utilized to propel the machine forward was a very popular form. This consisted of two wheels mounted in a suitable frame; the axle of the rear wheel being provided with cranks, one upon each end. The rider's seat was hinged so that it might oscillate up and down. Rods connected the seat with the cranks on the opposite sides of the machines so that movement of the seat would rotate the wheel. Stirrups were provided for the feet so that weight might be raised off the saddle upon the up stroke of the cranks and then brought to bear upon it again upon the down stroke. Both cranks revolved at the same time and no provision was made for the dead centre of the revolution with the consequence that the wheel was forced forward with a jerking motion.

The unicycle, or one wheeled machine, also met with much favor during this year. One prominent form of the same consisted of a wheel whose axle was extended on either side. Stout bars were mounted on the respective ends of said axle so that they might swing back and forth. These bars were connected at their upper ends by a saddle and were held in their vertical position by counter poise weights adjustably attached to their lower ends. The weights were, of course, of sufficient size to counterbalance the weight of any rider and by their adjustability could be arranged to counterbalance different sized riders by increasing or decreasing the leverage. This machine would no doubt not meet with favor with to day's rider who is constantly on the lookout to find some way to lighten his machine, as it was absolutely necessary for it

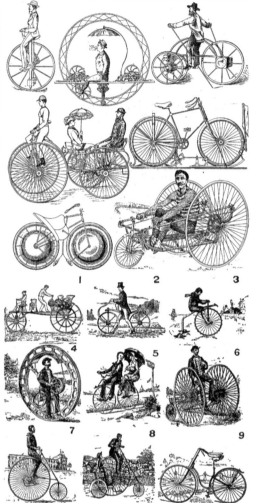

1   2   3

4   5   6

7   8   9

to carry a weight equal if not greater than the weight of the rider in order to keep him in an upright position.

After this year when cycling received its first impetus, it advanced steadily, but a great many of the inventors were still in the dark as to the practical principles of locomotion. This is plainly witnessed by a machine patented in 1877 by a Wisconsin genius. In form the machine was a tricyle, two of the wheels being arranged in the front and one to the rear with the seat supported between them. The front wheels were pivoted for guiding by means of the feet. The rider propelled himself forward by means of a couple of bars or sticks, one in each hand, and which he forced into the ground and pushed upon alternately upon the opposite sides of the machine.

In 1881 a patent was granted for a four wheeled velocipede for carrying three or more persons. This machine in fact was intended to take the

ing wheels capable of being depressed to engage the ground were mounted on each side of the machine. When the power wheel was operated the forward end of the frame was lifted by the forward friction wheel attempting to run up the inner surface of the main wheel and the weight of the frame and the rider was thus thrown on the main wheel forward of its contact with the ground with the consequence of a forward movement of the wheel. The rider was located entirely within the circumference of the wheel out of the way of all harm.

The rowing cycle of 1887 supplied a long felt want of some good out door exercising machine for oarsmen. This machine had three wheels and was provided with a regular sliding seat, the usual foot braces, and handles operating on slides to give as near as possible the movements of a pair of oars. These handles were connected to cords which passed about friction pulleys on the axles of the driving

elling over snow the runners would sink into the latter and thus cause the paddle wheels to strike and assist in propelling the machine forward.

In 1895 a bicycle was patented which could be used on either, land, water or ice. This combined machine had its wheels formed of hollow air-tight copper disks—were connected by a suitable frame and an ordinary crank and chain driving mechanism employed. The peripheries of the disks were provided with rubber tires for use on land. When the cycle was to be used upon the ice, the rear tire was removed, leaving exposed the toothed periphery of the disk. This periphery engaged the ice to drive the machine forward but the movement was a rolling, not a sliding one as in the cycle of 1894. The rear disk was also provided upon its sides with radially arranged paddle blades for use when the cycle entered the water. If the cycler desired a few miles spin upon the river some warm evening, he first released two weights attached respectively to levers

THE SIMONDS STEAM WAGON.

THE COPELAND MOTOCYCLE.

THE THORP SELF-PROPELLING VEHICLE.

THE OLSON CYCLE AND BOATING MACHINE.

place of the family carriage. The front axle of the machine was provided with cranks and the coachman seated over the same was supposed to propel the vehicle forward by operating said cranks. Other propelling devices were arranged under the rear seats so that the occupants of the vehicle might covertly assist the coachman on steep grades.

The epicycle of 1885 was a unique and interesting machine. It consisted of an annular ring or wheel having no central spokes or hub, and a frame supported within this wheel by small friction wheels that engaged its inner annular face. A power wheel provided with foot pedals was mounted on the frame and communicated with the forward friction wheel by and an endless belt. Suitable steer-

wheels to propel the machine forward. Clutches were provided between the friction wheels and the axles so that movement would be imparted to said axles in a forward direction only.

In 1894 a very successful ice bicycle attracted much attention. The most attractive feature of this machine was that any ordinary safety bicycle could be transformed in a few minutes into a perfect ice cycle. This was accomplished by securing runners to the front and rear wheels and in providing the rear wheel about its periphery with a series of projections that extended below the rear runner to engage the ice. Small paddle wheels operated by the main driving wheel were mounted on the respective sides of the machine. These wheels would not engage the ice but when the machine was trav-

or rods which were suspended from the front and rear axles. When he entered the water the air tight disks supported him, the paddle blades forced him forward, and weights hanging below the machine held it in an upright position in the water and prevented a capsize. A tourist mounted on this machine would not find it necessary to look for the bridge when he came to a river. He would simply release the weights without dismounting as he entered the water and after crossing the stream with as much ease as riding along a smooth road, would raise them again and continue his journey.

The latest novelty in the interesting art of cycling, is the bicycle skate. This comprises a body portion provided with securing straps or clamps for the feet and small ball-bearing pneumatic tired wheels mounted at each end of said body portion; said wheels being about three and a half inches in diameter.

It is claimed for this skate that a speed of from ten to fifteen miles per hour can be made after a little practice. It is not only intended for smooth surfaces but can be utilised on rough roads as well, both up hill and down. Uphill work in fact can be done with less exertion that riding a bicycle. It is also claimed that there is no danger of loosing control of the feet going down steep grades as one foot may be dragged to the rear in a cross direction and form an effectual brake.

A very recent patent has been granted for a bicycle which carries its own road although this is not the primary object of the invention. The rear wheel of this machine is peripherially grooved to partly receive an annular flange of greater diameter than said wheel and formed on a loose tread. Sufficient of the flange enters the groove to keep the tread always about the periphery of the wheel. The inventor says:

"When power is applied to the rear wheel to rotate it, it will tend to roll forward and upward on the inner circumference of the tread, the amount depending upon the power exerted and the resistance to be overcome, such movement tending to raise the rider and the rear of the frame. The centre of the power wheel is moved ahead of the centre of the annular tread and the weight of the rider and the frame is applied at the point of the tread intersected by a perpendicular passing through the wheel centre, tending to press that part of the tread toward the ground. As the power wheel is rotated the tread is thereby moved forward and the transfer of the point of application of the rider's weight is continous, maintaining the rotation of the tread and carrying the machine along." The inventor claims that he gains the weight of the rider as an additional impetus to throw the loose tread forward and thus propel the machine, as the power wheel is continually trying to ride up the flange on the tread forward of the contact of the tread with the ground. W. HARVEY MUZZY.

FROM "DANDY HORSE" TO 1889.

The illustrations (1 to 9 page 132,) show the evolution of the wheel from the "Dandy Horse" of Rochelle, France, about 1815, down to the period of the "safety," of 1890, now so universally in use and possessing many important improvements over the 1890 product. With what amazement the spectators looked upon the "Dandy Horse," with which a man could obtain the remarkable speed of 500 feet in two minutes—without falling off—and how little did any one dream in those days of the possibilities of 1896—the modern globe trotter on a 22 pound machine—and tandem racing against mile-a-minute express trains. Other curious horseless vehicles are illustrated on page 133.

AN ECCENTRIC BICYCLE.

The principal feature of this machine is the rear wheel, which is journalled eccentrically. that is, with its axle out of the center of the wheel. This is

calculated to give the rider a sort of gliding motion very similar to the ocean's swell. A choppy sea can be had by riding rapidly.

COMBINED WATER AND LAND BICYCLE.

This is the invention of a Chicago genius named Thore J. Olsen, and the illustration tells the story. The new hydrocycle is not much of a novelty when compared with this machine after all. What an elegant device for a fishing trip. The traveler may carry his necessary baggage, hunting and fishing tackle, yet the whole device weighs less than 100 pounds. On land it is designed to carry one person. but when water is reached a couple of friends can be accomodated. Nothing like it for genuine pleasure.

A NOVEL MOTOCYCLE.

In the fall 1889 there was exhibited in the streets of Washington a curious tricycle, or motocycle—the illustration herewith giving a fair idea of its

mechanism. The machine was built at Camden, N. J., the inventor being Mr. Lucius D. Copeland, of that city.

SELF-PROPELLING VEHICLE.

This is the invention of Thos. J. Thorp, of Chicago, Ill., and is so constructed that the vehicle body may be detached from the front wheels manifestly leaving a three wheeled vehicle upon which one may ride as on horse back, or the form of what here represents a horse being changed, the latter may constitute a suitable receptacle in which passengers may be seated as in an ordinary carriage. It can be run by electric or other power.

NEW SWISS BICYCLE.

The American consul at Geneva recently transmitted the drawing of a bicycle exhibited at the Swiss National Exposition. The principal feature of this machine is the back support, and a consequent utilization of an enormous force not applied

on any other bicycle—a "leg force " or "brace force," equal, it is estimated, to several times the weight of the rider. Greater speed, up-hill and on level ground, with less effort than in other "safeties" are superior points claimed for this Swiss machine.

THE SIMONDS STEAM WAGON.

This is an invention of real merit, by C. L. Simonds, of Lynn, Mass. The carriage weighs 437 pounds. The boiler is of the porcupine type, with 28 square feet of heating surface. The water tank holds ten gallons of water. The naptha tank holds five gallons, about what would be required for a 100 mile run.

Condition of Business in Patent Office.

The total number of applications awaiting action August 25th was 8,920, of which 2,527 have been pending between two and three months. The total number of patents, designs, trade-marks and prints issued was as follows :

Week ending August	4,		549
"	"	11,	515
"	"	18,	494
"	"	25,	485
Total for four weeks,			2,043

The Horseless Carriage.

The horseless carriage promises to be as pervasive in its social as well as its commercial influence as the bicycle. In Paris the dry good houses are all selling the horses that drew their delivery wagons, and using automobiles, and scores of electric dogcarts, and other horseless vehicles, are seen every evening taking their owners out to their homes in the suburbs, where land is rapidly going up in value. This is but the beginning of a great popular evolution. The horse in cities is practically forbidden to all except the rich. The horseless carriage is comparatively cheap, and a ride fed with oil or naphtha, at a cost of a few cents a day, will eventually put a carriage ride in the park within reach of any bookkeeper or clerk. The possibilities of this development are well shown by C. D. Lanier, in a late article. He says when a man earning $2,000 a year in New York City can maintain an equipage which will trundle him twenty miles away from his flat in an hour, a whole new class of citizens will become victims to the tennis, base ball, or golf habit, from which they are now sheltered by the mere inertia of time and space to be overcome. And with each advance in the art of moving rapidly there will be a corresponding increase in out-of-door sports, and a better opportunity to reach the fields and woods in short vacations allowed by the hurrying struggles of today.

The New Patent Law in Russia.

The Czar has sanctioned a new patent law in Russia, now in force. The salient features of this new law are as follows : Any new invention capable of being used industrially can be patented for fifteen years instead of the present periods, except arms, explosives, munitions of war, foods and chemical product ; but chemical processes and processes for preparing foods can be patented. By "new" is meant not previously published in print in any country, or published, worked, or patented by others in the realm. A patent in Russia, therefore, under this new law must be applied for on or before the issue of the corresponding United States patent. Old patents can be prolonged up to fifteen years by paying the corresponding tax to this period. Patents of importation will no longer be granted, but patents of addition on existing patents will be obtainable to expire at the date when the original patent would expire, the annual taxes on the original patent serving both. Taxes instead of being paid down at once will now be paid annually, beginning with a comparatively small tax and yearly increasing in amount. The cost of a patent of invention will probably be about £12 10s ($62.50), including translation of 1,000 words (extra translation 3s. (7.5c.) per 100), the applicant supplying the specification in English in duplicate, and drawings in duplicate, one Bristol board and other cloth, well executed in black lines, 14 inches by 8¾ inches or multiple of 8¾, including ¼ inch left blank at sides and bottom and 2 inches left blank at top (no marginal lines). An extra set of drawings should, if possible, be sent for use of the attorney. Any English words such as Figure 1, Figure 2, etc., should be inserted in pencil on the drawings, not in ink. The power of attorney must be in the Russian language. The annual taxes will probably be : Before the end of first year £2 15s ($12.50), before the end of the second year £3 ($15), and so on increasing in a gradually augmented ratio each year during the life of the patent.

Gas Motor Vehicles.

Any doubt which may have dwelt in the minds of engineers as to the feasibility of using gas as a motor for vehicular traction, must have been shaken, if not entirely removed, by the success which seems to have attended its application to the cars of the new tramway which has been formed from South Shore to St. Annes-on-Sea and Lytham. The gas is supplied by the Corporation, and it is said that the cost of running the cars will not exceed one shilling for every fifteen miles traversed. The motor is a double-cylinder gas engine of the "Otto" type, and it is placed in such position that all its parts are accessible for cleaning, oiling or repairs. The cars are self-contained, and are of the ordinary build in general use in the country. The machinery, although easily got at, is quite out of sight, and what is of more importance, there is said to be little noise, and no unpleasant sensation of heat or smell. This experiment in tramway traction by gas—the first we believe ever attempted in this country—will be watched with much interest.—*Trade Journals' Review.*

Andrew M. Carlsen, of St. Paul, Minn., sought to obtain a restraining order in the United States Circuit Court to prevent Anthony E. Lindstrom from manufacturing an animal trap which Carlsen alleged he invented and patented. Lindstrom claimed that his device was an improvement and not an infringement and the court so held.

The society of arts has awarded the Albert medal to Prof. D. E. Hughes, in recognition of the services he has rendered to arts, manufactures and commerce by his numerous inventions in electricity and magnetism, especially the printing telegraph and the microphone.

The crude petroleum production in the United States in the year 1895 was 52,983,526 barrels valued at $57,691,279, against 49,344,516 barrels in 1894, valued at $35,522,099. The total of production since 1855 is placed by J. D. Weeks at 709,713,403 barrels.

The report of the Chicago public library shows that 1,173,586 volumes were circulated during the year ending June 1, which breaks the world's record for free circulating libraries.

The longest artificial water course in the world is the Bengal canal in India, 900 miles ; the next is Erie, 363. Each cost nearly $10,000,000.

Claus Spreckels contemplates the erection of a beet sugar factory in California, which will consume 3000 tons of beets daily.

Palaces on Wheels.

The new vestibule "Limited" trains of the Northwestern line, running between Chicago and the twin cities of St. Paul and Minneapolis are the finest now running regularly on any road in the country. They are up to date, which in this age means ahead of anything before them. The train consists usually of an express car, buffet library car—solid comfort for gentlemen—private compartment sleeping car, a 16 section Wagner sleeping car and one or two 12 section sleepers added enroute from branches. Back of these are the ladies' day coach and general smoking car. All coaches are new and handsomely furnished and upholstered. The 16 section sleeping cars are gorgeously decorated and divided off into a series of salons or drawing rooms, each containing from four to six sections.

TYPE OF LOCOMOTIVE FOR NEW NORTHWESTERN FLYER.

The weight of these trains exclusive of the engine, averages 817,000 pounds. The putting on of these fast trains necessitates the use of powerful locomotives after the patern of the famous "999." These engines are the largest and most powerful ever run on the northwestern lines, capable of pulling these trains at the rate of 70 miles an hour. They stand 5 feet 6 inches high, which is to say that their drivers, inside the tires, are 66 inches in diameter, and including tires, 73 inches. The cylinders are 19 inches diameter by 24 inch stroke; weight on drivers, 81,000 pounds, and on engine truck, 45,100 pounds, or a total weight of engine proper, 126,000 pounds. The boiler is of the "wagon-top" type, 62 inches diameter, and has 281 flues, each 11 feet 6 inches long. The fire box is 96 inches long by 40 inches wide, and total heating surface is 1,665 square feet. The tender has a coal capacity of 8 tons, and holds 4,350 gallons of water sufficient for a run of 205 miles. The total weight of engine and tender is 226,900 pounds when loaded, 346,000 pounds. The wheel base of engine is one inch over 23 feet and of the engine and tender six inches over 47 feet. The total length is about 57 feet.

It is claimed that Mr. Tesla has at last succeeded in so reducing the cost of transmitting electricity that under his system electric power can be carried for a long distance with commercial economy on copper wire, even to the extent of 20,000 horse-power. The New York correspondent of the Philadelphia Press claims that leading capitalists have thoroughly satisfied themselves of the feasibility of the undertaking, and are now arranging under this system to carry electric power from Niagara to the city of New York for commercial use.

Experiments to test the relative efficiencies of plain and ribbed glass have shown most conclusively that much more light is thrown into the darker corners of rooms when ribbed glass is used. It would appear that the corrugations act to some extent as magnifiers or prisms, and the rays of light are transmitted with greater illuminating power and more general diffusion than in the case of plain glass.

A Joplin, Kansas, man has invented a tripoli filter that he thinks there is a fortune in. It is a simple contrivance intended to be attached to a hydrant, having two faucets, from one of which comes the filtered water which the pressure has forced through a cylinder of porous stone, and from the other a stream of unfiltered water as large as the size of the pipe will permit.

How Bicycle Tubing is Made.

Some drawn steel tubes have been made for years, for boilers and general use, but the great demand arose when the safety type of bicycle came into vogue, the diamond frame requiring the use of a greater length of tubing and necessitating that this should be as light as possible. There are variations in the methods for producing a cold-drawn steel tube but the principle of all is practically the same. Only a very high class of steel is suitable for the purpose, and that hitherto employed has been chiefly Swedish charcoal steel, containing a certain por²ion of carbon. She steel is taken in the form of a billet two feet long and about six inches in diameter. A hole is bored through the center and it is heated, annealed and rolled into the form of a tube about 1¼ inches in diameter, with wall of about 10 gauge. This is then drawn through a die and over a mandrel by means of a draw-bench until about 800 feet long, beautifully smooth and bright both within and without. This is not drawn at once, but in a number of operations and between each of them the metal has to be repickled an reannealed to prevent the crystallization to which the drawing process tends to give rise. The first drawings of the tube leave it about three-eights of an inch thick, but this gradually decreases until a tube is produced which is of the thickness of stout writing paper. This is the class of tube employed in bicycles and that imparts a strength and rigidity out of all proportion to its lightness.

About 12,000 workmen are employed in the logging industry of Minnesota. It is estimated that the total amount of white pine standing is 14,424,000 feet and of red or Norway pine 3,412,473,000 feet. In 33 counties there are 10,889,000 acres of natural forest, and in the whole state there are 11,800,000 acres of natural forest, not including mere brush and swamp land. The annual cut of pine for each of the last three years is estimated at 1,500,000 feet. The consumption of mercantile hardwood lumber in Minnesota is estimated at 100,000,000 feet annually.

In 1886 Hoffman determined the presence of the tuberculosis on the bodies of flies collected in the room occupied by a consumptive. Six years later, a physician of Switzerland, Dr. A. Coppen Sones, proved that infection can be, and actually is, carried not only by the bodies of flies, but also by their feet. Flies which have been infected with the bacilli were permitted to walk across the surface of sterilized potatoes. In two days' time numerous colonies of the bacillus prodigiosus made their appearance.

Herr Lilienthal, the engineer, who for many years has been experimenting in the building of flying machines, was killed August 11 near Berlin as the result of an accident to his apparatus. He was 48 years old, and had been engaged in his experiments along this line for 27 years. About two years ago he succeeded in "flying," 'or soaring, a fifth of a mile.

The American liner "St. Louis" has broken the record from Southampton, and it is now 6 days 2 hours 24 minutes, a cut of more than three hours. The average speed was 20.86 knots.

Largest River Steamboat in the World.

(Continued from first page.)

rooms, and four suites of parlors and bedrooms, with brass bedsteads. In addition there are 285 berths in the cabins and 120 berths for the crew.

For the illustrations and data herewith we are indebted to Smith & Stanton, publishers of Seaboard, the representative journal of the shipping interests of America. The smaller cut presents a comparative view of the various notable steamers that have plied on the Hudson since Fulton's Clermont came out in 1807, drawn to a scale, and appear in the exact proportion which they hold to one

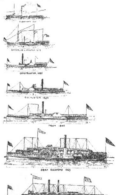

From Seaboard.

COMPARATIVE SIZES OF HUDSON RIVER STEAMBOATS, 1807-1896.

another. The Clermont could sleep about thirty passengers, while the Adirondack has sleeping accomodations for over 1,000; the Clermont ran at a speed of five miles an hour, while the big Adirondack glides along at a 20-mile gait. Following will be found the size and tonage of the boats in question:

Clermont, 1807, 133 ft. in length; tonage about 180.

Chancellor Livingstone, 1816, 157 ft. in length; tonnage 496.

Constellation, 1825, 149 ft. long; tonnage 275.

Rochester, 1836, 210 ft. in length: tonnage 491.

Troy, 1840, 294 ft. in length; tonnage 724.

Dean Richmond, 1865; 360 ft. in length; tonnage 2,555.

Adirondack, 1896, 412 ft. in length; tonnage 3,644.

Kite Flyers.

A convention of aeronauts will be held in Boston this month. It will be held under the auspices of the Boston Aeronautical Society, of which Professor W. H. Pickering of the Harvard Observatory is president. This is the only society of the kind in this country, but there are several in Europe, and the International Aeronautical Association, with headquarters in London. In order to promote public interest in the convention and to encourage kite designing and kite flying for the purpose of scientific experiment, the society proposes a competition for cash prizes. For the best designed and best flying kite a prize of $150 is offered, and a special prize of $100 will be given by Octave Chanute, ex-president of the American Society of Civil Engineers, for the best monograph on the kite giving a full theory of its mechanics and stability, with quantitive computations appended. Other and more important prizes will be bestowed upon successful inventors of aeronautical machines.

The Austin Current Motor.

The name "Current Motor" is applied to a mechanism that utilizes the power of the stream or current of a river in such a way that it can be applied to operate machinery. Several crude attempts have been made in the past to accomplish this, but it remained for the F. C. Austin Manufacturing Company, of Chicago, Illinois, to be the pioneers of a practical current motor.

Rivers have been utilized to run machinery by damming them and using a water wheel, but this necessitated the expense of the construction and maintenance of the dam, and was only available where the conditions were favorable, which necessarily very much limited their use. The current motor, however, can be used without a dam, in any river, in any location in the river where the water is deep enough to admit it.

There is almost no limit to the use to which the power generated by a current motor can be applied. For example, it may be used for generating electricity for electric lighting purposes, and for va-

THE AUSTIN CURRENT MOTOR.

rious machines used on a farm now run by steam, but its chief use will be in connection with irrigation and mining.

There are many localities in the West, and in fact all over the country, where the land adjacent to the river is too high to admit of the water being conveyed to it through ditches by means of gravity. Under such conditions the land is now generally not irrigated, particularly in the arid regions, except to a limited extent in those localities where vegetables and fruit are raised, where the value of the crop compensates in a measure for the expense of pumping water by steam or gasoline engine, both of which require fuel and an engineer. The capacity of the windmill is too limited to admit of its general use for irrigation on a large scale. To such localities the Austin Current Motor will prove a boon.

This motor has been experimented with and tested thoroughly and its practicability practically demonstrated. The only limits to the power, and consequently to the amount of water that can be pumped by the Austin current motor, are the dimensions of the paddles, the number of them and the force of the current.

The machine, which is anchored in the river, consists of a pontoon carrying two endless chains, to which are pivotally attached at suitable distances, reversible paddles having floats at their upper ends that buoy them up in their course through the water. By an ingenious device these paddles enter and leave the water in such a way that they do not detract from the efficiency of the machinery, but on

the contrary rather add to it. To the wheels put in motion by these cables is attached an elevator, carrying large buckets, which, when entering the water to fill themselves, add to the generation of power and are so arranged that when they reach the top of the elevator they discharge their contents with the least possible friction into a trough, there to be conveyed to the shore, where a reservoir has been constructed to collect the water and distribute it through canals and lateral ditches over the land.

The first outlay for a current motor of this description is comparatively speaking, not large, particularly when the enormous enhancement of the value of the land is considered, and the cost of running it is nominal. It works incessantly day and night without an attendant, and if the capacity of the reservoir is sufficient a large volume of water is accumulated to be used at the proper season.

The manufacturers, F. C. Austin Manufacturing Company, of Chicago, Ill., will furnish any further information that may be desired.

Wind Power.

A paper by E. O. Baldwin, published in the "Canadian Electrician," deals with the author's at-

tempts to utilize the power of the wind for producing current. They are strong at windmills in Holland, where they use them for pumping purposes, but the difficulties in the way of getting a useful supply of current are much greater than when the only work required is to lift water a short distance and speed regulation is of no importance. Mr. Baldwin's regulating apparatus consists of an additional coil on the field magnets to act as demagnetizing coil, and an electro-magnet which attracts a pivoted iron arm against the resistance of a spring and throws the demagnetizing coil into the field circuit whenever the voltage rises above the desired point. There is a very rapid vibration of the arm, and Mr. Baldwin states that he can adjust the regulation so that lamps can be run directly off a dynamo belted to the windmill without any perceptible flickering. Normally, however, a storage battery is used.—*Electrical Engineer.*

The death of Otto Lilienthal the German inventor of a jumping or soaring flying machine, is announced. He was experimenting with his invention and when up some distance from the ground the machine failed to work and the inventor fell to the ground, receiving fatal injuries.

Twenty-four governments, including the United States, nearly all the European nations, Japan, China, Persia and India, have officially notified the French Government of their intention to exhibit at the Paris International Exposition in 1900.

Death of Dr. George Brown Goode.

Dr. George Brown Goode, Assistant Secretary of the Smithsoian Institution and Director of the United States National Museum, died at his residence on Lanier Heights, Washington, D. C., Sunday evening, September 6, after a brief illness.

His most distinguished services to science have been in the department of ichthyology in association with the Fish Commission and his comprehensive work in the organization of the United States National Museum. Of these, as well as of his beau-

tiful private life and his great interest in all that pertains to good citizenship, a preciative eulogies have been made.

In other connections it might not be known, however, to those who are devoted to inventing that Dr. Goode was deeply interested also in their work.

The department of Arts and Industries in the Museum was organized entirely on the basis of the inventor. The collections were divided into classes according to the industry or activity on which they were based.

Especial notice is directed to the music collection, the collection of fisheries, of naval architecture, of animal products, of graphic arts and of physical apparatus. The specimens connected with each one of these classes were not upholstered for effect. They were not divided nationally neither historically, any further than such a concept was helpful to that other thought, which was ever dominant in his mind, of the evolution of humanity through its expression in the arts themselves. For instance, in his music collection he at once observed that the first effort in the world in this direction was the use of simple objects to produce rhythm with or without a continuous musical note and, so, rattles of all kinds leading up to drums and other substance occupied the lowest rank.

In the next class he quickly observed, as others have done, that musical tones are produced by the vibration of a string or a reed and by the vibration of a column of air in a tube, so the xylophone, the banjo, the clarionet and the horn were placed at the beginning of four classes in which by evolutionary processes the instruments were developed to higher and higher perfection.

In the Fishery Court he began with the simplest devices for picking up the products of the sea and carried his investigations through all the various middle steps to the latest elaborate devices for deep sea dredging.

In the class of Naval Architecture the raft, the ball-boat, the birch canoe and the dugout, very rude and simple at first, are shown to be the beginnings of our modern cruisers in which are made to co-operate all the thoughts of mankind upon this subject.

In the Animal Product collection it was Dr. Goode's intention to show how many species of animals, how many parts of the animal and in how many ways the animal kingdom could be made subservient to the wants and happiness of man. In this art also the first efforts of course were exceedingly rude and by a series of inventions, perfecting always that which had gone before, the whole animal kingdom at last is brought into the service of our race and not a particle is lost.

In these sections and departments of the Museum for which Dr. Goode did not assume to be personally responsible, the influence of his thought was felt and other series than those mentioned were also arranged upon the inventive idea. It is, therefore, with great pain that the INVENTIVE AGE records the death of one of the noblest citizens as well as one of the foremost students of human culture on the line of evolutionary invention and extends to all scientific workers and inventors its heartfelt condolence.

How Carnegie Rose from Obscurity.

Andrew Carnegie, in a communication to the Youth's Companion some time ago gave some interesting particulars regarding his early experiences in money-earing. The article was headed, "How I served my Apprenticeship as a Business Man." A similar and no less interesting article, "How I Became a Millionaire," appeared in a recent issue of Cassell's Magazine, London. It tells of the struggles of the Carnegie family, in the cottage in Dunfermline, Scotland, a view of which appears herewith through the courtsey of Iron Trade Review. The father was a master weaver in Dunfermline, who emigrated to America when Andrew Carnegie was a mere boy, and entered a cotton factory here. The son became a "bobbin boy" at 12 years of age, receiving $1.20 as weekly wages. At thirteen, he was set to fire a boiler in the cellar of a bobbin factory, where "the responsibility of keeping the water right and of running the engine, and the danger of my making a mistake and blowing the whole factory to pieces, caused too great a strain, and I often awoke and found myself sitting

up in bed through the night trying the steam gauges," At 14, he obtained a situation as messenger boy in the telegraph office at Pittsburg, where he became an operator after a time, and earned $1 a week extra by working evenings on telegraph reports for the newspapers. He attracted the notice of Thomas A. Scott, of the Pennsylvania Railroad, whose clerk he became at $35 a month. Mr. Scott himself earned at that time $125 a month. He put the young man on the scent of good investments, which ultimately led him to occupy the unique position which he holds to day—that of being not only a millionaire, but one of the greatest manufacturers in the world.

A New Illuminant.

The London correspondent of the Manchester *Courier* publishes a remarkable account of a new illuminant, which, if all that is said of it is true, will push both gas and electric light very hard. For its production no machinery is required save that contained in a portable lamp neither larger nor heavier than is used with colza oil or paraffine. This lamp, it is declared, generates its own gas. The substance employed is at present a secret, jealously guarded by some inventive Italians. The cost is declared to be at most one-fifth of that of ordinary gas, and the resultant light is nearly as bright as the electric light and much whiter. A single lamp floods a large room with light. The apparatus can be carried about as easily as a candlestick and seems both clean and odorless.

New Bicycle Boat.

A new bicycle boat was recently tested by its inventor, Mr. Weston Clark, of Berkley, Cal., and pronounced a success. The boat is a canvass-covered canoe, 14 feet in length, weighing 140 pounds with sail and mast, and with a beam of 33 inches. In the stem and stern are three air-tight compartments. The operator sits in the stern and works his boat by pushing with his feet upon two iron shoes, which revolve a propeller extending to the rear

Norfolk is to have a new $500,000 hotel. Sylvanius Stokes of Baltimore is principal promoter.

Chicago has 270 miles of electric railway—and more to follow.

The Silver Question.

Some of the most lucid and convincing arguments on the currency question are contained in a letter recently written by Mr. George D. Boulton, who has long held the responsible position of manager of the foreign exchange department of the First National Bank of Chicago, and considered one of best posted men on the question of finance in the country. The letter was written in reply to an enquiry from a business friend in Barrie, North Dakota, and is as follows:

There are a good many leading points on this question which I think can be briefly expressed and which appeal strongly to my side of the argument. One of the most urgent motives of the silver party is that they want cheap money. By that I suppose they mean money they can borrow cheaply or earn cheaply. Now the cheapest money in the world is in the strongest gold country—viz., England. The dearest money in the world is in the silver countries. For example, money in London today is 2 per cent per annum, while money in Mexico, China, Spain, India, and in fact in all silver countries in the world, commands a loaning value of from 12 per cent upward. In the other gold countries of Europe, while money is not so low as in England, the rate varies from 3 to 5 per cent to the borrower. I may cite as a good example of the two currencies two States adjoining one onother in South America—one Guiana, a gold country, with money at 4 to 6 per cent per annum, the other Venezuela, with like soil and climatic conditions, a silver country, where interest rules at 10 to 12 per cent per annum.

Should we depart from a gold basis Europe would undoubtedly send in all the currency securities—that is, securities that may be paid in anything but gold—to us, requiring an export of either gold or its equivalent in trade. It it takes gold it takes that much of our money circulation. If it takes merchandise it takes that at a largely reduced value. The consequence would be that shrinkage in money circulation would run into very large figures, while we could not put out silver or certificates sufficient to take their place for many months or years, so during the next three four years, instead of the circulation increasing, as silverites hope, it would materially decrease. After a lapse of time, no doubt, by putting their printing presses and mints to work, they could largely inflate our currency with new issues. Currency depletion means low prices for labor and everything else. Currency at a fair rate per capita means prosperity. Currency inflation means danger again.

Going back into history we find Europe using largely silver and gold together. With the expansion of trade one country after another found by sad experience their inability to keep the two values on a parity. England was the first to depart from this custom, then Germany, then France, Holland, Belgium, Italy, Austria, and, last of all, Chile. It was from no prejudice on their part, but from the requirements of trade, that this course was taken.

We can only have one standard, be it of gold, silver or anything else, and the experience of the world has been that gold was the best. Again, where the country is most sound on its currency question you will find the highest civilization. Where money is debased, or is other than the recognized standard of the world, civilization is on a much lower plane. We can find at the present time no silver country in the world, I think I might say without exception, that is in a prosperous condition, whose Government securities command respect and full prices in the markets of the world. To this statement our friends from the West will probably take exception, and cite as an example of a silver country being prosperous and in good condition the case of Mexico; but they will find it difficult to support their assertions. The writer had occasion last month to buy in the City of Mexico $50,000 of bonds issued by the Mexican Government. These bonds were bought at the rate of 48 cents on the dollar in silver, the net cost to the purchaser being $24,170 in Mexican silver. As the money to pay for these bonds came from this country, the amount of American funds used in the purchase of $50,000 Mexican Government securities was $13,012.11, or about 26 cents on the dollar. Now it seems impossible for any country to be in a sound and prosperous condition whose securities are so heavily discounted as in the above case.

Looking at the matter from an intellectual standpoint, we find arrayed on the gold side the high intelligence of England, France, Germany, Italy, Holland, Belgium, Norway, Sweden and Canada. On the other side we find an inferior grade of intelligence, an absence of public schools, and a lower plane of morality, as in Spain, Portugal, South American States, Mexico, China, etc. On which side shall we array ourselves?

Of course you understand it is not the intention of the party in power, or the gold party, to disturb the present silver circulation of the country, which is now $500,000,000. There is no desire to demonetize that. On the other hand, the whole contention is that all of our circulation shall be kept on a parity with the gold standard, and that this $500,000,000 instead of being reduced in value will remain equal to gold anywhere.

Borrowers throughout the country will have to recognize the fact that undoubtedly they will have to pay more for loans with silver ruling than they now do with gold. Again, if gold remains the standard, and we give our indorsement of the principle that we believe it the only standard for us, the money markets of the world will be open to us, and instead of having to pay a high rate for money borrowed the chances are we will have to pay a very much reduced rate—less than that which even now prevails. It is estimated that London alone has many hundred millions of idle money in its banks waiting for this matter to be settled, which will undoubtedly be released and used to a large extent on this side, if we commit ourselves unequivocally to the recognized standard of European nations. This course of events will be, if we make the change in accordance with the platform of the silver people, that in November, as soon as the silver President is elected, there can be no doubt at all but Europe will return our securities in large amounts. For these we have to pay gold or its equivalent.

This will entail a large export of the gold we now hold or of commodities. Gold will at once advance to a substantial premium. No legislation can probably be made by Congress until well along in the summer of 1897, during which period our circulation will be very largely depleted by export and hoarding. The return of our securities have got to be at very much below the present valuation ruling on our stock exchange—probably 15 per cent to 25 or 50 per cent.

If we can avoid a serious panic during such a crisis we may regard ourselves as fortunate. Under the most favorable circumstances we must look for great disturbances in value to all classes, a disorganization of labor and a hardening of money and general financial trouble, which will be felt by all classes, whether the farmer the laborer the mechanic or capitalist. Capital can always take care of itself and will feel the trouble the least, as it can largely unload its burden onto others. Now, legislation in favor of silver, when it comes, must be at least from nine months to a year off, and at the best it cannot do anything which will speedily restore our circulation to its normal amount per capita, as it takes time to coin silver, the capacity of our mints at present being only about $5,000,000 a month, or $60,000,000 a year.

The following can almost be taken as axioms:

No silver country is prosperous.

No silver country has a stable and firm government.

In no silver country is general labor well paid.

No silver country has its government securities at par.

No silver country has good public school facilities.

Cheap Telephony.

Switzerland may be justly described as a model Republic. It has no revolutions as in South America; no fierce political squabbles as in the United States. It goes its calm course undisturbed by the wars, and rumors of wars, by which its surrounding neighbors are constantly agitated, and, when two nations, who have almost come to blows, about some important matter, agree to arbitration, the disputants, as a rule, come to a settlement on the peace-inducing banks of the lake of Geneva. In education of all kinds, commercial, classical, technical and physical, Switzerland has, for many years, been in the front rank. The administration of the country, in all departments, is carried on most thoroughly, and at a minimum of cost, the President receiving, for his services, only a few hundred pounds per annum. Its railways, posts and telegraphs are admirably conducted, and at very moderate rates. Switzerland, having had a half-franc (*5d*) telegram long before England, which has a density of population three times greater, thought it practicable to reduce its minimum to *6d*. The smaller country has now taken another step in advance of other nations, on which she is to be congratulated. It is now possible to telephone from the smallest village to any place in the country, at a fee of from 1*d*. to 4*d*. for the most distant points (about 200 miles as the crow flies) on instruments *on which one can hear with perfect distinctness*, and which are kept in perfect order. The italics are ours.—*Indian Textile Journal.*

Japan's Diet voted $45,000,000 for the construction of railroads, telegraphs and cables at its last session, and $97,000,000 for the construction and purchase of war materials and ships. Since January, 1895, $600,000,000 has been invested by Japanese in banks, railroads and other companies.

Government Telegraphy.

[Abstract of Argument made by Mr. Delany before the United States Senate Committee.]

It is a curious fact that while great improvements have been made in telegraphy, and notwithstanding that telephony has come and covered the earth with its wires within the past twenty years, the arguments put forward against postal telegraphy have not changed in the slightest degree.

It is significant that none of the improved methods of telegraphy now in use originated within the controlling telegraph organization, all having come to it by purchase of competing lines, or from individuals outside. I refer to these matters because the opponents of government telegraphy have invariably advanced the argument that goverment control would discourage invention and improvement in systems. With the exception of its own Wheatstone system, all the great improvements used by the British post office are importations.

According to the report of the British Postmaster General for the year ending March, 1895, 66,189,000 messages, averaging 15 words each, were transmitted at an average cost of 15 cents, and, in addition, there were transmitted 5,400,000 press dispatches, averaging 120 words each, at a cost of 9 cents, or nearly 14 words for a cent. Nor is this all. There were 1,600,000 railway messages, averaging 25 words in length and representing 25 cents in value, transmitted free.

The press rate in this country averages about one-half a cent a word.

Now, as the deficit for 1895, in England, including insterest account, amounted to a little over $2,500,000, it will be evident that had the free messages been paid for at regular rates and press dispatches paid at a rate that such dispatches are charged for in this country, the British telegraphs would show a good balance of profit. In 1870, when the government took control, the business amounted to but 7,000,000 messages, an increase of more than ten fold in twenty-seven years, while the rates have been reduced from a maximum of about 4 shillings to a uniform rate of 6 pence.

The British operator has had two increases of pay since 1881, while his American brother has had four reductions, and to day the British operator is better paid for the same amount of work, and by his environment occupies a higher plane of comfort and contentment, than the American operator. Good behavior and diligence in his duties warrant him a life position, from which the whim or caprice of none can drive him. His increasing years of service are taken into account in various beneficial ways. He has his yearly vacation. He is not cut off in sickness, and, most important of all "he is not turned down" in old age, but retired on a pension, porportioned to his years of service. I can not conceive of a stronger incentive to a government system of telegraphy in this country than the example of thorough efficiency and success presented by the British post office.

No telegraph operated by the government in any country is conducted with a view to pecuniary profit. The aim is to spend all surplus earnings in improvements and extensions, and if none are necessary, then a surplus is prevented by a reduction of charges.

No one acquainted with the executive officers and heads of departments of telegraphs in this country can charge them with incompetency or lack of clear vision. They are experienced and able officials, familiar with all the details of their business. It is, therefore, strange that the wealthy few who control and dictate the policy to be pursued have not deferred more to the opinions of these managers, and been satisfied with a slower inflation of their shareholdings. The ability to so manage a vast concern as to earn dividends on a capitalization at least double what it should be, and this in times of great depression in business, commands admiration, but surely the wealthy owners have cut out a hard task for these men. They have been so hampered as to warrant the conclusion that the most primitive methods were the conditions most desired, and that difficulties in the way of cheap telegraphy should always be encouraged in order to maintain a great discrepancy between the cost of sending a message by wire and one by train. It has always seemed to me that the natural desire for great profits could have been met much better by a policy of encouragement of improvements, warranting cheaper rates and insuring an increase in business which would more than compensate for the reduction in charges. I doubt very much whether telegraphy could be carried on any cheaper by the government than it is now conducted by the companies if the same methods of operation are to be retained. Expertness of the operator has reached its highest development. Machine methods are as old as hand manipulation, but in this country they have not been used to any considerable extent and their development taken advantage of. A wrong start has been adhered to persistently, owing in a great measure to overconstruction of competing lines and multiplication of wires; and so long as one company gathered all the others in as fast as they came along, there were wires to spare, and therefore, as those in control argued, there was no use for increasing speed. Besides, wires afforded a basis for stock issuing. If this convient mine could have been ignored, it would have paid the oompanies much better to have abandoned the poorly constructed lines and concentrated traffic on a comparatively small number of well-constructed lines of high conductivity operated by machinery. The companies can hardly be blamed now for not taking down thirty poor iron wires and putting up one good copper conductor in their stead, even though it is now entirely practicable by machine working to make the single wire carry more messages for average distances than the thirty hand-worked wires, even when quadruplexed, so that four messages may go simultaneously.

The argument of those whose interests compel them to defend them by disparagement of improvements, which must surely render their great network of wires unnecessary and useless, is that machine working would be slower than hand working. With plenty of wires, no practical telegrapher will deny that a single short message can be sent by hand in the same time that it takes to perforate or prepare it for transmission by the machine system. The average layman is by this fact frequently deceived into grave error and by not pushing the comparison further. If the message be a long one, or if there are a thousand messages to transmit, it might take two days to get them off by hand, whereas, if there are a sufficient number of perforators, the whole lot could be transmitted in a few minutes. A perforating operator will prepare messages at the same rate of speed that a Morse operator can transmit them by hand, and a transcribing operator will typewrite them as rapidly as a sound-reading operator can receive, while the machine transmitter will send the dispatches as fast as 70 to 170 perforators can prepare them, or afford on an average, according to length of circuit, the same carrying capacity as 70 to 170 circuits worked by the present Morse system. One man could send them as fast as seventy men could prepare them.

Mr. chairman, without machine transmission, telegraphy, under the provisions of this or any other bill would not be practicable, and the rates stipulated may be seriously questioned on the basis of present hand methods of operation. In expressing this opinion I have no fear of contradiction from any disinterested telegrapher. I do not think that any government management could possible conduct a telegraph service on a cheaper basis at the present capitalization and by the present methods than it is now conducted, without the use of a machine system.

Twenty years ago the highest average of transmission over a single wire was, by the quadruplex system, about fifty words per minute; the telephone was only thought of for local use over distances of a few miles. Now it is practicable to telegraph 2,500 words a minute between Washington and New York, and 1,000 words a minute between New York and Chicago, while the telephone carries speech 15,000 miles.

The Wheatstone system has an average of about 150 words a minute. The English Wheatstone system is used between New York and Chicago. The average by hand is about fifty words a minute. In view of what has already been done. I do not think that any telegraph electrician will dissent from these propositions: That with machine transmission and chemical recording by the method referred to, over a copper wire weighing 850 pounds to the mile, and with an ordinary current power such as is used for quadruplex working, 1,000 words per minute can be plainly recorded over a distance of 1,000 miles, or, say, from New York to Chicago, and that over such a line 2,000 words per minute can be plainly recorded from New York to Washington and between other points throughout the country in the same ratio, according to distance. Last October, over an actual line, having but 130 pounds of copper to the mile (Philadelphia to Harrisburg and return), 216 miles, 940 words per minute were plainly recorded in dots and dashes, the current used being but 120 volts. This trial was conducted in the presence of a board of well-known electrical experts. With this system 8,000 words per minute have been recorded over an experimental line.

Here Dr. Delaney submitted estimates of cost and earnings of a line of two 1 ohm per mile copper wires between New York and Chicago operated at 1,000 words per minute.

I think generally, with reference to my experience in regard to this matter, that $25,000,000 applied to wires such as I have described in this statement would give greater facilities for carrying telegraph dispatches in this country than is now furnished by the wires of the Western Union Telegraph Company. It might not reach all the offices now reached by the Western Union Telegraph Company, but it would cover the large points which would take more than three-fourths of the traffic. I am unable to understand why Congress has not long ago authorized the Postmaster General to fix a maximum rate for telegraphic letters, and contract with the lowest bidders for their transmission, especially between cities separated by any considerable distance.

Engineering Tools at Pompeii.

Under the title of "Things of Engineering Interest Found at Pompeii," Professor Goodman gave his inaugural lecture in the engineering department of the Yorkshire College, Leeds. The lecturer remarked that he had recently visited Pompeii, and was not only charmed by the great beauty of the works of the ancient Romans, but also by their extreme ingenuity as mechanics—in fact, it was a marvel how some of the instruments and tools they were in the habit of using could possibly have been made without such machinery as we now possess. After explaining the situation and destruction of Pompeii by showers of ashes and mud, not lava, as is usually supposed, in the year 79 A. D., Professor Goodman showed a series of about fifty lantern slides, prepared from photographs taken by himself in Pompeii last Easter. The streets, he explained, were used as waterways to carry off the surface water, and probably sewage, from the houses. The pavements were raised about a foot above the streets, and stepping stones were provided at intervals for foot passengers. The horses and chariot wheels had to pass between, and in many places deep ruts have been worn by the chariot wheels in the stone-paved streets. The water supply of Pompeii was distributed by means of lead pipes laid under the streets. There were many public drinking fountains, and most of the large houses were provided with fountains, many of most beautiful design. The amphitheater, although a fine structure, capable of seating 15,500 people, was small compared with many in Italy. The bronzes found at Pompeii reveal great skill and artistic talent. The bronze brazier and kitchener were provided with boilers at the side and taps for running off the hot water. Ewers and urns have been discovered with internal tubes and furnaces precisely similar to the arrangement now in modern steam boilers. Several very strong metal safes provided with substantial locks have been found. The locks and keys are most ingenious and some very complex. On looking at the iron tools found in Pompeii one could almost imagine he was gazing into a modern tool shop, except for the fact that the ancient representatives have suffered severely from rust. Sickles, bill-hooks, rakes, forks, axes, spades, blacksmith's tongs, hammer, soldering irons, planes, shovels, etc., are remarkably like those used today; but certainly the most marvelous instruments found are the surgical instruments, beautifully executed, and of design exactly similar to some recently patented and reinvented. Incredible as it may appear, yet it is a fact, that the Pompeiians had wire ropes of perfect construction.

Drifting Bottles to Indicate Ocean Currents.

Among the scientific investigations conducted by he Hydrographic Office is one to determine the direction and strength of the principal ocean currents, by observing the drift of bottles, which are furnished by the office to the steamship companies to be thrown overboard. The office has secured some valuable results from these experiments. The time they were dropped into the water and the exact locality are noted on canvas attached to the interior of the bottle, so that the direction taken in the drift and the distance covered since last sighted can be accurately calculated. All ship captains are requested, on picking up a bottle, to inform the Hydrographic Office of the fact, and to indicate on the canvas the longitude and latitude, together with the date. The last chart issued by the office shows the general direction taken by several of these bottles, the distance covered, and the average drift per day. One bottle in four years has coverd 5000 miles of water and still drifts, although it is now bound over the same course originally taken by it. The experience of the office so far indicates that when one of these bottles gets in a current it will remain there and continue to float until it reaches the place where it was first cast adrift. The course of the hundreds of bottles serves to clearly indicate the two main features of the general surface circulation of the waters of the North Atlantic.—*Iron Age.*

Japan is making marked progress along all lines. In the last decade her exports have jumped from $75,000,000 to $300,000,000.

Photographing Over a Wire.

Modern science has yielded many wonderful things during the last quarter of a century, and along with the telephone, the phonograph and the other marvelous applications of electricity it is not improbable that before many years telephotography will take its place.

Dr. A. E. Kennelly, who is associated with Professor Edwin J. Houston in Philadelphia, recently declared to a New York Herald reporter that "seeing over a wire, or photographing over a wire, is not in itself more improbable or impracticable than hearing over a wire—that is, telephony—was thirty years ago, and since telephony has become so complete a success he must be bold who would say that seeing over a wire is not in the near future.

"The problem is, however, still unsolved, and appears to be entirely beyond our grasp at the present moment; although, strange as it may seem, the electric current transmitted along a wire is known to have properties which are essentially similar to those of a beam of light moving along a wire."

In this connection it should be said that applications for patents have already been made by inventors covering schemes of apparatus for seeing over a wire. While some commercial success might be obtained from a system of seeing over a wire, it is in every way doubtful whether photographing over a wire could be made a commercial success.

"As I have said," continued Dr. Kennelly, "there is no means of photographing over a wire in the ordinary sense of the term. It is possible, however, and has been for some little time, to produce in Philadelphia a picture specially prepared at New York city, at the other end of the line.

"This is done by having two pens running up and down sheets of paper at the same speed, one pen at New York and one at Philadelphia. The picture at New York has been so prepared by suitable processes that the bright portions, the lights and the tones, are non-conducting, and the dark portions, or the shadows, are conducting, or vice versa.

"As the pen runs over the conducting portion the electric current is sent through the pen by the wire at Philadelphia, and causes the pen to make a straight line over the portion which corresponds to the conducting portion of the line on the New York picture; the pen misses, or fails to make a mark, for the remainder of the line in New York.

"By running up and down synchronously over the line several times in succession the two pens will have covered the entire picture, and the Philadelphia pen will have made marks corresponding to the conducting portions of the plate at New York. A picture will therefore be reproduced in black and white at Philadelphia corresponding to the conducting and non-conducting portions of the New York plate."

American Locomotive Plant for Russia.

An entire locomotive making plant is to be taken to St. Petersburg from Philadelphia in a few days on the British steamer Laleham, which has been chartered for the purpose. The plant is to be erected at Nijnii-Novgorod, the commercial metropolis of the interior of the Russian Empire. Contracts for machinery for the plant, amounting to over $500,000, were awarded to American manufacturers, most of them Philadelphia concerns.

The plant is to be built for the Sarmova Works, an entensive establishment engaged in manufacturing cars, steamboats, steam boilers, and employing 5,000 hands. The locomotive plant will have a capacity for building 200 engines a year, and will employ about 1000 hands. All of the foremen and engineers will be Americans. The buildings have been completed and are now ready to receive machinery.

The Czar has given valuable eucouragement to the enterprise. Nearly 85 per cent of the railways in the empire are operated by the government, and the new company will get a great share of the work for them. The company will be known as the Russian-American Manufacturing Company. The consignment will aggregate over 3,000 tons.—*Age of Steel.*

The Court Protects the Inventor.

In 1888 John B. Russell, a young Detroit man, revolutionized the art of making gelatine-covered pills. Heretofore the drug to be dipped was made into plastic pills and impaled on rows of fine needles held in place on thin strips of wood and dipped in melted gelatine until they were covered. The trouble with pills so treated was that each one had a small puncture which did not exclude the air nor prevent the drug from oozing out, to the discomfort of the consumers.

Mr. Russell substituted a row of small tubes on a hollow stick, and by means of a simple air pump created a suction through the tubes. This stick was first dipped into a box of uncovered pills, and each tube readily picked up a pill. They were then dipped in gelatine and all covered except the lower part at the tube. When the top of the pill was sufficiently hardened, another row of suction tubes picked them up and the lower part was dipped. In this way thosands of perfectly covered pills were made in the same length of time that hundreds could be made the other way, and at the same time the work was done better.

Some time ago Frederick Stearns & Co. adopted Mr. Russell's idea and made a similar dipper, the principle of which was the use of suction to hold the pills while being dipped. The result was an infringement of patent suit before Judge Swann in the United States court. Stearns & Co. contended that suction is an old principle that has been in use ever since mankind has been on earth. Cain and Abel first introduced it in the garden of Eden.

Judge Swann readily conceded that suction is an ancient and time-honored principle, but he claimed that Mr. Russell had made a new application of it, in his pill-making device, and permanently enjoined Stearns & Co. from using their machine.

Electrical Dial Telegraph.

A new electrical dial telegraph, invented by G. E. Painter, of Baltimore, is a unique and practical arrangement for transmitting unwritten knowledge to those at a distance. This instrument was devised for use on men-of-war in transmitting orders from the pilot to engineer. It resembles in appearance the ordinary mechanical telegraph now in use, but as the latter indicates only eight or ten different signals, the new device improves upon it by transmitting, not only the ordinary orders to the engineer, such as "slow," "full speed," etc., but the exact number of revolutions the engines are required to make, ahead or astern; such as "9 rev. ahead," "143 rev. astern," etc., in order to keep exactly in line when drilling or manoeuvering.

The instrument is so devised that the transmitting lever or handle turns forward or backward at will, and the receiving instrument follows in like manner, its every movement. It is capable of moving at the rate of 100 indications per second. Take for instance, a ship whose engines are running 150 rev. ahead; only three seconds would be required to transmit an order to reverse and run 150 rev. backward. This is in reality, three hundred different orders.

One great advantage is the absence of springs or weights for adjustments, consequently there is nothing to get out of order. It is often the case that instruments work very nicely on shore, but tossed about when the ship is at sea, the result is they refuse to work at all on account of the position of levers controlled by springs. The instrument in its best form has both transmitting and receiving instruments in one case with an illuminated dial, for use at night.

In sending an order, it is only necessary to turn the handle with pointer attached, either way to the desired order. The one receiving this order then turns the handle of his instrument in the same manner, until the pointer stands opposite to the order received; the other instrument will then act in like manner, showing the sender that the message was received correctly. The handle is supplied with an automatic switch, by which the instrument acts as desired, and at the same time rings a bell to call attention to the order, and when released, cuts itself out.

Only four wires are required for each set of instruments, and its simplicity renders the apparatus not costly. It may also be used in transmitting a great many different kinds of messages in quick order, without resorting to speaking tubes.

This apparatus is only one of the complete ship equipment which Mr. Painter controls.

Self-Lighting Oil Lamps.

A new electrical self-lighting attachment for oil lamps is now on thd market. A small, dry electric battery, hidden in the base of the lamp, does the lighting in the most convenient and perfect manner. The arrangement is simple, reliable and safe. It will light the lamp once or twice a day during six to nine months without renewal, and it then costs but sixty cents to replace the battery. It is not liable to get out of order; a child can light the lamp without danger, and there are no complications whatever. The oil chamber may be removed and the lamp taken apart for cleaning as easily as in the case of a lamp not provided with the attachment.

A little device of wire has been invented for holding the cob so that corn can be eaten therefrom without soiling or burning the fingers—and yet some say that the achievements of this period cannot equal the past.

Changes in Mexican Patent Law.

The Government of Mexico has recently made some important amendments in the patent law of the country, which affect article 33 of chapter 5 of the law of June 7, 1890, on patents of invention, and also the transient article of that law. Article 33, as amended, reads as follows: "The owner of a patent of invention or a patent of improvement is compelled to prove before the Department of Fomento at the end of every five years of the existence of his patent, and in order to keep it in his possession for the following five years, that the payment of an additional tax has been entered into the Federal Treasury, as follows: At the end of the first five years, $50; at the end of ten years, $75, and at the end of fifteen years, $100. All these payments must be made in Mexican dollars. To prove that these payments have been made at the end of the said period of five years a term of two months is allowed, which cannot be extended." The Transient Article, as amended, reads: "The persons concerned who, on the date of the publication of this law, have incurred the forfeiture established by section 3 of the 37th article of the law of June 7, 1890, may avail themselves of the dispositions of the present law, to be exempted from the forfeiture penalty, provided they make the due payment of the taxes within three months after its publication, it being understood this concession will take place only in case that a third person will not suffer in his rights after the forfeiture may have been established."

From New York to Brooklyn by Tunnel.

Active preparations are being made to tunnel the the East River between New York and Brooklyn. A company has been chartered under the title of the Columbian Company, with a capitalization of $20,000, which sum will be increased as the necessary privileges are obtained from the Brooklyn Board of Alderman. The New York Board of Aldermen on July 14 passed favorably upon the new scheme, and it is expected that the Brooklyn Board will soon follow. A great portion of the preliminary work, such as determining the character of the material to be excavated, the required depth, the approaches and terminals, ventilating shafts, etc., has already been carried through. Operations will be commenced in October, and it is intended that the tunnel shall be completed within a year. The total length of the proposed line will be about 8700 feet, the exact length depending upon the terminals in New York and Brooklyn. The maximum grade will be about 4 percent. The Greathead system of tunneling, employing a steel hood, will be used. Two shafts will be sunk for excavation purposes, one located on each side of the river. Three million five hundred thousand dollars is the estimated cost for the entire tunnel, terminal facilities and property rights. The present plans contemplate a tunnel 24 feet in height, with a flattened arch, and about 28 feet in extreme width. A double track is to be used, and ample ventilation and illumination provided for. It is thought that the trip by electric cars from one end of the tunnel to the other can be made in three minutes. Twelve thousand passengers, it is thought, could be carried each way per hour during the busy part of the day.

The New Silver Certificates.

The new $2 and $5 silver certificates have been printed and are ready for issue. Like the $1 certificates, they are a striking departure in money making. Black ink is used in printing the front of the notes, while the back is of the conventional green. The $2 note was designed by Edwin H. Blashfield, of New York, and the face contains an allegorical representation of "Science Presenting Steam and Electricity to Commerce and Manufacture," and consists of five partly nude female figures in graceful poses.

Walter Shirlaw, of New York, designed the $5 note. The face has an allegorical picture representing "America Enlightening the World," a beautiful female, partly nude, holding in her right hand, uplifted, a lighted incandescent lamp. Reclining at her feet is a female figure of Fame with the traditional trumpet.

The notes are beautiful specimens of the engraver's handicraft, and in that respect will compare favorably with any work in that line.

A remarkable record was recently made at the works of the Youngstown Specialty Manufacturing Company, Youngstown, O., at which time 42,660 tin fruit cans were turned out ready for shipment in 12½ hours.

London and Berlin are soon to be connected by long distance telephone.

Recent Patent Decisions.

In the case of the Taylor Burner Co., Lt. vs. Diamond—U. S. Circuit Court, western district of Pennslyvania—the court held that a mere accidental use of some of the features of an invention, without recognition of its benefits, does not constitute anticipation.

* * *

It is held in the case of the Thomson-Houston Co. vs Kelsey Electric Railway Specialty Co.—U. S. Circuit Court, District of Connecticut—that "in a suit for preliminary injunction the burden of proof is on complainant to show an intention on the part of defendants to infringe. Making and putting upon the market an article which of necessity and to their konwledge is to be used to infringe complainant's patent is contributory infringement on the part of defendants." Motion for preliminary injunction granted.

* * *

Injunction suit of American Cereal Co. vs. Eli Pettijohn Cereal Co.—U. S. Circuit Court, northern district of Illinois—decided in favor of defendants—had right to use trade mark "Eli Pettijohn's California Breakfast Food.

* * *

In the injunction case of Cook & Bernheimer Co. vs. Ross, et al—U. S. Circuit Court, southern district of New York—it was held that "complainant is entitled to an injunction *pendente lite* against the further use, by defendants, of a square-shaped bulging necked bottle, as a package for Mount Vernon whiskey, when it appears that the complainant was the first to use a bottle of this shape as a package for whiskey and as a means of identifying this bottling of Mount Vernon whiskey from all others."

* * *

Mullen et al, vs. King Drill Co. et al—U. S. Circuit Court, District of Indiana—suit in equity for infringement of patent No. 355,462 for improvements in grain drills. Held that claims 1 and 2 were valid and usual decree entered for plaintiff.

* * *

In the decision entering a decree for plaintiff, in the case of American Soda Fountain Co. vs. Green, et al, tne principle is laid down that "the fact that all the elements of a combination may be found partly in one structure and partly in another is unsafe ground for overturning a patent."

* * *

The case of Beale, et al, vs. Spate, et al—U. S. Circuit Court, Southern District of New York—action based upon Letters Patent No. 363,695, for improvements in stair-pads, the principle is set forth that "the law is imperative that the language of the description and claims, when clear and unambiguous, must be adheared to, even though the court be of the opinion that the patentee has unnecessarily restricted his claims."

* * *

In the case of Harrison vs. Morton, Court of Appeals of Maryland, it is held that "an assignment of an unpatented invention which does not contain a request that the patent issue to the assignee does not convey a legal title to the patent when granted, but gives only an equitable interest."

The Invention of Envelopes.

The invention of envelopes is within the memory of middle aged persons, and was the result of a Brighton, England, stationer's endeavor to make his store look attractive. He took a fancy for ornamenting his store windows with high piles of paper, graduated from the highest to the smallest size in use. To bring his pyramid to a point, he cut cardboard into very minute squares. Ladies took these cards to be small-sized note paper, and voted it "perfectly lovely." So great was the demand that the stationer found it desirable to cut paper the size so much admired. But there was one difficulty. The little notes were so small that when folded there was no space for the address, so after some thought the idea of an envelope pierced the stationer's brain. He had them cut by a metal plate, and soon, so great was the demand, he commissioned a dozen houses to manufacture them for him. From such small beginnings came this important branch of the stationery business.

The new 800 foot lock at Sault Ste. Marie, Mich., was finally opened on the 3d ult.

The English houses of parliament are partly lighted by 40,000 electric lamps, which number is being constantly increased. Fifty experienced electricians are employed to keep the system in order. But there is still a yearly gas bill of over £2,000.

Pineapple Cloth.

The finest pineapple cloth comes from the Phillipines, but very good tissues are turned out wherever there are Malays, and of late years by Mongolians and other communities. The thread is obtained from the pineapple leaves in some curious way which separates the fine filament from all the other vegetable tissues. It is then partially dried and bleached in the sun, and is then carded and spun. After its spinning, and before it is thoroughly dry, it is woven on the old-fashioned looms, which are busy today in Asia. The technical skill possessed by the spinners and weavers is truly admirable. Men are too clumsy for the work, and women have a practical monopoly of it, but even among them there are many whose eyes and fingers are not quite delicate enough to distinguish between the thickness of one thread and another. The weaving is done within doors and usually in a Malay house, whose bamboo framework, walls made of leaves and heavy thatched roof, keep the interior quite dusky and damp. When produced the cloth is plain in color or else made according to an order, or according to Malay tastes. The finest quality of cloth is so fine as to be practically translucent, and some tissues which were worth more than their weight in silver would stand successfully the test of the Indian rajah, who would accept no cloth unless he could draw the whole roll through his signet ring.—*The Tradesman.*

The Throne of England.

The throne of England, so spendid in its rich trappings of silk, velvet, and gold wire lace and tassels, is simply an old fashioned, high-backed chair. It has been in use for more than 600 years, but the early history of the old oaken relic and the name of its maker are both unknown. The wood which composes this "throne" is very hard and solid, as may be imagined when it is known that the chair has been "kept in the dry" and well covered with rich cloth of various kinds since the days of Edward I. The back and sides of the chair were formerly painted in various colors The seat is made of a rough sandstone. This stone, which is believed to possess talisman powers, is twenty-six inches in length, seventeen inches in breadth, and nineteen and a half inches in thickness. Nevertheless, legends are told in connection with this wonderful stone, but the truth brobably is that it was originally used in Scotland as a "coronation stone," upon which the Scottish kings were seated while undergoing the ceremonies connected with being crowned "king of the realm of Scotland."

The $1 Edison Motor.

This cut shown herewith represents one of the most satisfactory and perfect of all small dynamos, capable of running toys, fans, light models and for

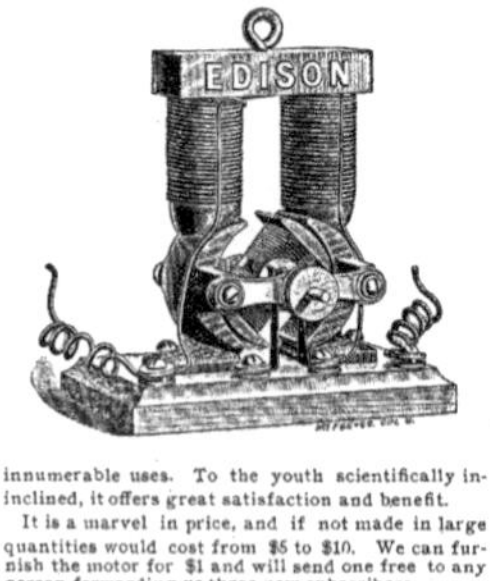

innumerable uses. To the youth scientifically inclined, it offers great satisfaction and benefit.

It is a marvel in price, and if not made in large quantities would cost from $5 to $10. We can furnish the motor for $1 and will send one free to any person forwarding us three new subscribers.

The immense Ellswick works of Armstrong, Mitchell & Co., New Castle, England, it is announced, are about to be further extended in several directions. They now give employment to between 17,000 and 18,000 men.

Ore Output of the Lake Superior Region.

The output of iron ore in the Lake Superior region, says "Age of Steel," has reached the 100,000,000 tons mark. The last decade has seen the largest part of this total mined and made into iron and steel. As the quality of these ores are of a grade necessary to the manufacture of Bessemer steel, the importance of these deposits to the development of the iron and steel industries of the United States cannot well be over-estimated. By the aid of the latest and most improved labor saving machinery, the mining of this ore has its own economical advantages. Electricity has been generously utilized and harnessed to special tools and machinery and the greatest results are obtainable at the least cost of labor. In the drilling for ore, and in the raising and crushing it, in the efficiency of pumping machinery, and in facilities for shipment, everything that money could do, or mechanical and engineering skill devise, has been done in an able and masterly fashion; Electric haulage in some instances costs as low as five cents per ton, and ore is raised at the rate of half a mile a minute in loads of 4 to 6 tons, and in one instance the efficiency of a modern pumping apparatus is given at 8000 gallons of water moving 1500 feet a minute. For the direct and rapid transportation of ore the Rockefeller interests have a railroad and docks, making the loading and shipping of 4,000,000 tons a year but an easy task. The same company are building a fleet of steeel ships of great capacity to carry their own ores. The product of the various mines tributary to the lakes is thus tabled in an admirable report of the same in a contemporary. It covers the period from the date of exploitation to the close of July of this year:

	Tons.
Marquette range	44,500,000
Menominee	22,000,000
Gogebic	19,500,000
Vermilion	7,500,000
Mesaba	6,500,000
Total	100,000,000

The capital invested according to the most reliable figures obtainable is thus tabled by the same authority:

Capital in mines	$97,500,000
Docks at upper lake ports, etc	16,250,000
Railroads, mines to docks	35,500,000
Vessels in ore trade on lakes	50,000,000
Lower lake receiving docks	16,000,000
Railroads to furnaces from docks	28,000,000
Total	$243,250,000

Across the Ocean on Wheels.

One of the most curious specimens of naval architecture ever placed on the water is the so called "roller" steamer launched last week at the Cail shipyards, at St Denis, near Paris. This vessel is built on a novel principle invented by M. Bazin, a well-known French marine engineer, who has embraced the theory that vessels can more easily roll over the water than cut through it. The boat just launched, and which is to be used for experimental purposes in the English Channel, is described as "a large rectangular iron box, about 120 feet in length, 40 feet wide and 5 feet high. It is mounted on six lenticular disks, or rollers of sheet iron, shaped like double converse lenses, 30 feet in diameter. In the sides of the box is the machinery, which is 750 horsepower. The propelling power is a screw hanging in brackets from axles of the rear rollers and submerged much as ordinary steamer screws. In the upper part of the vessel, between the disks, which pierce the box and extend beyond it about 6½ feet, are cabins fitted up with the accustomed sea going comforts. M. Bazin's vessel has a displacement of 280 tons. This displacement is made and the burden borne by the disks, which are sunk in the water about 10½ feet, which brings the bottom of the iron box about the same distance from the surface of the water."

It is claimed, as the results of tests made with a model, that the power necessary to keep the rollers at work is only one-fourth of the power that is required to keep the screw going, while by an extra expenditure of power, amounting to another fourth, the speed of the vessel is doubled. Thus, the inventor points out, with the use of rollers, the length of voyage may be greatly diminished, while the consumption of coal will be lessened, resulting in a great saving of time and cost. Moreover, it is asserted that the stability of the rolling boats will be far greater than that of the steam vessels now used. When ready for sea, the Bazin steamer will descend the Seine to the English Channel and cross to London. The inventor expects her to make at least 30 miles an hour in the Channel. M. Bazin, it is reported, has prepared designs for a large ocean going steamer on the roller principle, which he estimates will make the voyage from Havre to New York in four days. The innovation is a bold one, and the results accomplished by the experimental vessel will be watched with much interest.

Aftermath.

The Waterbury Watch Company shut down for a month on account of "no business."

Owing to the dullness of trade nearly one-half of the Fall River cotton mills have shut down.

Queen Victoria has conferred the Knight Grand Cross of the Royal Victorian Order on Lord Kelvin.

The closing down of the Spanger Steel and Iron Works at Sharpsburg, Pa., throws about 900 men out of work.

Professor Alexander Graham Bell has received the honorary degree of L.L. D. from Harvard University.

The Chicago Consolidated Iron and Steel Company, at Harvey, Ill., has made an assignment—assets and liabilities about equal—$300,-000.

The Indiana Bicycle Works at Indianapolis, have shut down. Overproduction and unsettled financial outlook are given as the cause.

The 61st meeting of the New England Cotton Manufacturers' Association will be held at the Profile House, New Hampshire, Sept., 21-24.

The Hungarian Minister of Agriculture estimates the wheat crop of that country at 140,-000,000 bushels, against 142,000,000 bushels last year.

Poor business is given as the cause of a 10 per cent reduction in salaries all around among the employes of the "Buckeye Engine' Works, Salem, Ohio.

The second order for American steel rails for Japanese railroads, placed recently with the Carnegie Company by Kenro Iwakura, the New York representative of Mitsui & Co., of Japan, for 9,000 tons, to be delivered by the end of the year.

The extensive additions to be made at the Altoona shops of the Pennsylvania Railroad will make them one of the largest industrial establishments in the world. Hereafter all the locomotives for the Pennsylvania Railroad and the Pennsylvania lines west of Pittsburg will be built at Altoona and only repair work will be done at the other shops.

To satisfy a judgment of $2,810, Frederick W. Duston's attic and a half experimental Boynton bicycle road, which was built at a cost of about $30,000, including power house and plant, was sold under foreclosure proceedings on July 20, by referee William G. Nicoll, and was bought in by W. H. Boynton for $500. The Long Island Boynton Bicycle Railroad Company, the New York and Suburban Investment Company and the Bellport Construction Company were named as defendants in the proceedings.

Knowledge Up to Date.

A small work of uncommon interest and value is The Living Topics Cyclopedia, which now costs, complete to date the small sum of $1.00. It is a unique publication, and its free specimen pages are worth sending for. Its latest issue gives the most important facts, "up to date," concerning, among hundreds of other important subjects, such titles as Cuba, Currency (a "living topic," indeed), Debts, national and foreign, East Africa, Egypt, Electricity, England, Engineering, France, German Empire, Gold, Greece, also concerning the States Delaware, Florida and Georgia. In general the object of the work is to answer the questions you would seek to solve by consulting your' cyclopedia, were it "up to date," which no cyclopedia is or possibly can be, because of its magnitude and cost. The Living Topics, being a small work, and chasing only of "living" topics, is continually in process, of revision, a new edition being published every month. After you have paid for one edition you are allowed to purchase later ones, within a year thereafter, at about one-third price, and thus keep your knowledge "up to date" at trifling cost. Address the publisher, JOHN B. ALDEN, 10 and 12 "andeveraat' St., New York.

Any person sending us the names and addresses of five inventors who have not yet applied for patents will receive the INVENTIVE AGE one year free.

For Mchanics and Engineers.

Every mechanic and engineer should have a copy of "Shop Kinks and Machine-Shop Chat," by Robt. Grimshaw. It is the most complete work of the kind ever published. If any " kink" is discovered in the machine shop—it can be straightened out—after consulting this valuable work. Sent to any address for $2.50.

TO INVENTORS.

We make a specialty of manufacturing goods for inventors. Make models, dies, do experimental work. Acquire special tools, machinery, etc. INVENTOR'S HANDBOOK FREE.

Cincinnati Specialty
Manufacturing Co.,
Third Street, Cincinnati, Ohio.

IDEAS DEVELOPED. Absolute secrecy. Send for particulars. Advice and suggestions given free when asked for. Correspondence and sample orders solicited. 25 years in business. Models, patterns and all kinds of castings and machine work.

GARDAM & SON,
96 John Street. New York.

ASSIGNMENTS OF PATENTS.

Partial List of Patents Assigned During the Past Month and the Consideration Therefor.

Patent No. 520,027. Hiram S. Atkins, inventor to Warren J. Atkins, Grinding Mills; All rights for U. S. and territories. Consideration $5,000.

No. 562,108. Albert Barber, to Jobs Lehman of Springfield, Ill. Extension Bedsteads; an undivided ½ of whole right. Consideration $1,200.

No. 363,720. Thomas Brett to the Brett Piano Co. of Geneva, Ohio' Siting Instruments; all rights for the U. S. Consideration $10,000.

No. 556,381. Lawrence D. Boyce and Joseph C. Thompson inventors. Arthur M. Cummings, assignor, to Geo. A. Baker of Harvede-Grace. Md. Ribbon Attachment for Type Writing Machines. All right' and title in said patent. Consideration $1,750.

No. 526,368. Elliott D. Barling, to Frank Otto of Pontiac, Mich., Smooth Wire Fence. All right, title and interest for the state of Illinois, except eight counties. Consideration $1,500.

No. 444,685, Horace H. and John B. Barnes, inventors, Oscar G. Wilmoth, assignee, to Barnes and Linton, Detroit, Mich., Tack Driver. All rights, title and interest. Consideration $3,000.

No. 542,517. John S. Burton, to the John S. Burton Separator Wood Work Co., of Merphsesboro, Tenn. Cable Ornaments for Buildings. Consideration $10,000.

No. 551,754. Daniel W. Bromley, inventor, W. R. Banon, assignor, to Geo. W. Weymar Huntington; W. Va., Dough Kneading Machine. Assigns all right, title and interest. Consideration $3,200.

No. 575,902. Alexander M. Bollinger, to Geo. S. Getty, of Greensburg, Pa., Rotary Secretary and Writing Desk. A one-half right, title and interest. Consideration $20,000.

No. 566,983, John W. Cook, to Herbert' Lee, and Geo. F. Miller of Portland, Orgeon. Clothes Pins and Lines. All right, title and interest for the states of Tennessee and Kentucky. Consideration $3,000.

No. 550,617. Charles W. Colony inventor, Wm P. Sanford, assignor to the Hise & Sanford Wooden Ware Co., Presses for Shaping Plates from Veneer. All rights. Consideration $5,000.

No. 513,094. Charles M. Coates and James L. Spragne, to Albison Little of Portland, Maine, Washing Machine. Assigns all rights for the states of Maine, New Hampshire and Vermont. Consideration $1,500.

No. 495,307. Frank J. and Mead C. Coon, to F. J. Coon, Washing Machine. All right, title and interest. Consideration $6,000.

No. 597,461. Joseph O. Clark inventor, John B. Baidge, Assignor to the Victor Manufacturing Co., of Maine. Character'ing Machines and Redirectedly Characterized Instruments. All rights. Consideration $19,997.

No. 553,673. Carl R. Evertz to the Derby Razor Co., of Derby, Conn. Razors. All rights. Consideration $5,910.

No. 553,986. Fred Girtanner to Charles R. Campbell of Sedalia, Mo., Straw Burner. Assigns all right, title and interest for the state of Missouri except one county. Consideration $3,500.

No. 558,901. Adolphus D. Goodwin to C. W. Sarfant and Sidney Shaltman of state of Virginia. Protecting Fruit Trees from Vermin. The right to Manufacture and vend in the U. S. except six states. Consideration $8,000.

No. 414,809, 444,219. David Grove inventor, F. H. Murphy, Wm. M. Thomas, assignors, to J. E. and T. M. Henderson of Covington, Ind., Reel for Winding Wire. All right, title and interest for the state of Texas. Consideration $8,000.

No. 541,238. Jasper N. Jennings inventor, Armstrong Glover assignor, to Henry A. Stevens of Stockton, Cal., Tire Tighteners. Grants the exclusive right to Manufacture and vend in the state of California. Consideration $2,000.

Othmar Mergenthaler' inventor, Mergenthaler Linotype Co., assignor to the Mergenthaler Linotype Co., of N. Y. and Brooklyn, N. Y. Matrix Making Machine. Assigns entire right, title and interest. Consideration $200,000,000.

No. 362,097. Thomas P. Murray to Benjamin F. McCauly of Washington, D.C. Burglar Alarms. Entire right for Virginia and W. Va. Consideration $2,500.

Geo. W. Marks to Harry M. Stanley Henderson, Ky., Stopper Necks. An undivided half of his right, title and interest. Consideration $2,000.

No. 545,179. Charles Spofford inventor, Wm. J. Hise, assignor, to the Hise & Sandford Wooden Ware Co., Process of and Apparatus for Making Dishes of Wood and Wood Pulp. All right to manufacture and vend for the U. S. Consideration $5,000.

No. 533,315. Wesley Sibbrel inventor, Louis C. Hedinger, assignor' to Charles Hedinger of St. Louis, Mo. Sleigh Runners. Assigns all rights. Consideration $2,500.

Benjamin F. Sammons inventor, W. F. Jackson. assignor, ½ except Buhl. Traction-enginer' and undivided one-half interest. Consideration $60,000.

No. 566,252. John F. Swift to Samuel M. Conant of Central Falls, R. I. and Adam Sutcliffe of Pawtucket, R. I. Tag Tying and Bunching Machine. All rights for the U. S. Consideration $2,500.

Oliver W. Whitehead to Warren E. Baeghly of Dayton, Ohio. Fence Stays. An undivided one-half of his right, title and interest for the U. S. Consideration $1,000.

STANDARD WORKS.

"The Electrical World." An illustrated weekly review of current progress in electricity and its practical applications. Annual subscription $3.00

" Electric Railway Gazette." An illustrated weekly record of electric railway practice and development. Annual subscription $3.00

" Johnston's Electrical and Street Railway Directory." Published annually $5.00

Dictionary of Electrical Words, Terms and Phrases. By Edwin J. Houston, Ph. D...$5.00

The Electric Motor and its Applications. By T. C. Martin and Jos. Wetzler..........$3.00

The Electric Railway in Theory and Practice. The first systematic treatise on the electric railway. By O. T. Crosby and Dr. Louis Bell.........................$2.50

Alternating Currents. By Frederick Bedell Ph.D., and Albert C. Crehore, Ph.D.......$2.50

Gerard's Electricity. Translated under the direction of Dr. Louis Duncan..........$2.50

Theory and Calculation of Alternating-Current Phenomena. By Charles Proteus Steinmetz......................$2.50

Practical Calculation of Dynamo Electric Machines. By A. E. Wiener..........$2.25

Central Station Bookkeeping. By H. A. Foster........................$2.50

Electric Arc Lighting. By E. J. Houston, Ph.D. and E. Kennelly, D.Sc.......$1.00

Electric Incandescent Lighting. By E. J. Houston, Ph.D. A. E. Kennelly, D.Sc....$1.00

Electric Motors. By E. J. Houston, Ph.D., and A. E. Kennelly, D.Sc.............$1.00

Electric Street Railways. By E. J. Houston, Ph.D., and A. E. Kennelly, D.Sc.......$1.00

Tables of Equivalents of Units of Measurement. By Carl Hering..............96

Copies of any of the above books or of any other electrical book published will be sent by mail, postage prepaid, to any address in the world on receipt of price.

THE INVENTIVE AGE,
Washington, D. C.

A large number of bicycle people will be interested in this issue. They will be interested in succeeding issues.

The Inventive Age

·AND INDUSTRIAL REVIEW·

·A JOURNAL OF MANUFACTURING INDUSTRY·
AND SCIENTIFIC PROGRESS·

Seventh Year. | No. 10.   WASHINGTON, D. C., OCTOBER, 1896.   { Single Copies 10 Cents. | $1 Per Year.

Motorcycles or Horseless Carriages.

The widely advertised motor or horseless carriage race at Narragansett Park, near Providence, R. I., Sept. 7-11, was not the important event its promoters anticipated, still it was probably in advance of all previous demonstrations of the kind in this country. In this experimental stage of development it is hardly proper to call these exhibitions "races." The Providence meeting, like its predecessors was more on the order of a practical demonstration of the status and prospects of the motor carriage than a test of endurance and speed. It is not surprising, therefore, that those who anticipated some of the excitement of horse racing were disappointed in this event.

Apart from demonstrating officially the phenomenal speed which motor vehicles in their present stage of development can successfully obtain the race cannot be considered a success, and it is doubtful if such races ever can be considered a success from a horse racing or spectacular point of view, unless machines of almost identical construction and equal horse power capacities should be used for this purpose, as the interest in a horse race depends, not only upon the time made, but also upon the fact that the horses come to the line at the finish at almost identically the same instant of time and therefore the excitement which is the distinguishing feature of a horse race finish is entirely wanting in motor vehicle races with the vehicles distributed from one end of the track to the other. The result, therefore, of placing on the track a number of vehicles designed for different purposes, with varying rates of speed and different horse power capacities, while it proved interesting to quite a large number of people who are looking forward to the early introduction of motor vehicles to replace horses, can hardly be said to have created much impression on the vast majority of the people present who were there merely to see feats of endurance and bursts of speed which are more or less unusual. Under the conditions prescribed by the Rhode Island State Fair Association there were to have been five heats of five miles each on five successive days. Owing to the rain, however, on two of the days, only three heats were run with the following results:—(There were practically only three contestants, viz: The

Riker Electric Motor Co., Electric Carriage & Wagon Co., and the Duryea Co., although the latter company was represented by six machines, making eight in all that started in the race.)

FIRST HEAT.

Riker Electric Motor Co.,	1st.	time	15:01½
Electric Carriage & Wagon Co.	2nd.	"	15:15½
First Duryea	3rd.	"	18:48½

SECOND HEAT.

Riker Electric Motor Co.	1st.	time	13:06
First Duryea	2nd.	"	13:14
Electric Carriage & Wagon Co.	3rd.	"	13:33

THIRD AND LAST HEAT.

Electric Carriage & Wagon Co.	1st.	time	11:27
Riker Electric Motor Co.	2nd.	"	11:28½
First Duryea	3rd.	"	11:59

The other gasoline vehicles were so far behind as to be virtually out of the race although they all made a creditable showing in going over the course at a rate of speed which is undoubtedly greater than there is any necessity for in actual service.

It will be seen from the above that while the Electric Carriage & Wagon Co., were beaten on the first two heats for first place, on the third heat they outdistanced all their competitors, making better time than was made in the previous heats and establishing the official five mile record for motor vehicles. This time was made without a flying start as the carriages approached the line at a speed not faster than a walk; the first mile being made in 2 min. 20 sec., the average for the five miles being 2 min. 17:4 sec., the fastest mile being made in 2:07 and the rate per hour 26:2 miles, which it is safe to assume has never been excelled or equalled by any motor vehicle ever constructed up to the present time.

When it is considered that this carriage is designed for use in the city streets and country roads and cannot be considered in any sense a racing machine, and when it is considered that it is only equipped with two 1½ H. P., motors as against two 3 H. P. motors on the Riker machine, and a 5 Brake H. P. engine on the gasoline machines, its performance is not only remarkable but amazing. It is fair to state, however, that neither the Riker or the Duryea machines were built for purely racing machines, although the construction of the Riker machine was such as would not permit it to run over rough streets or roads without serious danger of breaking down from "structural weakness."

The Providence Journal, the day after the first race, said:
"The vehicle which finished second was the manufacture of the Electric Carriage & Wagon Co., of New York and Philadelphia. It is after the design of Morris & Salom and is of the surrey shape. Two motors supplied the power, each of 1½ H. P. In general appearance this machine was by far the handsomest on the track."

The machine in question, which took part in the race, is precisely similar in every respect to the one which received the gold medal at Chicago, nearly a year ago, in the Times-Herald Motorcycle contest.

Further mention of other motorcycles will be made next month.

Purdy Campbell, a Chicago jeweler, has invented a clock that he claims will run forty years. The movement is geared so that the barrel wheel containing the main spring revolves once in two and a half years. When this wheel has made sixteen revolutions somebody will have to give the key seventeen turns. The clock will then be wound for another forty years. The first wheel from the barrel wheel crowds around at the rate of one turn a year. The dial plate is six inches in diameter.

Inventive Age

Established 1889.

INVENTIVE AGE PUBLISHING CO.,

8th and H Sts., Washington, D. C.

ALEX. S. CAPEHART.   MARSHALL H. JEWELL.

THE INVENTIVE AGE is sent, postage prepaid, to any address in the United States, Canada or Mexico for $1 a year; to any other country, postage prepaid, $1.50.
Correspondence with inventors, mechanics, manufacturers, scientists and others is invited. The columns of this journal are open for the discussion of such subjects as are of general interest to its readers.
Technical matter is particularly desired. We want practical information from practical men.
Nothing will be published in the editorial columns for pay.
The INVENTIVE AGE is thoroughly independent.
Advertising rates made known on application. Special facilities for furnishing cuts of any patented article together with descriptive article. Business specials 15 cents a line each insertion, 7 words to the line. No advertisement less than 25 cents.
Address all communications to THE INVENTIVE AGE, Washington, D. C.

Entered at the Postoffice in Washington as second-class matter.

WASHINGTON, D. C., OCTOBER, 1896.

Among the papers in the District of Columbia devoted to inventions and patents, none has credit for so large a regular issue as is accorded to the "Inventive Age," published monthly at Washington.—*Printers Ink.*

DURING the year ending June 30th there were filed 45,645 applications for patents in the United States Patent Office, including reissues, designs, trademarks, labels and prints, against 41,014 for the year previous.

IT must be remembered that the Patent Office does not assume responsibility for the acts of attorneys, and the selection of competent, and honest counsel should, therefore, be the first consideration of applicants for patents.

OF the 45,693,000 pounds of rubber exported from Para and the States of Amazonas in 1894, 24,753,000 pounds came to the United States. The wonderful growth of the bicycle industry has greatly stimulated the rubber industry and the exports of the United States exceed all of Europe combined.

FRENCH capital and cheap Chinese labor has resulted in opening up the almost inexhaustible coal deposits of Tanquin and the trans-Pacific rates are so low that the foreign product promises to drive the Pennsylvania anthracite from the market. The first cargo arrived in San Francisco last month and more are to follow.

MR. ELIAS E. RIES, one of the best-known electrical experts and consulting engineers in the country has moved his office from Baltimore to New York City—1031 Temple Court. Mr. Ries is an active member of the American Institute of Electrical Engineers and a contributor to the technical press on electrical and other subjects.

SOME prolific newspaper reporter has quoted Li Hung Chang as declaring the modern safety bicycle to have been invented in China something like 2,300 years B. C., and that its popularity extended over a period of 150 years. The women became so infatuated with it they neglected their household duties and the raising of children. Finally the emperor was obliged to issue a decree abolishing the wheel from the realm. Another newspaper critic who seems to fail to appreciate the hold the wheel has taken on the world in the closing days of the nineteenth century says that when the steamer St. Louis, with Li Hung Chang on board passed the Atlantic squadron, which had at vast expense, been gathered at New York to do him honor, the old fellow remained in his cabin and wouldn't look at it. He showed equal indifference to the naval exhibit at Philadelphia and to the immense enginery of the Cramp shipyard. But he has been tickled by the American bicycle, and takes home

with him an elegant one presented to him at Washington.

False Representations to Inventors.

The attention of the INVENTIVE AGE has been called to the wording of a recent circular letter sent out by Edgar Tate & Co., of New York, to patentees in this country, whose patents have been issued and the description thereof published in the Official Patent Office Gazette and elsewhere. They say:

"Notwithstanding the assertions of many uninformed attorneys, we stand willing to guarantee that we can yet obtain valid patents for your invention in the countries named below, viz: Canada, England, France and Germany."

Every well informed solicitor of patents knows that when an invention is published in a printed publication of any kind which is sufficient to enable others skilled in the art to which it relates, to make and use it, no valid patent can be afterwards obtained on application seven months from the date of filing his application in the United States to file one in England and France. This law does not apply to Germany, because that country has not yet complied with the terms of the Convention. To obtain a valid patent in Germany absolute novelty is essential. In France a valid patent cannot be obtained unless the application be filed before publication of the invention in any country.

Every patent as soon as granted by the United States Patent Office is published in the Patent Office Gazette which goes at once abroad, and unless the application for a patent is filed in France and Germany before the publication of the Gazette, there is no chance for obtaining a valid patent in those countries.

In England it is different. If the patentee deposits his application there before the Gazette, or any other printed publication containing a description of the invention, arrives in England, a valid patent can be obtained, because the law of that country says that the first to introduce an invention, whether the inventor or not, will be granted a patent. It is in very few cases indeed that the inventor has time after the grant of his patent in the United States to file an application in England and France before the expiration of the seven months from the deposit of his application at home, but the circular above referred to is sent to him just the same.

It was sent to one patentee whose application had been on file over one year. This man corresponded with Edgar Tate & Co., and they urged him to file his application in England, France and Germany, as his correspondence with them will show. We simply call attention to this matter as another of the many warnings given to our readers against being imposed upon by patent agents.

WHEN at the request of Congressman Cannon the senate receded from its amendment appropriating $64,000 to establish the classification division in the United States Patent Office, so strongly urged by the Commissioner and the American Association of Inventors and Manufacturers, it was tacitly understood and agreed that at the short session this winter an effort would be made to secure favorable action in the house. It is hoped the recommendations of the Commissioners will prevail this winter and that this division, wherein all applications may be properly classified according to principle, will be authorized. It is justified by all considerations of good service, and the monstrous sum to the credit of the Patent Office in the United States Treasury and the increased net receipts last year but emphasize the reasonableness of the demand.

As a part of its plans for practical work in the extension of the foreign trade of the United States, the National Association of Manufacturers is now preparing to establish an exhibition warehouse in Caracas, Venezuela, for the display and sale of American products of various kinds. Circular of Information, No. 10, which has just been issued by

the Association, gives the details of this enterprise and presents a large amount of interesting information about the trade of Venezuela. It is the opinion of the merchants of Venezuela that the following articles might be imported from the United States with profit, in addition to those that are now going in, viz: American building material, hardware, common glassware, cutlery, fencing wire, mining and sugar machinery, agricultural implements, carriages, cars, steam engines, lumber, cotton goods, certain kinds of wearing apparel, and all kinds of articles for home furnishing and decoration, carpets, curtains, rugs and novelties. Copies of the Circulars of Information of the National Association of Manufacturers can be obtained from the Bureau of Publicity, No. 1751 North Fourth Street, Philadelphia, Pa.

THE west seems to be a prolific field for patent selling humbugs and the Inventive Age has on several occasions called the attention of its readers to some of the ridiculous claims made by these clever parasites on the inventors of the country. We are in receipt of a letter from a well known gentleman in Muncie, Ind., who, in speaking of an extensively advertised patent broker in that city, says:

"He is swindling the poor patentees out of their $10, $15 and $30 right along. He makes all kinds of inducements to get their money and does them no good. I don't know of a patent he has sold in the last two years, except some of his own and he has the gall to mention these sales in his letter to patentees as extra inducements; of course he doesn't mention the fact that they are his own patents. He speaks (in the honied letters he sends out each week to inventors) of having the largest number of persons employed and this appears on his bonds he sends out ($1,000 bonds), when the fact is he has but one person employed in the patent business. This with five or six other claims just as ridiculously misleading are guaranteed under the $1,000 bond. The greatest point of all would be to compare his record of sales with the record of contracts taken, or patents taken on sale. This alone would convict him from the fact that he has not made half a dozen sales in the fifteen years he has been engaged in the patent business. I know this to be a fact.'"

THE encouraging outlook for the success of Wm. McKinley and the triumph of sound money principles in the forthcoming election is already having its effect on the business situation. The result is being discounted somewhat and as soon as all doubts and uncertainties are removed there will be unusual activity in all lines of business, especially in the extensions and improvements of steam, electric and street railways. A restoration of confidence means good times again all over the country.

THE offices of the Colliery Engineer Co., Proprietors of the Colliery Engineer and Metal Miner, Home Study, and the International Correspondence Schools, in the Coal Exchange Bldg., Scranton, Pa., were partly destroyed by fire Aug. 30. Fortunately the printing plant was in another building and the business was not seriously interferred with. The new offices are in the Mears building.

A cable dispatch to the N. Y. Herald states that the London Daily News has received a telegram from Odessa, Russia, announcing the discovery by M. Kildischewsky, an electrician, of an improvement in the telephone by the use of which distance has no effect upon the hearing. In a recent experiment between Moscow and Raitoff, on the Don, a distance of 890 miles, talking, music and singing were heard with perfect distinctness. For the purpose of the experiment an ordinary telegraph wire was used. The dispatch adds that M. Kildischewsky will go to London to experiment with his improvement on the Atlantic cable between London and New York.

The largest yacht operated by electricity yet constructed was launched at Upper Nyack, N. Y., on September 10, for her owner, Mr. John Jacob Astor. The yacht is 72 feet long, 12 feet beam and 4 feet draught. Four hundred and eighty cells of battery will be installed to operate two 25 h. p. motors, built by the Ricker Electric Motor Co., of Brooklyn.

It is reported that Newton F. Hurst a grocer's clerk in Buffalo, earning $5 a week, has just received an offer of $30,000 cash and a royalty on all manufactured, for a car coupler he invented.

Duties of Civil Engineers in India.

The strange duties falling to the lot of the civil engineers in the public works department in India have been the subject of many stories by Kipling and other Anglo Indians. An unusual work they were called upon to perform last year was the preparation of a valley containing several villages for inevitable destruction by a flood. A landslide at Gohna, in the northwest provinces, formed a natural dam across the valley of the Bireh Ganga, which rose about 900 feet above the bed of the stream. A lake commenced to form behind this bank, and it was evident that the dam would be submerged. Finally steps had to be taken to secure life and as much valuable property as possible in the valley below. The lake was rising so rapidly that there was not enough time to cut a breach through the dam, and it was necessary to let things take their natural course. A telegraph line was constructed from the lake to the various villages lower down in the valley, and substantial monuments were erected along the sides of the valley above the anticipated high-water mark, to which the inhabitants were ordered to proceed immediately on the announcement by telegraph of the breakage of the reservoir. At the same time all permanent bridges, except two, which were allowed to remain on the demand of the local authorities, were taken down and replaced by temporary rope bridges. Careful surveys were made of the lake, in order to determine about the date on which failure was to be expected, and the estimate was only ten days in error, very close work in view of the enormous volume of the impounded water, which was estimated at 16,650,000,000 cubic feet. The valley immediately below the dam has an average slope of 250 feet to the mile, so the opportunities for destructive results were very good. The failure occurred about midnight, a breach of 390 feet deep through the barrier. The flood swept away whole villages, the abutments of the bridges which had been removed, and scoured the valley clean, but owing to the precautions taken not a single life was lost. The average velocity of the flood wave was 26 feet per second for a distance of seventy-two miles, and the maximum velocity was forty feet per second. The depth of the torrent was about 260 feet in the gorge, immediately below the dam, and 160 feet some further down. The height and velocity were obtained by noticing the time when lanterns were swept away which had been placed at known elevations along the sides of the valley.

Typesetting by Telegraph.

The Long-Distance Electric Typograph Company, of St. Louis, has been incorporated with a capital stock of $100,000. Joseph Reifgraber, is the inventor. The electric typograph system is designed to enable a newspaper editor to do not only his own work of writing, but that also of the telegraph operator and typesetter, while the performance of all their duties is facilitated. The editor is to "pound out" a "story" on a typewriter-like machine which is electrically cnnnected with a telegraph wire system. His "story" will be automatically reproduced in every newspaper office in the country with his wire, and being furnished with the instruments which compose the electric typograph system. Mr. Reifgraber says no new principle is involved in his invention, which was patented about a month ago. "It is merely a new application of old principles," he says.—*Electrical Review.*

Telescope Hitching Post.

A Washington man, Isaac W. Lewis, is the inventor of unique hitching post that promises to become familiar and useful for the purposes intended. It comprises a ground tube with its upper end at or near the surface of the ground, inside of which is placed on adjustable post. This post can be lowered or raised as desired and kept in a rigid position. When not in use the post can be dropped down and covered with a cap plate on the tube.

London's Underground Electric Railway.

Work has been commenced on the underground electric railway in London. As now projected, the road will be only 10 miles long, but it will thread all the populous suburbs from Shepherd's Bush into the city, and the 14 stations will tap the great arteries of travel. The minimum depth of the tunnel will be 50 feet, and the maximum depth 100 feet, and access from above to the platforms will be had by monster elevators, each with a carrying capacity of 250 persons. Forty trains per hour are to be run. The fare will be four cents for the round trip. The schedule time for the 10 mile run, including 14 stops, is to be 15 minutes. To prevent any possibility of accident, there will be two distinct tunnels, one for the up and the other for the down traffic.

The Indian Patent Office Report.

The report of Mr. A. T. Pringle, Officiating Secretary under the Invention and Design Act, for the year 1895-96, shows a considerable increase in the number of patents over that of the previous year, but the increase is mainly due to the applications of foreign inventors. The number of applications, 417, against 375 of the previous year, is thus explained, and similarly the increased number of specifications filed, 320, against 294, is principally due to the foreign applicants, whose inventions had passed the exprimental stage before coming to India. Forty-two as against 32 natives of India were among the applicants, and the percentage of increase in that class since 1893 has been more than 90. This, by itself, the reporter considers satisfactory, but under the circumstances it of course points to a corresponding fall in the number of Anglo-Indians who have turned their attention towards improving the manufactures of the country in which they live.

Will Compressed Air Motors Succeed?

We have already expressed our belief that the compressed-air motor cars now being experimented with in New York city will not be a commercial success, and that we do not think an air pressure of 2,000 pounds to the square inch can be safely or economically handled. In this connection a correspondent states that not long ago he had a conversation with Mr. George Westinghouse, Jr., on compressed air as a motive power. As is well known, Mr. Westinghouse is an authority on air-brakes. He is quoted as saying that in the successful working of air brakes trouble enough was encountered with a pressure of 100 pounds to the square inch, and left it to be surmised as to the difficulties which would arise when the attempt was made to work with 2,000 pounds pressure.—*Electrical Review.*

No More Sea-Sickness.

Mr. James Goodwin of Lynn, Mass., has secured a patent on a self-leveling ship's berth which would seem to be the correct solution or remedy for seasickness. When the waves roll high and the ship is tossed about in a way calculated to disturb the stomach of the passengers, if one can seek the seclusion of his stateroom and there find a berth so supported and controlled by various mechanical devices that it is always level and wherein one may avoid the "motion of the vessel," the dread of an ocean voyage will have vanished. Ship builders should test this new device.

The *Railway Age* points out that Africa offers a wonderfully prolific field for railway building and one which is destined to witness great activity in that line. With an area of 11,500,000 square miles or nearly four times that of the United States, omitting Alaska, and with a population of 168,000,000 compared with our 65,000,000 the entire continent has only 8,000 miles of railways, against the 181,000 of the United States.

Prof L. J. Kimball of San Francisco, Cal., is the inventor of an electro magnetic apparatus which it is claimed will accurately locate subteranean deposits of oil water and minerals. If half the inventor claims is true it will cause another "revolution" —in mining circles—and do away with speculaton as to extent of "paying" veins, etc.

Dr. Nansen, the arctic explorer, reports that his stanch ship, the "Fram," withstood the strains of the ice floes with perfect success. She was equipped with an electric light plant run by a windmill. The plant fulfilled every expectation. The lowest temperature Dr. Nansen found was 62 degrees below zero.

The longest bridge in the world is the Lion bridge, near Saugong, China. It extends 5¼ miles over an arm of the Yellow Sea, and it is supported by 300 huge stone arches. The roadway is 70 feet above the water and is enclosed in an iron network.

It is said that Thos. A. Edison is greatly interested in flying machines and aerial navigation and that he has several novel ideas to apply to an experimental machine soon to be produced.

A rain-proof umbrella with a transparent cover enabling the user to see where he is going has been invented in England.

A Submarine Torpedo Boat.

There is now under process of construction in the yard of the Columbia Iron Works at Baltimore, a submarine torpedo boat, which is likely to rival the visions of Jules Verne. It is expected that this craft, which will be launched soon, will be able to maintain a speed of eight or nine knots per hour when submerged to a depth of 60 feet. A correspondent to the New York Sun states that the motive power consists of a boiler heated by petroleum, supplying steam to three tripple-expansion engines, each driving a propeller and capable of developing 1800 horse-power, a generator set and a battery of accumulators. When the vessel is about to go under water the smoke stack is housed and the top covered over. The engines will be kept running at full speed, and by turning the rudder she submerges to the desired depth. The speed is afterward maintained by the dynamos running as motors and supplied by the accumulators which were previously charged. It is expected that the battery will enable the craft to maintain a speed of 12 knots per hour for short distances. The fighting tube will be the only part visable when the boat comes near the surface. This contains a telescope and prism at the top and enables the maintaining of the proper course. As the prism is fully protected by Harveyized nickel steel armor, it is not likely to suffer much injury due to an occasional stray shot.

Air for the crew, which will consist of about 12 men, four officers and eight seamen and machinists, will be stowed in reservoirs and may be replenished by a pump connected to a rubber tube terminating in a float on the surface of the water. The armament will consist of two torpedo tubes, one forward and one aft, which will discharge automobile torpedoes. Should all of the driving machinery fail, she can remain below for an indefinite time. Water ballast will be used, thus enabling the maximum depth desired to be attained, but automatically controlled to prevent an excessive submergence.

The National Capital.

In 1892 Stilson Hutchins and Joseph West Moore published one of the most interesting works ever printed regarding the establishment, progress and development of the National Capital—the City of Washington. It contains about 300 pages and over 200 illustrations, by a corps of artists under Mr. Sid H. Nealy. The story of the settlement and location of the National Capital at this point, Washington's diplomacy, early speculations in town lots, the career of the unfortunate French engineer who made the plans of what is now the "Paris of the New World," the invasion of the British troops in 1812 and the destruction of the public buildings, together with sketches of Mount Vernon, the home of Washington, and other points of interest and various facts and interesting details regarding the different branches of government—all these features combine to make this one of the most interesting books ever published. It is bound in heavy, ornamented board, originally sold for $2 but is now out of print. The INVENTIVE AGE has secured the last lot of them and will furnish them to its readers for 50 cents each. To each new subscriber or renewal, $1.35. That is, the INVENTIVE AGE one year and "The National Capital" to any address, $1.35.

His Patent Imperilled.

Attorny Garrett McEnery recently appeared as counsel in a case before a Justice of the Peace at Suinsun. McEnery found it necessary to make frequent objections to the evidence that opposing counsel was attempting to introduce. The Justice, whose first rule of evidence, "everything goes," looked first annoyed and then indigant. Finally he could contain himself no longer and, as a ruling on one of Mr. McEnery's objections, roared:

"Mr. McEnery, what kind of a lawyer are you, any way?"

"I am a patent lawyer," replied the attorney."

"Well, all I've got to say is that when the patent expires you will have a hard time getting it renewed. Go on with the case."—*San Francisco Post.*

Stone Soles for Shoes.

A German inventor has hit upon a method of putting stone soles on boots and shoes. He mixes a waterproof glue with a suitable quantity of clean quartz sand and spreads it over the leather sole used as a foundation. These quartz soles are said to be quite flexible and practically indestructible; they give the foot a firm hold on the most slippery surfaces.

The first electric railway to be built in Egypt was opened for traffic at Cairo on August 1.

Joint Inventions.

When two or more inventors combine their efforts to produce an invention, they open the way for endless trouble. All these combined originators of a joint invention must apply together for the patent on the invention. And after the patent issues to all of them as joint inventors, each one owns all of it, no matter how their interests are apportioned, to the extent of being able to use the invention and license others to use it in any part of the territory covered by the patent. Then suppose there are two inventors, and one has a 1-100 interest in the patent undivided, and the other has 99-100 interest, then the man who owns the undivided 1-100 of the patent can make use or sell the invention anywhere without the consent or approval or against the wishes of the owner of the remaining 99-100 of the patent. An "undivided" interest in a patent means an interest in the entire patent. An interest in a patent can be "divided" so as to make the interest cover only certain territory, as, for instance, a single state, or a single country, or a single township, or a single shop. But if even a single shop has a right to manufacture under a patent, that shop can sell its product anywhere, so that it can make a price to suit itself without regard to the wishes of any one else. It is the case of two cats over one mouse, which the proverb says never did agree, and joint patentees are pretty certain certain to quarrel over their patent. In all cases of joint invention the whole interest should be conveyed to one single party; then there is some possibility that somebody will make something out of the patent.

Joint patents are structurally weak, because a truly joint invention is an unlikely and improbable thing; the probability is that one inventor invents one detail, and another another detail; it is much better to have each one of the joint inventors apply for an individual separate patent on what he individually and separately invented, and in such a case all of the patents should be assigned to some one party as a trustee, and this trustee should be a business man competent to manage sales and should be given full power to handle the patent. Of course the trustee may steal all the proceeds, but the supposed beneficiaries of his trust can then go to law to obtain their shares of the proceeds, so it is just as well in the end as if there had been no trustee, and the joint patentees had stolen from each other and gone to law over the matter.

The New Phonendoscope.

The newly invented phonendoscope is designed to be used by physicians and surgeons for detecting the presence of disease by sound.

The instrument consists of a circular flat metal box or tympanum, having on its one surface two apertures for the attachment of the rubber ear tubes, while the other surface is formed by a thin disk which is readily thrown into vibration. The best results are obtained by simply applying this disk to the surface to be examined. By an ingenious contrivance a second disk can be superposed upon this one and a vulcanite rod attached to the former, so that the area of auscultation may be extremely circumscribed. The condition of the sounds is only slightly diminished by the use of the rod, which thus combines the principle of the solid stethoscope with that of the tympanum. The rod furnished with the instrument is about two inches in length, but it is stated that there are other rods of various lengths to enable the "phonendoscopist" to receive sound vibrations from the natural cavities which communicate with the exterior of the body. Altogether we consider the instrument highly ingenious, carefully and compactly constructed, useful as an aid to auscultation, and not likely to entirely supersede the use of the stethoscope. It may also be found usefull in class demonstration, since it would be easy by means of branched tubes to enable several persons to listen at the same time.

The instrument would be particularly useful for the following purposes: 1. The sound of the respiratory organs, of the circulation of the blood, and of the digestive organs in the healthy body as well as in the sick subjects. 2. The sounds made by the muscles, joints and bones. 3. The sounds of the capillary circulation. 4. The slightest sounds produced in an any diseased condition of the body; hence it is possible to draw on the body dimensions, the position or any alteration in the position of the various organs and of the fluids which have gathered in the most important cavities in the body. 5. The sounds in the eye, the ear, the bladder, the stomach and the intestines.—*Lancet.*

A new and novel "rope pull" is the invention of B. R. Hill of Locca, Ga., —a contrivance made so that cotton rope may be pulled from its original coil without tangling.

A Revolving Tower.

France built the Eiffel tower and turned up her nose at the world. England's retort was to lay the foundation of her Wembley Park tower, a stolid, stupid retort, for even if the new tower is a few feet higher, it will be a mere imitation of the French original.

American's reply to the Eiffel tower was the Ferris wheel. "Anybody," said America, "can pile steel beams one upon another. It is only a shade more intelligent an undertaking than heaping stone upon stone. But we have put up a structure as big as your tower, and it goes round, instead of standing still."

France stopped to think. England, bull headedly enough, built a wheel of steel bigger than ours, and further differentiated by the fact that it sometimes sticks instead of going round, and leaves peripheral parties of merry makers to spend a night in the air.

NEW REVOLVING TOWER.

All of this is an old story. But now we discover what France has been thinking about, and it is quite a new story. "Your big wheel that goes round," says France to America, "and the English bigger wheel that won't go round, are only fit to amuse country cousins. What do you say to a great lofty building that spins slowly like a majestic top? You sit in a splendid hall, under noble arches, surrounded by stately palms and festoons of flowered vines, and while you eat your dinner and drink your coffee and talk to your best girl and hear the band play, you look out of the big windows at a city which seems to move beneath your gaze like the cloth of a gigantic panorama."

The inventor is M. Devic, and he calls his big tower the "Palace of Progress." This extraordinary sort of structure is shown in the architect's perspective drawing. A complete revolution will occupy about two minutes, and the views of Paris and the hills and plains of the Seine and Marne country will change as rapidly as the scenery

changes when one is strolling slowly along a road. The rotary building will be only half the height of the Eiffel tower, but, as it is to be erected near the summit of Montmartre, the highest point within the fortifications, it will command a broad view, cut only by the tower of the new church on the apex.

The bearings are said by the mechanical engineers who have prepared the specifications to be so designed as to absolutely assure the absence of all sense of motion. When you are not looking out at the view you will be as tranquil as in any other building, but when you swing your chair so that you face the window you enjoy a serene motion and contemplate a constantly changing spectacle.

The motive power which will supply the force necessary to turn the structure will be hydraulic, and its cost has been calculated to be only 87:12 francs per hour, although each time the movement is checked the hydraulic pressure needed to give it new impulse will represent an expenditure of 282:80 francs. Rozier, the caterer and refreshment contractor, who has made a fortune out of buffet concessions at all the race courses in the neighborhood of Paris, is the largest shareholder in the enterprise; and Marchand, manager of the Follies Bergeres, and of two or three other less important variety halls, has underwritten a large block of stock, and will control the music and the vaudeville attractions, which are relied upon to assist in drawing pleasure loving Paris to this vortex of delights. The upper part of the building will be occupied by a public ball room, to be open from 11 o'clock in the evening until 2 in the morning, and the place immediately below this for an artificial ice skating rink, so that the allurements of the Palais de Glace on the Pole Nord will be added to those of the Moulin Rouge and the Casino de Paris.

The "Cassard" Sold for $1,000.

That strange little vessel, the Cassard, expected by her inventor, Mr. Robert M. Fryer, to revolutionize ocean travel, was sold at auction a few days ago in Baltimore for $1,000. It was built in 1890 in Baltimore and for a long time the steamer lay at the wharf in Alexandria where the promoters expected to build up a great ship building industry on entirely new lines. The Cassard is remarkable for her narrowness, being 220 feet long and 16 feet extreme beam. She has 18 feet 4 inches extreme depth amidships, and draws 8 feet forward and 10 feet aft. She has a turtle-back deck and the lines aud ship give it an elliptical shape. She is extremely pointed at the forward end. Mr. Fryer said that thirty-five miles an hour would be easily reached, but no trial trip of the vessel ever developed more than a fourth of that speed. Her displacement was reckoned at about 350 tons. The basis of the structure is a keel that weighs 68,000 pounds. It is made of a series of plates laminated and as homogeneous as though solid. This keel runs from nothing at the bow to a depth of twenty-five inches outside of the hull, at the stern, and comes up through the hull a sufficient distance to form the backbone for all the other parts of the ship—the frames, the bedplates for the engines, boilers, condensers and all heavy weights in the ship—and all of which are placed directly upon this rigid and indestructible keel.

The shaft is 92 feet long and 10½ inches in diameter, with a four-bladed screw 9 feet 8 inches in diameter.

It is said more than $125,000 was sunk in the building of the vessel, which was never a success. The Cassard came to public notice again last winter by being towed to Newport News, where it was understood, after some changes, if her speed developed she would be bought by the Cuban revolutionists. This failed and she started for Baltimore but before reaching port an accident occurred, and she had to be towed to anchorage at Baltimore.

Correction.

In the September issue under title the New Patent Law in Russia was printed in regard to specifications. "One Bristol board and other cloth, well excited in black linns 14 inches by 8¼ inches." It should have read—One copy must be on good drawing paper or bristol board, and one upon tracing cloth. The size of the sheets should be 8 inches wide by 13 inches high all inches high by 16 inches wide; or, 13 inches high by 24 inches wide. One inch from the edge a single margin line should be drawn. The usual designation of figures, i. e. "Fig" should be omitted.

It is said that runaway horses are almost unknown in Russia. No one drives there without having a thin cord with a running noose around the neck of the animal. When the horse bolts the cord is pulled, and the horse stops as soon as it feels the pressure on its windpipe.

DANGEROUS AND STARTLING PROPOSITION

To Amend the Patent Laws

NO PATENTS ON LABOR SAVING MACHINES.

The following is taken from the Chicago Commercial Journal.

Editor Chicago Commercial Journal: Will you allow me the use of your valuable columns for the purpose of directing attention to what I conceive to be an egregious wrong? I refer to the granting of patents for inventions that have no other merit whatever than what is called "saving labor"—really to throw numbers of men, or women perhaps, out of work. I am not such a crank as to condemn all labor-saving machinery in toto. This country could never have been what it is today had it not been for this same labor-saving machinery. In the early days of the republic vast acres had to be filled, mines developed, and an immense expanse of land and water navigated, and labor was very scarce then; a machine which could be manned by five men and do the work of 500 was a great acquisition to the wealth and productiveness of the country. But that day has now gone by; laborers are plentiful, and, as Li Hung Chang says, may be made still more plentiful if we will. Under such circumstances it seems to me downright stupidity for the government to deliberately throw, goodness knows how many more men out of work by so encouraging the invention of more labor-killing machinery as to grant patents to the inventors of such, giving personal right to make, sell, and use the same for a period of 17 years. Generally such inventors and patentees are not able for lack of means to take advantage of their so-called "rights," so they sell said "rights" to some capitalized corporation, who, owning the machines which displace so many men, actually own and control the products of the labor of the number of men displaced.

I want, Mr. Editor, to secure the influence of your valuable journal in getting an amendment made to the patent law so that no other patents may be granted for inventions that have no other merit than the mere saving of labor.

INVENTOR.

It may be very old-fashioned, but we must say that we are inclined to agree with the contention of our correspondent.—Ed.

If the above article against patenting inventions for "saving labor" had been published in Egypt, China or some other like nation ignorant of invention as a promoter of civilization one might not even take notice of it, because the effect thereof would have had no greater baneful results than the words of either Li Hung Chang, or the author of the above quoted article, "that the day when a machine could be manned by five men and do the work of 500 was a great acquisition to the wealth and productiveness of the country has gone by; laborers are plentiful and may be made still more plentiful if we will."

The more startling thing about the above article is the comment of the Editor of the "Chicago Commercial Journal." He says: "It may be very old-fashioned but we must say that we are inclined to agree with the contention of our correspondent."

Would the Editor be willing to stop labor saving improvements in printing machinery? Would he go back to the boys feeding the sheets as they were discharged from the cylinder? Would he go back to dip, mold or star candles, and cease patenting improvements in Electric lighting mechanism, whereby the electric light may be furnished as cheap or cheaper than gas or any of the best known lighting means? Would he cut down thousands of acres of wheat by the hand scythe, and thrash the same with the flail, as used in the days of Gideon; or as used on one side of the Nile at the present day while the most approved machines for thrashing are in use on the other side? Would he have every mother stitch her life out of her with the hand needle, instead of providing her with a sewing machine with which she can do, in a short time, pleasurably, all the work of the home, even if she has a number of childern.

Every intelligent person knows that fields for employment of labor are opened manifold by the manufacture of labor saving machines.

It has been estimated that many thousands of laborers must contribute their services in one way or another to the building of a railroad of the average quality and extent, but after it is built some one will say, "Yes; but the old stage coach also required a large amount of labor for its construction and maintainance." This is true, but while the population since the stage coach days has increased probably two thousand percent; the amount of labor required in mines, forests, offices and manufactories attributed to the existence of the railroad, has risen to a seemingly fabulous extent. And whom does the railroad benefit more than the laborer? Rich men could always afford to travel in the best conveyances and by the most rapid means, but the poor man in the days of the stage coach could seldom afford to travel at all, and had to stay at home and live out his life within the horizon which his eyes met when he was born; and to live too on just what was raised on the land immediately near here; for to fetch articles from neighboring districts would be too expensive for his purse. Now, on the contrary, he can travel just as far and fast as his more fortunate employer, although he does not elect to go in cars so well equipped, for want of as much money. It has been estimated that every average engine can do the work of 1,000 horses working on good roads, and consequently the laborer can have his goods brought to him much cheaper than in the old days of wagon trucking. But the Correspondent of the "Chicago Commercial Journal" would argue that much more work and higher wages would be obtained if there was no steam engine. It is hardly necessary to state that if there were no railroads, the people instead of doing the same amount and kind of work, would not do it at all. It was the railroad that has made people dream of doing work on so large a scale. It has been estimated by eminent authorities that with patented mechanical inventions the population of the globe can do in one day what would require one hundred times the population of the globe one thousand times a hundred days, if they had no machines; and, indeed, men can do some things by machines which never could by any possibilty be done without them in any given time whatever.

While the results of invention may be a greater increase of production without an equal increase of consumption on the part of the rich, the overplus will be placed within the reach of the masses which at one time could not get meat for dinner, and often went without wheat bread. A few figures of the results of invention illustrate this point.

Within 8 years after the invention of the cotton gin by Eli Whitney, the production had risen in America from 139,000 lbs. to eighteen million lbs; and up to 1870 to four billion five hundred million lbs. While this was vastly out of proportion to the increase of population, it enabled the masses to use cotton goods who never used them before, and others who had used them began their use in greater quantities.

Upon the intraduction of the sewing machine as a tolerably practical device in 1846, each machine cost about $500, and it was well out of the reach of the middle and lower classes. At this period Pessimists of the type of the Correspondent of the Chicago Commerical Journal, stated that the machine at even that high price would soon destroy the occupation of the sewing women. At this date good sewing machines can be purchased at retail for from $20. to $60 apiece, and they are to found in a majority of the households, saving a great amount of labor to the mothers of the world; also in the factories of the world, employing millions of the world's inhabitants, saying nothing of the labor employed in their manufacture. It may be asked how is this possible, and the answer is, because the masses are better off and are using better wearing apparel, and more of it than was ever the case in the times of our ancestors. The trouble is not with patented inventions and inventors, but mainly with the capitalists and trust-companies; such, in a measure, preventing the distribution of the products of labor and restricting the sale or exchange of articles which are produced by workmen with the aid of patented machinery.

It is resonable to believe that the time will come when a man can make all he can wish for in a few hours of the day, and the remainder of his time will be given to recreation, and such studies as will serve as the best means to pass away the time. Thus, indeed, work will become a pleasure, and much more time will be given to beautifying the surroundings and environments of even the poorest laborer who may inhabit the world.

It should be borne in mind that patents are not a monopoly, for they are granted for a term of only 17 years, and at the end of that period the public possess the right to freely make and use the invention which is described and claimed in a patent and this, indeed, is but a small reward to such benefactors of their race.

Let America be as generous and grateful to her inventors and patentees as England was to James Watt, the inventor of a practical steam engine.

In Westminister Abbey, where England honors her great men, with burial, and records their names and achievements, there stands a monument bearing an inscription as printed below, from the pen of Lord Brougham, who esteemed it one of his greatest honors that he was called upon to record the nations appreciation of the man in whose honor the monument was erected.

"Not to perpetuate a Name
"Which must endure while the peaceful arts flourish;
"But to show
"That mankind have learned to honor those who best
"Deserve their gratitude.
"The King
"His ministers and many of the nobles and
"Commoners of the realm raised
"This monument to
"James Watt:
"Who, directing the force of an original genius,
"Early exercised a Philosophic research,
"To the improvement of the
"Steam Engine.
"Enlarged the resources of his country:
"Increased the power of man, and rose to an
"Eminent place,
"Among the followers of
"Science and real benefactors of the World."

ROBERT W. FENWICK.
CONSELOR IN PATENT CAUSES.

Hear Yourself Wink.

You can hear yourself wink. A wonderful machine has been invented by which the hitherto impercepitible sound of the action of the eyelid is made clear and distinct to the ear. The invention is called the phonendoscope.

The new wonder was designed to be used by physicians and surgeons for detecting the presence of disease by sound. With its perfection it will be possible for the medical world to prepare a perfect chart which will enumerate the sound of every disease known, and to instantly detect that sound with an application of the phonendoscope. There will be no mistake or misconstructions in the matter of a diagnosis.

The delicate instrument, which those who are fortunate enough to catch a glimpse of, now handle as carefully as if it were made of diamonds, consists of a circular or tympanum box of highly polished metal. On one of its flat surfaces there are two apertures for the attachment of that number of little rubber tubes, each fitted with a tiny elbow nub to be placed in either ear. The other surface shows a thin and sensitive disk, which is readily thrown into vibration. The thin disk, being controlled by the compressed air of the circular box, is so sensitive that the faintest breath upon the disk sends a wave current through the box, into the rubber tubes and against the ear drum of the listening person.

Not only can this breath upon the disk be heard, but its force in transit is multiplied a thousand fold, so that when it reaches the ear it sounds like the roaring of a gale of wind at sea. Tapping the disk lightly with the fingers sends a sound into the ears like the tramping of many horses, and the wink of an eyelid gives forth a sound like the fall of heavy waters.

By the arrangement of the rubber connecting tubes it will be understood that the ends can be put into the ears so that any one can hear himself wink or hear the sound of his most gentle breathing.

Can be Seen 60 Miles.

The penetrating powers and ranges of powerful lights, such as are employed in lighthouse service, rapidly decrease as the ratio of their luminous power increases. For instance, a light of 5,000,000 candle-power in the British Channel has in average weather a luminous range of about 44 nautical miles, while if the light be increased to the power of 10,000,000 candles, the luminous range is only five miles more, or 49 miles. According to current practice, lights up to 200,000 candle-power are obtained by means of mineral oil lamps, while electric lights are used for higher powers, and almost any power may thus be obtained. The highest power yet attempted is about 36,000,000 at Penmark Point, in the department of Finisterre, France, which when completed will be the most powerful lighthouse illumination in the world. The height of the tower in which it is to be located is about 63 metres, enabling it to be seen during the day from a distance of 18 miles in fine weather. During the night this light will be visible for 60 miles. The rotundity of the earth will prevent the rays from striking the eye directly at a distance of more than 30 miles, but the sky overhead will appear illuminated for 30 miles more. The estimated cost of this lighthouse is about $120,000.

BURIED ALIVE.

Grewsome Inventions For Preventing Fatality From Premature Interments.

It has often been quoted, and with much truth, that the inventor is the benefactor of his race and we can hardly fail to concur in this when we realize that we owe every convenience we enjoy in life to him. He is not satisfied though in simply supplying the conveniences of life but with untiring energy he is even now seeking to rob death of some of its terrors by following us into the grave and providing for our rescue in cases of premature interment through mistake and where the supposed corpse is simply in a trance, cataliptic or other state of suspended animation.

The primary object of all these ghastly inventions is to provide devices that can be applied to an ordinary casket or its box in such cases as where doubt exists as to the death or condition of the corpse. Another and very important object is to provide against any tampering with the devices, after they are in place in the grave-yard, by malicious or mischievously disposed persons.

One of the simplest forms of these devices consists of a small metal tube screwed in to the top of the casket or its box and extending up to the surface, the lower end of the tube being provided with an auxiliary tube for discharging air in proximity to the mouth and nostrils of the corpse. The upper end of the tube projects above the earth and is provided with a cylindrical glass casing in which is mounted a bright colored valve. This valve normally closes the upper end of the tube so that air cannot enter the buried casket, and is mounted on the upper end of a slender rod that passes down into the casket and rests upon the forehead of the deceased. Should the victim revive, his first natural movement would be to raise his head and elevate the rod carrying the valve and thus admit air to the casket to prevent his suffocation. Clutches are provided to prevent the rod descending again, as the victim in his paroxysms of agony upon discovering his condition might grasp the valve rod and pull it down and thus shut off his supply of air. When the bright colored valve is raised by the rod it can readily be seen through the glass cylinder and thus indicates a movement of the corpse to the watchers who take immediate steps to rescue the victim. If no movement of the valve occurs within a predetermined time, the tube is drawn out of the ground and earth filled in.

A more complicated form of this grewsome apparatus comprises two vertical tubes connecting the casket with the outer world. One of these tubes opens into the casket directly over the face of the corpse and is provided with a small shelf and a reflector above the same so that a lamp placed upon the shelf will throw its light directly upon the face of the deceased. Friends of the supposed dead by looking down through the top of the tube can plainly see the face at all times and any changed expression would be instantly noticed. The other tube is connected at its upper end to a fan for forcing air into the casket. The fan is started by the releasing of a catch through the medium of a string attached to the hand of the corpse. The movement of the hand also releases a signal which rises above the first mentioned tube, thus indicating a movement of the corpse. An additional electric signal in the form of a bell is also provided and is adapted to be set into operation by the movement of the hand. The top of this casket is divided into two sections so that in exhuming a victim, time can be saved in digging by permitting the rescued to be removed through the forward part of the casket without removing the earth from above the whole length of the same.

A very similar apparatus is also used in which the hands of the corpse rest upon the cross piece at the lower end of a rotatable shaft that passes up out of the top of the casket to the surface. The upper end of this shaft is provided with an index finger that is adapted to travel over a cylindrical index plate. This index finger and plate are covered by a glass case so that they may be readily seem at all times. Air passes down through the shaft directly into the casket. Upon an interment being made the relative positions of the index finger and plate are noted and any movement of the hands of the corpse in reviving will turn the rotatable shaft and change the position of the finger on the plate. If the supposed deceased should turn in the coffin or make any other violent movement he would push the rotatable shaft upward and raise the index finger clear of the index plate. Either movement of the index finger would at once be noted through the glass case and a rescue commenced.

A hermetically sealed coffin is also employed and in this case the exterior box is connected by a tube with the surface. Each end of the casket is provided with an aperture opening into the outer box. These apertures are closed by suitable plugs so as to seal the coffin air tight. The plugs are connected by a cord which passes through the hand of the corpse. Upon any movement of the hands in reviving the plugs are simultaneously withdrawn, admitting air to the casket from the exterior box. The movement of the plugs also releases a fan like signal which projects above the top of the tube so as to be visible above the surface of the ground.

All these devices before mentioned depend upon an air communication between the casket and the surface and a positive movement of the corpse to set the devices in operation, but this is not the case with an attachment recently patented. With this greatly improved mechanism the most delicate movement of the pulse of the supposed corpse will admit air to the casket and sound an alarm above the surface. A tank of compressed oxygen or other life sustaining gas is attached to the casket. A delicate bracelet of aluminum is placed about the wrist of the supposed corpse and is provided with contracting points connected to the wires of an electric circuit. The wires pass up to a suitable alarm above the surface. The contracting points are mounted on the most delicate springs and the faintest movement of the pulse will cause them to come together and close the circuit thereby sounding the alarm and opening the valve leading into the oxygen tank to admit air. This improvement is a very important one as the reviving victim sometimes has not strength enough to move any portion of his body and the before mentioned devices would therefore be useless in such a case.

Still another invention consisted of a long sharp knife set in the coffin lid and so placed that when the lid was closed it would pierce the heart of the corpse and thus make his death a certainty.

It is perhaps not inapt to here mention the grave torpedo for preventing the unauthorized resurrection of the dead. This consists of a metal cylinder fully charged with powder and balls and located at the head or other portions of the casket and adapted to be discharged by a spring actuated hammer to the injury or death of the desecrators of the grave. The trigger for exploding the torpedo is connected to the body of the corpse by suitable cords and any attempt to withdraw the dead from the casket will immediately operate it. The torpedo is concealed in the trimmings of the coffin and the plunderers of the grave are not aware of its presence until it explodes sending its deadly missles flying in all directions except into the casket.

W. HARVEY MUZZY.

The Great Wall of China.

"China abounds in great walls," remarks a Pekin correspondent in a recent letter; "walled country, walled cities, walled villages, walled palaces and temples—wall after wall and wall within wall. But the greatest of all is the great wall of China, built 213 years before our era, of great slabs of well-hewn stone laid in regular courses some twenty feet high, and then topped out with large, hard-burned brick, the ramparts high and thick and castellated for use of arms. It was built to keep the warlike Tartars out—25 feet high by 40 feet thick, 1200 miles long with room on top for six horses to be ridden abreast. For 1400 years it kept those hordes at bay, in the main, and is just as good and firm and strong as when put in place. How one feels while standing on this vast work, scrutinizing its old masonary, its queer old cannon, and ambitious sweep along the mountain crest. In speechless awe we strolled or sat and gazed in silent wonder. Twelve hundred miles of this gigantic work, built on the rugged, craggy mountain tops, vaulting over gorges, spanning wide streams, netting the river archways with huge, hard bars of copper; with double gates, and swinging doors and bars set thick with iron armor—a wonder in the world, before which the old-time classic seven wonders, all gone now, save the great pyramid—were toys. An engineer in Seward's party here, some years ago, gave it as his opinion that the cost of this wall, figuring labor at the same rate, would more than equal that of all the 100,000 miles of railroad in the United States. The material it contains would build a wall six feet high and two feet thick straight around the globe. Yet this was done in only twenty years, without a trace of debt or bond. It is the greatest individul labor the world has ever known. You stand before it as before the great Omnipotent—bowed and silent."

A new bullet, for which great destructive power is claimed, has just been furnished by an inventor at Anderson, Ind., says an exchange. The bullet has a hole ⅛ inch in diameter extending nearly through its whole length. This chamber, the inventor says, gathers air under strong pressure, caused by the rapid flight of the bullet, and the air expanding when the ball comes in contact with anything, causes an explosion of reat destructive force.

Steal Other People's Inventions.

Just why inventive genius and gullibility should go together it is hard to say. Certain it is that inventors are the most guileless individuals in their dealings with others on business matters, and fall easy victims to the spiders who lie in wait for such flies. The list of clever men who walk today, while those who ride owe their luxury to the other man's genius and their own shrewdness, is an interesting one. Here are a few cases picked haphazard from the chronicle of inventions that failed to benefit the inventor, or, at least, produced for him merely a little of what was his due.

It is not necessary to be very old to remember when hooks were first put in men's shoes in place of holes, in order to save time in lacing the shoe at the top. This was the brilliant idea of an inventor to whom it should have brought a fortune. It would have done so had he been a shrewd business man. Being merely an inventor, he hadn't sense enough to keep his idea to himself until the patent-office padlock had secured it against theft. In the innocence of his nature the inventor confided the idea to a friend while crossing the North River ferry boat and the friend hardly waited for the boat to tie up in Jersey City before he excused himself, started back to New York, and went on a dead run for a patent lawyer in order to have the idea secured for his own special benefit. Another man is known today as the inventor of the lace hooks. He owns a splendid house and is wealthy. The confiding inventor got nothing.

The inventor of a patent stopper for beer bottles, something that had long been wanted by the trade, sold the invention for $10,000 to a man who recognized its great money-making value. The purchaser is now worth $5,000,000, all of which he made from the sales of the patent stopper. Out of the goodness of his heart he presented the original owner of the patent with $30,000, so that this man got $40,000 in all for his $5,000,000, idea. To give some notion of the value of the patent rights on this bottle stopper the price came down from $1 to 6 and 7 cents a gross, and even at this enormous reduction a good profit could be made.

This last inventer was treated with princely generosity, however, in comparison with the genius who devised a pocketbook clasp in the shape of interlocking horns, with balls at the end which snapped shut with a single pressure. The idea was afterwards applied to gloves and became very much in favor. The inventor relinquished his prize for the magificent reward of a kidney stew dinner and 50 cents, the latter having been advanced by the purchaser to pay the inventor's expenses from Newark to New York. The man who secured the idea and patented it, after treating the inventor in the royal manner mentioned, made a big fortune by his shrewdness. What became of the inventor is not known.—*New York Recorder.*

New Ideas.

New ideas! Who has them? Nobody and everybody. Contradictory, isn't it? Yes, but true. An idea is new or old according to the language in which it is clothed, mostly. An idea, an old, old idea may come to us in such a form that it can very readily do duty for a brand-new one.

Again in the ability to express one's self at all is often found the chief merit of a so-called new idea. This we find in inventions. Putting forth an invention is merely crystallizing an idea which has been floating around in one's brain. How often we will find a half-dozen people using some simple appliance which they have contrived for themselves. It is so simple and so obvious that the idea of patenting it never occurs to them, and outside of the few in the immediate neighborhood the world hears nothing of it. Suddenly, in another part of the country, some bright inventive man sees the want, and turns out something to fill it. The value of the idea is at once recognized by all in that special profession or calling and by many outside of it, and the inventor is praised for the "newness" of his idea, while those who have been doing pretty much the same thing for years on their own account look with disgust on the man who makes a fortune out of what they regard as a pretended new idea. But, then, you see the latter knew how to present his idea to public notice, while it never occurred to the former to do this, even had he been capable of doing so.

It is precisely the same way with expressed language. The same idea may be presented a half-dozen times in as many different ways and each time through its new mode of expression start an entirely new train of thought.—*Philadelphia Call.*

A Chinese encyclopædia of 5,020 volumes has been added to the library of the British Museum.

Some Observations Regarding Inventions.

Not many years ago nearly all the inventive genius of the country was located in New England and the Eastern States, but during the last two or three decades the West has developed its share of mechanical ingenuity, though Connecticut still leads with the largest number of patents in proportion to the population, with Massachusetts second, the District of Columbia third, New Jersey fourth, Montana fifth, Rhode Island and New York sixth and seventh, Colorado eighth, and several other Western States well up in the list. Several of our well-known American millionaires owe their fortunes to valuable inventions. Of these, perhaps the four most conspicuous are George M. Pullman, Alexander Graham Bell, Cyrus H. McCormick, and Thomas A. Edison. On the second thought we believe the latter should have been named first, except for chronological sequence. Whether Mr. Edison's tangible fortune at the present time equals that of the others it would be difficult of determination; but, if the value of his holdings in all the various companies and inventions bearing his name could be computed, probably his wealth would exceed any of them. He is not, however, an investor and financier like some of the others, his attention and abilities being concentrated upon the intricate problems with which he is constantly wrestling. He is the most prolific inventor in the world, having secured mo:e patents for his own inventions than any other man, as shown by the record.

Alexander Graham Bell was born in Scotland, and came to the United States in 1872. For several years previous to the invention of the telephone he had been known as a writer on scientific subjects, but not as an inventor, no other patent having been granted to or applied for by him previously.

The protracted controversy in the courts instituted by inventor Daniel Drawbaugh, of Pennsylvania, who claimed to have filed the first application for a telephone patent, and that his discovery was stolen in some manner in the Patent Office, was finally settled in favor of the Bell patent, and which has been one of the most profitable ever issued. The first public exhibition of this telephone was made in Philadelphia in 1876.

An inventor not so well known, but perhaps more honored than all the rest in that his invention stood more for philanthropy than for possible profit, is Joseph Francis, who invented the lifesaving boat used in the Government service. The Fifty-first Congress voted him a medal of pure gold valued at six thousand dollars. It is the largest and finest ever given by this Government to any individual, and was presented to Mr. Francis at the White House in 1890 by President Harrison, with apporopriate ceremonies. On the occasion of Mr. Francis' last visit to Washington, in 1892, he donated the medal to the National Museum, where it is now on exhibition, together with his original life car, which saved two hundred and one lives from the wreck of the "Ayrshire" on the coast of New Jersey in 1847.

When we look further along the line to the names of Howe, Morse, Colt, Hoe, Whitney, Blanchard, Ericsson, and Eads we find that examples are not wanting for the inspiration of the American inventor, and more recent events prove that he is fully equal to the occasion and that his achievements will fully maintain the brilliancy of the banner of progress.—*New Ideas.*

The New Monster 16-Inch Gun.

What is expected to be the most powerful cannon in the world is now being constructed at Watervliet Arsenal for the United States coast defense service. When completed it will be mounted at Fort Wadsworth. The gun will lack only a small fraction of being fifty feet long. It will have a range of sixteen miles, and will be able to penetrate twenty-seven and one-half inches of the best steel armor at a distance of two miles. The weight of the gun will be 125 tons. It will throw its solid armor piercing projectile, weighing 2,370 pounds, with a velocity that can hardly be conceived. When the projectile leaves the muzzle of the gun it will travel at the rate of 2,000 feet a second. If brown powder is used in firing the gun the charges will weigh 1,060 pounds.

Largest Propeller Ever Cast.

The largest cast steel propeller ever made is one recently made at Chester, Pa. for the new iron steamship John Englis, soon to be put in commission beween New York and Portland, Me. There are four blades measuring from tip to tip 16 feet, and its hub where it fits on the shaft is 26 inches in diameter. The thickness of the metal at the hub is 6 inches. The casting weighed 15,000 pounds. Some idea of the strength and power of this huge screw may be gathered from the fact that the tensile strength of the steel is 71,000 pounds, with an elongation of twenty-seven per cent and a reduction of thirty-three per cent. When placed in position on the steamship John Englis, and driven at the full speed of the 4,000 horse power triple expansion engines, this propeller will make ninety revolutions a minute.

Report of Commissioner of Patents.

Under date of Sept. 5th Commissioner Seymour submitted a report to the Secretary of the Interior of the business of the Patent Office for the fiscal year ending June 30, 1896. Summarizing the various tables, there were received in the fiscal year ending June 30, 1896, 41,660 applications for patents, 1,641 applications for designs, 84 applications for reissues, 2,460 caveats, 2,064 applications for trade-marks, and 171 applications for labels. There were 22,791 patents granted, including reissues and designs; 1,782 trade-marks registered, and 11 prints registered. The number of patents which expired was 11,466. The number of allowed applications which were by operation of law forfeited for nonpayment of the final fees was 4,014. The total receipts were $1,307,090.30; the receipts over expenditures were $209,721.45, and the total receipts over expenditures to the credit of the Patent Office in the Treasury of the United States amount to $4,776,479.18.

On the 30th of June, 1896, all but four of the examiners had their work within one month of date, two were between one and two months, and the other two were between two and three months from date. At the close of the fiscal year there were 8,943 applications awaiting action on the part of the office.

Device for Suspending Stove-Lifters.

There are a thousand and one useful little devices in the kitcken—of every day use—that might have been patented and although sold for a few cents each, would have brought the inventor handsome royalties. The kitchen is still behind in the "march of invention." Scores of little labor-saving devices might be added. A recent one is that of James L. Kerstetter, of Milton, Pa.,—device for suspending stove-lifters. It consists of a small spring drum or cylinder suspended from the ceiling over the stove. Attached to the drum is a cord fastened to the "stove-lifter" and adapted when unwinding, therefore, to wind the spring, and vise versa. Thus the "kitchen maid" always has the "stove-lifter" in reach—never hot or "out of sight"—and always up and out of the way when not in use. Here is a simple, inexpensive device that doubtless will meet with universal favor.

A Chinese Pile Driver.

Piles were being driven in one of the new buildings for a foundation for a punch. They were 8 inches in diameter and 14 feet long. The staging was bamboo, and so was the frame for the hammer, which was a round piece of cast iron, with a hole in the centre for a guide-rod.

Attached to the hammer block were twenty-seven ropes, carried up to the top of the frame and down on the outside, looking very much like the old-fashioned Maypole. Twenty-seven women had hold of the ends, and with a sing-song, all together, pulled down; up the rod, four feet, traveled the hammer; then, at a scream, all let go, and down it came on top the pile, which was unprotected by a band or ring. The women were paid 20 cents in gold per day. This May-pole driver is in general use throughout Japan and China.—F. F. Prentiss in Cassier's for October.

Electrical Patents.

In twenty-five years the total number of United States patents rose from 98,460 to 568,619. Of the latter number, electric generators claim 3,117; electric railways, 2,019; electric lighting, 3,622; electric power, 1,183; telegraphy, 3,205, and telephony, 2,459.

The possibility of telegraphing through space, which is maintained by Tesla and other electricians has been practically demonstrated in Ireland in maintaining communication between the Fastnet lighthouse and the mainland, where a non-continuous system has been established with success by W. H. Preece, the electrician to the British Post Office Department. Formerly the difficulties at this place of carrying a telegraph cable up an exposed rock, where it was subject to constant chafing, were almost insuperable. The non-continuous system is now used and works admirably. The cable terminates in the water 60 yards off, and the electric currents sent from the shore find their way through this distance to two bare wires that dip into the sea from the rock.

The Pneumatic Tire Back in 1846.

Most people are probably aware that Mr. Dunlop was anticipated in his invention of the pneumatic tire by a Mr. William Thompson, who took out a patent for the device about 1846. The curious thing and one which is less generally known, is that the invention was very thoroughly tested, and shown to have great advantages, so far as the reduction of tractive resistance is concerned, over the tires ordinarily in use. Thus, in the "Mechanic's Magazine" of March 27, 1847, an account is given of a number of experiments made in the presence of the editor of that journal, which conclusively showed the superiority of the pneumatic system. A car fitted with the pneumatic wheels was drawn over a length of road in Regent's Park, half of the distance being firm and smooth, and the remainder covered with newly broken stone. The tractive resistance on the smooth portion of the road was 28 pounds with the pneumatic wheels, and 45 pounds with ordinary wheels. On the newly macadamized portion of the road the pneumatic wheels showed a tractive resistance of 38½ pounds, and the ordinary tires one of 120 pounds. It was also shown that the wheels stood wear very well, as the car with which the experiments had been made had been driven more than 1,200 miles over every kind of road, the wheels remaining in excellent condition. These wheels had an inner tube of soft rubber, and an outer one of leather, to take the wear; but in 1849 the leather was replaced by rubber, which was thickened on the tread, giving a tire apparently identical in all essentials with those in use today. It is most remarkable that an invention so thoroughly tested, and shown to have such advantages as this, should have been allowed to lie absolutely dormant for so many years. The very existence of the patent seems to have been forgotten till the re-invention of the device by Mr. Dunlop. Perhaps the fact that it was the horses and not human beings who would have profited at that date by the easy traction may have had something to do with it, as even now, when the pneumatic wheel is practically universally used for cycles, it is but seldom that horse vehicles are fitted to it. How far the existence of these old patents invalidates those since granted for the same invention, is a point on which it is, perhaps, best for an engineer not to express an opinion, as the action of the courts of law in patent matters is difficult to predict. The fault does not lie wholly with the lawyers, as the average expert witness aims at anything but telling the truth, the whole truth, and nothing but the truth.—*Engineering.*

Foundations of Fortunes.

Senator Farwell began life as a surveyor.

Cornelius Vanderbilt began life as a farmer.

Wanamaker's first salary was a $1.25 a week.

A. T. Stewart made his start as a school teacher.

Jim Keene drove a milk wagon in a California town.

Cyrus Field began life as a clerk in a New England store.

Pulitzer once acted as a stoker on a Mississippi steamboat.

"Lucky" Baldwin worked on his father's farm in Indiana.

George W. Childs was an errand boy for a bookseller at $4 a month.

J. C. Flood, the California millionaire, kept a saloon in San Francisco.

P. T. Barnum earned a salary as bartender in Niblo's theatre, New York.

Jay Gould canvassed Delaware county, New York, selling maps at $1.50 apiece.

C. P. Huntington sold butter and eggs at what he could get per pound and dozen.

Andrew Carnegie did his first work in a Pittsburg telegraph office at $2 per week.

Whitelaw Reid did work as a correspondent of a Cincinnati newspaper for $5 a week.

Adam Forepaugh was a butcher in Philadelphia when he decided to go into the show business.

A medical authority says celery is a cure for rheumatism, and asserts that the disease is impossible if the vegetable be cooked and freely eaten. The fact that it is almost always put on the table raw prevents its therapeutic powers from becoming known. The celery should be cut into bits, boiled in water until soft, and the water drunk by the patient. Put new milk, with a little flour and nutmeg, into a saucepan with the boiled celery, serve it warm with pieces of toast, eat it with potatoes, and the painful ailment will soon yield.

According to Seaboard the traffic of the Erie canal this year up to date exceeds the whole year of 1895 by 200,000 tons, and the midsummer grain delivery at Buffalo has been the greatest on record.

Recent Patent Decisions.

In the case of *ex parte* McFarlane, the Commissioner decided:—

This Office is not a proper place in which to try the issue as to ownership of a patent. It has no judicial function for the determination of private rights. The numerous questions pertaining to ownership of an invention or a patent are for the courts to settle. It is not even mandatory on the Commissioner to issue a patent to the assignee of an inventor.

* * *

In the case of National Harrow Co., vs Quick et al, U. S. Circuit Court of Appeals, seventh district. Letters Patent No 201,946, issued April 2, 1878, to De Witt C. Reed, for improvements in harrows, construed and held void for want of invention. To constitute anticipation of a device it is enough that such a device has been in well-established use, whether it originated in design or by accident.

* * *

In hearing, Loewer vs. Ross, Commissioner decides:—

Although a device was not commercially perfected and did not do as efficient work as later devices, yet, as it showed every feature of the invention in controversy and was adapted to perform the work, for which it was intended and actually did such work, it is held that such a device was a reduction to practice.

Where, after the issue of an original patent and before an application for the issue of the same, it was shown that other parties were using the subject-matter not claimed, held that the patentee has abandoned his right to a reissue to claim such matter by his laches in not applying for the reissue before the rights of other parties accrued.

* * *

In infringment case of Johnson vs. Brooklyn Heights Railroad Co., U. S. Circuit Court, eastern district of New York. Claims 1 and 2 of Letters Patent No. 454,214, granted June 16, 1891, to Tom L. Johnson, for a life-gaurd for electric and cable cars construed in view of the prior art and held to be valid and infringed.

* * *

There has been pending in the Patent Office for about three years a contention between W. S. Scudder and Ottmar Mergenthaler as to priority of invention in line-casting machines in regard to, first, the matrice bar having a series of type matrices on its edge independently usable; second, composing mechanism for assembling the type matrices in line and a series of stops to arrest the matrice at its proper place in the line; third, mechanism for selecting and conducting the matrices to a place of assemblage and adjusting them individually in order into a common line.

The Commissioner of Patents finds all these issues covering also a few minor points in favor of Scudder, and closes a long opinion and decision as follows:

"And I find not only that Scudder conceived the invention as an organized product of the intellect as early as May, 1890, but that he proceeded thence, under all the circumstances of the case, with reasonable diligence to the construction of his complicated, costly, but entirely successful second machine in October, 1892, and therefore priority of invention is awarded to Scudder, and the decision of the examiner-in-chief is reversed."

This decision does not effect the Mergenthaler patents as applied to the present machines, the patents covering them having been granted in 1884 and 1885.

* * *

In the suit of the American Graphophone Company against John R. Hardin, receiver of the North American Phonograph Company, for infringment of patents in the manufacture and sale of the so-called Edison phonograph, and for injunction, damages, accounting, etc., a decree in favor of the American Graphophone Company was entered in the United States circuit court for the district of New Jersey by the consent of the defendant, who also pays the American Graphophone Company a substantial cash sum in settlement of damages. This case is interesting to the public in view of the widespread belief that Thomas A. Edison is the inventor of the commercial talking machine of today. Such is not the case. He was the inventor of the tinfoil phonograph, which lacked three essential features of the graphophone, a perfect record, a permanent record and a record which can be detached from the machine and used at any other point and at any distance of time. It remained for Alexander Graham Bell, Chichester Bell and Charles Summer Tainter to discover, after long experiment and research, in the Volta Laboratory, in Washington, D. C., the principle of sound record making and reproducing, as set forth in the graphophone patents, which, in the present suit, are again vindictated. In part payment of the expense incident to the experimental work in the graphophone Alexander Graham Bell donated the Volta prize of $20,000 given to him for the invention of the telephone.

"Patent or No Pay."

Is the policy of "patent or no pay" a proper one, and are the interests of the inventor best subserved by the employment of an attorney who has adopted this principle? This is a question, propounded by a reader of the INVENTIVE AGE in Prophetstown, Ill., and it is frequently asked. There are, of course, exceptions to all rules but we believe the concensus of opinion among the leading practitioners before the Patent bureau and among those who have held official positions in the Patent Office is adverse to the "patent or no pay" system. In this age when the inventor is beset on all sides with appeals from unscrupulous and inefficient patent attorneys, the most plausible argument in favor of the "patent or no pay" counsel is the representation that inasmuch as the attorney's fee depends upon his ability to obtain a patent for his client, the preliminary search to ascertain the state of the art and probable patentability of the invention will be more thorough and the client will not be urged to put up first government and attorney's fees on an uncertainty. Opposed to this theory—and outweighing it, in the opinion of many—is the theory that as the fees of the attorney depend upon the obtaining of a patent a *patent in name*, but of doubtful value, will be obtained on insignificant claims. Thus it is that many inventors who think they have something of value in the seriously worded document with the big seal from the Patent Office, really have only an invitation to a fruitless contest in the courts or the allowance of claims so inadequate as to be entirely impractical and worse than worthless.

The establishment of an American Patent Bar, with rules and regulations covering this and other features of the practice has been frequently urged by leading practitioners and was the subject of much discussion and favorable action at the last annual meeting of the American Association of Inventors and Manufacturers, and of the American Bar Association. Commissioner Seymour heartily endorsed these recommendations to congress and in his last annual report embodied some suggestions that should be carefully read by all inventors, as they bear directly on the question of "contingent fees" and the general qualifications of attorneys. Here is what the Commissioner says:

In a former report I had the honor to recommend that authority be given to the Commissioner to frame rules, subject to the approval of the Secretary of the Interior, for the admission of practitioners before the Office to an organization to be designated the patent bar. Further reflection has led me to attach increasing importance to this measure, and has induced a belief that a noticeable evil from which inventors and the public suffer would be mitigated, if not almost removed.

It would be a manifest injustice to establish a patent bar to be composed exclusively of lawyers. Some of the most capable and honorable solicitors practicing before the Office have chosen other professions than the law, and it not infrequently happens that the mechanical expert is as well qualified to conduct the work of a solicitor before the Office, as those who conduct at the same time a patent-law practice before the courts. No suggestions now made or that have been heretofore made, had any thought of excluding men of this class from any of the privileges of the proposed patent bar.

It not infrequently happens that a practitioner, perhaps *under the stimulus of a contingent fee*, has so presented a meritorious invention that no adequate claims are found to be allowable, while yet one letter claim has been allowed. It is obvious, on careful inspection, that claims might be drawn which would cover the invention and thus secure to the inventor that which was his own, and which would be allowed in a latter action on the case; but instead of bestowing the requisite attention and labor upon the case, the unfaithful solicitor promptly directs the cancellation of all claims objected to, and thus puts the application into condition for allowance with an inadequate claim. Again, the unscrupulous solicitor takes an appeal, or a series of appeals, upon a case that is squarely and indubitably met by the references, for no assignable reason other than to secure from some unlettered client the *appeal fee;* for in such cases, though the appeal be regularly taken, no appearance is made before the appellate tribunal no brief is filed, and the inventor is left in no better state, but somewhat the worse for this petty system of plundering. Another class is not infrequently seen about the elevator in the Patent Office, lying in wait for those who have traveled hither for the purpose of presenting their inventions in person. They are without a known place of business other than the attorneys' room in the Patent Office; and against all these the establishment of a patent bar would provide a barrier, and it would, to some extent, insure to inventors a guarantee that those who have a right to practice here are men of recognized integrity and of ascertained fitness and capacity.

As matters now stand, any one who has not been proved before the Patent Office to have retained the money of his client or to have been guilty of other gross misconduct, and so refused recognition under section 487 of the Revised Statutes, is permitted to practice before the Office, whatever his attainments may be, whatever his character or capacity.

It is thought that a conservative regulation upon the subject of admission to practice before the Office is necessary for the protection of inventors against the greed of some who, as patent agents, solicit their business, and against the incompetence of others.

I therefore respectfully recomend that authority be given by legislation for the enactment of rules and regulations concerning the admission to a solicitors' bar of those entitled to practice before the Patent Office.

New Application of the Ratchet.

EDITOR INVENTIVE AGE: I have found a new application of the ratchet among the Alaska Indians which I do not think has been brought to the attention of inventors.

The halibut forms a staple food from the Alent peninsula far southward. The hook with which it is captured is made of two pieces of wood joined together in form of the letter V, the angle being securely lashed with tough spruce root. One limb of the hook serves as shank the other as fluke. The former is carved in form of some animal or being whose good offices are necessary in persuading the halibut to bite. In the example figured will be seen the head of a rapacious bird and of an otter. Through the middle of this shank a tough rope of

shredded spruce root is fastened, causing the hook to hang on its side. The fluke has a sharp nail run through its outer end pointing inward and backward for a barb, and is covered with split spruce root. A good sized herring is run over this limb of the hook, covering both wood and iron. The hook is let down near the bottom of the sea, the halibut seizes the herring to gulp it down and continues to make fresh spasmodic efforts which open its mouth wider and wider, mean while the iron barb takes up the slack in the fish's jaw, allowing forward motion but no retreat, permitting the mouth to open wider and wider but not to shut. The poor creature is soon drowned, hauled to the surface of the water and knocked on the head. If the canoe be large the fish is dragged aboard, but the Alent riding in his eggshell Kaiak, fastens his game astern and tows it ashore. Those who have read Renleaux essay on the principle of the stop or ratchet in machinery will be glad to see this primitive application of it.

O. T. MASON.

Engineering Magazine for August contains an article showing the increasing importance of the cyanide process in gold extraction, and pointing out that in the patent for the use of cyanides and in the one covering the use of zinc to precipitate the gold, the patentees are claiming discoveries that were published long prior to their patents.

At last the government has something substantial to show for the enormous expenditures on the Galvaston harbor. The next government map will show a depth of 24 feet on the bar and vessels drawing that much water can enter the harbor.

The Mines Department of New South Wales offers a substantial reward of $2,500 for the discovery of a payable gold field.

Protection to Engravers Wanted.

While politics is just now the chief topic under consideration it might be timely to begin the agitation of a subject political that is of vital interest to process workers, but which has not been brought to their attention, and that is, the injustice done the whole engraving business by the present copyright law. It is a grievance that can only be remedied through politics, and now is the time for engravers to acquaint themselves with the wrong being done their business and set to work to right it. They can have the matter adjusted by the next Congress, providing they begin at once either through their organizations or individually to correspond with their congressmen and demand that engravers receive the same recognition from the Government that is accorded lithographers and machine typesetters,

When the present copyright law was being formulated an association of oublishers, both American and foreign, sent agents to Washington to see that the law was fixed to suit them. The typographical and lithographers' unions had representatives there. The engravers, not being organized at that time, were ignored, and the result was their business received a blow that it will never recover from until an omission in that law is rectified. What that carefully planned omission is can be understood by a careful reading of the following extracts from the existing copyright law. First, is defined the person and subjects entitled to the protection of copyright:

"Sec. 4952. The author, inventor, designer or proprietor of any book, map, chart, dramatic or musical composition, engraving, cut, print, or photograph or negative thereof, or of a painting, drawing, chromo, statue, statuary, and of models or designs intended to be perfected as works of the fine arts"—

Now comes the discrimination. Note carefully the deliberate omission of the words "engraving, cut and print" from the proviso in the same law which reads:

"Provided, that in the case of a book, photograph, chromo, or lithograph, the same shall be printed from type set within the limits of the United States, or from plates made therefrom, or from negatives or drawings on stone made within the limits of the United States, or from transfers made therefrom. During the existence of such copyright the importation of any book, chromo, or lithograph, or photograph, so copyrighted, or any edition or editions thereof, or any plate of the same not made from type set, negatives, or drawings on stone made within the limits of the United States, shall be, and it is hereby prohibited."

No sooner was this law approved than the publishers rushed to Europe their orders for steel engravings, copper plate engravings, photogravures, photo-engravings and engravings of all kinds to illustrate their books and periodicals. Business with foreign engraving houses became brisk. In some lines, however, notably photo-engraving, their work could not compare with ours in quality, so photo-engravers were engaged from here to introduce American methods in foreign cities. The majority of the more skilled of our American engravers were shortly after compelled to give up the business by which they had brought credit to this country. Wood engraving, for instance, which had reached with us a higher degree of perfection than it was heretofore considered capable of, then received its death blow, and will soon be known as a lost art.

The magnitude of the engraving business now done abroad for American houses cannot be estimated, but take Harper's Bazaar as a single example of this loss. Every engraving of any importance in this publication is not only designed abroad but it is also engraved abroad, and every portion of this work is protected by United States copyright, because its illustrations are engravings. Were they lithographs they would have to be made in this country to be entitled to the privilege of copyright.

One need but turn to the nearest print shop to find that all the photogravures and engravings of every description on sale are foreign made. Further than this, the subjects of the pictures are foreign to our life and to our institutions, so that for patriotic as well as for business reasons we should see to it that this loophole in the copyright law is closed.

The late celebrated engraver, Ritchie, he who engraved such plates as "Martha Washington's Reception" and "Sherman's March to the Sea," told me that it would never be profitable to print from American plates while the present law stands. Still another blow to the engraving business is a provision in the later law permitting all prints having a name scribbled in pencil on their covers to be entered as "Artist's Proof," free from duty. So that now nearly all foreign engravings are alleged "Artist's Proofs," and can be purchased for less than the paper and printing would cost of an American engraving.

The Hon. Josiah D. Hicks, of Altoona, member of Congress for the Twentieth Pennsylvania District, introduced a bill to amend the inadequate proviso in the prevailing copyright law. This paragraph of his bill read as follows:

"Provided, that in the case of a book, photograph, engraving, etching, chromo or lithograph, the same shall be printed from type set within the limits of the United States, or from plate made therefrom, or from negatives, or from engraved or etched plates or drawings on stone made within the limits of the United States, or from transfers made therefrom; and the importation of the same is prohibited."

Then the publishers inspired editorials saying that this bill was a blow at them, and, of course, they killed it. Mr. Hicks wrote me at the time that if he could but get the support of the engravers of the country he could carry the bill through. There was not at that time a single publication giving as much space to the engraver's art as Anthony's Photographic Bulletin now does, consequently few even knew that a bill in their interest was before Congress.

This is the first time that the subject has been brought to the attention of process workers, and if they but interest themselves now by writing to their Representative they can secure remedial legislation during the coming Congress that will bring a revival in all the higher grades of process work.—*Stephen H. Hargan in Anthony's Photographic Bulletin.*

Mr. Bryan Was Right.

The cheapening of the products of labor contemporaneously with an increase of wages is the most beneficent fact of modern civilization. It has placed within the easy reach of the comparatively poor innumerable articles, both of use and ornament, that only the rich could enjoy in the days of our grandparents. It is the habit of the most extreme protectionists to claim for their policy the bulk of the credit for all this. Because the era of maximum prosperity—of greatest growth in wealth and population—was coincident with the operation of the protective policy they are in the habit of asserting that the relation of cause and effect exists between the two facts or conditions. We have no doubt that protection has been a great contributor to our national development. Without its aid our manufacturing industries could not have grown to large proportions; and in the absence of that factor, agriculture, mining, and transportation would have had much less development.

But the most important agency in reducing the cost of necessities and luxuries and enabling the poor, the men who work for wages to support their families and educate their children in a manner possible only to the rich of the old times, has not been a protective tariff. Mr. Bryan knows what is the cause of low prices, or did know it on the 16th of March, 1892. On that day he said, in the course of an eloquent speech in the National House of Representatives: "You must attribute it to the inventive genius that has multiplied a thousand times, in many instances, the strength of a single arm, and enables us to do today with one man what fifty men could not do fifty years ago. That is what has brought down prices in this country and everywhere."

That is just as true now as it was four years ago, but Mr. Bryan appears to have changed his view of the matter. We do not question his right to get rid of one conviction and take up another in its place. These be rapid times, and the transformation act is readily performed by statesmen and politicians of all parties. Mr. Bryan now asserts that the gold standard is the cause of low prices, but what he said four years ago has the great advantage of being in accord with universal experience and being confirmed by the observation of intelligent men everywhere. Reputable economic scientists in Europe and America, bimetallists and monometallists, agree with the Bryan deliverance of 1892. And the experience of nations proves also that wages have made the greatest advancement, both in amount and the purchasing power of every dollar, shilling, or franc, under the gold standard, which Mr. Bryan now asserts is responsible for the reduction of prices.

We cannot help believing that Mr. Bryan was right in 1892, but we do not particularly blame him for ignoring all that just now. It could not be helpful to his Presidential campaign for him to assert in 1896 what he asserted in 1892, notwithstanding the fact that the experience of mankind, the observation of intelligent people, and the conclusions of economic writers all agree with the opinion of '92, and do not generally agree with the '96 theory.

Inventive genius, stimulated by the patent system, has, next to our incomparable and immeasurable natural resources, been the greatest factor in our growth as a nation. Other nations have shared the benefit of our patent system, and we, in a less degree, have profited from their patent laws. Our prosperity really began with the establishment of the plan for the encouragement of invention, and The Post has shown that our national growth has been commensurate with and largely the result of the growth of our patent system. We find in Mr. Bryan's tribute to inventive genius, which is a positive though an indirect indorsement of our method of stimulating invention, a deliverance that can be heartily approved by some millions of his fellow-citizens who will vote against him on the 3d proximo.—Washington Post.

BOOKS AND MAGAZINES.

APROPOS of Mr. Cornelius Vanderbilt's disagreement with his son, and the latter's marriage to a woman very much older than himself, the editor of The Cosmopolitan, in the September issue, seriously discusses the education most useful to modern life, and substantially, if not in words, ask "Does modern college education educate?"

COMPRESSED AIR is the title of a new candidate for popular favor in the realm of technical journalism. It is a monthly, $1.00 a year—publication office at 26 Cortlandt street, New York.

"STATISTICS of the Colony of Queensland" for the year 1895, being a compilation from official records in the Registrar-General's office, has been received. It covers the period of development from 1860 to 1895—a fund of information obtained from no other source.

THE Wisconsin Engineer is the name given to a new quarterly magazine published by the students of the college of engineering, University of Wisconsin. It is designed primarily to "make known the results of original investigations by students and others connected with the University and to publish communications of general interest from graduates who are engaged in the practice of their profession." The first number contains much interesting technical matter and a special feature is the complete index to periodical engineering literature. Issued quarterly at Madison, Wis.,—subscription price $1.50 a year.

Lee's Home and Business Instructor even a rapid examination of this handsome little volume is sufficient to demonstrate its novelty and its usefulness. The very fact, quite self-apparent, that each of the ten departments included within these 400 pages is the work of an American specialist in that particular line, gives the book a value that no reprint or extract can possess. Here Penmanship, Bookeeping, Letter Writing, Banking, Law, Social Forms, etc., etc., are taught, by pen and picture, with a directness and accuracy found only in works carefully planned from the start and conscientiously executed all through. We do not know of any other American or foreign book of self instruction that can be justly compared with this excellent compendium, devised and realized by our home enterprise. 16mo, Russia leather, full gilt, $1.00; extra silk cloth, marbled edges, head-banded, 75c. It forms the fifth volume of Lee's Pony Reference Library. Laird & Lee, Publishers, Chicago, Ill.

AIR CASTLE DON.—The author of "Tan Pile Jim" through the famous publishers, Laird & Lee of Chicago, has again placed all old and young readers under obligations to him for giving them the profitable pleasure of reading a new book called "Air Castle Don." Throughout this handsomely printed book Mr. B. Fenman Ashley carries his readers blissfully from "Dreamland to Hard pan and the wealth of information that an intelligent reader can absorb from this unique work assumes something of the proportions of an encyclopedia.

It is hoped that Mr. Ashley may give the reading world many other examples of so deft a pen. "Air Castle Don" belongs to the Young America Series which are finding a warm reception every where and the publishers deserve the thanks of the reading public for giving them such a wealth of entertaining and instructive matter as is found in these publications. Air Castle Don may be obtained of the publishers Laird & Lee, Chicago, or of any first class book-seller. Price one dollar.

William T. Underwood and Leroy C. Underwood, of Itasca, Tex., have invented and patented a furnace or camp stove that can be carried from place to place with ease, and is useful for cooking and heating. The stove is a hollow cylindrical body provided with a door opening and a series of inwardly projecting studs for grate-rests arranged midway between top and bottom. The grate, to be used when food is to be cooked, is removable, so that when the stove is required for heating alone, large fuel can be used.

Those who have camped out, hunting, etc., know the comfort of a good fire and the pleasures of a good cooking arrangement.

THERE is big money to the inventor who can discover a means of treating the clay so that brick when burned and laid in the wall will not effloresce from dampness or other causes.

IT is announced that Russia has obtained absolute freedom of trade in northern China.

The daily consumption of coal by the St. Paul on her record-breaking trip was 315 tons.

C. W. Hicks of Melrose, Ga., thinks he has struck the right principle in a flying-machine.

The number of locomotives in the world is estimated to be 105,000, representing a total of 3,000,000 horse-power.

The third series of sales of ostrich feathers occurred in London Aug. 10-12. The quantity of feathers offered was large, over 58,000 pounds, against 49,000 pounds at the previous sale in June, and 66,700 pounds during the corresponding time last year.

The Bliss School of Electricity.

The success of the Bliss School of Electricity, (incorporated), is only another evidence of the fact that coupled with good management, the city of Washington is an ideal location for an institution of this nature. The environments of this city, its peculiar and superior advantages for education in all branches are especially attractive. This is not only an inventive age but the drift of human intellect is more and more in the line of specialists. The aim of the knowledge-seeker today is to excel in some special branch or feature of a profession. After mastering the fundamental principles of education, the youth of today, if he is to successfully compete in the commercial strife of life, must turn his attention to a profession or a branch of science with a view to its mastery. What more promising field presents itself than electricity? Rapid as has been the developement of electrical science and wonderful as have been the achievements in electrical engineering during the past twenty years the signs point to still greater strides and more wonderful discoveries in the future.

It is claimed for the Bliss School of Electricity, now entering upon the fourth year under the presidency of Louis Denton Bliss. that it is the only institution in the country where practical electrical engineering is exclusively taught, offering to students a complete course in one year. The institution is located in the shadow of the dome of the the Capitol of the nation, so to speak, for it faces the public grounds on the north. Thus almost within a stone's throw, and free to every student, is the Congressional Library with its 1,600,000 volumes—soon to be removed but a couple of blocks further

away to one of the most spacious, costly and architecturally perfect "home of books" in all the world. It is the design of this school to give practical rather than theoretical instruction; to turn out electrical experts rather than mere electricians or "wire workers." The student begins at the very foundation of electrical science and if he is apt, studious and persevering he rises to a successful if not the highest plain. It can truly be said that by restricting the course of study to electricity, pure and simple, and omitting everything that does not directly bear on the science, this school graduates its students thoroughly versed in the practical application of Electricity in all its forms.

The electrical field is a facinating abode of inventors, the records of the Patent Office showing each year an increased number of patents in that line, until now they exceed any other single class.

The faculty of the Bliss School and the personnel of its directory are calculated to inspire confidence and command respect. Louis Denton Bliss is president: John A. Chamberland, B. S. Dean and Professor of Mathamatics and Draughting; Chas. B. Pardoe, A.M., E. E., Professor of Electrical Engineering; Chas. M. Emmons, M. D., Professor of Electro-Therapeutics; Sergeant Thos. Lippincott. U. S. A., Instructor, Military Corps of Electrical Engineers; Frederick C. Schofield. Instructor Armature Windings and Electroplating; Edward G. Niles, B. S., LL. B., Lecturer, on Electrical Jurisprudence; Chas. C. Baden, Laboratory Superintendent; T. Jno. Newton, Secretary and Treasurer.

The illustration presented herewith, shows the exterior, and three inside views of the building, which promises to become an institution of national rather than local importance.

The Announcement Catalog for 1896 '97, has just been issued.

What is said to be the largest dynamo ever constructed is now being built at the Westinghouse Electric Company's works, at East Pittsburg, Pa. It is for the Allegheny County Light Company, and when completed will weigh 90 tons. The field or base has been cast in two sections, each weighing

56,000 pounds. Its base is 17 feet 6 inches, so that the armature will be 16 feet 5½ inches in diameter. The dynamo will be of 1500 kilowatts power. Four of them are to be built.

Pioneer Pill Machine.

On the 29th of last September a patent was granted to Parker J. Noyes of Lancaster, N. H., for the first automatic machine ever invented for coating pills or tablets by compression. This machine permits the application to a pill of an effervescent coating, a feat hitherto considered impossible. This coating is soluble at once on entering the stomach and allows the medicine to take effect without delay. Mr. Noyes has a patent on that idea as well as the machine. A full set of machines have been built and successfully running in Mr. Noyes' establishment during the past year. They are a marvel of ingenuity and their work is the acme of perfection. The medicine is placed in one hopper of the machine, the coating in another, and after the machine is once started the pill is formed, coated and droped out all finished and ready for use at the rate of several thousand a day.

New Water Lifter.

Mr. R. W. Farris, of Rockvale, Tenn., is the inventor of a windlass or water lifter which promises to be a useful device in districts where wells are deep, and also where it is necessary to raise large quantities of water. The machine consists of a cylinder two feet in circumference supported on an axle horizontally, and at the end of which is a cog wheel, and is rotated by means of a lever which has an arc of cogs attached at one end which fit into the cog wheel, turning it at every stroke, thereby revolving the cylinder. The mechanical arrangements are such that every stroke, of the lever causes four revolutions of the cylinder, thus winding in at each stroke eight feet of the rope which is drawn through a pulley above and wound around the cylinder. The machine has its advantage over a pump for raising water in that a bucket of water can be raised much quicker and there is no time lost in pumping off the warm water which stands in the pipe of a pump.

No Friction in Ball Bearings.

By recent experiments it is found that in comparing the ball bearings with babbitt bearings under 300 pounds pressure per square inch and 800 revolutions per minute babbitt bearings 2½ inches in diameter, lubricated with twenty drops of oil per minute and having a half inch lateral play, heated badly. Under the same pressure ball bearings were run 2,600 revolutions, or more than three times as fast, without signs of warming.

The Working Man's Dollar.

In reply to a request from the American Manufacturer that he should say a few words to workingmen upon the silver question, Andrew Carnegie briefly and pointedly responds:

"The question, as far as concerns workingmen and all others receiving salaries, lies in a nutshell.

First. Today they are all paid in dollars as good as gold, worth 100 cents everywhere. These gold dollars being worth nearly two silver dollars of the silver standard countries, buy two dollars worth of the tea, coffee, etc., used by workingmen. This is the reason that these articles are now so cheap in gold.

Secondly. The present dollars paid workingmen, being gold dollars, buy a full gold dollar's worth of any article in all countries which are gold standard countries.

Therefore, if American workingmen are paid in silver, the cost of everything they use from silver-bearing countries will be nearly double. This includes tea, coffee, sugar, spices, etc.

Thirdly. The cost of articles purchased by his gold dollars from gold using countries will also be doubled, as the silver dollar, containing about 53 cents worth of silver, will be taken by the gold using countries only at its value in silver. Today our dollars are taken at 100 cents. The Mexican dollar buys in Brazil 53 cents worth of coffee; it buys in London 53 cents worth of anything: but the American dollar today buys 200 cents worth of coffee in Brazil and 100 cents worth of anything in London. No silver advocate can dispute these facts. If free silver coinage comes, which Mr. Bryan desires, the wages of the workingman will be thus nearly cut in half. The question he has to consider is, will wages under silver nearly double? If not he is a loser by silver.

I answer, No! Here is the fact upon that subject. Wages in silver using countries have not advanced as the value of silver has fallen. The workingmen in Mexico, Brazil China, Japan and India get just the same amount of silver, which is now worth 53 cents on the dollar, as they did when silver was worth 100 cents on the dollar.

Now, the workingman who desires his revenue to be reduced nearly one-half should vote for Mr. Bryan; the workingman who does not wish this will make a great mistake if he votes so.

It may be asked why employers are not in favor of debased money, when they could pay their workingmen in it and thus save about one-half the cost of labor. The reason is that employers know only too well that debased money has always resulted in disaster to business. It shakes confidence, and business is based upon confidence. Employers are never prosperous unless the workingmen are prosperous. It is when labor commands the highest wages that profits are highest. When labor can be obtained at very low rates, because many men are idle, the employer makes no profit. He always loses.

I should like workmen to look around and consider the situation. We read of millions of spindles standing idle in the textile factories, of the iron mines closed around Lake Superior, of the furnaces blown out, and almost every manufacturing concern in the country running with reduced forces. There is only one reason for this paralysis in business, and that is the threat of reckless, ignorant men to lower the standard of money.

The day that the republic decisively declares that its standard of value will remain what it has been for so many years, gold—will see the return of genuine prosperity. Until that day comes we have nothing to look for but such depression as now prevails, and even worse.

The question is to be decided by the workingman and those who receive salaries. If they vote for gold, they vote for prosperity; if they vote for silver they will have themselves to blame for reducing the value of wages, and for the hard times certain to follow.

This a question above party, and in this contest I am neither Republican nor Democrat, but I shall vote and work for the maintenance of our gold standard as a patriotic duty. Labor in this republic shall never be reduced to the level of China and Japan by any vote of mine.

In renewing a subscription for the Inventive Age, Mr. Farrelly Alden, Agent for the Cleveland Stone Company, adds, " Send it as long as I live; am now 74." Mr. Alden represents one of the most extensive stone companies in the country, their shipments of grindstones alone amounting to over 1,000,000 tons.

Recent Inventions.

A. P. Crell, of Ionia, Mich., has invented an electric mail car which will automatically pick up or deliver mail pouches at designated stations.

An automatic coal-stoker is the invention of William H. Hannan, of Syracuse, N. Y., and on which he has secured a patent, and which will interest users of steam-boilers.

The longest commercial distance at which the long distance telephone is now operated is from Boston to St. Louis, a distance of 1400 miles. This line is more than twice as long as any European telephone line.

Among the inventions recently deposited for patenting purposes in France is "an alarm for trains in case of an attack by brigands, having for its object to enable engine drivers to warn passengers in all the compartments of the train when he sees the brigands.

A dispatch from Odessa speaks of a Mr. M. Kildischesky, having invented an improvement in the telephone, by the use of which distance has no effect on the hearing. He expects to telephone across the Atlantic. Electricians generally discredit the story.

A new machine for cleaning cotton seed, the invention of J. Howard McCormick, of New Orleans, was recently tested and pronounced a success. The cotton seed is fed into the machine from the top and after passing through the machine is perfectly cleaned of lint, while the lint is separated and cleaned of dust and dirt. This machine has a capacity of seven tons a day, and is simple in construction. It is now run by an electric motor, but in practice can be operated by steam or other power.

A useful office appliance has been patented in Europe in the shape of an apparatus for attaching stamps to envelopes. A rectangular box is fitted to hold 200 stamps piled one on the other, gummed side down, with two little hooks at the bottom, holding the stamps in place. A downward pressure of a verticle handle fixed to the side of the box releases the two hooks and forces down a stamp on the moistening pad. It is claimed that with this apparatus envelopes can be stamped at the rate of 1,200 to 1,500 an hour. For general use a special stand is constructed to carry boxes for three or more values of stamps, with the moistening pad in front.

A company for the manufacture of enamel, to be applied to metal and other substances, has begun operation in South Baltimore. It controls the patent of Theo. Zwermann, which provides for the application of enamel to iron castings of all kinds; also of what is known as Majolica enamel, which is used in decoration on iron and steel castings. Stoves, ranges, kettles and other utensils are lined with the enamel, which is composed of a combination of borax, silicates, soda and metal oxides. The compound is prepared by a cooking process in two large kilns, which each turn out 1000 pounds of enamel every three hours. After preparation it is cooled in a large vat built especially for the purpose, when it becomes a hard substance, transparent in appearance. It is then ground in mills and ready for application. The metal work is treated by two processes, the wet and dry. The articles to be enameled are placed in a furnace and heated redhot. They are then sprinkled thoroughly with the powdery substance, which forms a paste over the surface, and, it is claimed, cannot be removed by extreme heat or cold.

American Bicycles Abroad.

The American wheel is driving bicycles of domestic build out of the market in Germany. The American make is being given the preference by the nobility as well as the people generally. The German manufacturers have become so desperate that they have induced their publications to refuse the advertisements of American builders. The machines built there are much heavier and are no stronger than those manufactured here, and the lightness and neatness, combined with staunchness and durability, of the American wheel was a revelation to the Germans of the superior skill of our mechanics. The cycle industry in Germany has been giving employment to 100,000 hands. As one of the leading cycling papers of that country appeals to its readers not to ride American wheels, the inroad being made must be very serious.—*Traffic.*

The Hamburg-American Steamship Company's new 20,000 ton twin screw steamer "Pennsylvania" which was successfully launched from the Harland & Wolf shipyards at Belfast, Ireland, on September 10, is the lagest merchant vessel now afloat. She is 585 feet long, 62 feet beam, and 42 feet deep, and is designed to make an average speed of 14 knots an hour. With 20,000 tons displacement, the vessel will have a carrying capacity of 13,000 to 14,000 tons of cargo, besides a large amount of passenger accommodation.

Lithographing Stone Discovery.

Those who know how rare and costly such stones are will appreciate the discovery of large quantities in Custer County, South Dakota as a company of Omaha men has just been organized for quarrying and marketing same, for the utility of the stone discovered has been fully demonstrated by practical men. Bavaria has long supplied the world with lithographing stone, and now Omaha is to become the base of an industry which must reach enormous proportions. It is also reported that a fine quality of lithographic stones has been found in the hills north of Santa Barbara, California.

The Umbrella.

There is probably little doubt that Jonas Hanway, the eccentric philanthropist, was the first person to carry an umbrella in London. He did not lack courage, and being a Quaker, was not afraid of sneers and jeering remarks. This was during the reign of George III. These first umbrellas were made of oiled canvas with cane ribs, and while large, clumsy, and heavy, were not over strong. The French used umbrellas at a somewhat earlier period, and they probably originated in the far east, where they were used as a protection from the sun.

Will the Telephone Supplant the Telegraph?

The progress of the telephone is shown, says the American Machinist, by the fact that while only 65,000,000 telegraph messages are sent in this country each year the number of telephone conversations is 750,000,000. These figures include, of course, the messages sent over public lines only, as in the nature of the case the figures cannot be given for private lines.

In spite of the hard times and the paralized condition of business The Mergenthaler Linotype Machine Co. paid out to its stock holders on the first of this present month of October the munificent sum of $400,000 in dividends. That makes the last of four dividends paid by the Company this year, making in all $1,600,000 clear profit to the stock holders. We are informed that the company still has a large balance in the treasury and will be ready for another quarterly dividend the first of next January.

The Inventive Age is a thoroughly independent magazine, published in the interest of inventors. It costs but $1 for a whole year, and while we have received hundreds of complimentary letters we have yet to receive the first complaint. It is published at the seat of government—its building, erected especially for the publishing business—being one of the handsomest blocks in Washington and located just one block from the Patent Office. We extend an urgent invitation to all inventors to make our office their headquarters when in the city and whenever we can serve them in any way we are at their command.

William A. Eddy, of Bayonne, N. J., has succeeded in making several distinct photographic views of Boston from a great height, by means of a camera supported from kites. The kites were of the tailless type used at the Blue Hill Observatory, where an altitude of 7,441 feet was obtained, and were six and seven feet in diameter. Four to eight of these kites were required to support the camera, depending upon the strength of the wind. Distinct views were obtained of the Common, Beacon Street, Commonwealth Avenue, Charles River, and the outlying suburbs, and Mr. Eddy estimates that in one of the views the camera was, at the moment of exposure, 1,500 feet above the pavement.

The third series of sales of ostrich feathers occurred in London Aug. 10-12. The quantity of feathers offered was large, over 58,000 pounds, against 49,000 pounds at the previous sale in June, and 66,700 pounds during the corresponding time last year.

As a result of the purchase of the Mesaba range of iron mines by John D. Rockefeller and associates, it is now stated that Mr. Rockefeller will proceed to build the biggest blast furnaces and Bessemer steel works in the world.

The cause of milk souring is the rapid growth of bacteria, and it has been discovered that it is the warm, sultry conditions existing just before a storm, rather than the electricity or thunder, that is the cause of souring.

One of the latest uses of the bicycle is the organization of fire brigades.

Machinery vs. Brains.

The skilled workman of today is as much a product of the times as the machines he is so largely occupied in tending, and though the operative of today may be individually as intelligent, and undoubtedly far better educated, and an all-round better citizen than his predecessor of a previous generation, yet so far as his actual work is concerned, brain power is being practically superseded by steam power. I make this rather sweeping statement deliberately and with a sincere conviction that, disguise it how we may, while our machines are becoming almost human, nay, almost superhuman in their powers, the workingman himself is, as a matter of fact, more and more nearly approximating to the condition of an automaton—a wage earning machine. He is living upon the brains of dead and gone inventors, pioneers of mechanical industry, like Maudsley, and Bramah, and Whitworth. The mechanic of the present day gets through his day's work without the necessity of exerting himself to more than the most trifling extent: either bodily or mentally. I quite admit that this is a gain, on the whole, for all the parties concerned; for the workman, because though his work is lightened, he earns better wages, and is at liberty to devote his intelligence to doing good for himself in other ways—his status is raised; he has time for rest and rational recreation; his day's work is not, as formerly, his day's sole occupation. It is a gain too, for his employer. A single man's production is now from five to fifty times what it was in the early days of mechanical engineering. It is good for the community-at-large—directly because the increased facility of production lessens the price at which good articles may be purchased, and, indirectly, because the man, relived from exhausting toil, is a more valuable citizen than he who has no leisure to devote to the improvement of himself and his fellows. I must not, therefore, be understood as conveying a reproach when I assert that the mechanic of the period has his mental as well as his bodily labor performed for him.—W. D. Wansbrough in Cassier's Magazine.

The $1 Edison Motor.

This cut shown herewith represents one of the most satisfactory and perfect of all small dynamos, capable of running toys, fans, light models and for

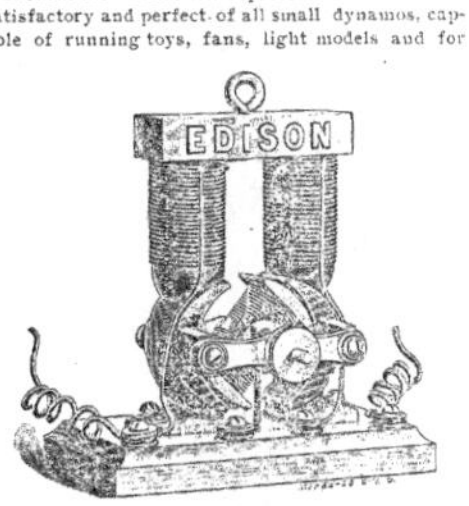

innumerable uses. To the youth scientifically in-inclined, it offers great satisfaction and benefit.

It is a marvel in price, and if not made in large quantities would cost from $5 to $10. We can furnish the motor for $1 and will send one free to any person forwarding us three new subscribers.

Pavements made of granulated cork mixed with asphalt have proved successful after two years' trial in London and Vienna. They are never slippery, are odorless, and do not absorb moisture, besides being clean, elastic, and lasting. Near the Great Eastern station in London, the wear in two years amounts to about one-eighth of an inch.

The death is announced of the veteran Italian meteorologist, Prof. Luigi Palmieri, at Naples, at the age of 89 years. He was widely known in connection with the observatory on Mt. Vesuvius, of which he had had the control for 36 years.

In the Pabst brewery at Milwaukee is a machine which corks, wires, and caps sixteen thousand two hundred bottles a day, automatically.

In France when a railroad train is more than ten minutes late the company is fined.

NEW PATENTS FOR SALE.

Advertisements inserted in this column for 20 cents a line (about 7 words) each insertion. Every new subscriber sending $1.00 to THE INVENTIVE AGE will be entitled to the AGE one year and to five lines three times FREE. Additional lines or insertions at regular rates.

FOR SALE.—Entire interest or Royalty Canadian patent 52,794 issued July 1896. Smith's Mucilage brush, absolutely prevents clogging of the bottle; a rare chance, investigate. Address J. F. Smith, Anacostia, D. C. 9-12.

FOR SALE.—OUR TWENTY-FIVE CENT ARTICLE is ready for the market. State Agents and Canvassers wanted. Safety Kettle Cover Holder Co.; Gobleville, Mich. 9-12

WANTED.—A party to furnish the expenses for Foreign Patents for a valuable Bicycle Novelty, will give in return one-third interest in the same. The U. S. Patent, to be issued, also for sale. Address, J. C. INVENTIVE AGE, Washington, D. C. 9-12

FOR SALE.—My Patent, No. 554,065—Sand-Paper Holder—both for the United States and Canada. A very useful device. For further information address, A. Hornig, Sandusky, Ohio. 6-8

WANTED.—Some one to join as co-partner with me to bring out Link and Pin Car Coupler No. 485,333. Rev. J. W. Poston, Holly Springs, Miss. 5-7

FOR SALE.—Good patent. Easy and cheap to manufacture. Thoroughly tested. Many in use. Recommeded by all using it. Ready seller. Price reasonable. Good investment. Pat. Jan. 9th, 1894. For references and price, terms etc., address P. O. Box, No. 1072, Tacoma, Wash. 5-7

FOR SALE.—Patent, 492,875, Folding Hoe, entirely new principle. Will sell State or County rights cheap. Any live man can make money right in his own state or town handling the hoe. Inexpensive to manufacture, and sells on sight. For full information, address Dr. F. A. Rice, 62 Main street, Lockport, N. Y. 2-7

FOR SALE.—Patent No. 517,919 A Chimney Regulator. Best ever invented. Is sure to be a blessing to every home. Will take some land in exchange. Address Jorgen Jurichs, Danforth, Ill. 7-9

FOR SALE.—A half interest in my "Hair Grower," I have made a discovery whereby I can grow a full head of hair on the baldest head; cure any case of dandruff or scalp disease to dead certainty. I want a partner with money to put this new discovery upon a large scale. A fortune for some one. Sample bottle sent upon receipt of $1.50. Address, Geo. W. Schoenhut, Eldora, Iowa. Box 98.

FOR SALE.—A complete set of Patent Office Reports and Official Gazette of the U. S. Patent Office from 1790 to 1894. H. B. Dickinson, Ravenna, O., Adm'r of Estate of Bradford Howland. tf.

FOR SALE.—Patent No. 52,264, Canada issued May 15, 1896. United States, No. 566,997, issued Sept. 1. 1896. (Boiler presure indicating alarm) alarms with a bell. For particulars address Daniel C. McAulay. Port Moriea Cape Breton, Nova Scotia. 10-12

FOR SALE.—Patent No. 561,789, dated 9th June last. The most amusing toy out ("Pugnacious Roosters") will make you laugh to look at them. Offers received. Address Samuel B. Field, Scotch Plains, N. Y.

BUSINESS SPECIALS.

Advertisements under this heading 20 cents a line each insertion—seven words to the line. Parties desiring to purchase valuable patents or wanting to manufacture patented articles will find this a valuable advertising medium.

WANTED.—Reliable manufacturer to manufacture on royalty, or will sell at a moderate figure. Caliper Adjustment. Described elsewhere. Address, Arthur Munch, 653 East 5th, st., St. Paul, Minn. 4-7

WANTED.—Reliable manufacturer for my Window Ventilator for Churches, Hotels, Factories, Schools and Public Buildings, Theatres, Private Residences and Railroad Coaches. Simple in construction. Easily operated. Address Geo. B. Pattison. Rose's Valley, Lycoming Co., Pa. 10-12

WANTED.—A Partner to patent my Sash Lock. Superior to any in use. Beats weights. Simple, cheap, durable, efficient and ornamental. Address, O. C. Billman, Smithville, O. 9-12

Improve
Your Spare Time

You can successfully study at your home, under our direction, Mechanical Drawing, Perspective, Lettering and Mathematics. Send for catalog; free. Penn Correspondence Institute, Philadelphia, Pa.

Aftermath.

The Northwestern Millers' Association, known as the Flour Trust, has disbanded.

The Air Pressure Cylinder Company of New York City has been incorporated, with a capital of $1,000,000.

There are about 6,000 stockholders in the American Bell Telephone Company. Their average holdings are 39 shares each.

The Chicago Edison Company has closed a sale of $1,200,000 new five per cent 30-year gold bonds. The proceeds will go into improvements.

It is the intention of the Post Office Department to extend the delivery service in remote districts by means of letter-carriers mounted on bicycles.

For two weeks, beginning January 25, 1897, there will be held in Madison Square Garden, New York city, an exhibition of gas apparatus and appliances. It is the first affair of the kind attempted in this country.

The Ohio Acetylene Gas Company has been formed with a capital of $5,000,000; and it proposes to supply acetylene in steel cylinders of sufficient capacity to provide light for a house of eight to thirteen rooms for six or eight months.

The death of Charles L. Chapin, the oldest telegrapher and electrician in the country, occurred at Philadelphia on the 14th ult. He had attained his 66th year and given a complete half century of his business life to telegraphy, having been initiated into its mysteries by its most prominent originator, S. F. B. Morse

The Argentine Republic has granted manufacturers in the United States the privilege of importing their products into that country, under bond, free of duty, until sold. All goods must be sent in through the regular ports and may be taken to any city of the republic. The Venezuelan government was the first to issue such an edict. The Argentine law passed that congress on June 3, and official notification of the action was made to the Argentine minister in Washington on Aug. 20.

One of the most unique buildings in Chicago will be the banking house now being constructed for the Illinois Trust and Savings Bank. In that city of wonderful sky-scrapers this building will nestle down with only two stories to its foundation, as the owners have determined to restrict the edifice to the sole use of the bank. Mr. D. H. Burnham is the architect, and it is safe to say that no financial institution in the world will have a building more perfectly appointed for their special needs or more beautiful in artistic design and proportion. This will be one of the notable buildings in that great western city.

THE Chicago Drainage Canal ranks as one of the greatest works of constructive engineering in the world. In a previous issue the Inventive Age published an illustrated article dealing more particularly with that feature or nature of the work requiring the invention and use of new and novel labor-saving machines. A work has just been published by the Engineering News Co., of New York, giving a complete description of all the machinery used and methods of work adopted in carrying on this stupendous enterprise. It is profusely illustrated and is edited by Charles S. Hill, C. E., associate editor of Engineering News. The book will be appreciated by all engineers and others who take an interest in the details of great feats in engineering and inventions. Contractors engaged in similar public works can gain from this review valuable information, while as a history of a monstrous undertaking that ultimately means a water-way for large vessels from the Great Lakes to the Mississippi River as well as the perfect drainage of one of the largest cities in the world, the work becomes one of unusual interest to the unscientific readers as well.

Denmark Enters the Union.

The accession of Denmark to the International Union for the Protection of Industrial Property is officially announced—took place Oct. 1. 1894, and includes the Faroe Islands, but neither Iceland, Greenland, nor the Danish West Indies.

Specifications and Drawings

Of any patent sent to any address on receipt of ten cents. Where entire sub-class list covering any particular field of invention is desired, 5 cents each.

Applications Pending.

At the close of business in the Patent Office Sept. 15th there were pending, awaiting official action 9,622 applications for patents, classified as follows:

In arrears under one month	2,168
Between one and two months	4,369
Between two and three months	2,403
Between three and four months	682

There were 124 designs and 57 trademarks, under one month in arrears, awaiting action on the same date.

The total number of patents, designs, trademarks l, etc., issued for the four weeks ending Sept. 22, was 1,774.

Attorneys Disbarred.

The following attorneys have been disbarred from practicing before the United States Patent Office:

Wm. R. Stringfellow, doing business under the name of the "New Orleans Patent Agency."

Oscar B. Barber, Warren Mills, Wis.

C. D. Zerbe, formerly of Chicago, address now unknown (patent shark.)

Charles W, Springer, Springfield, Ill.

ASSIGNMENTS OF PATENTS.

Partial List of Patents Assigned During the Past Month and the Consideration Therefor.

Patent No.—— Charles Armstrong, inventor, to E. Butler and H. G. Potter, of Miles City, Montana. Ball Bearings. One-half interest in said invention. Consideration $1000.

No. 556,240. Joseph Askins, inventor, to the New York Sproket Wheel Co., of New York, N. Y. Locks for Bicycles. All his right, title and interest. Consideration $99,500.

No. 564,062. George H. Evans, inventor, Frank McLaughlin, assignor to the Risdon Iron and Locomotive Works of San Francisco, Cal., assigns all his right, title and interest. Consideration $1000.

No. 299,546. George E. and Edwin R. Hovenden, inventors, David C. Cutler' assignor to the American Reminder Clock Co., of Illinois. Electric Fire Alarm. Assigns all his right, title and interest. Consideration $5000.

No. 538,500. Edwin B. Hutchinson, inventor, J. W. Buckley, assignor to the American Office Supplies Co., of Chicago, Ill., Account Book. Assigns all his rights. Consideration $8,980.

No. 569,108. Henry D. Lorash, inventor, to Chris Hilgemeier of Indianapolis, Ind. Dumping Wagons. One undivided half of his interest in said invention. Consideration $1000.

No. 549,019. John L. Rappleyea, inventor Charlotte A. Rappleyea, assignor to Gilbert E. Jacobs of Newburg, N. Y. Dustless Coal and Flour Sifter. All right, title and interest for 12 states and the District of Columbia. Consideration $8000.

No. 506,942. Patrick S. Ryan, inventor, to S. B. Judd and others of Aurora, Ill. Milk Coolers and Aerators. All his right, title and interest for the states of Wisconsin and Iowa. Consideration $1,900.

No. 575, 841. Wm. J. Ross, inventor, H. P. Roberts, assignor to J. C. Frederick, of Brockton, Mass. Anti-Friction Ball Bearing Sprocket Wheel for Bicycles. One undivided half interest for the United States. Consideration $1,200.

No. 581,412. Leroy S. Pfouts, inventor, to Frank M. Wyant of Canton, Ohio. Hot Air Furnaces. Exclusive right in said invention. Consideration $1000.

No. 356,561. Chester R. Thompson, inventor, to L. C. Rice, of Macon, Ga.. Addition Pencils. Assigns all his right, title and interest for the United States. Consideration $300.

No. 564,938. Amos J. Scritchfield, inventor, to John M. Whitehead, of Janesville, Wis. Fountain Pens. Assigns his half interest in said invention. Consideration $250.

No. 564,539. Conrad S. Schwarz, inventor, to the Standard Hat Cleaning Machine Co., of New Jersey. Hot Cleaning Machine. Assigns his entire right. Consideration $5000.

No.s 441,594, 595. Ambrosa C. Spicer, inventor, J. W. and Frank M. Hays, assignor's to Oscar B. Moss, Burlington, Iowa. Window Shade Holders. Entire rights for certain counties in the state of Iowa. Consideration $6000.

No. 565,513. Chris C. Schuler, inventor, to J. A. Bower, of Cedar Rapids, Iowa. Clevis. Assign's his entire right, title and interest for the United States. Consideration $10,000.

No. 588,089. William T. Smith, inventor, to Ben Walmsley of Bolton, England. Boring or Cutting Tool. One undivided third of his right, title and interest. Consideration $500.

No. 544,938. Herbert M. Sturgis, inventor, to W. H. Wood of Shelby Co., Tenn.. Adjustable Window Curtain. Assigns all his rights for 12 states. Consideration $50,000.

Lewis H. Slaght, inventor, to Roxa Slaght, of Waterford, Canada. Sharpener for Lawn Mowers. All his right, title and interest in said invention. Consideration $500.

No. 530,163. Charles A. Cypers, inventor, to William Nice, Jr., of Philadelphia, Pa., Incubator. One undivided half of his entire right. Consideration $500.

No. 506,980. John W. Cook, inventor, to Geo. F. Miller, and Herbert Lee, of Portland, Oregon. Clothes Pin. Assigns his right for the state of Ohio except twelve counties. Consideration $2000.

Cornelius Callahan, inventor, Cornelius Callahan, Company of Maine. Assignor's to the Puritan Fire Supply Co., of Massachusetts. Cut off for Valves. Assigns all his right, title and interest. Consideration $19,000.

No. 562,215. Frank L. Carr, Jr., inventor, to William A. Aiken, of Norwich, Conn., Display Stands. Assigns his entire right. Consideration $200.

No. 570,368. Thomas H. Coakley, inventor, to E. Madison Mitchell, of Baltimore, Md., Self Locking Can Tops. Assigns one-fourth interest. Consideration $10,000.

No. 559,845. James N. Crabb, inventor, to the Good Manufacturing Co., of Indianapolis, Ind. Wash-basin and Bathtub Stoppers. Assigns his entire right, title and interest. Consideration $3000.

Charles M. Clinton and James McNamara, inventors, to DeForest Van Vleet of Ithaca, N. Y. Typewriting Machines. Assigns two undivided fifths of the whole right, title and interest. Consideration $3000.

No. 552,228. Walter A. Barrows, inventor, to Jennie E. Westlake, of Chicago, Ill. Music Leaf Turner. His entire right for the United States. Consideration $4000.

No. 521,927. Dozier T. Bently, inventor, to Edward L. C. Ward, of Fulton Co., Ga. Syringe. One-third of his right, title and interest. Consideration $1000.

No. 487,821. William H. Betts, inventor, Amelia Schwerin, assignor, to William S. Allen, of New York, N. Y. Mop Wringer. Entire right for the state of New Jersey and New York except seven counties. Consideration $3000.

No. 549,484. Henry A. Bierley, inventor, to W. T. Curtis and M. T. Humble. Bed Spring All his right, title and interest for the state of South Dakota. Consideration $1,600.

No. 593,900. Cornelius A. Burnham, inventor, to Henry W. Holmes and Lorenzo Streeter, of Holly, Mich. Device for Twisting Stay Wires in Fences. Two undivided thirds of his right, title and interest in said invention. Consideration $2,250.

No. 583,719. Virgil W. Blanchard, inventor, to the Odorless Gas Stove Co., of West Virgina. Gas Mixers and Hydro Carbon Furnaces. Exclusive right, title and interest in said invention. Consideration $4,999,500.

No. 537,263, Robert H. Avery, inventor, Frederick R. Avery, assignor, to Avery & Rouse Steam Thresher Co., of Peoria, Ill. Threshing Machines. All the right, title and interest in said invention. Consideration $3000.

No. 505,965. James Q. Dickinson, inventor, to Edward L. C. Ward, of Fulton, Co., Ga.. Adjustable Beds. One undivided half of his entire right, in said invention. Consideration $500.

No. 353,311. John W. Dudley, inventor, Dudley Packing Co., of Ill., assignors to the Cincinnati Dudley Metalic Packing Co., Piston Rod Packing. Said company grants exclusive right, to manufacture in the states of Ohio and Kentucky. Consideration $10,000,

No. 593,331. George W. Deane, inventor, Mrs. Nellie Deane, assignor, to J. F. Jackson. Seed Planter Attachment for Plows. One-half interest in said invention. Consideration $500.

No. 406,709. James W. Dawson, inventor, to Christen A. Ervig and George Johnson, of Minneapolis, Minn. Invisible Burglar Alarm. Electric Mats. All the right, title and interest in said invention for twenty-six counties in Minnesota. Consideration $2,500.

No. 384,311. John H. Doub and Fabius Robbins, inventors, to Thomas F. Jones, of Walnut, Kans. Washing Machine. One-half interest to all the right, title and interest in said invention for the United States except six states. Consideration $1000.

No. 553,899. Eli Catlin, inventor, to J. E. Farber of Post Oak, Texas. Well Drilling Machine. Assigns his entire right. Consideration $900.

Knowledge Up to Date.

A small work of uncommon interest and value is The Living Topics Cyclopedia, which now costs, complete to date the small sum of $1.00. It is a unique publication, and its free specimen pages are worth sending for. Its latest issue gives the most important facts, "up to date," concerning, among hundreds of other important subjects, such titles as Cuba, Currency (a "living topic," indeed), Debts, national and foreign, East Africa, Egypt, Electricity, England, Engineering, France, German Empire, Gold, Greece, also concerning the States Delaware, Florida and Georgia. In general the object of the work is to answer the questions you would seek to solve by consulting your cyclopedia, were it "up to date," which no cyclopedia is or possibly can be, because of its magnitude and cost. The Living Topics, being a small work, and treating only of "living" topics, is *continually* in proces, of revision, a new edition being published every month. After you have paid for one edition you are allowed to purchase later ones, within a year thereafter, at about one-third price, and thus keep your knowledge "up to date" at trifling cost. Address the publisher. JOHN B. ALDEN, 10 and 12 Vandewater St., New York.

"Bubier's Popular Electrician"

Is the name of a monthly publication which contains a vast amount of valuable information on all electrical subjects. Its department of "Questions and Answers" will be appreciated by students and amateurs desiring information or instruction on any problem that may arise. THE INVENTIVE AGE has made special arrangements whereby we can supply that popular dollar journal and THE INVENTIVE AGE—both publications one year—for $1.50.

TO INVENTORS.

We make a specialty of manufacturing goods for inventors. Make models, dies, do experimental work, make special tools, machinery, etc. INVENTOR'S HANDBOOK FREE.

Cincinnati Specialty Manufacturing Co.,
Third Street, Cincinnati, Ohio.

IDEAS DEVELOPED. Absolute secrecy. Send for particulars. Advice and suggestions given free when asked for. Correspondence and sample orders solicited. 25 years in business. Models, patterns and all kinds of castings and machine work.

GARDAM & SON,
96 John Street, New York,

The Inventive Age
AND INDUSTRIAL REVIEW
A JOURNAL OF MANUFACTURING INDUSTRY AND SCIENTIFIC PROGRESS

Seventh Year. No. 11.   WASHINGTON, D. C., NOVEMBER, 1896.   Single Copies 10 Cents. $1 Per Year.

The Chicago Tower.

Nothing is too stupendous for Chicago to undertake. She excelled in the matter of World's expositions; she surpasses New York in voting population and now the Eiffel tower is to be eclipsed. From the altitude of 1,150 feet the United States flag will be flaunted in the breezes when the city tower is completed. This tower is to be constructed by the City Tower Company, and covers an entire block corner of Congress and Loomis streets.

The projector of this great enterprise is D. R. Proctor of Chicago, and the architect is Francois La Pointe, a well known Canadian, who came to Chicago in 1887.

The tower will rest on four four-cornered supports, each 50 feet square, and meeting in an arch 200 feet wide and high. On top of these arches will be a vast landing 250 feet square and capable of accommodating 20,000 people. There will be six other landings, with a capacity of 20,000 more. The second landing will be 150 feet square and 450 feet from the starting point. The third landing, 75 feet square, is to be 675 feet above the ground; the fourth, 50 feet square, 1,000 feet above the foundation, and two others will be at elevations of 1,040 and 1,080 feet. The last will be provided with a powerful searchlight and telescope. Sixteen elevators will carry passengers to the first landing, eight to the second, six to the third, and from there four cars will take passengers to the 1,000-foot landing. A theater, restaurant, booths, and exhibits of every description will occupy the landings.

In magnitude and design the tower will out-rival any structure of its kind in this or any other land. At the highest points electric lights will be arranged to produce artistic effects. Thirteen flags will fly from the tower. At sunrise, by pressing a button they are to be unfurled.

At sunset pressure of another button will furl them and they disappear for the night. Electrical displays will be made from the top landing at night. Many other features will be added for the amusement of persons visiting the tower. The estimated cost of the structure will be $800,000. Only the finest steel will be used in its construction. The annual income its stockholders expect to derive will exceed $300,000, while they estimate the total cost of management at $50,000.

Bathrooms, barber and florist shops, delicacy booths, and convention rooms are devised in the plans as a means of revenue. On each landing will be telegraph and telephone stations. Firework exhibitions will be given and the concert halls are expected to afford considerable income.

W. W. Duffield, superintendent of the United States coast and geodetic survey,

says the tower would be of value for meteorological observations in the upper currents of air, and Prof. Willis Moore, chief of the government weather bureau, gives the following opinion: "We have an example of observations at the Eiffel tower, 300 meters (984 feet) in Paris. At that tower remarkable results

PROPOSED STEEL TOWER AT MONTREAL, CANADA.

have been obtained as to wind velocity. At the earth's surface the wind dies down until it reaches a minimum at sunset.

"This is not true, however, at the lake crib, Chicago, for the minimum is reached there about midnight. The time of maximum wind at the earth's surface is usually about 3 p. m. At the Eiffel tower these conditions are almost exactly reversed in the colder months. This is a very interesting phenomenon for study. Such a tower would enable us to determine how much sooner a cold wave will strike a point 1,000 feet high than at the earth. It would give a wonderful place for obtaining storm and wind velocity. "It is safe to say a wind of 100 miles an hour might be expected there.

"The most important observations of all on such a tower would be those relating to atmospheric electricity. There is hardly a point regarding diurnal change, abnormal change, or seasonable change of meteorological elements that would not be aided by records from such a tower."

D. R. Proctor, the Chicago man who designed the tower and organised the company, was born in Gloucester, Mass., Oct. 10, 1830. He has been a resident of Chicago for 13 years, is a mechanical engineer, and an inventor of some note. He is not a little proud of his accomplishments up to the present time, and predicts that nothing can interfere with the successful completion of the tower within a year.

"When it is finished," he said, "Chicago will have a structure with which not even the Colossus of Rhodes could have compared."

Francois LaPointe, the Canadian architect, who represented the government at the world's fair, is associated with Mr. Proctor. He has elaborated the designs of the tower and will superintend its construction. He is a native of Montreal, and has built many handsome churches and government buildings in his own country. He came to Chicago in 1887 and has since lived here.

He is also the architect of a similar structure to be erected at Montreal in Mont Royal Park. This tower will be but 600 feet in height but will be placed on an eminence 750 feet high. The base of this structure will be 250 feet wide; there will be six landings and the tower will accommodate 20,000 people—only half the number, however, contemplated in the construction of the Chicago tower.

Of the architect, Mr. Francois LaPointe, Le Courrier L'Quest, of Chicago says:

Major Francois LaPointe was born in Montreal. May 28th, 1848. He is the eldest son of Francois Audet LaPointe, of Chicago, formerly constructor and lumber dealer in Montreal. He is related through his mother, to the honorable Judge L. A. Ouimet of Montreal, also to M. Gideon Ouimet;

The Inventive Age

Established 1889.

INVENTIVE AGE PUBLISHING CO.,

8th and H Sts., Washington, D. C.

ALEX. S. CAPEHART. MARSHALL H. JEWELL.

The INVENTIVE AGE is sent, postage prepaid, to any address in the United States, Canada or Mexico for $1 a year; to any other country, postage prepaid, $1.50.

Correspondence with inventors, mechanics, manufacturers, scientists and others is invited. The columns of this journal are open for the discussion of such subjects as are of general interest to its readers.

Technical matter is particularly desired. We want practical information from practical men.

The INVENTINE AGE is thoroughly independent.

Advertising rates made known on application. Special facilities for furnishing cuts of any patented device together with descriptive article. Business specials 15 cents a line each insertion, 7 words to the line. No advertisements less than 25 cents.

Address all communications to THE INVENTIVE AGE, Washington, D. C.

Entered at the Postoffice in Washington as second-class matter.

WASHINGTON, D. C., NOVEMBER, 1896.

Among the papers in the District of Columbia devoted to inventions and patents, none has credit for so large a regular issue as is accorded to the "Inventive Age," published monthly at Washington.—*Printers Ink.*

A comparison of the various new industries established in the South during the present year shows quite an increase over the year 1895.

THE Department of Promotion and Publicity to the Tennessee Centennial and International Exposition at Nashville has been placed under the charge of Mr. Herman Justi, who is a gentlemen of great executive ability and well and favorably known throughout the South.

NICOLA TESLA'S new triumph demonstrates the scientific possibility of creating brilliant illumination by means of vacuum tubes which are not in contact with the electric source, the device being so constructed as to make something like 1000,000,000 vibrations a second. The result is a bright white light of greater brilliancy than arc illumination.

A PEKING cablegram of Oct. 22 says: An American syndicate will advance 30,000,000 taels for the construction of the Han-Kow and Peking Railroad. The line will be 700 miles long and will cross 27 rivers, including the Hoang-Ho, all of which will have to be bridged. The entire work will be transferred to the syndicate, but the shares of the company will ostensibly he held in China.

AT an expense of something over 3,000,000 the government has given to the port of Charleston an excellent harbor. According to a recent survey the entrance to the harbor consists of a safe channel of eighteen and one-half feet at low water and twenty-three and one-half feet at high water. The greater portion of the channel is three to five feet deeper than this. The prospects of soon having a twenty six foot harbor fills the merchants of Charleston with joy.

Conserative old London is going ahead of any American city in horseless carriage progress says Electrical Review. It is announced that the Road Car Company, which operates 900 omnibuses in London, intends to eventually do away with the use of horses and to run motor omnibuses. It is said that 100 of the new vehicles will be put in service during November and 300 more in January. The business enterprise of New York and the hustle and push of Chicago are thrown in the shade by this project of British enterprise. Where in the world is there a more advantageous field for horseless carriages than the level boulevards and avenues of Chicago or the picturesque park drives of New York City?

The longest commercial distance at which the long-distance telephone is now operated is from Boston to St. Louis, a distance of 1,400 miles. This line is more than twice as long as any European telephone line.

American Association of Inventors and Manufacturers.

The Sixth Annual Meeting of this Association will be held at Washington, D. C., Tuesday, January 19, 1897. The hour and place of meeting will be announced hereafter. Officers will be elected for the coming year and other important business will be transacted.

The Executive Council has sent out a schedule of subjects on which discussion is invited and members are free to select other subjects if they desire. The Council's list of subjects is given below:

1. Patent Laws, their growth and origin.
2. Benefits that flow from the American Patent System.
3. The necessity of increasing the facilities of the Patent Office.
4. Should a Court have power to decree the payment of license fees instead of granting an injunction?
   (*a*) Where other parties are licensed.
   (*b*) Where the invention is not in use under the patent.
5. How ean Sec. 4887 R. S. be amended to the best advantage?
6. Can Interference Proceedings be simplified or improved?
7. Have any illegal or detrimental actions in the Patent Office come to your knowledge recently?
8. Have you any suggestions for improving the working of the Patent Office?
9. How can the classification of inventions in the Patent Office be made more convenient, complete, or effective?.
10. Should not the law department of the Patent Office be furnished by the Courts having jurisdiction of patent cases with a copy of the record in each cause decided?
11. In the present system of removing or reducing clerks in the Patent Office from a higher to a lower position for the best interests of the service?
12. Does the progress of invention tend to enlarge the rewards of labor?
13. Is the introduction of labor saving machinery a direct benefit to laboring men?
14. Does the security of property in inventions by law promote progress and civilization?
15. The relation of manufactures to agriculture.
16. The effect of invention upon travel and commerce.

While the foregoing subjects are deemed to be proper ones for discussion by the Association it is not intended to discourage the presentation of papers on other subjects relating to patents; and papers are invited from members of the Association and others upon any subjects within the Association's recognized field of work.

A full report of the proceedings will be printed in the INVENTIVE AGE.

Chinese Pile Driving.

However far the Chinese were ahead of us in many mechanical inventions, there are still very numerous appliances regarded as indispensable in western civilization, and long in use in Europe, of which the Orientals seem to have no knowledge or experience. The following description of Chinese pile driving, given in an interesting article in "Cassier's Magazine" for October, illustrates this very pointedly. Mr. Prentiss says that piles were being driven in one of the new buildings for a foundation for a punch. They were 8 inches in diameter and 14 feet long. The staging was bamboo, and so was the frame for a hammer, which was a round piece of cast iron, with a hole in the centre for a guide rod. Attached to the hammer block were twenty-seven ropes, carried up to the top of the frame and down on the outside, looking very much like the old fashioned Maypole, Twenty-seven women had hold of the ends, and with a sing song, all together pulled down. Up the rod four feet travelled the hammer; then, at a scream, all let go, and down it came on top of the pile which was unprotected by a band or ring. The women were paid 20 cents in gold per day. This Maypole driver is in general use throughout Japan and China. But we do not need to go so far East to see such primitive arrangements. We have seen exactly the same appliance at work by government employes in Turkey, with this difference only, that no women are there employed in out-door labour.—*Trades Journal Review.*

One of the uses of skimmed milk is in the manufacture of an artificial ivory, which in most respects resembles the original. The milk is mixed with borax and subjected to a high pressure. The product is well suited for combs, billiard balls and pipe mouth-pieces.

Machinery Compared with Muscular Power.

Speaking of prime movers before the Association for the Advancement of Science, at London several years ago, Sir Frederick Bramell drew an interesting picture of the puny thing that muscular power, whether animal or human, really was when compared with the vast efforts exerted nowadays by machinery. Contrasting a galley, for example—a vessel propelled by oars—with a modern Atlantic liner, and assuming that prime movers were non-existent and that this vessel was to be propelled after galley fashion, he proceeded thus: Take the length of the vessel as 600 feet, and assume that place could be found for as many as 400 oars on each side, each oar is worked by three men, or 2,400 men, and allow that six men under these conditions could develop work equal to one horse power. We should then have 400 horse power. Double the number of men and we should have 800 horse power, with 4,800 men at work, and at least the same number in reserve if the journey is to be carried on continuously. Contrast the puny result thus obtained with 19,500 horse power given forth by a large prime mover of the present day, such a power requiring, on the above mode of calculation, 117,000 men at work and 117,000 in reserve, and these to be carried in a vessel less than 600 feet in length. Even if it were possible to carry this number of men in such a vessel, by no conceivable means could their power be utilized so as to impart to it a speed of 20 knots an hour.

This illustrates how a prime mover may not only be a mere substitute for muscular work, but may afford the means of attaining an end that could not by any possibility be attained by muscular exertion, no matter what money was expended or what galley-slave suffering was inflicted. Take again the case of a railway locomotive, in which we have from 400 to 600 horse power developed in an implement which, even including its tender, does not occupy an area of more than 50 square yards and that can draw us at 60 miles an hour. Here again the prime mover succeeds in doing that which no expenditure of money or of life could enable us to obtain from muscular effort.—*Cassier's Magazine.*

Exportations of Copper.

Copper has become one of our most important exports, the quantity sent abroad showing a large increase. For the first seven months of 1896, the export of fine copper was 142,476,307 pounds, as against 78,110.642 pounds in the corresponding seven months of 1895. There was, besides a considerable shipment of copper matte, amounting to 11,019 tons, representing 16,585,408 pounds of fine copper, and carrying the total shipment of the latter up to 159, 061,715 pounds. This is at the average rate of 22¼ million pounds per month, an export unprecedented in the history of American copper production. The value for the seven months was $16,546,172, or at the rate of more than 2¼ millions of dollars a month. The United Kingdom, France and Germany are our chief buyers, but the other countries of Europe took nearly one-third of the whole, their purchases being largely increased over 1895.

The Most Refractory Substances Known.

M. Moissan expresses in the *Annales de Chim. et Phys.* His opinion that the most stable compounds yet known to science disappear in the electric furnace, being either decomposed or volatilized. The only exceptions are contained in the series of perfectly crystallized compounds discovered by him, consisting of borides, silicides, and especially carbides of the metals. M. Moissan intimates his intention of publishing a description of these compounds within a short period. He regards them as being probably among the original constituents of the globe, and as still existing in some of the stars,

At Klausthal, Germany, a bolt of lightning instantly melted two wire nails 5.32-inch in diameter. To melt iron in this short time would be impossible in the largest furnace now in existence, and it could only be accomplished with the aid of electricity, but a current 200 amperes and a potential of 20,000 volts would be necessary. This electric force for one second represents 5,000 horse-power, but as the lightning accomplished the melting in considerable less time, say 1-10th of a second, it follows that the bolt was 50,000 horse-power.

A suit has been instituted in the United States Circuit court at San Francisco by Herman Cramer claiming $5,000,000 from the Singer Sewing Machine company, profits alleged to have been made by the company in selling machines infringing on Cramer's patent.

The Chicago Tower.

(Continued from first page.)

formerly superintendent of Public Instruction and now legislative counselor in Quebec. He received his first instruction at the Brothers of the Christian Doctrine, took his classic course at the college of Terbonne where he received his first drawing lessons. Then he entered the Polytechnic school of Montreal, and took a course in the school of arts and trade. He associated himself with his father under the firm of Francois La Pointe & Son. Married in 1869 Caroline Milaire, of St. Eustache. He opened an architect office and built the bridges on the aqueduct canal of Montreal, the hydraulic press, the slaughter house at St. Henrie, the handsome buildings of the exposition in Montreal, 1880—1884, the colleges of Notre Dame, St. Laurent, St. Cesaire etc.; The churches

FRANCOIS LA POINTE, ARCHITECT.

of St. Laurent, St. Huges, et of Aurora; the residence of the mayor of Valparaiso, Indiana, and directed a number of other enterprises of the highest importance. In 1886 he established himself in Chicago. In 1866 he entered the military college and left it with brilliant certificates, acquired by rigid examinations.

He took part as officer in the expedition against the fenian invasion at Lacolle, Pidgeon Hill and Niagara. On his return, he took the command of Company No. 6 of the 65th battalion C. M. R. as Captain. 1880 he was commissioned to take possession of the Northern Railroad in the name of the government of Quebec. Major LaPointe had 450 men under his command, seized at the point of the bayonet all the stations of the road, from Montreal to Ottawa. Only one fatal accident occurred and the manner in which Major LaPointe conducted this expedition brought him the congratulations of the honorable Joly, then first minister of the government of Quebec. He was chief superintendent of the association of St. Jean Baptiste of Montreal in 1884, at the occasion of the Golden Wedding of that society and the entire press was unanimous in praising the executive qualities of Major LaPointe. He is father of 13 children, of which 7 survive, the eldest, Francois, Raoul, et Alfred, distinguished themselves by their talents and occupy enviable situations. All three are artists as inspectors of the departments in photography and engraving in the best establishments of Chicago.

Major LaPointe is now partner of D. N. Proctor, one of the most distinguished engineers of the United States. Both direct the grand projects of construction of the towers of Chicago and Montreal.

A new fire-proof paper, made in Berlin by L. Frobeeu, is reported to be capable of resisting even the direct influence of flame, while it may be placed in a white heat without harm. It consists of ninety-five parts of the best asbestos fiber, which is washed in a solution of permanganate of calcium and then treated with sulphuric acid, and five parts of ground wood pulp, the entire mass being placed in the agitating box, with the addition of some lime water and borax. After thorough mixing the materials is pumped into a regulating box and allowed to flow out of a gate into an endless wire cloth, where it enters the usual paper-making machinery.

Inventions by Women.

An interesting feature of the Tennessee Centennial Exposition will be an exhibit of inventions of women. Mrs. G. H. Ratterman is in charge of the department and is now so fully advised as to warrant the statement that this will be the greatest collection of inventions by women ever secured. Since last January the chairman has sent out many thousands of folders and writtern hundreds of personal letters urging the women inventors all over the world to become interested in the great enterprise. The great newspapers and magazines have contributed to the agitation and Mrs. Ratterman believes that no feature of the great exposition will be of more interest.

Regarding the work in hand Mrs. Ratterman writes as follows:

"So far the patents promised are almost twice as many as there were on exhibit at Atlanta, and almost as many as there were at the World's Fair. It has been a very difficult task to get the inventor's address. When I was appointed Chairman of the Committee I did not know that woman had had the time to think out the many improvements, which come in the scope of her household labor, as well as articles for the personal use of herself and childern. Since August, there have been one hundred and ninety patents granted to women. Is this not an evidence, that woman posessesa inventive genius, and also that she is using it? Time was when any such departure from the old stereotyped beaten track would have been frowned down upon, but in the advance of women, by sheer force of character and invidiuality, she has 1mt dignity to every position she has aspired, and disarmed all criticism.

Fifty years ago there were only two positions to be held by women—seamstress and teacher—but what a field for them today to use their energies. Upon entering on my duties as Chairman the perplexing question was, "how will I reach our women inventors?" First I applied to the Commissioner of Patents at Washington for the index of women inventors to whom the United States government had granted patents. There I found city and state addresses but not street nor number of residence, and as our most important inventors reside in the large cities, as New York, Chicago and Boston, it is almost impossible to reach them. I have found great assistance through the mayor's of the cities. I am also pleased to state that I have received valuable assistance from the press. I then sent a circular letter to the women of our country asking them to loan us their inventions as an exhibit in the Womens Building, showing the public what women are doing with their brains; also to let the public know of the patents in existence. Many have answered. The Commisioner of Patents has promised me the Government exhibit of women's inventions, containing 150 models (this was not at the World's Fair) if Congress recognizes the Tennessee Centennial Exposition.

The walls of the W. C. & I., room will be decorated with wall paper from the New York Society of Art. I desire if possible to have an exhibit on a dummy

horse, of everything that woman has invented for horses—as fly net, horse brush, saddles, rein protector, a spring to detach vehicles from runaway horses; a bicycle outfit invented by women—saddle tire, and a bicycle suit; a sewing cabinet, with everything pertaining to that art, etc. On each article will be a white silk banner inscribed with gilt letters, name of inventor, and when patented, and if exhibitors so desire, I will have a placard on articles stating, that orders for sale can be filled in the salesroom in the Woman's Building.

When we remember there are five thousand of these patents extant, I feel confident of success and hope to be able to represent several hundreds. Many women have turned their attention to car fenders, snowplows, windmills, carts and street-sweepers and they are inventing many electrical appliances. The firemen's hat and the typewriter for the blind I think deserve special attention. A woman has invented artificial marble, also, a deep sea telescope. Woman was the *first inventor*. Woman was the first to bring food to the family, the first to weave, the first to dress skins, the first to make pottery, and to encourage art and religion —according to Prof. Mason's statement. I believe the world will be astounded, after seeing this exhibition, at the inventive genius of women."

☞ W. R. Lee, a Nashville harness maker, has invented an adjustable halter which is to be used in the United States calvary.

THE CHICAGO TOWER.

PICTURESQUE WASHINGTON.

Brief Description of the More Important Points of Interest.

The city of Washington has been called the Paris of America by foreigners and there are those who see in the environments of this city even greater beauties than belong to the gay Parisian Capital. Its broad avenues, public squares and parking system, the architectural beauty of the numerous buildings dedicated to governmental affairs, its cleanliness, the cosmopolitan character of its people, the numerous monuments and points of great historical interest combine to make this city the mecca of meccas for Americans as well as foreigners. Readers of the INVENTIVE AGE who have never had the opportunity of visiting the Capitol city as well as those contemplating a visit sometime in the future will be interested in the following brief description of a few of the leading points of interest.

One of the most beautiful structures in the world is the new Congressional Library, which, with its magnificent golden dome, has become as familiar a sight as the Capitol itself. It is built of the purest

State Hall and New Congress

The Capitol

white granite, covers over three acres, and is lighted by nearly two thousand windows. The beauty of its decorations is seen in every piece of carving, especially the heads carved upon the keystones of its window arches, thirty-three in all, representing as many races of man. The vestibule is lined with rare Italian marble, and adorned with lofty Corinthian columns exquisitely decorated. The staircase with its white marble balustrades has newal posts surmounted by two bronze lamp bearers, while the upper staircases are ornamented with twenty-six marble figures representing various arts and sciences. Beyond the vestibule is the reading room an octagonal one hundred feet in diameter, and designed to accommodate two hundred and fifty readers, furnishing each with a desk, and four feet of working room. Here will be found eight colossal figures, representing Art, History, Philosophy, Poetry, Science, Law, Commerce and Religion, and around the galleries of the rotunda are bronze statues of men representing these subjects. On each side of the reading room is a magazine repository. The lenght of shelves provided for would extend forty-three miles, and accommodate 1,800,-000 volumes. Above the reading room is the Art Hall and Map Room, and on the ground floor there are four corridors, lined with marbles of rarest beauty.

As we approach the Capitol and climb the magnificent stone staircase on the east front, the first objects which attract our attention are the famous Bronze Doors, by Rogers, with their high reliefs representing scenes in the life of Columbus. As we pass through these we find ourselves in the Rotunda directly under the dome, surrounded on all sides by its fresco decorations, and awe-inspiring paintings of history. Turning to the left we find ourselves in Statuary Hall, where we may stand in certain positions and whisper to each other through long distances, the form of the arched roof reflecting the sound from one end of the hall to the other. Going a little farther to the left we find ourselves upon the floor of the House of Representatives. Surrounding the House are committee and reception rooms, but we leave these and, ascending the magnificent marble stairs, take another glance at the House from its Gallery, and then, crossing to the opposite end of the building, visit the Senate, and the beautiful reception rooms of the President,

Vice-President and Senators. As we return we can not help stopping for a few moments before the splendid picture of Perry's victory on Lake Erie. Then we go down the stairs again to the old Senate Chamber, now the Supreme Court, and crossing the Rotunda once more, we pass through the old Congressional Library, which still contains the thousands upon thousands of volumes. which have not yet been transferred to the new building. Going out upon the broad Portico on the west side of the building, where the view across the wide green lawns, and of the city with its avenues and parks, is hardly surpassed for beauty, we proceed to some of the other points of interest.

We will glance first into the Pension Bureau, which comes into especial prominence once in every four years. Here is held the Inaugural Ball. This structure is fireproof, being built entirely of brick, even to the steps, and is the largest brick building in the world. All around the building, on the frieze outside, march the Boys in Blue, and inside among the clerks will be found many an old soldier who marched with these same Boys in Blue more than thirty years ago.

From the Pension Bureau we will go to the Patent office, two blocks west, and, if you are an inventor hours, and even days, may be pleasantly employed in examining the thousands of models and innum-

THE PATENT OFFICE.

erable patents here displayed. Crossing the street we come to the Post Office Department and take the elevator to the Dead Letter Museum, where we find everything under the sun which could be sent through the mail, and some things that couldn't. We can hear try to puzzle out the blind addresses deciphered by experts, and going out on the gallery we can overlook the work of the clerks, who handle 18,000 letters a day, in the Opening Division. On our way to the next Department we stop for a few moments at what was formerly Ford's Theatre, where, on the 14th of April, 1865, President Lincoln was shot, and afterwards carried to a house just opposite, where he died. A marble table set in front of the house commemorates the event.

Taking the cars, we will proceed to the State, War and Navy Building, which, with its five hundred rooms and two miles of marble halls, claims to be the most magnificent office building in the world. Let us visit the War Department first, noting first the portraits of its Secretaries and the group of the three great War Generals,--Grant, Sherman and Sheridan. The three frames are draped with the stars and stripes. In the hall above we find papier-mache figures, on which are displayed the uniforms which are, and have been, used in the American army at various periods. Next we will visit the Navy Department with its beautiful models of war vessels, and then proceed to the State Department Library; but it is not to see the books that we come here, although many of them

are rare and very valuable. But perhaps the dearest relic in Washington to patriotic hearts is Jefferson's first draft of the Declaration of Independence. Near it on the wall there hangs a fac-simile of the Declaration itself, the original copy of the latter being in an indestructible steel safe. There are other things of interest, too: the desk on which Jefferson wrote the Declaration of Independence; Washington's Sword; Commodore Isaac Hull's testimonials, etc.

Leaving the State Department, we walk through the White House Grounds, and are soon in the central vestibule of the Executive Mansion, the President's residence. The most noticeable things to be seen upon entering is a beautiful screen of stained glass mosaic, which separates the vestibule from the central corridor. We pass through another corridor to the East Room with its immense chandeliers of cut glass, its eight carved mantels surmounted by mirrors, and the full-length portraits of Washington, Martha Washington, Jefferson, and Lincoln. Peculiar interest attaches to the portrait of Washington because it, as well as the original of the Declaration of Independence, was saved by Mrs. Dolly Madison, who carried them with her across the Potomac, when the British burned the White House, in 1814. The East Room is the only one usually shown to visitors, but some times people

are more fortunate and you may possibly see the Green, Red, and Blue Rooms, the first of which contains a portrait of this same brave Dolly Madison.

One of the most interesting places to visit, and which must be seen before two P. M. however, as it is not open to visitors after that hour, is the United States Treasury Department. First we will visit the Cash Room, with its beautiful marble walls, which make it one of the costliest rooms in the world; but surely that is fitting here, if anywhere, for in this room the daily transactions run into millions of dollars.

A visitor himself may have a part in the business, by presenting at one of the windows a national currency bill, for which he will receive coin for its face value. The system of making new money, exchanging new for old, and destroying the old is what one sees at the Treasury. The manufacture of paper money begins in the Bureau of Engraving and Printing, which we will visit next. Every morning at nine o'clock, the currency, a million dollars a day, is brought from the Bureau of Engraving and Printing to the Treasury, in a large steel wagon, attended by a force of guards. The money is delivered to the Division of Issue, where it is again counted. Having received its final count, it is turned over to the sealing clerk, who wraps the packages, containing from $4,000 to $4,000,000 each, in plain brown paper and seals them with the Treasury seal, after which they are deposited in the Currency Reserve vault, where they remain two or more months. As one new lot is added each day, another lot is taken out and put into circulation. Down in the basement we will find the macerator, a huge spherical receptacle made of steel. It contains water and is fitted inside with 156 closely set knives which, as they revolve, grind the contents exceedingly fine. The lid is secured by three Yale locks, each with its own key. The key of one lock is held by the Treasurer; the second by the Secretary, and the third by the Comptroller of the Currency. These three officials, or their deputies, with a fourth representing the banks, meet at one o'clock, at stated periods, and place in the macerator, its tribute, a million dollars. The keys are then turned and the machine set in motion. At the end of four or five days the maceration is complete, and the committee of four unlock the valve and the liquid pulp flows out. Most of it goes back to the Bureau of Engraving and Printing, where it is made into cardboard, but samples of the million-dollar pulp are made into souvenirs and sold.

Through the bars of the inmost doors, we may look into the Silver Vault, which contains 103,202,000 silver dollars put up in sacks of $1,000 each.

A second vault contains beside its silver nearly three million dollars in gold coin.

The Smithsonian Institution, a beautiful specimen of Norman architecture, and the National Museum will be found of great interest to the specialist or scientist. In the Smithsonian Institution will be found an enormous collection of rare and beautiful birds, while in the Museum we find over 3,000,000 objects classified in 24 departments. Here may also be found the uniform worn by Washington as Commander-in-Chief, and the tent which he used during the Revolution, while the halls are filled with porcelains, minerals, rare and beautiful tapestries, with many other things of historical and scientific interest.

Last, but by no means least, let us look at that magnificent shaft, the Washington Monument, towering to a height of five hundred and fifty-five feet. Notwithstanding its plain appearance from the exterior, it is found upon examination to be a wonderful and truly beautiful piece of architecture. Only by climbing 900 steps can the memorial stones, with which the interior is set, be seen. One hund-

One Thousand Pounds Steam.

A correspondent in the *Engineer* has the following to say concerning high pressure steam :

I see that Hamilton Fraser & Co., of Liverpool have built the steamer "Inchoma," which is to use 260 pounds steam—365 pounds on boiler—which is a very bold advance on the 180 pounds steam of the "Campania" and "Lucania" of the Cunard Line. I venture to hope that some other firm will before long go in for 1000 pounds steam. Jacob Perkins in his steam gun—invented in 1825—used steam from 700 to 1200 pounds pressure, and continued it for many years, although his "generator" was only two flat iron plates bolted together with a very thin space between them. That space was filled quite full of water and then heated up to 500 or 600 degrees F. The gun was fired by letting out a drop of hot water at one end and forcing in a drop of cold water at the other end. Now, why should not that system be utilized at sea, where economy of coal and also economy of space are of great importance ? One long pipe of small diameter, with a force pump at the lower end, and connected at the upper end to the valve chest of the first cylinder, would supply 1000 pounds steam with ease and perfect safety. The pipe being so thick would be absolutely "unexplodable," but of course would have a safety valve loaded, say, to 1100 pounds. The "Campania" crosses the Atlantic in 130 hours, or less, and she uses 180 pounds steam. Such steam expanded 19 times down to 8 pounds into a vacuum produces, "theoretically" 15, 847, 184 foot pounds from each cubic foot of water boiled off at that pressure. But steam of 1000 pounds, if expanded to 8 pounds into a vacuum—94 expansions —would produce, "theoretically" 24 375, 225 foot pounds from each cubit foot of water boiled off at one thousand pounds pressure. The "Campania" is said to burn three thousand tons of coal in crossing from Liverpool to New York, with 180 pounds steam. But if she were fitted with boilers and engines suited for 1000 pounds steam she would cross at same speed and use much less coal in the ratio of the above figures, say, 24, 375, 225, : 15, 847, 184 : 3000 : 1950 ; that is to say, she would save more than one third of her coal bill and also save one third of her stokers, and one-third of her coal trimmers, and be able to carry 1050 tons more cargo.

A Washington, D. C., chemist claims to have discovered a process for making chemically pure whisky by means of electricity.

Invention and X-Rays.

When electro-magnetism was first discovered, only a little practical utility was even hoped for, much less did any one prophesy the many wonderful and money making inventions that have since been created by inventors. Depending almost exclusively upon this discovery, are the invention of the telephone, telegraph, electric railways, dynamos electric motors, and indirectly, electric lighting, electric printing machines, electric stock tickers, and many miscellaneous electrical inventions.

In looking over Edward P. Thompson's new book on "Roentgen Rays and Phenomena of the Anode and Cathode," it is apparent that invention has already started in the direction of applying the discovery to useful purposes, and especially at present to the means for applying the X-rays in a more effectual manner. Some of the more prominent inventions brought forth are:—

The standard tube, which permits the use of one tube for currents of different sources from high efficiency for generation of X-rays.

The sciascope, more generally called the fluoroscope, is the result of direct invention, which permits one to investigate with the fluorescent screen much more readily than by the means first employed by Prof. Roentgen.

Calcic tungstate is a well known chemical, but unless by a direct attempt to surpass the qualities of barium platino cyanide as a fluorescing material under the action of X-rays, this chemical has never before been known as possessing six times the fluorescing power exhibited by the platinum compound employed by Roentgen.

The focus tube, as now made, is also the result of invention, although a tube of a similar kind and used for other purposes by Crookes, had been known.

Several inventions that have been of use in connection with photography have been revived in connection with this art, as for example, what are called atops or perforated holes for cutting off superfluous X-rays, thereby obtaining darker shadows. The book referred to contains sciagraphs of snakes, fish, frogs, etc., made by such means.

A minor invention, but all the same, a true invention, is that in which the glass bulbs are made with a very thinly blown portion through which the

THE WASHINGTON MONUMENT.

DUST PORTRAIT ON GLASS.

red and seventy-nine of these are from 40 states and 16 cities, from battle-fields at home and historic spots abroad, from schools, societies and lodges, all over the land. But no tablet can equal in beauty or historic interest the magnificent view which lies spread out before us, when we reach the top. The beautiful public grounds, the splendidly paved streets and avenues, with their 70,000 shade trees, and beyond the stretches of green hills, are seen while across the river lies Arlington with its silent army. The Potomac blue in the summer sunshine, narrows to a silver thread in the distance, and then is lost in the hills, beyond which lie—faint shadow mountains against the horizon—the far-away Blue Ridge.

The annual statement of the Mergenthaler Linotype Machine Company shows a net profit of $2,243,523, over and above expenses during the year ending October 21st 1896.

A nickel-in-the-slot machine, which takes an X-ray photograph of any object you choose, for the modest sum of five cents has been invented by a resident of Hartford, Conn.

X-rays pass in order to reach the outside, based upon the property of X-rays which shows that they pass through glass with difficulty.

A late invention is that of a universal tube, which is provided with means for adjusting the vacuum, which invention was invented upon the principle that necessity is the mother of invention, for X-rays are very particular about the degree of vacuum. If so low that only striae are generated, the rays are exceedingly weak and in fact can only be detected by the most delicate means. If the vacuum is so that a current will not pass through it, then, of course, X-rays are not generated at all. There will be plenty of room for improvements in the means for regulating the vacuum, as it is evident that some ways can be invented that would be worse than not having any.

Some remarks may be made in regard to focus tubes, as the best form has not yet been made, because there are so many different kinds advertised under the names of different inventors, not only in the United States but in several foreign countries.

A remarkable invention is that which assists in reducing the time for taking X-ray photographs, which are called in the book, sciagraphs. In the early X-ray experiments, from ½ to 2 hours were occupied in producing the pictures, whereas at the present time only a few seconds are needed, for example, for taking the picture of the skeleton of the hand. The idea is the combination of the two properties of the X-rays, one producing action upon the photographic plate, and the other producing

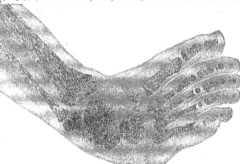

SCIAGRAPH OF CLUB FOOT OF CHILD.

light by acting on a fluorescent screen. The invention consists in combining a photographic plate with a fluorescent screen, and, therefore, the negative effect is produced, not only by the direct action of the X-rays, but by the action of the light into which the X-rays have been converted by the fluorescent screen. There is room, of course, for development in this respect.

Many difficulties were experienced in the early experiments, and they have not yet been removed. The author has often noticed one direction in which difficulties arise, and that is in respect to the "timing." He has seen plate after plate ruined because left, either too long or too short, or because the rays were too strong or too weak, or because the fluorescent screen was not of the right material or strength.

A very ingenious invention is that recently invented and consisting in using a camera in conjunction with a fluorescent screen. The camera takes a photograph of the picture which is exhibited by the screen, and to this end the sciascope is attached to the camera. The operator looks into the camera and when the focus is adjusted, exposes the film.

Laying aside the consideration of mechanical inventions as necessitated by the discovery of X-rays, many other illustrations of inventions might be cited relating to the apparatus for generating X-rays, and for carrying on the experiments in the numerous subordinate discoveries. For example, in testing the action of X-rays upon electrically charged bodies, several have shown much genius. In order to prove the source of X-rays, it is simply wonderful to notice how many different ways were invented. The result of all was conclusive as to the exact source of X-rays. It reminded one of a

patent suit where an attorney exercises a great deal of ingenuity in hunting facts which may bear upon the case, and which may prove the matter one way or the other. The grand result was the establishment of the source of X-rays without any doubt, whereas if only one of them had been relied upon, and formally accepted, no one would be absolutely satisfied that the result was correct.

Inventions and Inventors.

The genius of invention is not always allied with the practical business knowledge necessary to make it a commercial success. This has been illustrated in many cases where the work of one man's brain becomes the wealth of some one who may not have more of grey matter than the law allows. Inventors are seldom business men, nor are they all gifted with discretion in speech. A discovery with some men is an incentive to boastfulness. The misfortune is that the crowing is done before the egg is hatched. Nor are men scarce who take advantage of this weakness and lose no time in practically appropriating to themselves what is or ought to be the property of another. This is the reason why in so many cases inventions and inventors are so wide apart in the commercial results of some new

and brilliant device. One man sows the grain and another gets the harvest sheaves. Men are to be found in comparative poverty in whose ideas others have found a mine of wealth, and while it is true that some of this injustice is due to the inventor himself, it neither excuses or condones the larceny of those who by ingenuity or design contrive to secure the rights and fortunes of others.

There is no doubt a mass of unwritten history behind the invention of which we jeatly boast, that while not detracting from their services to man would exhibit a species of rascality at present unsuspected. Invention has made some men rich and many poor, and is likely to still carry out the old programme. It is probably past the stage of public annoyance and abuse. The modern Jacquard is not likely to see his loom made into a bonfire, nor the new Hargreaves to be chased out of his native town. There may be differences of opinion as to this or that phase of the labor saving question, but practically none as to the rights and the protection of the genius behind invention. The inventor, however, will continue to be sponsor for his own actions. The same care he takes of his watch must be taken of his idea. If he gives it away it is his own act, and if he lets it be stolen it may be chargeable to his own neglect. Ignorance as to the methods of securing his rights can hardly be justified in these days when instruction can be had for the asking, care being taken that advice is asked of the right party. It is very often simply a lack of common sense, judgment and prudence that places the inventor at the mercy of men without conscience, and in the unfortunate condition of losing the benefits of his ingenuity, another man securing both his rights and his fortune.—Age of Steel.

Edison's New Kinetoscope.

Mr. Edison is now working upon an improved vitascope, which instrument he calls the "kinetoscope," which will make negatives covering as many as six hundred figures in motion, so that immense spectacles may be displayed by it, filling the whole space of the stage of a theater. An entire pantomime could thus be reproduced, or a yacht race, a sham battle, a parade, or a horse race, with a large part of the onlookers in the grand stand. A 1000-foot ribbon will run for about thirteen minutes, a sufficient length of time to answer the length of a sensational or thrilling theatrical scene, while in the case of a waterfall, dashing waves, wind-stirred forest and similar effects, a repetition will serve all purposes of realism and scenic delight. Speaking of it Mr. George Parsons Lathrop, in the North American Review, says:

Where, for instance, it is desired to show a waterfall in the background, or a seashore with waves rolling in on the beach, or a storm at sea, there can be no question that the vitascope would represent these things, taken absolutely from life, with a thousand-fold more effectiveness and pleasure to the audience than anything in the line of most skillful stage device with which we are now acquainted. Marine views, with sailing vessels, steam yachts, and boats in motion—the dimpling and rippling of the water, and people embarking disembarking—could be rendered with delicious and genuine vividness. Flags fluttering in the wind canvas awnings shaking over house windows; passing clouds, mist and sunshine in the sky, and puffs of dust in the air could all be conveyed to us without a flaw. A crowd in a background or street; a busy scene in the markets; the coming and going of vehicles—all these could be shown to perfection. It is easy to see further, that spectacular effects of distant multitudes, of armies advancing battling and retreating could be placed before us as nothing else can place them now. Sir Henry Irving achieves a great stage point with his masqueraders trooping through the scene in his production of "The Merchant of Venice." Imagine how much more brilliant and vivacious the effect might be with vitascope figures for auxiliaries! Then, too, in any out-door scene in "As You Like It," for example or "King Lear" or "Macbeth"—would it not be a great enchantment to have the tree boughs waving and sunlight and shadow flickering precisely as in nature? In the matter of panoramic scenery furthermore, where it is desired to give the impression that the actors are passing through a long stretch of landscape, either on foot or in boats, the vitascope offers facilities which the next generation will probably enjoy to the full.

Medals for Engineers.

Mr. Robert Grimshaw has ordered a series of silver and bronze medals to be awarded to those engineers, firemen and others, (1) Who stand the best $e_xam_ina_tion$ on boilers, pumps and indicator practice, and (2) whose engines, after being one year in their charge, are in the best condition. The best details of competition, names of judges, etc., will not be announced until the medals are delivered in the various cities, in each of which the medals is to be competed for.

Another Smooth Swindler.

An exchange says that a smooth patent right swindling scheme is being worked in some of the States. A man comes to a farmer with a patent wagon-tongue for which great claims are made. He has only that one county left, and will sell the right for two hundred and fifty dollars. A few days later, while the farmer is thinking over the matter, another man comes along, who has learned that the farmer has the right to the valuable invention, and offers him four hundred dollars for it paying him $10 down. The farmer at once goes and closes the deal with the first party, giving his note or cash for the two hundred and fifty dollars. In the meantime the second man disappears and the two schemers meet and "divy." If a note, it is discounted at the nearest bank.

A substitute for stained glass is found in fector'ium, a galvanized iron web covered with a gelatinous substance. Experiments have been made with it in Europe, and it is said to be tough, durable, a bad conductor of heat, and easily manipulated.

A French chemist has made a blue soap, which will render unnecessary the bluing in the laundry. In ordinary soap he incorporates a solution of aniline green in strong acetic acid. The alkali of the soap converts the green into blue.

A Study in Car Refrigeration.

BY JOSEPH W. BUELL.

"The world does not stand still. Changes come quicker. New ideas new inventions, new methods of manufacture, of transportation, new ways to do almost everything will be found as the world grows older and the men who anticipate them, and who are ready for them will find advantages as great as any their fathers or grandfathers have had" —Phillip D. Armour.

Practical inventions add rapidly to the economies of any business, and their influence is particularly noticeable in the field of railroad transportation where the operating expenses have been greatly reduced in the past twenty years, until now under favorable circumstances a loaded freight car can be propelled a mile with one pound of coal. This economy in result has been attained by gradual changes in mechanical methods and devices which have greatly lowered the cost of construction as well as the operation of railways.

To test the necessity of a change is to find out where the relative percentage of cost exceeds that of the return of work; or what reduction can be anticipated by the introduction of a new device or new method?

A study of the subject of Car Refrigeration brings out some interesting facts to show that it is an exceedingly wasteful process as effected in the well established systems of chilling by the use of ice, when we come to consider the different items of expense that are influenced or included within the cost of said method, which might be obviated by the adoption of another refrigerant. But this will hardly be remedied by improvements on the present method as ice refrigeration has undoubtedly reached its limit in point of efficiency and economy, as the result of the many plans and innumerable slight variations of method, that have aimed at the proper arrangement of the ice and the manipulation of the currents of air that flow over it. A full understanding of the advantages of a change, however, makes necessary an inquiry into the various items included or affecting cost, as now practiced.

One of the first considerations is for a method that will lessen the number of delays to cars, en-route, experienced in the present system, and thereby lessen the anxiety of consignor as well as consignee, as to the risk of loss or damage, by removing the need of attendance at certain points along the route, necessary to replenish the ice boxes, as well as avoiding the making of numerous contracts to furnish ice, which must be subject to change every year owing to the fluctuation in prices and which vary slightly according to the location of the line of road and the relative location of the points on the road to supply markets.

As the transportation of fresh meats requires expensive rolling stock which is used solely for the shipment of meat in car load lots in one direction, while taxed for mileage in both—as the cars must be returned empty—it is important that the volume of shipment should be large; and for these reasons space within a car is precious and should be reserved solely for the cargo of meat to be transported. Under the present methods this space is encroached upon by the stock of ice stored for the purpose of preserving the meat, and which takes a large portion of the available shipping space of said car, thus making the car too small to accomodate return car load shipments of any other commodity, while it allows of swashing of the water of melting ice, thus throwing out moisture and seriously affecting the contents of the car while the meat that is close to the respective ice boxes, situate at each end of the car, is often covered with frost, which must render that portion of the cargo a less marketable product. The consumption of ice for each car per trip under the severest service of a summer's heat amounts to four thousand pounds (4000), between Chicago and New York City, while it is customary to stop every twenty-four hours to replenish the stock necessary to keep the melting down to a minimum. This in total amounts for the entire trip to handling six thousand pounds (6000) of ice, and an average haulage of four thousand five hundred pounds (4500) upon which a rate of 45 cents per hundred (which is the freight rate charged to points east of Buffalo and Pittsburg) thereby adds an expense of $20.25 in freight without giving a return in marketable material at the point of unloading, while further deducing from the profit of the meat shipped per car on account of the cost of the ice used and the attendance in replenishing its stock, which at the least estimate must amount to $2.00 and it would be safe to say, $10,00 more per car. The consequent saving of space, by the doing away of ice, would allow of the addition of seven or eight more beef, which accomodation per train load would amount to betweed two and three cars in saving—which reckoned at 20,000 lbs would be a saving in freight of $90.00 per car or from $180.00 to $270.00 per train. This would count up as quite an item in the course of a year's shipment of beef from Chicago, when we stop to consider that there is a daily outgo from that city of 300 to 600 cars a day.

This is a fair statement of the losses that are experienced in the above method of refrigeration. The writer proposes to suggest a method of refrigeration that would greatly reduce the cost, seemingly entirely feasible in its practical application and, that will do the required work, in a more satisfactory manner.

To this end it is proposed to use compressed air as the refrigerating medium.

Attempts have been made to apply this method to car refrigeration, but one of the chief difficulties encountered in this direction is the providing of power to drive the compressor, without occupying any space on the car for power apparatus, or without increasing the demands on the locomotive, for live steam, out of proportion to its required supply for propulsion. Then, again, attempts have been made to belt and gear from the axle of the car which have so far proved fruitless, as the movements of the trucks in relation to the body of the car are so varying and the changes of motion so sudden both transversely and up-and-down, as to break any belt, from the shocks received, or to wear down gearing in so short a time as to soon render it useless. But the greatest of all difficulties has been to meet the contingency of side tracking, and preserve the contents of the car after the train has come to a standstill. This problem can however be practically solved and it is the purpose of this statement to show in a general way its practicability.

It is not the purpose in this paper to describe in detail the arrangement of the apparatus to be used although it should be mentioned that the apparatus employed is compact and located beneath the floor of the car. It is the purpose herein to set forth the plan and scope of this method considered commercially and to show that as a policy it is within practical bounds.

The chief consideration therefore is whether we can derive power at a minimum cost without disturbing the efficiency of the locomotive, in its legitimate work. It is proposed to take power from the axle of the wheels of a moving car, to drive the air compressor, to effect the refrigeration, and in order that the refrigeration may equal that of the present ice method we will need to appropiate 3.86 h. p. per car from the axle.

No attempt will be made here to describe the manner of doing this, except to state that the apparatus is simple, requires no attention, and is positive in its operation. The practical inquiry we have to make, however, is whether this proposed demand to be made upon the locomotive, by adding train resistance, is an exorbitant one, or one that would be or could be objectionable to a railway corporation, or which would be an objectionable policy.

As this is such an important consideration we will have to examine in detail the demands made upon a locomotive in running with a freight train from Chicago to New York City, and ascertain if it would be a wise policy to abstract any portion of the available power at any time for this purpose or whether it would be admissible under the conditions of work.

Freight locomotives have power available, ranging from 500 to 1000 h. p. according to the size and type. In order to calculate the power consumed during a trip, as compared with the amount generated that is held in reserve and that is necessary, we will have to note the variance in load in a train of given weight, travelling at a given speed. For this purpose let us select a locomotive of the ten wheel type, as found on the Pennsylvania Road, in the fast freight service,—stem cylinders 21 inches in diameter, stroke 26 inches, weighs 133,000 pounds, of which 101,000 pounds are on the drivers (drivers 68 inches diameter) and which with tender weighs 205,000 pounds. Such an engine is reputed as being able to move slowly 2,725 tons on a level, 655 tons on a grade of one foot in a hundred; 355 tons in one foot in fifty. It will be found by computation that such a locomotive travelling at a rate of 20 miles an hour with an M. E. P. of 90 pounds in the cylinders (that is a fair average) can yield 826 h. p. while having a tractive power of 15,060 pounds. Running at this rate such a locomotive will haul 66 cars, of 22 tons each (that being the weight of the average 28 ft. refrigerator car loaded) on a level tangent with an expenditure of 12 effective horse-power per car.

But as the train resistance, or effort in hauling varies with the grade of a road as well as the degree of curves encountered, the speed being the same, the number of cars that can be hauled over a certain line of road will depend on these charateristics.

On the Pennsylvania Road the average mixed freight train load is 18 cars. By dynamometer readings it is found that resistance of cars and tenders moving very slowly on a level amounts to from 6 to 9 pounds pull 8, per ton weight of train, which varies according to certain conditions, as the size of rails and irregularities of track, size of journals, etc. But on the Pennsylvania Road it averages 7 pounds pull per ton for all rolling and axle friction, while for the engine it amounts to 8 pounds per ton, which includes in addition to the rolling and axle friction that of the machinery.

This resistance is increased by a rise in grade that amounts to the number of feet rise per mile multiplied by 3,787, the product being also pounds pull per ton. Resistance due to speed depends somewhat on the state of the weather and the calculation should be considered an approximate one only. The rule generally followed is that the resistance increases with the square of the speed divided by 171, the result being pounds per ton of train. While to ascertain the resistance due to curves we multiply the degree of a curve by .5; the product will be the resistance in pounds per ton due to the curves on a road.

While following the above method for computing difference of load, we are enabled to arrive at the output of power as well as the amouut of pressure and volume of steam necessary most of the time, in overcoming stiff grades and curves which are liable to be encountered periodically along the route.

To be continued.

Postal Card Magnet.

No doubt you've all made a rubber comb pick up bits of paper by first rubbing it briskly on a rough coat sleeve, but did you ever hear of a postal card that could be turned into a magnet? Balance a walking stick on the back of a chair and tell the spectators that you are going to make it fall without touching it or the chair.

Having thoroughly dried a postal card, preferably before an open fire, rub it briskly on your sleeve and hold it near one end of the stick. The stick will at once be attracted to the card, and will follow it as if it were a magnet. As it moves it will soon lose its equilibrium and fall from the chair. Of course you understand the principle of the experiment. By rubbing the card you awaken electricity in it, and it thus becomes a sort of magnet, with the power to attract light bodies. Do not try the experiment in damp weather.

Child Inventors.

That many children have great ingenuity of mind in fashioning toys of various kinds is well known. That they have very, frequently turned this quality to good use in the invention and construction of some of our most useful mechanical appliances is attested by the following instances:

The children of a Dutch spectacle maker happened to be playing one day with some of their father's glasses in front of the shop door. Placing two of the glasses together they peeped through them, and were exceedinly astonished to see the weather cock of the neighboring steeple brought within a short distance of their eyes. They were naturally puzzled, and called their father to see the strange sight.

When the spectacle maker looked through the glasses he was no less surprised than the children had been. He went indoors and thought the matter over, and then the idea occurred to him that he might construct a curious new toy which would give people a good deal of amusement. Not long after the telescope was an accomplished fact.

A poor Swiss, named Argand, invented a lamp with a wick fitted into a hollow cylinder, thus giving a supply of oxygen to the interior as well as exterior of the circular flame. At first Argand used the lamp without a glass chimney, the invention of which important adjunct would doubtless have been delayed for some time had it not been for the thoughtless experiments of his little brother.

One day when Argand was busy in his workroom, and sitting before the burning lamp, this boy was amusing himself by placing a bottomless oil flask over different articles. Suddenly be placed it upon the flame of the lamp, which instantly shot up the long circular neck of the flask with increased brilliancy. Argand did not allow such a suggestive occurence to escape him. The idea of the lamp chimney almost immediatly came into his head, and in a short time his invention was perfected.

The Tramps of "Outer Space."

In the book where is found so much wisdom, promise and poetry, we are told that "The heavens declare the glory of God." And surely when night has wrapped her sable mantle round our world and the starry hosts appear, beaming in little points of light, in large splendid stars and beautiful constellations, we can in a measure understand the sentiment of the sacred writer, who saw the Creator's glory in the mysterious worlds marching through the skies obedient to the divine law that set their unchanging course from the dawn of time.

How often, as we look into the interminable above, turning from world to world, from beauty unto beauty, we see darting athwart the sky, rifting the curtain of night, the startling meteor, which, like a rocket from the hands of Jove, speeds earthward with a fiery train, bursts and leaves us wondering whence it came. Ever since man first turned his enquiring eyes towards the heavens, he has been thus wondering. Today he is but little nearer the solution of the mystery than were the ancients—only we do not now look upon the heavenly visitors as miracles.

If one desires a good idea of the size, shape and chemical constituents of meteorites, a better place than the National Museum cannot be found for such an object lesson. There on pedestals and arranged in cases, are three hundred specimens, which have wandered from "outer space" to find a terestial resting place. The most of these are small, but there are a few among them that are astonishingly large. One battered all over, as if it had encountered many obstacles in its flight, weighs three tons; another, with all the middle portion melted out, forming an irregular ring, tips the scales at 1400 pounds, and a smaller, with claims to size distinction, possesses just 999 pounds of almost pure iron.

Nearly all the meteorites shown here have an uneven brownish black surface; many of them are covered with little indentures, and all show the glazing made by the intense heat as they plowed through the atmosphere. It would seem that the heat generated by a "sky stone" moving through the air at the rate of twenty miles a second would be sufficient to completely melt it; and this could not fail to happen if the air did not continually blow off the melted portions—which form the luminous train. It is a surprising fact that while a meteorite in motion is outwardly blazing hot, its inner portion is intensely cold. This can be more readily believed when we understand that the cold of outer space, through which the meteor comes, is far below that of freezing mercury.

The Museum's sky stone exhibit which is in charge of Mr. Wirt Tassen shows some beautifully polished sections of these fiery wanderer, which are composed of metal pure enough to be worked directly into useful articles. One small stone of about 50 pounds weight has the appearance of pure silver, and others exhibit various markings caused by the presence of different chemical compositions. These usually consist of iron, nickel, phosphorous, sulphur, carbon, oxygen, silicon, magnesium, calcium and aluminum, and are generally found in combinations—iron alloyed with nickel, phosphorous as phosphides, sulphur, as sulphides, etc. Neither gold or silver have yet fallen from the skies; but the microscope has revealed many minute diamonds hidden within the glassy envelope of the sky-stone. These little gems belong to the carbon department of the meteorite, and are no doubt the result of chemical action set up when the atmoshperic heating process began.

There have been a number of theories advanced in regard to the presence of meteorites in space, and their fall. One idea is that they are thrown out from our planet by great volcanoes, and return to earth after voyaging for a time in unknown depths. Another and perhaps better theory—is that they are the results of volcanic action on other planets. But the most wonderful and startling idea in connection with these visitors, makes them portions of "dead world's," which, after losing their vital forces, like our cracked and barren moon—were rent asunder and their fragments cast into the boundless realms of space; some to wander perhaps forever, and some to come within the atmosphere of Earth and other planets, to blaze and burst and scatter in still finer pieces the remnants of once active spheres.

Sometime a mighty stone may fall and bring within its frozen heart a fragment of architecture, an implement or some other object which will tell that a people similar to earth's inhabitants fashioned it—people who have long since fulfilled their destinies, leaving their world ages ere it grew cold and dropped from the lights above, a mere grain of sand in the desert of space.

From ancient days, Christian, Jew and Pagan, have turned to the skies in speculative wonder or "looking for a sign," and any thing that comes from the void above is sought with deep interest. Livy tells of a shower of stones, which fell about 652 B. C., and impressed the Roman Senate to such an extent that a nine days solemn festival was declared. The Phrygians and Phoenicians worshiped a meteorite as Cybele, "The Mother of the Gods;" and the Moslems still turn in prayer toward the kaaba, in the mosque at Mecca, where is the Black-Stone (a meteorite) that "fell from Paradise with Adam," and marks the direction—as it did many hundred years ago—toward which the faithful must turn in their devotions.

The Japanese worshiped, in the temple of Ogi, one of these visitors from the heavens; the East Indians included one in their religious ceremonies at a Buddhist temple, after decking it (the stone) with flowers and covering it with butter and powdered sandalwood; and a Christian church in Europe, something over four hundred years ago, had suspended from its ceiling a portion of a meteorite, whose fall was looked upon by the superstitious natives as a miracle.

The recent newspaper report, that a meteorite covering two acres of ground had fallen in a western state, was a very large mistatement; no such quantity of "sky matter" has ever come to earth at one fall. But there is good scientific authority for the fact that over ten tons of this mysterious product came down in a lump in Arizona. Another recorded fall occurred in South America, where 30,000 pounds of maleable iron fell from the sky and sank to a great depth in the ground.

There would be a greater number of large meteorites found, if, instead of bursting, they could come to the ground whole. But their flight is so swift and the intense heat so suddenly generated by atmospheric friction, that the rapid expansion causes, in the majority of cases, an explosion, which casts the fragments, many miles apart.

It has been suggested that "sky stones" are parts of those erratic wonders, comets, which for some cause have gone to pieces, not from atmospheric friction, for in their field of operation there is no atmosphere; and their luminous trains are not composed of fiery particles brushed from the speeding nucleus by air.

What are meteors—embryo worlds, that fall ere they are large enough to maintain their positions among the "outer lights?" Are they born of mighty mountains above, whose awful volcanic throes rend the rocks and send them whirling into profound silence and darkness? Or do they come from lands where once green fields and pastures smiled beneath a sun that rose and set, bringing gladness to millions of human beings, until its waning power, growing less and less was insufficient to sustain life on the planet, now a dead world, which without solar attraction, and receiving no radiant energy from its sun, with one last terrific quiver broke into innumerable pieces and disappeared.

J. E. Price.

Knots Tied by Machinery.

If inventions continue to multiply at the present rate, the day may speedily come when man will have to sit with folded arms, while his work and even his pleasures are turned out for him. Science has lately given us a marvel in the shape of a card-counting machine. Two of these most interesting automata ? now working are used for counting and tying postal cards into small bundles. Two of the machines are capable of counting 500,000 cards in ten hours and wrapping and tying the same in packages of twenty-five each. In this operation the paper is pulled off a drum by two long "fingers" which come up from below, and another finger dips in a vat of gum and applies itself to the wrapping paper in exactly the right spot. Other parts of the machine twine the paper around the pack of cards and then a "thumb" presses over the spot where the gum is, and the package, tied with the paper slip, is thrown upon a carry belt ready for delivery.

Swiss Citizenship.

Consul Germain, of Zurich, under date of August 20, 1896, sends the following information relative to Swiss citizenship:

Once a Swiss, always a Swiss—this is practically the case with the children of this Republic. Anyone born of foreign parentage in Switzerland might remain all his life in the country and never become a citizen thereof, unless he chooses to purchase that citizenship, which is the only mode of Swiss naturalization.

German Scientist has invented a mirror of celluloid, which accurately reflects every object. Great care is taken to select celluloid without a flaw, and with a backing of quicksilver it is as perfect as a mirror of glass. The celluloid mirror is unbreakable, is cheaper than glass and lighter.

Boats over 4,500 years Old.

It would be hard to say when the first boat was built. The material of which such articles were mainly composed is so perishable that one could not expect them to be preserved so well as ancient pottery and metal-work have been. Nevertheless, a story comes from Egypt, by way of Chicago, which declares that five boats belonging to a period 4,500 years ago (2600 B. C.) have recently been exhumed from the sand near the pyramid of Gizeh. They measured about 30 feet in length, 8 feet in breadth and were 4 feet deep. Cedar had been employed in their construction. So graceful was the shape that it was easy to perceive that marine architecture had at that time reached a fairly high state of development. The boats, it is thought, had not been intended for service in war or ordinary commerce, but for some exceptional and sacred use, like bearing the bodies of the dead, perhaps the royal dead, to their places of burial, or carrying the perfumes, spices and gums that were offered in homage by surviving relatives. The boats had evidently been buried intentionally and not by accident, being at some distance from the adjacent river Nile, and near the pyramids. It seems probable that these barges, if they may be so called, were propelled by oars or paddles; but it has been impossible to discover any rowlocks, thole pins or other fulcrums on the gunwales. Of the five boats, three were transferred to the Gizeh Museum, and one of these, says "The Chicago Journal," has already been taken to the metropolis of the West. Two boats were merely propped up and suitably covered near the spot on which they were originally found. In such an excellent state of preservation were these barges that it was not necessary, in repairing them, to add more than one-twentieth of the amount of wood originally employed to insure safety in transportation. Light iron braces were introduced here and there. Until this discovery was made, the oldest boat known to be still in existence was a viking craft found in Norway, but dating back only 900 years, as nearly as could be guessed.—*New York Tribune.*

American Machinery for China Mints.

Bridgeton, N. J., has carried off the contract for making the new machinery for the Chinese mint, consisting of fine coining presses, with attachments and dies; two punching presses, with feed attachments and other requisite machinery, making a coining equipment complete and up to date in all respects. We are likely to get our share of the numerous contracts which China, in her political and industrial regeneration, will have to distribute among the Western nations, and there is not the slightest doubt that we shall be able to hold our own both in the matter of price and quality. Besides the equipment of her mint, we are setting up a locomotive plant for her, and when she comes to lay down her railroads the American rail and the American car and the American engineer will all be there. In fact, our business prospects with that great country are extremely promising, and, as we have never cheated or bullied her, nor tried to do either, we ought to stand better with her than some of our competing nations, who have done both.—*New York Tribune.*

The Montgomery (Ala.,) Advertiser reports that the commission appointed by the Secretary of the Navy to test a teredo proof paint invented by Thomas J. Childerson, a painter at the Pensacola navy yard, has concluded its labors. On March 16 four pieces of solid heart pine wood were sunk at the navy yard. One piece was unpainted, and the others had one, two, and three coats of the teredo-proof paint respectively. At noon on September 15, the three members of the commission assembled, and had the four pieces of wood raised. The piece that was not painted was literally honey-combed by the teredo, fell to pieces. The other three pieces were not touched by the insects and were perfectly dry on the interior. The commission considers the invention a perfect success, and one that will be of vast interest to the government and ship-owners. Their report has been forwarded to Washington.

A contract for twenty-five Air Compressors and twenty-five Air Receivers, of medium and small sizes, has been closed by the Clayton Air Compressor Works, Havemeyer Building, New York, with one Company, delivery of the entire order to be made within six months from date. They also report sales of five Air Compressors of standard pattern during the first week in November, and the indications point to a decided revival of trade in Air Compressors, many orders having been held in abeyance, pending the result of the election.

Fortunes in Puzzles.

Sam Loyd is a remarkable man in more ways than one, says the Brooklyn Eagle. He is known the whole world over as a mathematical genius, and to try to tell just when he was not something extraordinary would be a difficult task. At nine years of age he was champion of the New York Chess Club, and for the past forty years he has been recognized as a leading problemist, his problems reaching into the thousands. For many years he was editorially associated with Paul Morphy, the latter being the player and Sam the problemist.

The latest puzzle which has come from a head that has devised thousands of them is a circle of thirteen "Chinks," or Chinamen, each having a red jacket, blue trousers, pigtail, sword and wooden-soled shoes, and by simply giving the wheel a one-eighth turn one of the Chinamen, pigtail, sword, and all, disappears, and the question is where does he go? For with the same turn back he appears again in all his gorgeous array. In his boyhood days Sam was a sleight-of-hand performer, and disturbed somewhat the staid residents of Philadelphia, where he was born on Jan. 20, 1841, by his precocious genius on the lines named, adding thereto an unusual gift of mimicry. Mr Loyd now resides at 153 Halsey street, Brooklyn, he having been a resident of that city for twelve years, and he told a reporter who sought him out that his puzzle-making no doubt grew out of his mechanical bent and his love of Chess, although he does not remember the time when he was not making something, but he is desirous of having it understood that puzzle work is only a sort of diversion. He is considered in many quarters, judging by testimonials, as one of the best mechanics living. He is an inventor and has been granted many patents, including among them those of steam engines.

Mr. Loyd has in possession a Jorgenson repeating watch which is worth $1,000 and no one is allowed to touch it but himself when it needs repairing. He is also the inventor of a bicycle, has travelled all over the world, speaks fluently a number of languages, has been an artist and engraver all his life, a newspaper writer for twenty-five years, having been connected with most of the New York dailies.

Mr. Loyd owns up to the great sin of having invented the "15 block puzzle" and to which he solemnly avows there is no answer. It was freely stated at the time it was playing havoc with the brains of the country that he made $1,000,000 out of it. He says nobody made a cent. One large dry-goods firm in New York sold 100,000 at three cents apiece, and it cost more than that to make them. Millions of them were sold, however. Mr. Loyd says he served on the Grand Jury shortly after the "15 puzzle," became the rage, and it was necessary to visit the jails, almhouses and insane asylums and on a day when he was at one of the latter institutions the doctor gravely told him having been previously informed that he was the inventor, that there were 1,500 persons there who had become violently and hopelessly insane through trying to solve that awful puzzle.

By the "pony puzzle," Mr. Loyd says, he made more money than on all other puzzles put together. More than a thousand millions of them were disposed of for advertising purposes. He is also the inventor of "Parcheesi," which has been a source of great comfort to country people on long winter evenings, and with which many city people kill time. Twenty-eight years ago he invented "La Petite Bagatelle," which latter was brought out by another man as "Pigs in Clover," and which also had a paretic influence. To sum it all up, Mr. Loyd is a remarkable man, and puzzle concocting is only incidental to a mind possessed of wonderful mechanical bent.

A new signal light, called the "rat-trap" light, is now being tested in the French navy, which is said to throw out an electric beam which cannot be seen either to the right or to the left of the ship, and can only be discovered under conditions known to those who are seeking for it. It is claimed that signals can be exchanged between vessels of the fleet without much risk of discovery by an enemy; and it is argued that with such a light a squadron could be guided in its course by a single ship, and almost any kind of weather, without its being known to a hostile fleet lying within a mile or two of the course which is being taken. The value of such an invention can be readily understood, as it would do away with rockets and other signals which are used in time of war. In a recent experiment a squadron which had left port in the night three hours ahead of a pursuing vessel was quickly found by means of this "rat-trap," which was the only light carried on the squadron.

The new state law prohibiting the employment in factories of children under 14 years of age went into effect in New York on October 1.

Behr's High-Speed Railway.

Considerable attention is being given by the English technical press to a system of lightening express trains invented by Mr. F. B. Behr. The method proposed provides for a speed of from 100 to 150 miles per hour, and it has also been before the public for several years, but without meeting with any practical application heretofore. It is now about to be put on trial at Brussels, where a line about three and one-eighths miles long and oval in form will comprise one of the attractions of the 1897 exhibitions to be held in that city.

A cross section of the car and general arrangement of the track is shown in the illustration. It will be seen that four guide rails are necessary, and the top rail which bears the weight of the train is supported at from 40 to 60 inches above the ground. Each carriage is supplied with its own motive power, so that little strain is put upon the couplings.

On the Brussels line the guarantee of 95 miles per hour has been given for running over the curves of only 550 yards radius. The mechanical stress when running at 95 miles per hour on the above mentioned curves reaches an alarming amount. Thus, the centrifugal force tending to throw the cars outwards reaches the respectable figure of 20 tons

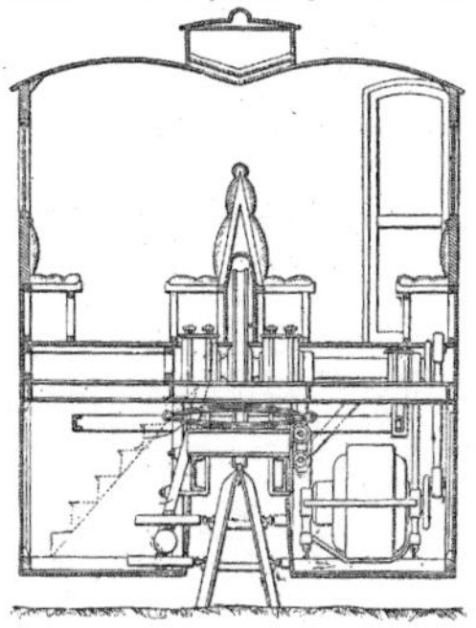

This stress is to be taken by the guide wheels on the inside.

The following are the improvements claimed by Mr. Behr, the mention of which may be of interest to our readers.

Firstly, the carriages have a considerable total length to enable them to pass readily round sharp curves. This is effected by constructing the carriage in two or more seperate parts, joined together by a pivot or universal joint. A flexible enclosure covers a platform between the two adjoining parts, from which access is gained to each part. Each half is supported on the rail by two driving wheels, placed as close as practicable together so as to obtain a comparatively small fixed wheel base.

Secondly, the center of gravity of the carriage is kept below the level of the line of rail, even when the entire floor for the passengers is raised above the line of rail. This is effected by placing the motors that propel the carriage at the bottoms of the two lower parts of the vehicle, and so arranging them that although they partake of the vertical motion of the carriage body upon its springs, the distance between their driving shafts and the driving wheel axles, by which they are connected by driving gear, shall always remain constant.

Thirdly, the carriage body, though having a vertical motion on springs, is so guided that it cannot move laterally; this is effected by constructing the framing carrying the driving wheels and bogie entirely separate from the framing of the carriage body and guiding the former in a vertical position by means of guide wheels running on lateral guide rails on the line rail supports. The framing of the carriage body, which rests upon the wheel framing with springs, is prevented from lateral motion by providing the wheel frame with verticle rails, against which bear rollers attached to the body.

Fourthly, two guide rails are provided, situated one above the other, against which bear two corresponding guide wheels on each side of the wheel frame of the carriage. In order to insure that both wheels shall always be kept automatically in contact with both the guide rails so as to distribute the lateral strains on the carriage uniformly over both, the guide wheels are mounted upon vertical arms projecting upward and downward from a horizontal axis that can turn in bearing in brackets on the wheel frame.

Double Electric System.

The application of electricity to the Manhattan Elevated Railroad in New York is on a double reciprocal principle—that of the directly flowing current from the power-house through the third rail, and that of the storage battery. The third rail carries the average current needed by the motor. When the motor, though going down grade, or other reasons, cannot use the average, the excess is taken up and conserved by the storage batteries. When the third rail current is inadequate, as on an up grade, the storage batteries supplement it. The motor was built by the Electric Storage Battery Company upon the trucks of an ordinary locomotive. It is eighteen feet long, eight feet high. The motorman's cab is in the center of the structure, and in front and behind the cab are compartments for 256 storage batteries. In addition to the ordinary machinery for operating the motor, there are two tanks containing air for the brakes. The initial trial was made on Monday on the Third Avenue line, and was sufficiently successful to lead Russell Sage to consider it at least a mechanical success. Its economic value is yet to be determined. The substitution of electricity for steam on the New York Elevated system will prove a boon to the traveling public and to the occupants of buildings along the lines. For years these have been subjected to all the dirt and grime and discomfort inseparable from the use of coal, and its banishment will be a cause for general rejoicing.

Jail Cells Made of Water Pipe.

A new idea in jail construction has recently been successfully tested in Boston. In brief the scheme is to construct the cells of hollow pipes and fill them with water. When a pipe is severed the water escapes, and by a system of registers in the office of the jail the fact is made known, as well as the particular cell where the pipes have been attacked No attempt is made to have the pipes particularly hard. Common gas pipe is as good as any, and will answer every purpose. The water is kept under a high pressure, so that will be sure to give the alarm when the pipe is severed.

Under the usual system of jail construction it is aimed to make the bars so hard that saws will not effect them; or at lest so hard that cutting would be a slow process. But convicts in jail are just as clever as the men who construct jail cells, and methods have been discovered for taking the temper out of the hardest steel. Nitric acid will do, and so will a common candle. If the flame of the latter is kept for several days close against a bar of chilled steel it will become so soft that a common steel saw will cut it. Solid steel plates have been eaten with acids and escape made possible. The filling of hollow pipes with water seem to be a good thing.—Chicago Chronicle,.

The Great Siberian Railway.

The Great Siberian Railway is now completed to Kransnoyarsk. During the season of 1895, 918¼ miles were built. This gives a direct route from Petersburg to the Yenessei river, a distance of 3,056½ miles. The proposed length of the Great Siberian Railway from Clubliabinsk to Vladivostock on the Japan Sea is 4,547 miles, of which more than one-third is now completed. A large amount of work has also been done on the branches. There are now engaged upon the actual work of construction over 70,000 workmen, beside engineers and officers. Up to 1896, $32,488,000 had been expended. The plan of building across the mountains and canyons on the south of Lake Baikal which was the most diffcult feature of the whole enterprise, has been abandoned and trains will be ferried across the lake by transfer steamers, a distance of about twenty miles.

When soft coal is thrown upon fire, there is an immediate disengagement of volatile hydrocarbons, which will pass of unconsumed, and create a dense smoke, unless there is a large quantity of air present in such a condition that the temperature is not reduced below that of ignition. A western inventor has development an arrangement whereby the act of opening the furnace door to fire starts into operation an arrangement for injecting heated air into the furnace, and this injection continues for a predetermined time, being shut off automatically by the attached mechanism at the end of a period, the length of which is determined by experiment, which will allow of the combustion of the volatile constituents of the fuel.

Patent Brokerage Again.

One of our correspondents writes us as follows:

"Referring to what you said in a recent article about patent selling agencies, I would remark that your condemnation of them seems very strange to me, for it appears as if there is a necessity for just such agencies, because inventors are not business men as a rule and hardly any of them can be in a position to dispose of a patent to the best advantage. A party centrally located and keeping in touch with inventors and manufacturers and with a familiarity in the business and the right ability I should suppose would not only be of assistance, but an absolute necessity.

What there is about the business that should attract only fakes is beyond my comprehension, unless it is that the whole patent business is disreputable. I have had letters from a large number of them and several have offered to see my patents and sell them with out asking an advance fee. All the business is done between the parties whom they interest and myself and their commission is not to be paid until the sale is made and the money paid over to me. I am totally unable to see where there is any opportunity for sharp practice, as no money is paid them and they handle none. They can make nothing only upon effecting a sale. I have written two of them for references and they referred to all the National banks in their respective cities, to Bradstreet's, to the American Express Co., etc. Would fakes go to the trouble and expense of all this? A man in my employ knows one of the firms who have written and he says that they do a large business and are strictly honorable and reliable in every respect."

The INVENTIVE AGE is only too happy to hear of a patent brokerage firm which is "strictly honorable and reliable" for when such a firm is successful in disposing of patents, it encourages the inventor and promotes the welfare of all concerned in inventions. It would bring more subscribers to the INVENTIVE AGE, more advertisements and increase scientific progress generally. Honorable and fair dealing men have engaged in the patent brokerage business heretofore, and they may be engaged in it now, but they seem to fall within the old maxim that "the good die young."

We know of one man in Chicago who was highly recommened by many of the leading patent law firms of that city. He undertook to sell patents for inventors without requiring any advance fees for advertising or other purposes. His plan was to first receive from the inventor a copy of his patent together with a full opinion of its scope and validity prepared by some well-known and reliable patent law firm. Clothed with this information he placed the invention on his books and probably did what he could, together with numerous other cases, to make a sale. We do not know whether he was successful, but his plan was certainly legitimate and harmless, if not successful for the inventor.

Another man located at Media, Pa., concieved the idea of selling American patents to Europeans, and vice versa. He would take over to Europe a lot of what seemed to him to be salable inventions, and endeavor to exploit them in Europe. While there he would also take up European inventions of seeming merit and bring them into the United States, and endeavor to exploit them here. He generally made the round trip once a year. Patent solicitors were much pleased with his scheme and he received a great deal of encouragement from them. The last time he appeared at the INVENTIVE AGE office he had with him a large number of letters from eminent patent soliciting firms in Washington and elsewhere giving him information as to valuable patents procured for their clients and which were awaiting development. We saw this gentleman twice, but have not seen or heard from him during the past three or four years and do not know whether his scheme succeeded. If it had been a success we doubtless would have heard from him again.

Largely advertised firms have undertaken to sell patents without an advance fee of any kind, but meeting with no success, they soon fell into the ways of the transgressor by requiring an installment in advance for printing and advertising, and became in a little while so callous and bold that they merely received the installment, placed the inventor's name and invention on their books and reached out, by extensive crculars paid for with the money advanced by the captured inventor, to secure more advance fees with which to keep the ball rolling.

Like our logical correspondent the inventor or owner of a patent sees no reason why a legitimate patent brokerage business cannot be carried on with success for the inventor besides the broker. Solicitors of patents of forty or fifty years standing will tell you that such a thing as a successful patent selling agency is unknown to them. When we say successful, we mean is actually selling inventions and helping the inventor. The patent brokerage business is as delusive as free silver. There are hundreds of thousands of people in the United States, who cannot see why free silver wont benefit the country, in face of the fact that other nations have tried it time and again without success. It is contrary to the laws of trade.

It is the old story that humanity likes to be humbugged and is willing to pay well for the privilege.

One of the greatest obstacles to success in selling patents on a large scale, is that the value of a patent cannot be foretold. It is just like any other business scheme, and unless managed with integrity and ability will not succeed. The mangement of a patent requires the undivided attention for from one to five years of some person with means, business tact and ability. The placing of a patent in a long list in a broker's office just like a list of houses in a real estate agent's hands, has never, to our knowledge, been of any practical value to the inventor. The work requires the undivided attention of some one or more individuals to that particular invention, just as the carrying on of a business scheme requires the undivided attention of one or more men.

An eminently successful inventor remarked to us the other day that a machine which would coin solid 18 carat gold dollars out of scrap iron would be a total failure unless it had the proper push and management behind it. The same letter which comes to you who have valuable inventions awaiting development, also comes to the men who have the most ridiculous devices. The other day a man received a patent on an improvement in incandescent lamps. His patent was one of a series on the line of inventions which he had been developing. It was of no earthly value to any one under the sun but himself, or the owner of all of his patents. Still he received a circular form a Cincinnati firm soliciting the agency for the selling of the patent on the ground that the agency believed it to be worth $125,000.

To those who read the INVENTIVE AGE and cannot be convinced of the error of patronizing the patent broker we would say, give him a trial as the experience and satisfaction may be worth the money.

The Developemnt of China.

The recent visit of an illustrious Chinaman to the principal industrial centers of Europe and America has, undoubtly, awakened expectations which, there is good reason to believe, are not unwarrented. We all know, or ought to know, that China is one of the most promising fields of commerce, and that recent events have induced many eyes to turn eastward in anticipation of the dawn of that prosperity which is sure to blossom in that quarter. China may be regarded as a virgin soil, the surface of which has only been scratched; and yet the foreign trade, even now, is of the annual value of £30,000,000 sterling. What may it not be expected to yield in the future now that so many barriers have been removed? The progress that Japan has made within a recent period is known to all the world. Is there any reason why China should not rise out of the stagnation that has so long depressed her, and become developed an reanimated in a similar manner? A powerful bureaucracy, banded together in defence of time-honoured abuses; misgovernment; the stupidity of the official classes in disdaining foreign aid either in money or management; the rigid execlusiveness that closed the doors which it was not obliged, by treaties, to leave open—these have operated to keep China far in the rear of her smaller but more enlightened neighbour. But events have moved rapidly within the last two years. The late war has shattered many miserable impostures, and dispelled many absurd superstitions that kept the Flowery Land outside the pale of the civilised world. The treaty of Simonoseki opens up a vast field of industrial enterprise. It is imposible to exaggerate its importance as regards the future. All obstacles to the importation of machinery are now removed. British goods of all kinds are now admitted without let or hinderance. Foreign capital and foreign organization will now pour in, and if British traders are true to themselves they are bound to reap a profitable harvest.

China is greatly superior to Japan in the natural resources which favour the growth of national wealth. We know what she has been as a grower of teas and silks, both of which are better quality than those of Japan. She grows her own cotton, while Japan has to import it. The cultivation of sugar and tobacco might be developed and increased to an enormous extent. There is hardly a crop that does not lie buried under its surface. China only wants to be opened up as Japan has been. Only let enterprise have full scope there, in equal degree, and the foreign trade may be estimated at £200,000,000 per annum. And there is nothing to prevent this being realised.

No nation is better equipped than the British for carrying of a large share of the trade which must spring up with the further developments which are imminent in China. We have at least 60 per cent. of the carrying trade by sea. It was the British nation that first broke down the barriers of Chinese exclusiveness. British trade has already taken root, and it is certain to expand and flourish. We have obtained a good footing there, and we must see that we are not supplanted by other nations in the race for the coming harvest. It is a field that is worth struggling for, and as long as our spirit of enterprise does not fail, nor our energies relax, we shall have no cause to fear the result.

In his conversation here with certain ironmasters, Li Hung Chang showed the interest he felt in the development of railways. This is a project evidently uppermost in his mind. We know that a largely increased navy, to retrieve the losses sustained in the last war, has become, to China, a necessity of the first importance, She is not yet in a position to supply her own needs in such matters. She will have to come to Europe, or to America, to find those who can build her battle-ships, and those who will lay out and construct her railways. Iron bridges, steel rails, locomotives, railway cranes, and rolling stock, will be required in immense quantities; and the question here will be, who will supply them? Which of the countries, through which the ex-Viceroy has passed, will be favoured with the orders? Great Britain, France and Germany will all, doubtless, expect a share of the work. But what is to be said of America? China is disposed to be friendly with the United States, as was evident when she chose an American to negotiate terms of peace with Japan. We may expect that American manufacturers will exert themselves to the utmost to get the lion's share of the spoil. America is more favourably situated geographically than Britain for dealing with China; and we know that Trans-Atlantic engineers can now compete with us in the race for railway and iron contracts. In shipbuilding, they have no chance against the Germans, and we may reasonably count upon being able to hold our ground as builders of warships, and in all other departments of trade, although we may expect to have keen competion. This question is a most important one at the present time, when, owing to the enormous productive power of Great Britain, she is obliged to look around for new markets. We cannot but think that Englishmen hardly realise what an immense volume of trade may, in the near future, emanate from the Chinese Empire; and it behoves us to bestir ourselves and make preparations for maintaining our industrial supremacy there as elsewhere, and our reputation for that excellence in wares which has so long distinguished the British trader.—Trades Journal Review.

A New Naptha Engine.

Charles P. Willard & Co., 197 S. Canal Street, Chicago, are offering a new naptha engine, illustrated herewith, which presents some entirely new features in a boat engine, which render it especially desirable as a motive power for small boats and launches.

Heretofore, we believe, all naptha launches have required the presence of a flame burning in the gen-

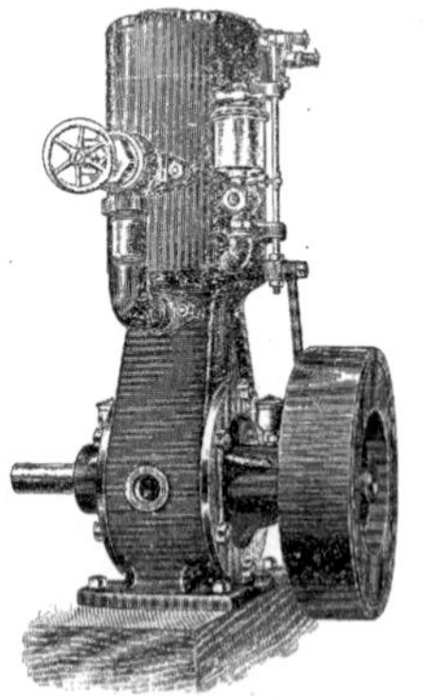

erator while it was in operation, to vaporize the naptha, and also required the presence of several gallons of naptha in the boat at all times. By the entirely different system upon which this engine operates, there is no fire in the boat whatever.

An electric spark from a sealed battery ignites the gas inside a closed iron cylinder and furnishes the motive power. Only a small quantity of naptha need ever be carried in the boat at one time.

A Multi-Millionaire's Practical Benevolence.

Mr. James Creelman, correspondent of the New York World, in a dispatch to that paper from North Carolina says:

All that North Carolina has needed, in union with her sister Southern States, is capital to develop her wonderful natural resources and employ her labor.

Here George Vanderbilt has established himself in a vast domain of 145.000 acres consecrated to science, agriculture and forestry. All this is for the sake of the public. No king, no emperor, has ever set his hand to such a vast scheme of usefulness, Intended almost exclusively for farmers. He has already spent $10,000,000 on the estate, and is giving something like $1,000,000 a year. He employs more men then the Department of Agriculture at Washington does. From every land he has brought trees, plants and flowers, noble herds of cattle and rare breeds of fowl. He allows the farmers of the country to breed from his bulls and stallions free of expense. He sends the eggs from his costly flock to be hatched out in the farms of South Carolina. He has gathered the experts of Europe and America to work out in this matchless place the practical problems which confront the American farmer and stock-breeder.

And when he dies this wonderful organization, with all its property and equipment, will be given to the government of the United States for the benefit of the very men who are now blindly following political demagogues bent on setting the poor against the rich.

A Suggestion for the Patent Office.

EDITOR INVENTIVE AGE: The official Gazette of the Patent Office under the present condition of affairs, is proving to be rather more of a curse to inventors and patentees than a benefit to them. This journal is published every Tuesday at Washington and contains an illustration and short description of each patent issued during the previous week, and also gives the name and address of each patentee. This would prove of great advantage to the inventors providing the Gazette fell only into the hands of honest inventors and manufacturers, but this is not the case by any means.

Its columns are open to the inspection of evil minded persons as well as to honest ones. Swindling patent attorneys, fake advertising agents, fraudulent patent selling companies, organizations, promoting concerns and patent sharks of every description take the Gazette simply becapse it contains every week the names and addresses of a new supply of probable victims. To every address, so conveiently furnished them each week by the Patent Office, they mail their vile, misleading, and fraudulent circulars, booklets and literature, replete with glittering offers and inducements, false assertions and propositions that end only in extorting more money from the too credulous inventors.

As a simple means to protect inventors from these schemes and swindlers, I would suggest that all patentees addresses be omitted from publication in the Gazette. The inventors name may be printed as usual, but his Post Office address should not be given. In the Patent Office let matters be so arranged that only the *names* of the various patentees will be visible to publication, and their addresses can only be ascertained by the purchase of patent copies from the Commissioner at the regular price of 10 cents per copy.

This would prevent nefarious mailing business of the swindlers to a great extent, for to obtain anything like a complete list of addresses in their manner would prove to be a too expensive plan for their adoption, therefore the greater number of patentees would be entirely free from the false luring literature which is now being scattered broadcast over the land to every inventor whose name and address appears in the Gazette as a patentee. The absence of the patentee's address in the paper would not in any way interfere with the probable sale of a patent. Manufacturers and investors promote or purchase patents for their own personal benefit, therefore when they observe an invention of probable value to them in the Gazette they will not hesitate to send for a copy of the patent which interested them. At any rate the one great evil thus remedied would more than balance any possible disadvantage of the system. Inventors would certainly prefer to receive no offers of any kind, than to receive false ones or a mixture of honest and dishonest ones so gotten up that it would be utterly impossible for an inexperienced inventor to distinguished one class from the other.

This simple plan if it could be adopted would not only afford greater protection to inventors, but the extra quantity of patent copies disposed of, would considerably increase the revenue of the Patent Office.

ERNEST LAWRENCE.

Malone, N. Y.

ASSIGNMENTS OF PATENTS.

Partial List of Patents Assigned During the Past Month and the Consideration Therefor.

No. 506,980. John W. Cook inventor, to H. V. Adix of Mount Tabor, Oregon. Clothes Pins and Line. Entire interest for the state of Missouri except five counties. Consideration. $2,500

Toliver L. Chisholm inventor, to Robert M. Grubbs of Atoka Indian Territory. Hinges. One undivided third of his right, title and interest. Consideration $1,000.

No. 555,683. Lemi B. Denton inventor, to Orin A. Ward of Grand Rapids, Mich., Caster. An undivided half interest. Consideration. $2,000.

John B. Davids inventor, E. M. C. Davids and others Assignor's, to The New Haven Toy Game and Novelty Co., Game Board. His entire right, title and interest. Consideration. $10,000.

No. 290,323. Henry Dickson inventor. Wesley I. Craig Assignor, to Edward J. Dickson of Pittsburg, Pa., Apparatus for Drying Green Brick. Entire interest except the states of Illinois and Indiana. Consideration. $1,000.

Loury A. Gray inventor, to Edward Lipps of Baltimore, Md., Apparatus for Preventing Habits of Self Abuse in Stallions. One undivided half of his right, title and interest in said invention. Consideration. $2,000.

No. 452,037. Jacob F. St. John inventor, to Henry Sperry of Nashville, Tenn., Artificial Stone Composition. An undivided half interest for the state of Tennessee. Consideration. $5,000.

Hercules Sanche inventor, to the Animarium Co. of New York City. Electric Connection. Assigns his entire right in said invention. Consideration $100,000.

No. 556,782. Max Sussmann inventor, the Exploitation Co., Limited. Assignor's, to The Sussman Electric Miners Lamp Co. of London, England. Secondary Batteries. All their right, title and interest. Consideration. $10,000

No. 527,131. Benjamin O. Smith, inventor, to Samuel S. Rush, Hame Fastener. His entire right, title and interest for New Mexico. Consideration $900.

Ira C. N. Sweet, inventor, Oramel C. Ainsworth, Assignor, to William Radley of Sandwich, Ill., Pumps. Assigns his entire right. Consideration $920.27.

No. 560,239. Theodore L. Stewart, inventor, to Mary J. Stewart, of Brooklyn, N. Y., Metal Working Tools. All his right, title and interest in said invention Consideration $5,200.

No. 597,992. Christian A. Salzman, inventor, to John A. Robbins, of Hamilton, Ohio. Combination Tools. An undivided half of his interest in said invention. Consideration $5,000.

No. 516,616. Joseph C. Somers inventor, The Columbian Amusement Co., Assignor's to Samuel Barton of Atlantic City, N. J., Merry-Go-Rounds. Exclusive right to construct and operate in the states of Pennsylvania and New Jersey. Consideration $1,750.

No. 490,688. Darwin Smith, inventor, The Smith Insect Exterminator Co., to Wm. F. Stolts of Des Moines, Iowa. Insecticides. Assign their entire right, title and interest in said invention. Consideration $1,010,21.

No. 604,912. Herbert L. Mitchell, inventor, to Howard L. Meyers, of Pittsburg, Pa. Coffee Pot. One undivided fourth interest. Consideration $500.

No. 592,962. George W. Huber, inventor, Morrison C. Nell, assignor to Thaddeus S. Krause, of Philadelphia, Pa. Machine for Compressing Powdered or Granular Substances. One undivided half of the entire right. Consideration $1000.

No. 556,537. Thomas F. Hagerty, inventor, Wm. E. Hoyt assignor to Chas. L. McCrea and Albert L. Wilson, of Dayton, Ohio. Sad Irons. Exclusive right to make use and vend in 22 counties in the state of Ohio. Consideration. $1000.

No. 563,141. Charles F. Hitchcock and Frank H. Edwards, inventor's, to Harry C. Johnson, of Los Angeles, Cal., Lawn Sprinkler. His entire interest in said invention. Consideration $500.

No. 561,708. William L. and Albert Hensel, inventors, to Henry C. Evert of Pittsburg, Pa. Folding Boxes. Assigns his entire right in said invention. Consideration. $3,758.

Clarence D. Hammon, inventor, to Ward W. Swift and James C. Cogin, of Minneapolis, Minn Tire Tightner. Assigns all rights in said invention Consideration $1000.

No. 476,548. John Nixon, inventor, to Ellerson J. Marsh and Abner C. Thomas of Sistersville, W. Va. Depurator. All rights for the county of Allegheny in the state of Penn. Consideration $1000.

No. 544,963. James L. Cummings, inventor, to W. H. Garmany. Coffee Pot. Entire right for the states of Alabama and Georgia. Consideration $2,000.

No. 571,968. Anson D. Gole, inventor, to Barclay S Smith, of Philadelphia, Pa. Attachments for Preventing Refilling of Bottles. An undivided half of his right. Consideration $1000.

No. 521,459. Thomas Carter, inventor, to W. T. Ward, of Ellsey, Miss. Brake for Side Bar Vehicles. Assigns all rights for the state of Mississippi except six counties. Consideration $700.

No 517,916. George W. Chandler and John C. Dale, inventor's, to J. H. Brodus. Car Brake. Entire right. Consideration $1000.

No. 562,151. Joseph H. Pierce, inventor, to J. B. Elliott, of Denver, Colo. Water Filter. Assigns all his right, title and interest for the state of Kansas. Consideration $1000.

Frank A. Pulhemus, inventor, to Lucius C. Harris, of Grand Rapids, Mich., Printing Presses. One undivided half interest in said invention. Consideration $1000.

No. 449,124. Charles Porter, inventor, Albert C. Morton, assignor to Solomon Smucker, of Philadelphia, Pa. Wire Bracket Shelf. Entire right for the state of Pennsylvania. Consideration $500.

No. 571,268. Nicholas P. Perkins, inventor, to James O. B. Palmer and others of Roanoke, Va. Tobacco Cutter. To each one undivided fourth of the right, title and interest in said invention. Consideration $1,500.

No. 556,154. Julius Lewis, inventor, Jeannette Hilgers. Assignor to The Elastic Tread Horse Shoe Manufacturing Co., of New York. Elastic Tread Horse-Shoes. All right, title and interest in said invention. Consideration $24,250.

No. 543,879. Samuel W. Ludlow, inventor, The Ludlow National Fire Alarm Co. Assignor's, to Walter S. Ludlow. Fire Alarm Apparatus. Assigns all his right, title and interest for the United States and several counties in the States of Ohio and Indiana. Consideration $10,000.

No. 551,009. Charles H. Tebbetts, inventor, Herbert F. Daggett, assignor to Frederick D. Berkeley, of New York City. Advertising Cabinet. Assigns all right, title and interest for the state of N. Y. Consideration $1,250.

Patent No. 605,910. Harry J. Buell inventor, to Jacob Benedict of Fort Wayne, Ind., Supporting Attachments for Bicycles. His undivided one half interest in said invention. Consideration $1000.

David Boyle inventor, Margaret H. Boyle. Assignor, to the Pennsylvania Iron Works Co., Ice Machine. All right, title and interest for the U. S. Consideration $20,000.

No. 445,768. Charles O. and Lucien Barnes inventor's, George Pope, Assignor to D. S. Wegg of Chicago, Ill., Car Coupler. All the right title and interest. Consideration. $2000.

No. 545,839 George W. Barfield inventor, to the Barfield Car Coupler Co., of Hamilton, Tenn., Car Coupler. All his right, title and interest. Consideration. $190,000.

No. 411,774. Rudolph C. Baird inventor, to A. C. Bothell of Sheldon, Iowa, Holdbacks for Vehicles. All the right, title and interest for the state of Minnesota. Consideration $4000.

No. 562,595. George C. Esclin inventor, to Thomas Wanless of Chicago, Ill., Oil or Gas Burner for stoves. Exclusive right to sell or vend in the state of Iowa. Consideration. $3000.

No. 531,306. Stephen Essex inventor, to Helen E. and Mary C. Essex of Providence, R. I. Conveyer Chain or Link Belt. All right, title and interest. Consideration $500.

No. 536,618. James P. Bickerstaff inventor, to Albert M. Beaner of Beaver Falls, Pa., Stoves and Ranges. Exclusive right, title and interest in said invention. Consideration $2,100.

No. 458,013. Arthur W. Brock inventor, Mulford Gray Assignor, to Loren D. Rodman of Oshkosh, Wis., Heating Drum. All rights for the state of Wisconsin. Consideration $1000.

No. 552,228. Walter A. Barrows inventor, John B. Brittain Assignor, to Jennie E. Westlake of Chicago, Ill., Music Leaf Turner. Assigns all right, title and interest. Consideration. $2,500.

No. 606,634. Francis Abel inventor, to the Kansas City Excelsior Mfg., Co., Excelsior Cutting Machine. The entire right, title and interest. Consideration, $2,500.

No. 555,670. Henry D. Adell inventor, to V. L. Weakly of Columbus, Ohio. Elevator Shaft Closing Mechanism. One fourth of his entire right, title and interest. Consideration $750.

No. 504,192. John Ammon inventor, to Alfred Dodge & Son, of New York City. Piano Hammer. His entire interest in said invention. Consideration, $2,500.

No. 411,705. Charles R. Adams inventor, to Jacob A. Ulrich of Dayton, Ohio. Policeman's Club. Assigns his etnire interest fort he U. S. Except the New England and Middle States. Consideration. $5000

☞ We furnish Inventors
with fine line cuts and
illustrate and describe
new inventions.

*Our building, one block from Patent
Office, cor. 8th & H sts.*

Special Offers to New and Old Subscribers.

THE INVENTIVE AGE one year and Edison's Famous Dollar Motor. $1.75

THE INVENTIVE AGE one year and Zell's Condensed Cyclopedia—reprint of former $6.50 edition; 800 pp.; cloth binding. $2.00

THE INVENTIVE AGE one year and Stillson Hutchins' work—"The National Capital"—very interesting; 300 pp.; 200 illustrations, $1.35

THE INVENTIVE AGE one year and a list of 50 firms who manufacture and sell patented articles, $1.00

THE INVENTIVE AGE one year and three copies of any patent desired, or one copy of any three patents, $1.00

THE INVENTIVE AGE one year and a five line (35 words), advertisement in our "Patents For Sale," or "Want" columns, three times. $1.00

THE INVENTIVE AGE one year and "The Mechanic's Complete Library" the best work published for mechanics and inventors; 576 pp., $1.75

THE INVENVIVE AGE one year and "Edison's Cyclopedia of Useful Information;" 520 pages; cloth, $1.15

We can also furnish any magazine published at from 10 to 25 per cent below the regular subscription price, in connection with the AGE. Premium offers are good for renewals as well as new subscribers.
Address,
THE INVENTIVE AGE,
Washington, D. C.

Royal 8vo, cloth, 68 Diagrams, 43 Half tones. Price $1.50.

ROENTGEN RAYS
And Phenomena of the Anode and Cathode.

Principles, Applications and Theories.

For Students, Teachers, Physicians, Photographers, Electricians and others.

By Edward P. Thompson, M. E., E. E.

Mem. Amer. Inst. Elec. Engineers, Amer. Soc'y Mech. Engineers, author, Inventing as a Science and an Art. Assisted by LOUIS M. PIGNOLET, N. D. C. HODGES and LUDWIG GUTMANN, E.E. With a Chapter on Generalizations, Arguments, Theories, Kindred Radiations and Phenomena, by Professor WM. A. ANTHONY, formerly of Cornell University, Past. Pres. American Inst. Electrical Engineers.

THE BOOK INVOLVES

Researches of Spottiswoode, Lichtanberg, Karsten, Hammor, Puggendorff, Gassiot, Plucker, Crookes, Goldstein, Hertz, Lenard, Kowalski, Roentgen, Righi, Varley, Elster, Geisel, Thomson, (J. J.) and Elihu; Lodge, Swinton, Salvioni, Rowland, Edison, Tesla, Bergmann, Pitchford, Mestnes, Chas. Henry, Brauty, Stokiow, Pupin, Stine, Dafour, Sylvanus P. Thompson, Terry, Scribner, M'Berly, Rice, Mischin, Appleyard, Bugost, Bood, Mayer, Murray, Lafay, McKay, Perrin, Thomson (Lord Kelvin,) and many other eminent Physicists and Electricians.

On receipt of a copy of this work Dr. W. C. Roentgen wrote as follows: "I express to you my sincere thanks for kindly sending me your book "X Rays," which I have read with great interest."

INVENTIVE AGE PUBLISHING COMPANY,
WASHINGTON, D. C.

CONTENTS.

CHAPTER I.—Anode and Cathode phenomena in open air, compressed gases and Low vacua.
CHAPTER II.—Action of the magnet upon the cathode and anode columns of light, and other kindred occurrences in the discharge tube.
CHAPTER III.—Electric images, electrographs, anode and cathode dust pictures, photo-electric pictures, and portraits, bas-relief fac-similes and other curious pictures based upon discharge of cathod.
CHAPTER IV.—Anode radiations, motions, effects, strata, velocity and kindred occurrences, such as heat striae and sensitive state.
CHAPTER V.—Cathode rays in high vacua, inside of discharge tube.
CHAPTER VI.—Cathode rays outside of discharge tube.
CHAPTERS VII, VIII, IX, X, XI, XII.—Roentgen Rays. Properties, laws and principles of Applications, Instructions on electrical apparatus for generation. Construction of discharge tube. Difficulties experienced and how overcome. Miscellaneous phenomena.
CHAPTER XIII—Roentgen Rays in diagnosis.
CHAPTER XIV.—Generalizations, Theories and kindred radiations.

"Bubier's Popular Electrician'

In the name of a monthly publication which contains a vast amount of valuable information on all electrical subjects. Its department of "Questions and Answers" will be appreciated by students and amateurs desiring information or instruction on any problem that may arise. The INVENTIVE AGE has made special arrangements whereby we can supply that popular dollar journal and THE INVENTIVE AGE—both publications one year—for $1.50.

Improve Your Spare Time

You can successfully study at your home, under our direction, Mechanical Drawing, Perspective, Lettering and Mathematics. Send for catalog free. Penn Correspondence Institute, Philadelphia, Pa.

An Offer of Interest to Every Reader.

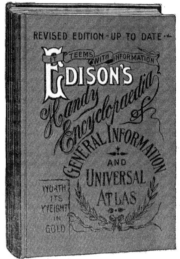

ntive Age

TRIAL REVIEW

CTURING INDUSTRY
AND SCIENTIFIC PROGRESS

ON, D. C., DECEMBER, 1896.

Single Copies 10 Cents.
$1 Per Year.

uncovered and studied by anti-
of these reveal a knowlege of the
sics of which many of us supposed
ignorant.

g article written by W. T. Bonner
dences of an adaptation by the
rly centuries of the laws which
of the household apparatus of the
e the story of his investigation in
e.

recent paper on "Water Tube
y the writer at a meeting of Gen-

In the center of the first hall in the National
Museum, at Naples, containing the bronze relics
of Pompeii, are preserved two apparatus for heat
ing water.

Photographs and drawings fully illustrating
their detail construction, were furnished me by M.
Francisco Milone, a Neapolitan engineer of considerable reputation, at the request of the directors of
the National Museum, to whom my communication
to the Minister of Public Instruction was referred.

There is a remarkably close analogy between the
ancient Pompeiian boilers and the water-leg

The Inventive Age

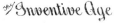

Established 1889.

INVENTIVE AGE PUBLISHING CO.,

8th and H Sts., Washington, D. C.

ALEX. S. CAPEHART. MARSHALL H. JEWELL.

The INVENTIVE AGE is sent, postage prepaid, to any address in the United States, Canada or Mexico for $1 a year; to any other country, postage prepaid, $1.50.

Correspondence with inventors, mechanics, manufacturers, scientists and others is invited. The columns of this journal are open for the discussion of such subjects as are of general interest to its readers.

Technical matter is particularly desired. We want practical information from practical men.

The INVENTINE AGE is thoroughly independent.

Advertising rates made known on application. Special facilities for furnishing cuts of any patented device together with descriptive article. Business specials 15 cents a line each insertion, 7 words to the line. No advertisements less than 25 cents.

Address all communications to THE INVENTIVE AGE, Washington, D. C.

Entered at the Postoffice in Washington as second-class matter.

WASHINGTON, D. C., DECEMBER, 1896.

As a result of the election it is reported that Baker & Company, the extensive platinum dealers of Newark, N. J., have closed a purchase of nearly $200,000 worth of crude platinum.

THE J. H. McEwen Manufacturing Company, makers of dynamos and electrical machinery of Ridgeway, Pa., with offices in New York, were compelled to make a general assignment last month for the benefit of creditors. Readers of the INVENTIVE AGE will remember the extensive inventions of this firm shown in December issue just one year ago. Poor collections is given as the cause of the failure and the firm hopes to resume shortly, under the former management. The assignees are J. H. McEwen and C. A. McCauly.

C. H. BENJAMIN, an eminent professor of mechanical engineering, has been making some interesting experiments to find out exactly how much of the power evolved in the boilers of the ordinary factory is transmitted by the machinery and used in turning out work. He found that the average loss is about 75 per cent, that the loss is rarely below 50 per cent and is often above 85 and 90 per cent. And this amazing waste is going on all over the world wherever men are applying the forces of nature, stored in fuel, to the moving of machines.

Much of this waste could be saved by a more scientific use of our present machinery. But even at best the amount of the power generated that is actually used is very small. This shows that, skillful and ingenious as we are in making mechanical devices, the best of those devices are but clumsy efforts to take advantage of nature's bounty. Also it suggests the time when persistent and progressive man will make power and work very nearly equal in every machine, and thus enormously increase the efficiency of machinery and enormously reduce the cost of manufacturing, railroading and every other industry not exclusively dependent upon the physical labors of men and animals.

In Massachusetts machinery represents the labor of 100,000,000 men. If the loss is fifty per cent of the power on machinery this would show that in operating the machinery within the confines of Massachusetts' eight thousand square miles the labor of at least fifty million of men is lost every day. If this calculation were extended to the machinery operated throughout the country, what an incalculable loss of labor would be shown.

THE whole country, Congress included, knows how urgently the Patent Office, as far as the inventors' interests are concerned, hasbeen tested. Congress, instead of expending the $3,500,000 surplus Patent Office fund for the purposes of providing greater facilities for the transaction of its necessary business, has turned this surplus into the Treasury of the United States and positively declines to employ the force necessary to carry on the current work of the bureau. For four years now the INVENTIVE AGE has been trying to make Congressmen understand the relations of the Patent Office to the people, and to give that legislative body a clear conception of its duties in relation to that office. But the work so far has been in vain. The trouble is the inventors of the United States are not organized and cannot make themselves felt in the legislative halls, and unless Congress can be impressed with the fact that the inventors know their rights and are prepared to maintain them, nothing will be done to render them justice. The trouble is largely with the inventors themselves because they never have made a united effort to secure their rights. Several years ago during the Patent Centennial an organization was founded in this city called the American Association of Inventors and Manufacturers. Dr. Gatling the famous inventor of the Gatling Gun was made President and Mr. George Maynard, Secretary; and a number of the famous inventors of the country became members. But in spite of the fact that nearly every inventors in the country has been informed of this organization and its high purposes of working for the true interests of inventors, the great body of these men have not taken advantage of this opportunity to band together for their common interest while thousands of them have joined fake institutions and received the reward of disappointment. For some reason many, perhaps the majority of American inventors seem to drift naturally into the clutches of the fake patent broker. But when the membership in such an institution as The American Association of Inventors and Manufacturers is offered to them they do not think it worth the while, when the fact is that this association has done more in the last three years to promote the best interests of the Patent Office than all other societies combined. It is advisable that all inventors should join this association for the protection and advancement of their common interest, but if they do not wish to do this they should at least urge their congressmen to recognize their rights and give them decent facilities in the Patent Office and necessary reforms in the Patent Laws.

It is said that Secretary of the Interior Francis will recommend to Congress the erection of a new building to be used for the Interior Department and thus vacate and turn over the Patent Office to the uses to which it belongs, that is the important patent business of the country. This would be a wise move on the part of the government, as it would save enormous rents that are now being paid for a number of Bureaus that cannot find room in the present Patent Office building. Every inventor should write to his Congressman and urge the adoption by Congress of this plan. The inventors have a right to the Patent Office, but they will never get their rights unless they demand them in a respectful but positive way. Here is what the Washington Times has to say on this subject:

Secretary Francis is reported as harboring the intention of asking Congress to do what should have been done long ago—surrender the building now occupied by the Department of the Interior to the Patent Office, for which it was originally intended, and erect a new building for that department. The Patent Office has long been suffering for want of room, and the inconveniences imposed upon it on this account instead of diminishing must necessarily increase every year, because the mechanical genius of the American people and their progressive and enterprising spirit stimulates invention. Take, for instance, the one subject of electricity, and see how in the brief period of twenty years this field has been enlarged. Yet, to day we are but at the threshold of the achievements of this element, whose marvelous activity ranges from the ringing of a bell to the harnessing of a Niagara to the wheels of industry. The multiplication of invention in this branch alone makes constant demands for more room for clerical force and for the storing of models. And what is true of this is also true of every other field of mechanical activity.

The Patent Office has long had a right to demand that more room be given to its affairs. The inventors and all who have business with it have a right to demand it, for the fees they pay not only pay all the running expenses of the office, but make a large surplus to be covered annually into the Treasury. This surplus at last accounts amounted to something like four million dollars, a sum which would pay for a site for the new department building and leave a handsome balance for the structure itself. The department over which Secretary Francis presides embraces a number of important bureaus, some of which are squeezed into the Patent Office building, others occupy rented quarters, and only one, the Pension Office, has a respectable habitation of its own. The transaction of the public business is seriously hampered by these conditions and the only practical remedy is that which Secretary Francis advocates.

"Hard Times, Come Again No More."

The following is the full text of a circular issued by Barnhart Bros. & Spindler, of Chicago, just before election. It is as interesting now as then:

"Hard times! hard times! come again no more!"

We will sing it, we all hope it, but do we know what hard times are? We sing it while we eat beefsteak at twenty cents a pound, oysters at fifty cents a dozen, and three kinds of bread at the same meal; we say it while we smoke cigars, two for a quarter; we think it while we stretch our comfortable legs on Brussels carpet, before a blazing grate, with well-groomed boys and expensively-clad girls around us; we shout it to our neighbors across our smooth lawns, or through our plate-glass windows; we groan it as we read our morning and evening papers, our plentiful magazines and our costly libraries; we dream of it in our soft and springy beds, while our coal-fed furnaces keep the whole house warm; we maunder about it in our well-equipped offices, shout it through our telephones, ring the changes on it as we send telegrams and take expensive summer outings. We meet in our political, social literary and business conventions, and ring the changes on it while we are spending fortunes with railroads, hotels, restaurants and places of amusement.

And yet in these days we do not know what hard times are; we think we do, but we do not.

The writer knows of a time within his remembrance—and he is no patriarch—when, in one of the richest parts of one of the most favored States in the Union, the whole town of some 2000 inhabitants possessed altogether not over $300 in money; all exchange was by barter; there was no cash payment, because there was nothing to pay with. Among the best and richest families (and there were many who thought themselves well-to-do), beefsteak was a once-a-week visitor; round beef was a luxury, oysters were an unheard-of dainty, corn bread was the usual, wheat the rare food; the cheapest pipe tobacco was a dissipation; cold bedrooms, scanty wood fires, woolsey and calico were in the house; 6x8 window-panes were helped out by hats, old papers and rags; a weekly paper was an extravagance, and served several families. Ten books made a good, fair library; beds were slatted or corded; rag carpets were occasional, ingrain scarce and Brussels a tradition; the sole vacation was a ride to the annual picnic in the one-horse shay; nobody had time, money or heart for conventions or amusements. We men worked from 5 A. M. to 7 P. M. (the aristocrats shortened the time by two hours), and the women worked at all hours.

And yet it is doubtful if there was in those times such a universal spirit of unrest and discontent, such a concert of growling, as today.

Is it fair? Are we just? Can we afford to waste time in bewailing hard times, when times are easy on us, and treat us far better then we deserve?

A manager once said with a chuckle: "The mails have been good to our house today; the first one brought an order in every letter, and the last one brought a remittance in every letter." We all know (mayhap we are) persons who would have said: "Alas, we have fallen on evil times; the first mail brought no money, and the last mail bronght no trade."

Let us put aside these ugly tempers of ours; look towards the sun; smile at the shadow; all sunshine makes the desert; "it's a pretty world, senor," enjoy its beauties; let us borrow no trouble; shed light on our neighbors; quit us like men, and times will seem (as they are) good.

(Continued from First Page.)

the heating surface greatly increased, but the heating was rendered more efficient, thus showing that the ancients fully understood the principle of distributing across the furnace a certain number of tubes, in order to increase the heating surface and to aid the evaporation by means of a more active circulation of the water.

It has been suggested that this apparatus may have served at some time to heat wine as well as water, which suggestion appears reasonable, as many competent authorities agree that the Pompeiians made great use of hot drinks. Probably this urn, or boiler, was found in one of the "tepmpodi," or, in modern language, *cafes*, in Pompeii or some other city of the Campagna. At Pompeii were to be found several of these merchants or dispensers of hot drinks, but my informant states that he has not ascertained to what other city of the Campagna it may have appertained.

The photographs reproduced here are Figs. 8 and 9. Evidently the subject offered this layman little in the way of either romance or sentiment; consequently our record is deficient in that it lacks any accurate description of the articles illustrated in the photograph. The apparatus shown in Fig. 9 is the boiler referred to in the different descriptive articles which have appeared in the technical papers lately. Judging from its appearance it probably consists of an annular water space connected by horizontal water tubes; or perhaps simply an outer vessel containing an inner smaller vessel whose flanged rim rests on the upper edge of the outer vessel, after the manner of the ordinary double boiler so common in the modern household.

The group, Fig. 8, is said to represent merely ordinary kitchen furniture, used no doubt, in the culinary department of some large household. Certain features of this group brought out by the photograph might indicate that this apparatus was utilized for more important work, and it is unfortunate that we cannot have some further information as to its internal construction.

The description of the two boilers is somewhat meager and unreliable, but sufficient is given to establish the fact that the larger boiler, Fig. 9 at least, is provided with a water jacket, with some form of grating for supporting the fire underneath. A cock at one side, which appears to be very artistically decorated, served to draw off the heated liquid.

The photographs, reproduced herewith, were obtained from the gallery opposite the National Museum at Naples. Apparently little effort has been made by the authorities at the museum to trace the origin and history of the different relics I have described above, my informant stating that no one had interested himself in the matter, although some of the articles were discovered as many as 40 years ago.

We are indebted to the *Iron Age* for the cut and description.

Pneumatic Tubes for Mails.

The mention in Postmaster General Wilson's report, of the desirability but enormous cost of pneumatic tubes for mail service in cities suggests a scientific dream. As is well known, the sewers of Paris are great underground galleries, the main ones large enough to be traversed by men in boats, and in them electric wires are hung and all necessary tubes arranged. If that could be done in Paris, as it was many years ago, with the old fashioned costly appliances, why not much more readily in Philadelphia, with all the advantages of modern economical methods, and in a subsoil so well adapted to the work as we have here? There is no difficulty in conceiving a harmonious system of sewers under this city, fulfilling the same functions as those of Paris, solving the problem of overhead wires by taking them all underground, carrying gas, water, pneumatic and many other desired pipes and tubes in a position where they would be out of the way and yet would be always within reach, and there would be no difficulty in constructing the same without disturbing the surface, except perhaps for the smaller branches. It is only a question of money. Some day no doubt, the city will have the money, and will see its way clear to make this great improvement.—*Phila. Exchange.*

Skylight Glazing without Putty.

A new system for the construction of skylights and glass roofs has been designed to secure tightness and elasticity and to avoid the use of putty, as shown in the accompanying sketch. The skylight bars, A, are made of rolled iron or steel in the form of an inverted T, a long strip of spring sheet metal, B, is formed so as to fit over the vertical web of the bar and form a gutter on each side. the outer edge of which is inclined at an angle of about 45 per cent, leaving

enough above the edge of the bar to give a continuous cushion spring for the glass to rest upon. On top of the glass is placed a cap, C, which is pressed down upon it by brass nuts and bolts, E, which are firmly riveted into the web of bar, A. In cases where the glass is lapped a condensation spring, D, made of spring sheet metal is used, leaving the edge upon which the upper light of glass rests at a slight angle, making a spring support for the glass as well as a tight joint. The glass is free to contract and expand without danger of injury, thus enabling larger lights to be used than with other systems. This system is not affected by the vibration in factories due to the use of heavy machinery or other causes. The system has been patented by Mr. Chas. Escher No 143½ Freemont St., Jersey City, N. J.

Successful Inventions.

Under this head we will hereafter give a brief notice of those new and commercially successful inventions which are brought to our notice. So many patents are taken out on devices which have proved to be of no commercial value that we have thought it would be refreshing to our readers to get a brief account of the commercial successes of small inventions not generally heralded in the daily press. We believe that as large a percentage of inventions are successful as those business enterprises not based on patents. In the reports from Bradstreet's, Dunn's and other commercial agencies it has been estimated that ninety-five per cent of all business ventures fail, thereby leaving only five per cent successful. A man who conceives a new invention and believes it will become popular and successful is just as much justified in taking out a patent as the man who conceives a business scheme is justified in carrying out that scheme.

COLE'S CANOPY.

One of the first exceedingly useful and successful inventions which we will mention is that of Mr. Warren Cole, Jr., of Knoxville, Tenn. Mr. Cole is what one might call a universal genius as he has invented many practical and useful devices in widely divergent arts. The one which we have selected to illustrate is his canopy now manufactured on royalty by the Southern Canopy Co., of Knoxville, Tenn., and sold under the trade name of "Dixie." Although the patent on this canopy is only two years old, the invention has gone into extensive use under the management of the Southern Canopy Co., which took hold of it about a year ago. The management consists of J. C. Woodward, President; C. A. Nickerson, General Manager; and Walter Woodward, Secretary and Treasurer. The canopy is attachable to the headboard of any ordinary bedstead and consists of a substantially U-shaped frame of flexible steel wire having coils at the ends where it is attached to the bed. Its free end can be bent down by a cord and fastened to the foot of the bed to hold the frame in a substantially horizontal position to allow the netting to cover the bed. It is designed to supersede those canopies which are attached to the ceiling over the bed, and which are such a nuisance to housekeepers. In Mr. Cole's canopy there can be no injury to the walls as it is attached directly to the bed and moved about the room with the latter, and when desired to raise the canopy out of the way the holding string may be loosened, whereupon it automatically springs into a vertical position, the netting falling down gracefully over the headboard. This drapes the bed as beautifully as a lace curtain or handsome lambrequin. Many

ladies attach ribbons to the top of the canopy and bring them down diagonally across each other on the front of the netting when the canopy is raised. Bows are often attached at the top corners and at the point where the ribbons cross each other therebye dressing the bed most beautifully and artistically.

In the first illustration the canopy frame is shown lowered over the bed without the netting, the second shows it in the same position with the netting on, and the third shows the canopy drawn back in a vertical position as when out of use in the day time.

The Southern Canopy Co. was organized about a year ago with a capital of about $1,000,000 and during that time has sold thousands throughout the country, having supplied many of the handsomest hotels in the United States with them. Mr. John Wanamaker has already sold thousands of dollars worth of them, and the company has just received an order for hundreds of the canopies with which to equip the Florida East Coast Hotel System, owned by Gen. Flagler, and under the superintendence of C. B. Knott. This system in-

FIGURE 1.   FIGURE 2.

FIGURE 3.

cludes the princely Ponce de Leon, the Alcazar, the Palm Beach, the Royal Poinciana, and the Royal Palm Hotels. The Canopy Company will pay a 6 per cent dividend for the first year's business, and will probably pay from 10 to 25 per cent in dividends the second year of its existence. The inventor also comes in for an exceedingly comfortable income from the royalties.

Stoves in Shoes.

An effective means of warming the feet has been patented in Germany. The inventor calls it "heatable" shoes. Within the heel of the shoe, which is hollowed out, there is a receptacle for a glowing substance, similar to that used in the Japanese hands-warmer, says the Popular Science News. Between the soles, imbedded in asbestos covers, is a rubber bag, which is filled with water. The water is heated and as it circulates while the wearer of the shoe is walking it keeps the surface of the foot warm. A small safety valve is provided so that the bag cannot burst. The warmth given by this sole never rises above 60 degrees Fahrenheit, and will last about eight hours. The sole is slightly thicker than that of a wet weather boot.

Paper Matches.

The time-honored scheme of rolling up a piece of paper, and using it for a lighter has been utilized by an inventor in the manufacture of matches. The invention promises to revolutionize European match manufacturing, and is perfectly timely, because the wood for this purpose is constantly growing scarcer and more costly. The new matches are considerably cheaper than wooden matches and weigh much less—a fact which counts for much in the exportation. The sticks of these matches consist of paper rolled together on the bias. The paper is rather strong and porous, and when immersed in a solution of wax, stearin, and similar substances, will easily stick together and burn with a bright, smokeless and odorless flame. Strips one inch in width are first drawn through the combustible mass spoken of above. and then turned by machinery into long, thin tubes, pieces of the ordinary length of wood or wax matches being cut off automatically by the machine. When the sticks are cut to size they are dipped into the phosphorus mass, also by the machine, and the dried head easily ignites by friction on any surface.

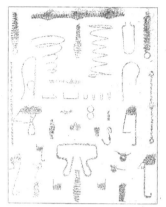

STOVER'S AUTOMATIC WIRE COILER.

together and the wire begins to feed through as before. The advantage of the machine is that all the work is done automatically, the wire being taken from a reel and fed through the machine continuously. By changing a cam a large variety of coils can be made. The machine is made by the Stover Novelty Works of Freeport, Ill.

The End of the Chesapeake.

An English journal contains the following item, for the truth of which we cannot, of course, vouch; but it is interesting if true:

"It is not by any means widely known," says the journal, "that the Chesapeake, famous for her historic encounter with the British ship Shannon in 1813, is in existence to-day, but is used in the somewhat inglorious capacity of a flour-mill, and is making money for a hearty Hampshire miller in the little parish of Wickham. After her capture by Sir Philip B. V. Broke she was taken to England in

sprocket wheels and all complete. It is intended solely for submarine use. It is painted with a waterproof composition which prevents rust.

The machine itself consists of the regulation shaped bicycle frame, two wheels, the usual running gear, with some additions, and two cylinders. These cylinders are realy the secret of the whole machine and its ability to travel below the ocean. They are about 8 feet long and 9 inches in diameter through the center. Their construction is a bit peculiar, in that while they are constructed of copper they are balanced with several hundreds of pounds of lead fastened to the under side. This is done so that when beneath the surface the cylinders will maintain their proper position. The cylinders are filed with air and serve to keep the submarine bicycle at the desired depth.

It is by means of these same cylinders that the

to operate t double.

As the pe revolve at t small cone- propeller to unique veh

Fixed in air gauge, amount of a pressure of the seat

The met rather odd. for one, bu is so evenly to one side and consequ

When it i forced from the submar

MODERN BURGLAR ALARMS.

What the Present Day House Breaker has to Contend With.—Electric Alarm.

To not a few alone the mere mention of burglars sends the cold chills stealing up and down the spine, and yet, with the assistance of a very small part of the modern anti-burglar improvements, a house can be made practically burglar proof and impregnable even to the most wily and skilled of house-breakers.

In a residence provided with these up-to-date devices, the midnight prowler's presence is known to the occupants, from the moment he enters the yard gate or vaults the fence and, if he returns alive and uninjured it is because he has been frightened away before completing his object.

The electric alarm takes precedence in all this great mass of anti-burglar mechanism, as its workings are least understood by the ordinary burglar and even if understood could not be defeated in their present state of perfection. It is not at all necessary that the occupants of the premises take any active part in resisting the burglary. They can lie quietly in bed and yet be made aware of every movement of the unsuspecting thief below.

Let us suppose that a burglary is to be committed upon a premises guarded with improved burglar alarms and foils. As the professional crook unthinkingly lays his hand for a moment upon the wire of the fence enclosing the grounds, he completes an electric circuit with the earth and, all unknown to him the first intimation of his presence is received in the bedrooms of the house and his exact location read upon the electric indicators. His next act of opening the gate also sounds the same alarm by completing another circuit through the medium of concealed spring contact points mounted on the gate and gate-post respectively. Suppose, though, by some means or cunning he has entered the yard without operating either of these alarms. He reconnoitres, and passes about the silent house in which every one seems to be wrapped in deep sleep. Nothing warns him of the death trap into which he is about to step. He stealthily mounts the porch steps but does not notice the slight sinking of one of the treads that controls another electric alarm circuit. Neither do his suspicions seem to have been awakened by the peculiar springy nature of the electric door mat upon which he has just trod.

Suppose though that this particular burglar is a very cunning one and has so far avoided all the alarms, as he supposes, by altering the electric circuits which he has located during the day. If he has so altered them he has reckoned without his host for little does he know of the delicate instrument within that is connected to all the circuits throughout the house and which sounds an alarm upon the faintest variation in any of them, such as would be caused by altering or tampering with the wires. Every thing to him though seems to be going well. He attacks the door but for some reason his keys and nippers fail to make an impression, so he sets to work to remove a portion of the glass panel to permit him to operate the lock from the interior. He unwittingly cuts through the gold letters of the sign painted upon the glass and which form part of an electric circuit. The breaking of this latter circuit indicates in the bedrooms that an attempt has been made to enter at the front door but the occupants feel no concern as they are aware that even though the burglar make an entry he can do no great harm, except to himself. He is now a house breaker in the eye of the law and is answerable for his crime, so the breaking of the last mentioned circuit also sounds the alarm at police headquarters to warn the officers that this particular residence is being burglarized and to summon them to arrest the intruder. Unless they are quick though they will only arrive in time to remove the battered remains of the unfortunate burglar.

If, instead of the glass, he had cut through the wood panels of the door his metal cutting tool would have encountered two separate sheets of thin metal and would thus have completed another alarm circuit. He now quietly opens the door and in so doing sounds another alarm. Before the door has opened sufficiently to admit him, it is brought to a sudden standstill by a patent door chain which he cannot reach to unloosen. Having failed to open any of the doors he next attacks the windows and is here more successful as he soon has the outer shutters open but not without again sounding the alarm and indicating his presence at the window on the electrical indicators in the bedrooms. After several attempts he at last opens the window thereby again breaking the electric circuit. There is apparently nothing now barring his entry but the window shade. He cautiously raises it but stops upon hearing the tinkling of a mechanical bell set in motion by the moving shade. All seems quiet though in the rooms above so he continues to raise the shade. He has almost got it up when the loud report of a dropping torpedo disconnected by the moving blind, startles him into flight. But he pauses and waits. He is puzzled for the alarm does not seem to have been taken up, but all remains quiet as before. If his suspicions would only keep him away now he would yet have a chance to escape. After a while he returns and again enters through the open window. He finds himself in a handsomely furnished library. He quietly tip-toes about in his felt slippers all unconscious that every step he takes on the electric matting sounds an alarm above. As he peers about by the aid of his dark-lamp his eyes light upon a rich dressing gown hanging on a hook in one corner. He removes the gown but does not notice though that the hook gradually moved upward upon being relieved of the weight of the garment and completed a circuit of another alarm and again indicated his exact position in the house. But still the deep silence is unbroken. The safe next attracts his attention and his drill is soon working through the metal. Suddenly he becomes aware of a peculiar hissing sound. This is caused by the escape of the compressed air from the air chamber caused by the escaping air which operates the contacting pieces of an electric circuit and sounds the safe alarm in the rooms above. But the peculiar action of the safe has frightened the burglar from his task. He suspects or fears something unusual in the behavior of the safe.

He next determines to try his luck in the bedrooms above. As he presses the heavy portieres aside in passing from the room into the hall the electric stairway alarm is sounded above. The slide is drawn across his lantern and in darkness he mounts the steps making not a sound. He stops to listen and finds all quiet as before. He reaches the first landing but does not see the delicate black wire stretched in his path just above the floor. This wire is connected to the deadly double barrelled swivel gun which is so mounted that any pressure on the wire will first swing the gun around to cover the point from which the pressure comes and then operate the hammer to discharge it. His foot strikes the wire and all unknown to him the delicately balanced gun is swung around noiselessly to cover his position. There is a deafening report as both barrels of the gun are discharged simultaneously and the unfortunate man falls dead without so much as a groan as he has been literally riddled by the scattered charges of balls. Simultaneously with the report of the gun every light in the house is lighted by the making of a circuit connected to the gun and the large alarm bell on top of the house set in operation to warn the neighbors.

The above mentioned alarms form a very small number of the great mass of improved burglar detectors.

The portable key fastener is a very handy little safe-guard for almost any one to carry, especially if travelling much. This consists of a piece of wire adapted to be passed over the knob shank above the key and with its two pendant ends stuck through the eye of the key to prevent its being turned by nippers or being pushed from the key hole for the insertion of another key from the outside. The alarm pocket knife is also a very unique little instrument. This is in the form of an ordinary pocket knife with the exception that it is provided with a spring hammer and a recess in one end for a cartridge. When in position against a door the opening of the same will operate the hammer and explode the cartridge, thus sounding an alarm. The door torpedo is also a very popular form of alarm from the fact that it can be instantly applied to any door. It consists of a weighted torpedo hung by a chain from the end of a thin blade. The blade is stuck into the crack above the door after the door is closed. Upon opening the door the blade slips out, thus removing the support for the torpedo and allowing it to fall to the floor and explode from the concussion.

W. HARVEY MUZZY.

Novel Telegraphy.

The celerity of telegraphic communications was well demonstrated during the recent Oxford-Cambridge boat race. The press-boat, as it steamed up the river paid out a cable during its entire course, and at every point in the race the news was immediately flashed by means of this line direct to London. Six minutes after the race was over papers could be bought in that city giving full particulars of the race and a description of the scene at the finish. The cable employed was a seven-strand conductor, insulated with vulcanized rubber, with wraps and braids of flax. From eight minutes after to thirty-seven minutes after 12 o'clock no less than thirteen complete messages were sent.

The Reason Why Patents are so often Declared by the Court not Sufficient to Protect Inventors.

So serious has the loss and disappointment to patentees been of late by reason of their claims being thrown out of the Circuit courts, and United States Supreme Court, that the whole system is being denounced by patentees and manufacturers, and by such capitalists as encourage inventors and patentees by investing money in buying their patents, or in manufacturing under them. The main reason for all this trouble is that many inventors at the start, employ cheap agents or solicitors, or persons of no legal reputation to secure their patents.

These agents very often, by reason of insufficient specifications and in accurate or insufficiently demonstration by drawings of the inventions intrusted to them for the purpose of securing patents from the U. S. Patent Office. Owing to this, applications for patents are rejected in part by the primary examiners of the Patent Office, and in order to avoid the trouble of contesting the leading claims of the cases before the Primary Examiners, the agents, solicitors, cancel these claims, and secure for their clients patents on merely the shells, rather than the kernels of their inventions. In time the inventions on which these patents are granted are found to be quite valuable, but alas, on consulting competent patent law counselors, the patentees are told, to their astonishment, that, while they were the original and first inventors of much more than is covered by their claims at the end of the specification and all the other novel things that might have been claimed and secured at the date of their patents, are now beyond their control because of their concession thereof to the public.

But cannot a "Reissue" be secured which will give these patentees what they really invented, and unjustly denied claims for at the date on which they filed their original applications? The Courts, Circuit and Supreme, have pretty strongly intimated that no patentee can come back to the Patent Office and by reissue secure what he or his agent has voluntarily surrendered at the time the case was up for examination in the Patent Office. The idea of the courts seems to be that an inventor, or his attorney dedicates to the public the subject matter embraced by the cancelled claims when he fails to contend therefor at the start, and that the patentee should not be allowed to change his mind at a subsequent period, even though the acts were committed by an agent or solicitor without his knowledge or assent. The courts feel that they stand as guardians of the public, as well as of the inventor and patentee.

While it does seem but right that the patentee should have a remedy by Reissue in many cases within 2 years after the grant of the patent it cannot be said that the government is unjust to most patentees, who employ incompetent counsel and dedicate their rights to the public with their eyes open to the fact that, if they voluntarily so dedicate and relinguish their claims when rejected [not contending for such claims, either by argument before the primary examiner, or an appeal to the board of examiners-in-chief, or the Commissioner of Patents in person, or the Appeal Court of the District and the Equity Court] they forever are stopped from again setting up these claims. The result is, if suit is commenced on such patents in the courts, it is decided by the judges that the patent only covers a very limited field and infringers go free and the patentee bears the loss of the suit with costs of the court. Many thousands of dollars are spent in litigating such patents, while if competent legal counsel had been employed to prepare the specifications and claims, only a sum of $40, $50 or $100, extra of the Government fee, would have been required to be expended in each case, and the rights of the patentee would have been protected and royalties from infringers collected.

Patentees should act wisely in the choice of those who are to prepare their specifications, drawings and claims, and not be induced by any false lights held out by cheap and incompent agents and solicitors, to entrust to them their inventions which may prove to be worth many thousands of dollars if properly secured by letters patent.

It is a well established practice when landed estate is to be purchased for prudent men to employ acknowledged competent counsel with respect to the title of such property, and this without regard to the fact that such counsel require the payment of their fees at the time of their employment, and this is the case even if such fees are much greater than ordinary lawyers charge. The fact that such acknowledged or reputable counsel were employed, gives the public confidence in the title of the party owning the landed estate. It would seem that such rules should hold good in respect to patent property which ofttimes exceeds in value landed estate.

ROBERT W. FENWICK.

A Study in Car Refrigeration.

BY JOSEPH W. BUELL.

Continued from November issue.

The route we shall consider extends from Chicago to Jersey City over the Pennsylvania Trunk Line, and we will calculate, with a small load, say, 12 Refrigerator Cars, which with the saving of space effected is equal to 14 or 15 cars as now loaded (a saving to the R. R. Co., of one breakman) and let the rate of speed be considered 20 miles an hour.

In passing over the first division, of that line going East we find that there is 65 miles of level road, 212.74 miles of descending grades averaging declivities of 17 feet, to the mile; 190 miles of ascending grades averaging 17 feet rise to the mile, and along this part of the system we find 348 curves, which for calculation we will approximate as averaging 8° curvature. The distance between Chicago and Pittsburg by this route amounts to 468 miles.

The average heavy work resistance on that section of the Road may be represented as follows:

Resistance of cars and tender due to	rolling and axle friction, 7	lb.
	grade 17 ft. × .3737	6.43 lb.
	Speed $\frac{20^2}{171}$	2.3 lb.
	curvature 8° × .5	4.0 lb.

Total resistance of cars and tender, travelling at said rate, without considering curves, on a level 9.3 pounds per ton, or 2662 pounds pull for a train of 12 cars and tender, 15.73 pounds per ton, or 3806.66 pounds per train, on the lengths of track having an ascent of 17 feet per mile, which latter is increased to 19.73 pounds per ton or 4774.66 pounds, with curve resistance added.

RESISTANCE OF ENGINE.

Weight, 66.6 tons.

Resistance of Engine due to	rolling, and axle and machinery friction,	8 × 66.6	532.8 lb.
	grade 17 × .3787 × 66.6		428.23 lb.
	Speed $\frac{20^2 \times 66.6}{171}$		155.84 lb.
	Curvature 8° × 5 × 66.6		266.4 lb.

A total resistance on a level road of 677.84 pounds, 1116.878 pounds in a rise of 17 feet to the mile and which is increased to 1373.278 pounds, when including curve resistance of the degree specified. Thus it can be seen that while on the level and descending grades the haul of 352.6 tons (Engine, 12 cars and tender) will require a pull of 3201.03 pounds, it will require the effort of 4913.53 pounds to haul that load up a grade of 17 feet to the mile, which would be increased to 6147.93 pounds if the curvature of the track on that grade amounted to 8°. Hence in the milage of 468, we have:

66 miles level pull at 3201.63 pounds requiring 117 effective H. P.

212.74 miles of descending pull &c., which will require less H. P.

190.14 miles ascending grade, pull 4913.53 lbs., requiring 272 H, P.

While the resistance on either grade would be increased by an 8° curue to 68 H. P,, more.

On the next division, that between Pittsburg and Philadelphia the grades encountered average 34 feet to the mile in ascending and the same in descending. There is:

72.91 miles level requiring 177 H. P. expended.

241.08 descending, requiring about the same expenditure.

209.72 ascending 34 feet with pull of 3495.82 pounds or requiring 4?6 H. P. Just west of Altoona Pa., the Maximum grade is reached, and for twelve miles it averages 1.5 feet rise to the hundred, or about 79 feet per mile, and in one point exceeds this a trifle. Let us call it 12 miles at 80 feet per mile, which with the load we are considering would require 782 H. P. (Effective).

On the third and last division, that between Philadelphia and Jersey City, the road is the finest in the country, presenting a level stretch most of the distance, with easy reverses, in fact it is considered the finest run in the country.

Thus it can be seen th[illegible] between Chicago and New York city the mo[illegible] power required to haul 12 refrigerator cars at a constanl rate of speed of 20 miles per hour, ranges from 177 effective H. P., to 782 H. P.

The entire haul is 1087.71 miles and the coal consumption 80 tons. the average being 148.5 pounds per mile run on freight trains on that road, which at the cost of $1.50 per ton on the tender (this exceeds the actual amount paid by that road) would amount to $121.00.

Assuming the economy of a locomotive engine in actual practice at 4.75 pounds of coal per H. P., per hour; that amount of coal consumption (80 tons) would yield 34.005 H. P.-Hours. As it would require 54.3 hours of constant travel at the rate of speed mentioned to cover the distance,-629 H. P., would represent the average available power to be found in the boiler from the consumption of coal fed, which when taken in connection with the above range of power consumption exceeds the average draught for power purposes and which will conveniently allow of a further train resistance of 31.2 H. P., most of the time without stalling or even discomoding the locomotiue for its required work, at the given rate of speed: and it can also be seen that the abstracting of 31.2 H. P., for the entire trip, the coal consumption represented would amount to but 10.00 dollars for the entire train, and which we will show latter on, will produce the same refrigerating effect, as the consumption of 4000 pounds of ice.

In the proposed change', the writer is advocating the putting of COAL AGAINST ICE, while effecting the same refrigeration at a less cost.

In order to demonstrate that the appropriation of 38.4 H. P., in train resistance will be sufficient, to drive the apparatus of a size required to perform the refrigeration, we will have to consider a few more measurements.

Now it is required in the problem we have to solve to find the dimensions of an air refrigerating machine to produce an effect equal to the melting of 41 pounds of ice per hour. The performance of a refrigerating machine may be stated in terms, of the number of thermal units, withdrawn in a unit of time. The melting of 41 pounds of ice per hour is equivalent to 5904 British Thermal Units per hour 98 British Thermal Units per minute. Let us assume that the pressure in the car is that of free air (14.7 lbs., to sq., inch) and the temperature low enough to preserve meat or say 32° Fah. Let the cut-off in the expanding cylinder be 29.4 pounds by the indicator; or 44.1 pounds absolute. Let the delivery pressure in the compressor be 39.4 pounds by the indicator, i. e., let the loss of pressure in the cooling passages be 10 pounds. Let the initial and final temperatures of the cooling medium be 60 and 80 f.

Assuming adiabtic compression and expansion:

$$T_4 = 492.7 \left(\frac{14.7}{44.1}\right)^{\frac{0.4}{1.4}} = 360\ ;\ \therefore\ t_4 = -100.7° \text{ Fah.}$$

$$T_2 = 492.6 \left(\frac{54.1}{14.7}\right)^{\frac{4}{1.4}} = 714\ ;\ \therefore\ t_2 = 254.2° \text{ Fahr.}$$

The air used per minute is therefore

$$98 \div (32 + 100.7) \times 0.2375 = 3.1 \text{ pounds.}$$

The horse-power of the compression cylinder with adiabatic compression is.

$$W_c = \frac{3.1 \times 778 \times 0.2375 \times (254.2 - 32)}{33000} = 3.86 \text{ H. P.}$$

If the compression curve may be represented by

$$pv\ 1.2 = \text{const.},$$

then the work of compression will be

$$W_c = 3.1 \times 144 \times 14.7 \times 12.4 \times \frac{1.2}{0.2} \left\{ \left(\frac{54.1}{14.7}\right)^{\frac{0.2}{1.2}} - 1 \right\} = 126050 \text{ foot lbs}$$

and the horse-power is therefore 3.8.

Therefore the horse-power of expanding cylinder is

$$\frac{W_e}{33000} = \frac{3.1 \times 778 \times 0.237 \times (32 + 100.7)}{33000} = 2.27 \text{ H. P.}$$

The net H. P., required is therefore 3.86 - 2.27 = 1.6 H. P., or 1.5 H. P., as the air does work, in expanding, which diminishes the amount of external power required to drive the compressor,-of course this would vary a little according to the manner of compression.

The volume of the compressor-piston displacement without clearance will be

$$\frac{3.1 \times 12.4}{120} = 32 \text{ cubic ft. at 60 revolutions per minute.}$$

The volume of the expanding cylinder under the same condition is

$$.32 \times \frac{360}{492.7} = .235 \text{ cubic ft.}$$

If the clearance of the compressor be assumed to be 0.02, the piston displacement should be

$$.32 \div \left\{ 1 + 0.2 \left(\frac{54.1}{14.7}\right)^{\frac{1}{1.2}} - 0.02 \right\} = .32 \text{ cubic feet.}$$

If the clearance of the expanding cylinder be assumed to be 0.05, the piston displacement should be

$$.235 \div \left\{ 1 + 0.05 \left(\frac{44.1}{14.7}\right)^{\frac{1}{1.4}} - 0.05 \right\} = .247 \text{ cu. ft.}$$

If further an allowance of ten per cent be made for imperfection, the dimensions of the cylinders may be: Compression cylinder 8 inch diam. & 12 inch stroke or equivalent. Expansion cylinder diameter 6.9 inches: stroke 12 inches.

These results computed upon the thermodynamic principles represent the theoretical efficiency of an air machine adequate to perform the required refrigeration and the power necessary to drive the same, but there is found in practice a large difference between the theoretical efficiency and the actual efficiency of an air machine. The principal cause of this large difference, is due to the fact that the temperature of the air leaving the expansion cylinder is very much higher than theory indicates for adiabatic expansion(-109°.7 while the actual temperature is-55°) This result is not due to any deficiency in the adibatic law of temperatures but is attributed to rapid absorption of heat in the air by its contact with the metallic parts of a machine, so that in computing sizes of cylinders and the required power we must include an allowance of nearly 50 per cent for losses, which includes the loss as above pointed out as well as that for friction of parts of machinery.

The cylinders, in practice, should be; compression 10 inches diameter by 15 inches stroke (or equivalent with shorter stroke) (Expansion, 8½ inches diameter by 15 inch stroke, or the equivalent,-and experiment has shown that they require 7.7 H. P., to drive the compressor and return 4.5 H. P., in the expanding cylinder. The net H, P., that would be required would be 3.2 according to the type of compressor used.

Thus it can be seen that a temperature as low as thirty-two degrees fah., can be maintained and a dry cold produced by a circulation of air which absorbs the excess of heat as well as moisture of the meats (the principal causes of their decomposition).

But to obtain the very best results in this method of refrigeration it is necessary to observe a number of details of method.

The object of all refrigeration is to obtain the greatest amount of refrigeration with the least expenditure of heat. The condition of the interchange of heat between the outside and the interior of a car, is dependent first upon the extent of surface exposed to the moving air and radiant heat of the outside, as well as the amount and quality of insulation that acts as a barrier to its transmission to the apartment within. Hence the heat to be removed should be the excess that permeates the insulating walls protecting the inner compartment from the external heat. Which we find under the present methods amounts to an ice melting effect of 4000 pounds per trip, for a compartment containing about 2000 cubic feet. But according to the plan and purposes now under consideration it is proposed to remove the heat as rapidly as it penetrates the outer insulating wall and before it reaches the interior of the car, which is an economy in work to be performed never before disclosed, and it may be briefly stated, that the practical arrangement consists of providing an air space that will surround the car, and that will be formed by leaving a space of about 4 inches in width between the insulating walls, thus forming an enveloping sheet of air that is to be rapidly withdrawn, by the air compressor expanded and chilled and returned to gather the entering units of heat, and as this is continuous, a circulation is established that removes the incoming heat that penetrates the outside shell of the car and thus prevents its ever reaching the interior of compartment, with decided advantages over present methods.

In all other methods the low temperature of a chilling room has to be handled over and over again to extract the increment of temperature that is dissipated therethrough, which is generally gradually raising the temperature of the whole,-in order to maintain its integrity,-and we can see that additional work has to be performed as the pressure has to be correspondingly raised to extract what little heat that has been added to the great bulk of cold air. Although in handling the same the engines apparently run easier and consume less steam, yet the refrigeration measured by the amount of heat removed eases off more than the work does, so that each H. P., expended produces less and less effect as the temperature falls.

It is not the purpose in this article (for legal reasons) to disclose the maner of meeting the contingency of the removal of power by the train stopping for a long time in some isolated spot where the attention would ordinarilly be required to preserve the contents of the cars but this problem has been successfully and positively solved and it can be cheaply demonstrated that, after the proper treatment of a car load of meat before starting, a side tracking of three days in the sun would not be followed by deteriortion of the meat within the cars.

It can be said, however, that it has nothing to do with the old methods and in no way applies to subjecting the meat to laborious and expensive treatment.

As a policy, the adoption of such a plan cannot be objectionable to the railway corporation, for a policy that best subserves the interest of their patrons, is unquestionably the best policy for them to adopt.

Flying Through Space.

The most interesting article ever published on Aerial Navigation was printed in a recent issue of the Washington Evening Star and was written by Frank G. Carpenter about Professor Langley's experiments. The following is a portion of the article:

With in the past few months an invention has

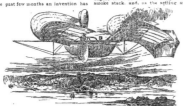

LANGLEY'S LATEST AERODROME.

been made here at Washington which promises to revolutionize the travel of the world. It may transfer the vessels of the ocean to the air and carry the locomotives among the clouds. The development of it will, in all probability, change the warfare of the world, and it may make war so terrible that the national troubles of the future will be settled by arbitration. I refer to Mr. Langley's aerodrome. The word means air-runner, and the machine is such that it runs faster upon the surface of the air than a horse can trot.

For sixteen years Mr. Langley has steadily pursued his work upon it. Engrossed as he has been, first in astronomical investigation and later in administer in the greatest of our scientific institutions, he has had only his leisure moments to devote to it, and now, after thousands of experiments and hundreds upon hundreds of failures, he has accomplished what scientists once declared to be impossible. Knowing that his work was done almost at the risk of his scientific reputation being questioned during the early years of it, he kept the object of his investigations to himself. Today the world knows practically nothing of them, and it was only last May, after persistent urging on the part of his friend, Prof. Alexander Graham Bell, that he allowed him to state the fact that he had succeeded.

Since then additional improvements have been made. A new and better machine than that which flew half a mile in May last has been tested. It has made a more successful flight, and today Mr. Langley permits me to give in my own words the first full description of his success to the public. I have spent several days with him upon the island in the Potomac river, about thirty miles below Washington, where his last experiments have been conducted, and on Saturday, November 28, I witnessed the most successful flight which has yet been made. I saw this machine, made chiefly of steel, weighing as much as a three year old boy, yet so large that it would just about fill the average parlor, moved by a steam engine which was a part of it, dart forth from the launching stage and fly in an almost straight line through the air a distance of more than 1,500 yards, or over three quarters of a mile. It flew almost as far as the length of Pennsylvania avenue between the Treasury and the Capitol. The flight was horizontal. There was not a quiver of the wings and the great bird-like aerodrome swam, as it were, upon the planes of the atmosphere. It first flew to the right, across the bay toward a strip of woods, and, as Mr. Langley and myself watched it, our hearts for the moment came into our throats, for it seemed as though it would dash itself against the trees. As it neared them, however, it gracefully swept round and downward, then turned and rose and as straight as an arrow flew across the bay where we were standing on toward Washington. It continued to fly in this straight horizontal line until the water which furnished the steam was exhausted, when it slowly but gracefully swept down and rested upon the water. It lighted so gently that not a bit of its machinery was injured, and had it not been that the evening shades

silken wings, and the white, silvery substance which bound the body containing its machinery, it seemed like a wonderful new species of bird. The great danger of losing the machine in the trees led Mr. Langley to put only enough water in it to allow it to fly about one and one-half minutes. It could have carried water for about five minutes, but as it was, it flew by two independent stop watches, one

PROFESSOR LANGLEY.

minute and forty-five seconds, being the only flight of any aerial-machine except itself which has ever lasted for more than a very few seconds. In this minute and three-quarters it flew a distance of almost a mile, going at the rate of over thirty miles an hour, and showing that if it had been fully supplied with water, it would have flown for two miles. As it was, its flight was only limited by the exhaustion of its steam, and there seemed no reason but that with more steam to run it, it might not have gone on indefinitely. With a machine ten times its weight, Mr. Langley told me, a condensing apparatus could be carried upon it which could use the water over and over again, and the same amount of water would carry it for hundreds of times its present flight. The machine flew against the wind. There was nothing of the balloon nature about it. There was no gas bags to uphold it. Its wings were immovable, and they merely steadied

were falling it could have been flown again. I have never seen any inanimate thing look so like a thing of life.

It was as graceful as any bird, and as it swam through the air, its propellers, which were going about at the rate of over a thousand revolutions a minute, made a whirring noise like the wings of a bird in rapid flight. The feathery smoke of the engine could be seen wreathing its way out of the smoke stack, and, as the setting sun caught its

it as it flew like a bird through the air. The force which carried it onward was generated upon it.

As I looked at it I could hardly realize the remarkable thing which Mr. Langley has accomplished. Let me repeat it.

The aerodrome is a machine made almost altogether of steel. A balloon floats because it is lighter than the air. This machine weighs more than one thousand times as much as the air through which it moves. The working parts of its machinery are of steel, and it carries a peculiar steam engine which forces it along through the air. In constructing this machine, the question of weight was an all important one, and everything had to be reduced to the minimum. The aerodrome weighing less than thirty pounds, carries about four pounds of water. This is about two quarts, and the little engine is so wasteful of it that its flight must be proportionately short, for when the water has been once converted into steam, the aerodrome must stop flying, as there is no more water to furnish steam to run it. The machinery of the air-runner is very light, indeed, but it requires a considerable force to move it in proportion to its weight. Its engine is equal to more than one horse power, and the movable parts of the machinery weigh twenty-six ounces. You could put all of its machinery into a peck measure. Now, a horse weighs a thousand pounds. Think of reducing the size of a horse to a peck measure, and its weight to that of a kitten, and you have some idea of Mr. Langley's aerial engine.

What does the aerodrome look like?

I have described it in flight. I examined it at rest and I have gone carefully over its different parts. It is about fifteen feet long and about fourteen feet wide from the tip of one wing to the other. The machine moves through the air on much the same principle as that by which the twin-screw steamer forces its way through the water. On each side of the aerodrome there is a sort of screw propeller or pair of blades in the shape of one cutting of a screw so hung upon a pivot that when the steam is on they fly around at the rate of a thousand revolutions a minute. They look, in fact, much like the wheels of an electric fan when in action. They cut the air so rapidly that you cannot see the blades, and they are in fact, a pair of wheels about four feet in diameter flying at this wonderful speed around through the air. As they move they screw the air ship onward, and this advancing motion keeps it up in somewhat the same way that a swift skater can be supported by thin ice.

The machinery is in a metal receptacle which ends in a smokestack. This is hung to a frame work of steel. The wings, which are stationary, are fastened to the upper part of the frame work, and they extend out above the body holding the machinery.

The machinery is wonderfully delicate, but it is as strong and at the same time as light as scientific investigation can make it. The fuel is gasoline, which is converted into gas before it is used, and which furnishes such an intense heat that it would melt the boiler in a second if there were not a special pump by which the water is kept flowing rapidly through the boiler, the intense heat converting some of the water into steam as it flows. Every part of the machinery is of the most practical nature and it has been constructed at an enormous expense of patience and experiment. It may be said that nearly every atom of the aerodrome as it is now put together is the result of experiment. The making of the boiler alone consumed months

ONE OF LANGLEY'S RUBBER MOTOR FLYING MACHINES.

of work. Every bit of the machinery had to be constructed with scientific accuracy. It had to be tested again and again. The difficulty of getting the machine light enough was such that every part of it had to be remade many times. It would be in full working order when something would give way, and this part would have to be strengthened. This caused additional weight and necessitated the cutting off of that much weight from some other part of the machinery. At times, the difficulty seemed almost heartbreaking, but Mr. Langley went on piece by piece and atom by atom, until he at last succeeded in getting all the parts of the right strength and proportions. Even after he had completed his model and had it ready for flight, he was confronted with an unexpected difficulty, which was, it seemed at the time, almost impossible to surmount

This was the launching of the machine into the air. One of the most difficult things that large, birds, have to contend with in flying is in getting a start. You know how difficult it is to launch a ship into the water. It is far more difficult to launch an air ship. Mr. Langley found that his machine had to be clamped down on the launching stage and to be arranged in such a way that the machinery could be started, so that it should receive a slight initial velocity and then be released with a spring. This looks easy. It was hard. But Mr. Langley at last succeeded in launching his machine by hanging it to a movable table, so that it could be turned to face the direction in which the flight was to be made, and so that the wheels of the table would carry the aerodrome straight out in a horizontal line and launch it off into the air. The launching apparatus which we used on November 28 was built on the top of a house boat, and the work of arranging the table was no small one. As I stood upon it and examined its construction Mr. Langley said:

"It don't seem to be much, but it is the result of five years of experiments."

I here asked Mr. Langley what first attracted his attention to aerial navigation.

"I can't tell when I was not interested in it," he replied. "I used to watch the birds flying when I was a boy and to wonder what kept them up. I afterward heard the theory that they possessed great muscular power. You know some scientific men have stated their belief that the muscular strength of birds must be enormously greater in proportion than that of men. But this it seemed to me could not be true. I could not believe what some French mathematicians calculated, namely, that an eagle must be nearly as strong as man. It finally occurred to me that there must be something in the condition of the air which the soaring birds instinctively understood, but which we do not. This idea I held for a long time, the flight of birds continuing to be a wonder to me. It is curious how an idea of that kind sticks to you. I seldom saw a bird flying that I did not think of it, and even lately I have watched them for hours, trying to understand how they could move about through the air, rising and falling, soaring up and sailing down without motion of the wings."

"But Mr. Langley I thought that birds used a great deal of strength to fly. They can't fly without moving their wings, can they?"

"The soaring birds can," replied Mr. Langley, and they do fly long distances with apparently very little exertion. Darwin once watched the South American condors, which, you know are immense birds, for hours. He says they ascended and descended, soared and circled about, with scarcely the movement of a feather. He could not detect a single flap of their wings."

"I remember," continued Mr. Langley, "how I stood one cold November day on the Aqueduct bridge that crosses the Potomac river above Georgetown and watched a turkey buzzard which was lazily soaring round and round watching something in the river below. The wind was blowing a gale. It was going at the rate of at least thirty-five miles an hour, still the bird moved about with the greatest ease, keeping generally on one level, but swaying a little as it went round and round. It was not more than sixty feet above me. I could see it perfectly and could not note the flapping of a wing, though I watched it for a long time. I stayed, in fact, until I got so cold that I had to leave."

A Patent Flycatcher.

A machine for catching flies off the backs of cattle, and so affording the animals relief and comfort, has been invented by a farmer in Madison county, Ky. The fly catcher is a kind of covered pen or passageway through which the animal must walk to secure relief. A few feet from the entrance there is a cupola or dome in the roof of the passageway, made of glass and arranged as a fly trap. Beyond this the passage is darkness. The animal walks through the machine, and just as it passes under the dome and enters the darker part a set of brushes sweeps off the flies, which naturally rise into the lighted dome, and the steer passes out the other side free of flies. The flies are retained in the dome trap. The inventor has experimented with his machine and finds that the animals soon learn the value of the machine and know enough to walk through it when the flies begin to bite. The device has been patented.

Inventors and Inventions.

One of the most deeply-rooted of popular fallacies is to the effect that a quick way to fortune, or at least, to competence, consists of inventing something useful; and the delusion is kept up by certain interested persons, who, at regular intervals, and by advertisement, invite the public to "sit down and think of something." It is, of course, understood that as soon as something is thought of, the thinker must go at once to the advertiser, and engage him to obtain a patent for the idea. If the idea is turbid, the advertiser will, for a consideration, assist in precipitating or developing it, in giving it a lift from the blooming fields of imagination towards the very stormy sea of commercial experience. The fact is, that the average man can no more invent a new and useful thing than he can write a new popular song. The two things are fairly comparable. Numbers of successful engineers never trouble themselves with attempting new inventions, Many of them have a positive aversion to the uncertainty that surrounds all new departures in physical science, and they insist everything they adopt shall bear the stamp of practical experience and ample trial. A decided preference for novelties, because they are new, is a very dangerous taste, and is only excusable in young men with limited responsibilities. It is only safe when controlled by an old or thoroughly experienced head with a strong critical talent

Any one who would revel in new inventions has only to buy a few copies of the journal published by the Patent Office in London. One of the recent weekly issues of this publication contains details of patents which number in all four hundred and fifty-eight, nearly all of which are illustrated. In order to thoroughly appreciate this list, a very extensive experience in inventions would be necessary; there is, however, ample subject of interest to the intelligent reader with a moderate knowledge of mechanics, who will probably come upon more than one old acquaintance under guise of a new invention. The volume before us has the usual variety of "notions," many of which are palpably useless, for the Government although it requires that, in order to be entitled to the privilege of a patent, the invention shall be new and useful, does not guarantee the novelty or utility of anything that is offered. There is a boiler for heating water for domestic purposes by means of a petroleum lamp. A hollow dish full of water is suspended below the boiler and over the lamp. No provision is made for removing calcareous deposit from this dish, which would in time be filled solid, and would spoil the boiler. A washing "dolly" is patented in which the twisting movement of the "dolly" is effected by an up-and-down pumping motion, which would at least double the labor of washing by increased friction without corresponding advantage. An ice store box is another example of useless invention, which, after enveloping the box in several non-conductors, exhausts the air in it for the better preservation of the ice. The trouble of making a durable air-tight joint alone would destroy the utility of the invention. The use of sheets of colored glass to be cemented to walls, instead of tiles, is an old idea that has been known for many years. An improvement in furnaces provides an elaborate washing apparatus for arresting soot. It, of course, destroys the draught of the furnace, and, further, demands a supply of steam to effect its object. The same result could be accomplished by a similar sacrifice of steam to render combustion more complete. This would dispense with the costly washing apparatus. A horse collar filled with compressed air is another old friend, which will continue from time to time to appear a novelty. It is an excellent cushion as long as it remains inflated, but if the air gets out, the horse is wounded before the accident is discovered. The pressure on a collar may at times amount to four hundred pounds, and sudden strains of this sort always wreck the bag some time or other.

For the jointing of lead or other soft metal pipes a screw is provided similar, but not quite so good or cheap, as one that was in use in 1870, on tin-lined lead pipes in England. A much more ancient idea is that of mounting a water wheel between two moored barges, anchored in a river or tidal estuary in order to obtain power to generate electricity or for other purposes. There are other inventions of undoubted merit, such as an apparatus for sizing balls or spherical objects. This operation used to be done by means of sieves, but when the spheres only varied by very small amounts, the sieves were rapidly blocked. The new apparatus consists of an inclined trough of convenient size and of a V-section; the two sides are made adjustable, so that a tapered slit may appear at the bottom, which will suit the largest and the smallest spheres of the lot to be sorted. Below the trough, at close intervals, are small funnels with pipes leading to different boxes. The use of the apparatus may be thus explained. Suppose a pound of mixed lead shot has to be sized, it is poured slowly into the upper end of the inclined trough, and, as the pellets roll down it, they drop through as soon as they reach a part of the slit that is wide enough, and each size falls into the box provided for it. Balls for bicycle bearings are thus sorted, and shot required for a percentage balance have been satisfactorily sorted by the same means.

Other inventions of interest are to be found for imbedding woven wire in glass, for the tuning of wind instruments, for silencing the noise of the exhaust from gas and oil engines, for weighing machines, lamps, sprayers for applying paint, preserving wood, pressure filters, knitting machines wood working machines, and the making of packing cases. Most of the inventions bear evidence of a certain amount of thought and experiment, and among them a lecturer on mechanics would find numerous examples of good and bad design. All of them represent work done without pay, for a prospective reward, and of this reward Mr. Carpenal's summary will go into a very good sentence. Not more than 5 per cent. of the inventions that are patented bring profit to the inventor. This does not mean that only 5 per cent. had any merit, but that only that number of inventors were fortunate in getting their inventions into good commercial hands capable of making money with them.—*Indian Textile Journal.*

A Great French Lighthouse.

The penetrating power and range of powerful lights, such as are employed in lighthouse service, rapidly decrease as the ratio of their luminous powers increases. For instance, a light of 5,000,000 candle-power in the British Channel has in average weather a luminous range of about 44 nautical miles while if the light be increased to the power of 10,000,000 candles, the luminous range is only five miles more, or 49 miles. According to current practice, lights up to 200,000 candle-power are obtained by means of mineral oil lamps, while electric lights are used for higher power, and almost any power may thus be obtained. The highest power yet attempted is about 36,000,000 at Penmark Point, in the department of Finisterre, France, which, when completed, will be the most powerful lighthouse illumination in the world. The height of the tower in which it is to be located is about 63 metres, enabling it to be seen during the day from a distance of 18 miles in fine weather. During the night this light will be visible for 60 miles. The rotundity of the earth will prevent the rays from striking the eye directly at a distance of more than 30 miles, but the sky overhead will appear illuminated for 30 miles more. The estimated cost of this lighthouse is about $120,000.

New Process of Gas Making.

The new and ingenious process of gas making invented by Prof. Riche, of the technical school at Lisors, consists simply in the distillation of the combustible, whether liquid or solid, and the forcing of the gases and vapors produced by this distillation through a layer of glowing coal. The plant is described as having one or more vertical retorts closed by a cover provided with a hydraulic joint and opened at the lower end, which dips into a water reservoir, and a pipe connects the lower part of the retort with the cooler. After the retort has been filled to three-quarters of its size with glowing charcoal, which in continuous working is the residue of the preceding charge, and heated to red heat either by means of a lateral firing or gas burners, or, finally, by waste heat, the cover is removed, the retort filled up with wood, and closed. The distillation commences immediately, and its products—as aqueous vapor, pyrolignic, methylic and carbonic acid—not being able to escape in an upward direction are forced to traverse the glowing charcoal, and are consequently transformed into carbonic oxide, hydrogen, methane, etc., which pass through the cooler into the holder and into charcoal, which takes the place of the charcoal contained in the retort at the beginning of the operation, the latter having been consumed by steam and carbonic acid for their reduction. In order to avoid the formation of an excessive quantity of charcoal, water is allowed to drop down form the top of the retort; the steam thus produced transforms the surplus carbon into carbonic oxide and hydrogen, which combine with the gases already formed.

SUN HEAT POWER.

When Properly Stored Will be a Mighty Force.

Thomas A. Edison said recently that some of the scientists have been speculating as to what the world will do for power when its supply of coal has been burned up. At the same time the commercial inventors of this age are working hard to find out practical solutions of this problem, and there is not the slightest doubt in my mind that more than one method for getting a great deal more power than we shall ever need will be practically and successfully worked out long before there will be actual necessity for decreasing coal consumption.

"So long as the sun-shines man will be able to develop power in abundance, no matter whether there are coal mines or not, and when its rays have been quenched power will no longer be needed. It is from the sun, in fact, that we get all our power now, for coal is only stored sunshine, and the rays that fall upon the earth's surface to day are as full of power as ever they were.

"John Ericsson showed by his experiments with the sun engine, on which he was working when he died, that each square yard of the earth's surface receives sufficient heat from the sun, when its rays fall perpendicularly through a dry atmosphere, to make enough steam to run an engine of one horse power. The only obstacle to the direct use of the sun's concentrated rays for the purpose of making power in this latitude lies in the fact that, excepting in midsummer, the sunshine does not fall vertically but at an angle. There are many days when the sun does not shine at all, and many more on which its rays have to pass through an atmosphere filled with an invisible vapor that absorbs a great portion of their heat before they ever reach the earth.

"On the desert sands of Africa, and in some parts of our own Southwest, where the sun shines uninterruptedly through dry air, the use of its rays in this manner would be more practical. Of course, the amount of power that is theoretically available in this way is almost beyond computation—enough could be got from the desert parts of the earth alone to operate the machinery of many worlds like this.

ALMOST PAST BELIEF.

"No one who has not familiarized himself with the experiments of the scientists along these lines can apprehend the possibilities of the scheme for getting power direct from the sunshine. As a matter of theoretical and demonstrated fact, it is possible to get enough power from the sun's rays that fall on a territory no larger than Manhattan Island to operate all the machinery of the world."

If, as Ericsson demonstrated, heat enough to make a horse power may be developed from each square yard, from one square mile enough may be collected to make as many horse powers as there are square yards in a mile, or 3,089,600. At the outside, the city of New York does not use more than 400,000 horse power, and this amount would operate all its cable, electric, horse and elevated cars, haul all its trucks and carts and wagons, run all its steam engines, all the tugs and ferry boats in its harbor and all its other machinery of every kind.

Chicago probably uses a little less, Paris about the same, and London about twice as much, or, say, 2,000,000 altogether. These four great cities, then, use more than a million lens horse power, altogether, than one square mile would produce.

Now, Manhattan Island contains twenty-two square miles, and such an area, according to the calculations just made, absorbs enough heat to make 69,971,200 horse power, whenever the sunshine falls directly and unimpededly upon it. Twenty-two square miles of territory on Sahara, or even in Arizona, could furnish all the power humanity has use for at this stage of its industrial development.

Although the greatest possibilities in the way of changing sunshine into power exists in the dry and sunny South, there are great possibilities, even in the latitude of New York and Chicago. Suppose, for the sake of argument, that there is an average of clear direct sunshine of only one hour a day the year through. That would be 300 hours, or thirty working days (holidays not counted) and a square mile of territory would therefore produce the equivalent of more than three millions of horse power for one month.

The power, however would be furnished intermittently. There would be days and days when the entire three millions odd horse power would be developed, and other days, and sometimes weeks, when hardly any power at all would be available.

The Belfast (Me.) letter carriers claim they are first in the country to use electric lights in delivering their mail. They use a small lantern fastened to their breasts, which is lighted by a small battery carried in their pockets. The lights last two hours.

The Efficiency of Compressed Air.

For some time past the engineering world has been put on the qui vive by the advocates of compressed air, who have with assiduity sought to prove that this agent is more economical as a power transmitting medium than is electricity, not only for the operation of stationary motors of all kinds, but for railway cars as well. We have no fault to find with the honest convictions of any one, but unfortunately for our compressed air friends, the works of daily practice are the best evidence that it is a delusion which has taken hold of them. By this we do not mean to imply that there are no uses for compressed air; quite the contrary. We believe and are convinced that for many purposes, too numerous to mention here, compressed air, as such, but not as a substitute for electricity, has a wide field open to it. Nor unmindful of the fact that under certain conditions compressed air may in certain cases, involving ventilation and fresh air, supply, make it preferable to electricity.

The successful compressed air transmission as compared with electric transmission has been the subject of numerous essays from the highest authorities in engineering, and the question is now to be found discussed in all text books on the subject, so that it is not necessary for us to go into detail. But an actual example is worth a volume of theorizing, and we are fortunate in being able to present in this issue an authentic account of a comparative test between the two forces under discussion. This test, carried out by Mr. Lewis Searing, a well-known Western mechanical engineer, showed that the total efficiency of a compressed air mine pumping plant was only 9 per cent; that is, it took 312 horse-power at the steam cylinders of the compressors to do the work equivalent to 28 horse-power as measured by the amount of water pumped against a given pressure. The electric plant installed accomplished similar work at an efficiency of 44 per cent., not-withstanding the fact that both the dynamo and motor were underloaded.

So far as the efficiency of the compressed air plant is concerned, we must confess that we are not surprised at the result, when one considers the length of transmission. What with the loss in compression, leaks at joints and loss of pressure due to friction in the pipes, together with the low efficiency of the compressed air pumps, the wonder is that the showing was as good as it turned out to be. It may be objected that it is hardly fair to single out an exceptional case like this and hold it up as a basis for the general condemnation of an entire system. Such is not our intent. What we desire to emphasize, however, is our belief that few compressed air plants can to-day stand a rigorous comparative test with electricity without showing such a decided difference in favor of the latter that further operation by air would be considered disadvantageous.

The test of Mr. Searing also throws some light on what may be expected from compressed air as a motive power for street cars. True the conditions are not absolutely identical, but the only essential difference is the absence in the case of street cars of the connecting pipe between the compressor and the air motor on the car. We leave it to our readers to make their own conjectures as to the allowance to be made for this difference and to calculate how much would have to be added to the 9 per cent. efficiency of the pumping plant to compensate for the removal of theconnecting pipe.—*Electric Engineer.*

The Dangers of Acetylene.

There may be a great future for acetylene as an illuminant, but the time when it can be used with tolerable safety has not yet arrived. In France several companies have been formed to exploit lamps and other appliances whereby the gas may be manufactured and used at home, and it has been supposed that as there is every probability of a great reduction in the cost of calcium carbide, a much cheaper and better illuminant than ordinary gas or electricity would be obtained. Scientists have approved of the use of acetylene if certain precautions were adopted, but in spite of all rules as to its use, public confidence was severely shaken a few weeks ago by two serious explosions, one at Lyons, and the other at Paris. In both instances two persons were killed, and several others injured. Whether acetylene is safer in the liquid or in the gaseous form is still a question among experts. It will not liquefy with a less pressure than 12 atmospheres, which, however, is only about a sixth of the pressure required to liquefy carbonic acid, and the latter has been used on the railways with remarkable immunity from accident. Many conflicting theories as to the nature and properties of acetylene are still exercising scientific minds, and it seems clear enough that the explosive character of acetylene in any form is not yet sufficiently understood to warrant its general use as an illuminant.

How the Ancients Used Rubber.

That the rubber industry might, in a moderate sense, be traced several centuries back, like any other modern industry, is not surprising. A contributor to the "Gummi Zeitung," R. Schuck, proves, from an old Spanish book, "La Monarchia Indiana," printed in 1723 at Madrid, that rubber water-proofs were worn in South America 200 years ago. The natives called the rubber-yielding tree, of moderate height, with large, ash-colored leaves ulquahuill. The milky sap was allowed to coagulate, thrown into boiling water, and kneaded into balls, called ulli, which were used in their games and otherwise utilized. When the Indians had no calabasher or other vessels in which to collect the rubber they made the sap run over their naked body, and stripped the water-proof cloak or tights off that would be the prototype of waterproofs. They also made armour of rubber, as no arrow could penetrate through it, thus anticipating the bullet-proof shields of a few years ago. The nobles sported sandals and shoes of rubber, and the court dwarfs and jesters had to tumble about on huge rubber soles for the common amusement. The Spaniards coated their hemp garments with ulli, but noticed that these water-proof materials could not bear the heat of the sun.

COLUMBIA CALENDAR FOR 1897.

A Very Useful Memorandum Block with Leaf for Each Day, Filled with Bright Thoughts and Graceful Pictures.

The twelfth annual issue of the Columbia Pad Calendar has made its appearance in more pleasing form than ever before, having scattered through its daily leaves many charming illustrations, with an appropriate thought or verse for each day in the year. Among the topics are bicyling, outdoor life,

and good roads. The cycling fraternity, to say nothing of the general public, has acquired a decidedly friendly feeling for the Columbia Calendar, and its annual advent is always looked forward to with interest and pleasure.

One feature of the calendar is its neat stand, so arranged that the block can either be used upon the the desk or hung upon the wall.

The calendar can be obtained for five two cent stamps by addressing the Calendar Department of the Pope Manufacturing Company at Hartford, Conn.

B. & O. Fostering Local Industries.

General Traffic Manager Randolph, of the Baltimore and Ohio Southwestern, is paying considerable attention just now to the development of the local industries along his line. For a number of yeurs the Baltimore and Ohio Southwestern has fostered its local industries and endeavored to place them on a paying basis, by promptly furnishing cars, &c., for their convenience, but still further efforts are to be made. Coal Traffic Manager W. W. Peabody, Jr., has been placed in charge of this department, in addition to his own, and it is expected that the Baltimore and Ohio Southwestern will add greatly to its revenue by securing new manufacturing industries to locate along its line of railroad in Ohio, Indiana, and Illinois. This road has great advantage in being able to furnish very cheap fuel.

A New Flooring Material.

The name of papyrolith is given to a novelty in flooring material which has lately been invented by Otto Kraner, Chemnitz, the article being a special preparartion of paper pulp which is in the form of a dry powder; when mixed with water it may be spread like mortar over stone, cement, or wood, where it dries quickly and may be smoothly planed, besides which it may be tinted almost any color, in this way adapting it for parquetery with variegated borders, or for panels and mosaics. Among the various advantages claimed for the inventor for the use of this product are freedom from crevices, deadening of noises, and poor conduction of heat, also considerable elasticity, safety from fire and remarkable durability. It may be employed, too, for wainscoting and other architectural purposes, as well as for flooring.

The New East River Bridge.

The bridge to span the East River and unite the cities of New York and Brooklyn will be constructed by a commission acting in accordance with authority granted in an act passed by the Legislature in May, 1895. The commission is composed of the Mayors of the two cities, together with six other persons, three appointed by each Mayor. A suspension bridge is to be built from a point near the foot of Broadway, Brooklyn, to a point near the foot of Grand street, New York. The cost of constructing the bridges is to be borne in equal shares by the two cities, the money to be raised by bonds issued as may be determined by the commission. A portion of section 5 reads as follows:

"But nothing in this act contained shall prevent said commissioners, in their discretion, from con-

35 inches deep by 30 inches wide, built up of plates and angles. The web system will consist of angles 1½ by 4 by 6 inches. The trusses will be placed 72 feet apart. The floor system will be composed of plate steel beams placed at each panel point, 20 feet apart. The plate stringers will be 2 feet deep and will run beneath the rail and roadways. The floor beams will be extended beyond each truss, and on the extended portions the roadways will be supported. There will be two elevated railroad tracks, four tracks for electric or other cars, two roadways 18 feet wide and two foot walks 12 feet wide. The width of the bridge will be 118 feet; the minimum height in the middle of the span for 400 feet will be 135 feet; hight at the pier head lines 117 feet in the clear; grade of approaches 3 per cent; grade of elevated tracks, 2 per cent; total length of bridge, 7200 feet. Wind pressure will be provided against by drawing each pair of cables together at the center, and also by a double system of lateral bracing on the top and bottom chords of the trusses, the latter

tracting with any corporation to operate a railroad across said bridge if said commissioners shall determine it to be in the public interest, and any such corporation is hereby authorized to sell the same at such a price as its directors may, by a vote of a majority of them, assent to, and upon such purchase such cities shall be vested with all the rights, powers and privileges of said corporation."

Upon completion, the care, management and control of the bridge "shall be vested in the said commissioners and their successors, who shall possess in relation thereto like powers as are at the time of the passage of this act vested in the trustees of the New York and Brooklyn Bridge."

The New York tower will be located at the foot of Delancey street, the anchorage two blocks in shore, or between Mangin and Tompkins streets, and the terminus in a plaza formed by two blocks bounded by Norfolk, Delancey, Clinton and Broome streets. The Brooklyn tower will be at the foot of South Sixth street, the anchorage between Kent and Wythe avenues, and the terminus at Broadway and Havemeyer street. This location places the bridge a mile and a half north of the present one.

The foundations of both towers will be carried to rock, which is found at about 65 feet on the New York side and about 85 feet on the other. Each foundation consists of two masonry piers rising 23 feet above high water and supporting the columns forming the two legs of the tower. Each column will be of the box type, about 4 feet square. The two halves of each pier will be placed 97 feet between centers, and the columns will rise parallel to the level of the bridge floor, and from thence will batter toward each other. The towers will be 335 feet feet in height above high water. The towers will be placed 1600 feet apart. The bridge proper will be supported by four steel cables 18 inches in diameter, and each composed, as at present contemplated, of 6800 wires, each 3-16 inch in diameter. The anchorages will be placed 570 feet back from the towers and will be of masonry about 150 feet square by 100 in hight. Their combined weight will be over 160,000 tons, and they will resist the pull of the main span, which will weigh 12,500 tons.

By referring to the engravings it will be noticed that the cables only carry the center or main span, the approaches being formed of independent dock bridges extending from the towers to the anchorages, and provided with an intermediate steel pier. The trusses suspended from the cables will be riveted structures, 45 feet in depth and having cords

being riveted to the floor beams and to the chords. Proposals were received by the commissioners on October 7, 1896, for furnishing the materials for and constructing the foundations of the New York tower.

Each tower foundation will consist of two stone masonry piers, each pier resting on a caisson sunk by the plenum-pneumatic process. The contractor must assume all responsibility for the difficulties

Invention and Industrial Progress.

The progress of industrial development, both here and elsewhere, has in many is not in most cases been pioneered by the genius of invention. It is only by a comparison of data and conditions that

we can realize even in an approximate sense, the march of industry. The changes made have been rapid and wide, and extended over the whole area of labor. Nothing seems to have been left untouched by the magic finger of inventive genius. In the field and the factory, from the plough to the loom; in the methods of transportation by sea and land, from the old coach to the modern palace car, and from the galley of the Roman to the greyhounds of the sea, in everything and everywhere, the wizard of genius has been busy at the greatest transformation scene in history. In looking on results as they loom up in their several groups or masses, we are apt to overlook the agencies that were both volitio-

nal and creative in bringing these things to pass. And yet not a roll of the wave onward, or a step of the foot forward, but had its inception and momentum in the brain of man before it became the work of his hands. We grow cotton and clip wool, and spin and weave and a million looms are moving to the hum of the spindle, but how few there are who know anything of Arkwright. We gridiron the planet with rails of iron and steel, and the stately locomotive crosses the plains and climbs the mountains. We send our treasurers from sea to sea on wheels that never tire, and can compass the globe with less trouble and less time than an old crusader could make his way from Castile to Judea. And yet how seldom does the average traveler think of Stephenson or remember Watt. The same may be said of everything of a like character: We appropriate their services, and boast of their use, but the thread connecting the machine with its inventor is to often missing. We appreciate the product, but forget the man: It is not likely that this will always be the case. History always brings what is best and most valuable into profile and prominence, and it may yet be that such benefactors of the human race as Edison and Fulton, will have a higher niche in public honor than the Cæsars and Napoleons once crowned with posthumous fame.

Be this as it may, the pickets of industrial progress are not warriors, but scientists and inventors. Many of these have perhaps failed in their efforts and have wasted time, energy and money, on what was practically of no value. The residue, however, has been of inestimable value to all mankind, and as a constituent in its development has become of its most vigorous and vital factors. The following figures give a forcible presentation of the activity shown in the field of invention in this and other countries from the earliest records to the close of 1895. The total number of patents issued is placed at 1,544,419. The United States claim 562,458 of the total, the rest 981,961 were divided among foreign countries.—*Age of Steel.*

Two New Coast Defense Patents.

Two new inventions intended to facilitate coast defense are announced.

One of the inventors is C. D. Haskins, of Boston a nephew of Clark C. Haskins, inspector in the Chicago electrical department in the City Hall. The invention of Mr. Haskins of Boston is an automobile torpedo, or a torpedo fitted with a device connected with its steering apparatus which will cause it when in motion to be diverted directly toward the object which has had a magnetic effect on the device.

The other invention is the creation of Francis B. Badt, secretary of the Siemens & Halske Company of Chicago. It is an electro-magnetic sentinel, to be used in connection with submarine mines placed for coast defence, or in any other place where it is desired to have an automatic contrivance to announce the approach of a mass of magnetic material such as is found in every armored vessel.

Both of these inventions depend upon the magnetic attraction of the iron hull of the approaching vessel for their proper action, and the inventors claim that tests made show their devices to be operative. If their expectations are not disappointed in more extended practical tests, these devices should prove valuable additions to present methods of defensive warfare.—*Electricity.*

THE Indian Textile Journal, published at Bombay, India, is one of the most representative of the textile and engineering industries of the far East.

It is now in the seventh year. The last issue illustrates a couple of motor cotton mills established at Shanghai, China, which are indications of the extraordinary activity in the industrial development of the Celestial empire.

The Harrison Telephone Company's Affairs.

A remarkable story comes from Chicago concerning the business methods of the Harrison International Telephone Company, a receiver for which was applied on Oct. 5, last. Mr. Edward M. Harrison was one of the defendants in the suit for a receiver, and in a long answer filed in the Federal Court in Chicago on Nov. 20 he makes some startling disclosures. It is stated that he practically admits all of the charges of fraud and wrecking and alleges that he was also a victim of President L. E. Ingalls and his associates on the Board of Directors.

Mr. Harrison alleges that the company is utterly insolvent, and that its assets have been squandered by officers and directors, who voted large salaries to themselves and money for mythical expenses. He joins in the petition to have the business wound up, and that the men who secured stock fraudulently be ordered to pay for it.

He says all the assets have been absorbed in the interests of the officers and stockholders defrauded. He asserts that the contract of May 4 last was not adopted at the annual meeting, transferring the telephone property to the International Construction Company, and he asked that it be set aside.

Mr. Harrison says that the men who were made directors were to receive 2500 shares at 20 cents a share for the use of their names, as part of a scheme to entice the public to invest in the $80,000,000 of capital stock, but as soon as they were elected directors they donated to themselves $2,000,000 each of the $50,000,000 treasury stock and voted stock to Mr. Ingalls without consideration. He says he was induced by the false representations of C. M. Ferree and A. L. Stone, the promoters of the corporation, to transfer to them a two-thirds interest in his patents and claims; that Ingalls was aware that the patents were worth only $300,000 when the company was incorporated with $80,000,000 capital on the pretence that a large sum was needed to fight the Bell Company, but really to delude the public into believing that vast sums of money had been invested by the directors.

Mr. Harrison charges Mr. Ingalls with falsifying the minutes of the May annual meeting, which he controlled by holding a big majority of shares.—*Electrical World.*

Book Reviews.

The Mastery of Books. By Henry Lyman Koopman, A. M., Librarian of Brown University. Cloth, 12mo, 214 pages. Price, 90 cents. American Book Company, New York, Cincinnati and Chicago.

The schoolboy needs such a book as this, and every student needs it. Few read much before the age of twelve, few read widely after the age of twenty. Within eight years one must select some few hundreds from the thousands of good books that await reading. The schoolboy needs at once restraint and encouragement, guidance and freedom of choice.

Mr. Koopman's argument for wide reading is so strong that it makes the mature reader regret opportunities missed. To the boy, its incitement will come in time. In the discussion of how to read, the author urges very cogently the necessity of close attention and conscientious diligence.

Mr. Koopman does not tell dogmatically just what to read. He tells simply how to select, laying down principles of guidance. He shows the right way, yet leaves freedom for individual choice. His discussion of newspapers and fiction is especially practical and tolerant.

The chapters on "Reference Books and Catalogues" and on "Memory and Note Taking" will appeal to every practical teacher.

The style is delightful. Mr. Koopman is an easy writer. Regarded merely as literary essays, many of these chapters take high place.

There is an increasing tendency among the Educational institutions of this country to give the German language a better opportunity to be heard. This is proper and right. The German nation is a mighty force in the world today and the literature of that great people is a mine of intellectual wealth. For this reason our educators must naturally be interested in all new German texts. One has recently been published by the Great American Book Company, New York City, entitled "First Year in German," and was written by J. Keller, Professor of German Language and Literature in the Normal College of New York.

The many high merits possessed by this work will, we feel assured, convince teachers that it is the realization of the long desired First German Book. Most beginning methods fail in one of two points: they either concentrate their attention almost wholly upon the difficult German inflectional and syntactical system, and lack in simplicity and in adaptability to young minds; or they sacrifice to simplicity a real mastery of grammatical knowledge. Some of our best text-books belong to the former class; they are far too advanced for high school use, and do much to convey the impression that German is a bugbear to be shunned on account of its inherent difficulty. The present book avoids both these errors; it is eminently simple, systematic, teachable. Everything essential for first year study is presented in a logical order—that is, the grammatical facts first needed for reading and most easily comprehended are those first presented.

A NEW TELEPHONE RIVAL.

Postal Telegraph Company Ready to Enter the Field.

A New York morning paper says—The withdrawal of telephone instruments from several Western Union Telegraph offices in Brooklyn gave rise to the rumor that the long-threatened fight between the telegraph and telephone companies had begun in earnest. President Cutler of the Metropolitan Telephone and Telegraph Company said last night that the withdrawals were in the usual routine of business and were of no especial significance. The long truce between the Western Union Telegraph Company and the Bell Telephone Company has come to an end. Each company is now free to engage in the others line of business. The agreement has not yet been renewed. No open warfare is to be inaugurated until the pending negotiations are closed. The Postal Telegraph desires, it is said, to ally with the telephone company to the exclusion of of the Western Union. The Postal prepared, it is alleged, to launch a new telephone company, backed by ample capital and controlling telephone patents.

Telephoning Through a Man.

A curious telephone incident occurred at Rock Dell last week. In some way the telephone wire had been cut, and previous to repairing it two young men of this vicinity undertook a novel experiment. One of them, Torger Anderson, went to the Rock Dell store, and the other, John Lindale, took an end of the broken wire in each hand. The two men set their watches alike, so there could be no mistake. Torger took down the receivers and rang up Oslo. Although Lindale received a severe shock, he kept hold of the wires and the message was clearly conveyed through his body. In like manner Torger Anderson telephoned to Dodge Centre and to Austin, and the message and replies were heard as clearly as though the wire wasn't mended with a human being. Lindale was four miles away from the store, and stayed there for fifteen minutes, in accordance with the agreement.

Daily Free Mail.

Congress has made available $40,000 for the purpose of enabling the Post-office Department to conduct experiments in daily free mail delivery in rural districts. The experiments are now being conducted in various states, and on the results of these a special report is to be submitted to Congress early this month. Much confidence is felt that the experiments will be successful, and the hope is expressed that rural free delivery may become an established part of the postoffice policy in the near future. There are few things within the province of government to supply that would do more to make farm life pleasant. A daily mail delivery in country districts, with the closer touch and intercourse with the outside world which that implies would mean a social transformation of considerable importance. The farmer would no longer be the isolated individual that he is. Thus one of the chief objections to country life, the feeling of being cut off from the world, would be overcome. The farm life would be rendered more pleasant, with the result that the sons and daughters would be less tempted to desert it for the city. More than that, many persons would find it more agreeable and feasable to follow their inclinations and leave the crowded centres of population for rural homes. The establishment of a system of rural free delivery would have the effect of raising the value of farm land. The reform is one which the postal department and Congress should institute at an early day.

The Second Annual Convention of the National Association of Manufacturers will be held on the 26th, 27th, and 28th of Jan., 1897, at Philadelphia.

It is expected that this convention will be one of unusual interest, as the President will submit a report of the first full year of practical work in the lines mapped out by the original Convention held in Cincinnati, Jan., 1895.

This report will describe the initiation of a practical movement in the direction of the cardinal principles originally adopted by the Association, to pass upon the work which has been done and to indicate a policy to be pursued by the President and Executive Committee during the next year; in which year it is confidently expected that existing conditions will enable the Association to make vigorous progress.

PAPER BOTTLES.

They Are Claimed to be Air-Tight, Water Tight and Better Than Glass Ones.

The days of the glass bottle are numbered. It is announced that in the near future bottles will be made of paper. A company has been formed to manufacture them, says the New York Journal.

The advantage claimed for the paper bottles are many. A glass bottle is extremely liable to break and, in the case of old wine, the breakage of a bottle in a bin causes serious loss. The paper bottle, it is claimed, cannot be broken, unless considerable force is used. Bottles have been made of toughened glass and jars have been covered with wickerwork, but still the breakage occurs. It is claimed that unbreakable paper bottles will stop this.

Paper bottles can be manufactured for about half the cost of glass bottles, and can be made water-tight as well as air tight. As brewers well know it is no easy matter to make a glass bottle that is air-tight when beer is the liquor it contains. All kinds of experiments have been made to accomplish this result, but none have succeeded. With the paper bottles the matter will be comparatively easy, as the paper will give when the cork is driven into the neck of the bottle, and will be sealed perfectly.

Glass bottles, too, will freeze and their contents spoil. In the paper bottles the liquid can defy the efforts of the Frost King. This will mean a saving in more ways than one. There is no occasion for the laborious packing in straw that has to be done in the case of glass bottles. The paper bottles, being practically unbreakable, there is no need for straw as a safeguard against rough treatment while in transit, and as the papier-mache will keep the contents warm, there need be no packing to keep the cold out.

The paper bottles are an American idea, but the trade in them will be carried to all parts of the world. No item of loss in ocean traffic has been greater than that caused by the breaking of bottles during the rolling of a ship in rough weather. On this account the paper bottles will be welcome in every quarter of the globe where liquor is shipped for export.

A Few New and Valuable Patents.

The editor of the Grafton (North Dakota) Record has been looking into the patent business a little, lately and he finds some new and novel inventions in the northwest not generally known to the eastern genius. He notes the following:

Storm doors for shot guns when handled by children will be found to be a great saving on window glass and doctor's bills.

Baby crying machine; cries while the baby is asleep and removes that lonesome feeling from the home; also imitates the hum of the mosquito, which will help to shorten up the window.

A contrivance for pulling tacks out of bare feet. Can also be used to remove dead mice from traps. Can be used in any climate.

A simple and effective device for removing the hole from doughnuts. Can be used on boarding house doughnuts same as on those your mother makes.

A collar button made of rubber that bounds up and hits you in the face when you drop it on the floor, instead of rolling out doors and falling into the well.

Machine for taking the fiber out of ox steak. By carefully following directions a thirteen year old hen can be tendered down to a dozen fresh eggs.

J. C. Chester, a deaf mute from Glendive, Mont., known as the walking telephone, is causing considerable merriment in and about the city. He has coiled about him several feet of insulated wire. In his hip pocket he carries a dry battery. Fastened to his ear is a receiver, while hanging from a hook on his side is a transmitter and receiver for the public. He talks through a tin tube. He is an inventor on his way to Washington to secure a patent on his portable telephone. He begs his way from city to city.

Some time ago a new electrical battery was announced as the invention of two Clydebank young men. It was to be worked with salt water and coal ashes, and would revolutionize the application of of electricity for motive power. It is now stated that a syndicate has been formed to exploit the invention.

In the motor carriage race between London and Brighton, England, recently, the American Duryea motor wagon was the first to accomplish the distance of fifty-two miles from point to point. The time occupied was four hours.

AUSTRALIA.

Fortunes are still often realized by Patenting American Inventions in that rich series of countries forming the Australian Colonies, viz., Victoria, New South Wales, South Australia, Western Australia, Queensland, New Zealand and Tasmania.

George G. Turri, Registered Patent Agent, of Melbourne who is known throughout the world as a Leader in his Profession, and has during the last 9 years obtained an unusually large number of Patents for American inventors and Patent Attorneys, may now be employed on lower terms than have ever yet been quoted. The rates have certainly never been tempting until now. George G. Turri places approved patents (no others) effectively before Australian capitalists, with whom he is in touch, and has negotiated many highly important and profitable sales on a purely commission basis.

As delay involves Fatal to Validity, send immediate instructions naming Colonies where patent is required, and enclose one copy of full specifications and drawings to GEORGE G. TURRI, Son Building, Queen and Bourke Streets, Melbourne, Australia. Cost (from $30 per patent.
See full particulars and obtain blanks (free) at INVENTIVE AGE Office, Washington. Special Rates to profession. Correspondence invited.

Inventors having patent attorneys are requested to forward instructions in all cases through them, that course being to Inventor's best interests.

"Your plan possesses a great deal of merit."— Hon. Chauncey M. Depew.

WHAT IS IT YOU WANT TO KNOW?

Write our Department of Research and Inquiry. Ten years established. Limitless sources of information throughout the world. Minimum fee, $1.00, must accompany each inquiry.

Associated Trade and Industrial Press,

Rooms 9, 10, 11, 12 and 13, 610 13th St., N. W.,
WASHINGTON, D. C.

ELECTRICAL INVENTORS

having new ideas, or in search of them, or who wish to learn what other workers in the same field of research are doing, will find

ELECTRICITY

an invaluable publication.

It is the ONLY Popular, Brightly Written, Non-technical, yet scientifically accurate, Electrical Journal made for Practical People.

Published weekly by

ELECTRICITY NEWSPAPER CO., 136 Liberty St., New York City.

Subscription, $2.50 per year.

Sample copies FREE.

ELECTRICITY and THE INVENTIVE AGE will be sent one year for $2.75

THERMOMETER & BAROMETER

COMBINATION.

Combination Barometer and Thermometer— Our well known leader. Blue and Gilt back with metal side posts. Fancy top and bottom and handsome enough to decorate the home of a millionaire. Extra well made and dependable on the weather changes in both weather and temperature. One of the biggest bargains in our list. A handsome thermometer, together with a glass tube containing a chemical preparation which rises and falls for action of the atmosphere, foretelling correctly all changes in the weather. This is a new style storm glass and the most reliable ever known, being in use in all signal service stations. The slightest change in the atmosphere in some way affects this glass. The thermometer is also of the best quality. Both are enclosed in a stand with walnut top and bottom and blue and gilt back. On either side is also a strong metal bar giving the whole great strength and adding to its appearance. Altogether it is a most remarkable combination for the price and we take pleasure in offering it to our patrons. This is but one of the many cheap combinations of which the market is flooded which are neither good for comment or use and can never be relied upon to forecast either the weather or in fact any thing else.

There is no reason for you to depend on the weather Bureau reports as every person who owns one of our Combinations can make his or her forecast and tell exactly when it will be fine, rainy and in fact every change in the weather. With this you can tell whether to put on rubbers and a mackintosh or a pair of summer shoes and a costly dress. Don't take any more chances about the weather but be prepared. Each one packed securely in a neat box so that they can be mailed up without any fear of getting broken or damaged. Postage 10c.   Price 25c. Each! $3.00 Doz.

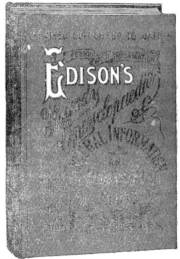

CPSIA information can be obtained
at www.ICGtesting.com
Printed in the USA
BVHW041533270119
538685BV00019B/81/P